BUILDING VENTILATION

BUILDING VENTILATION

The State of the Art

Edited by Mat Santamouris and Peter Wouters

from Routledge

First published by Earthscan in the UK and USA in 2006

For a full list of publications please contact:
Earthscan
2 Park Square, Milton Park, Abingdon, Oxfordshire OX14 4RN
711 Third Avenue, New York, NY, 10017, USA

First issued in paperback 2015

Earthscan is an imprint of the Taylor & Francis Group, an informa business, an informa business

A catalogue record for this book is available from the British Library

Library of Congress Cataloging-in-Publication Data

Building ventilation : the state of the art / edited by M. Santamouris and Peter Wouters.
        p. cm.
   Includes bibliographical references and index.
   ISBN-13: 978-1-84407-130-2
   ISBN-10: 1-84407-130-8
   1. Ventilation. 2. Air flow. 3. Dwellings—Heating and ventilation. I. Santamouris, M. (Matheos), 1956- II. Wouters, Peter.
   TH7654.B86 2006
    697.9'2—dc22

                                                                                        2005028612

ISBN-13: 978-1-138-98801-9 (pbk)
ISBN-13: 978-1-844-07130-2 (hbk)

Typesetting by Mapset Ltd, Gateshead, UK
Cover design by Yvonne Booth

Contents

List of Figures, Tables and Boxes

Figures

Tables

Boxes

List of Acronyms and Abbreviations

ACH	air changes per hour
ACS	adaptive comfort standard
AHU	air handling unit
AIVC	Air Infiltration and Ventilation Centre
AR	aspect ratio
ASHRAE	American Society of Heating, Refrigerating and Air-Conditioning Engineers
ASTM	American Society for Testing and Materials
BEMS	building energy management system
BMS	building management systems
C	Celsius
CAV	constant air volume flow rate
CBL	conventional boundary power law
CBPL	conventional boundary power law
CEN	European Committee for Standardization
CENELEC	European Committee for Electrotechnical Standardization
CFC	chlorofluorocarbon
CFD	computational fluid dynamic(s)
CFU	colony-forming unit
CBSB	Canadian General Standard Board
CIBSE	Chartered Institute of Building Services Engineers
clo	average clothing
cm	centimetre
CO	carbon monoxide
CO_2	carbon dioxide
CPD	Construction Product Directive
CPU	central processing unit
CW	closed window
°	degree
° Ch	degree hour for cooling
1D	one dimensional
2D	two dimensional
3D	three dimensional
D2	National Building Code of Finland
DAE	differential and algebraic equation
dB	decibel
DNS	direct numerical simulation
DOE	Department of the Environment (UK, US)
EC	European Community
ECBCS	Energy Conservation in Buildings and Community Systems
EEC	European Economic Community
ELA	effective leakage area
EnREI	Energy Related Environmental Issues in Buildings

EOTA	European Organization for Technical Approval
EPBD	European Energy Performance in Buildings Directive
ERH	equilibrium relative humidity
ET*	effective temperature
ETS	environmental tobacco smoke
ETSI	European Telecommunications Standards Institute
EU	European Union
f	farad
F	Fahrenheit
f/cm^3	farad per cubic centimetre
ft^2	square feet
g	gram
g/m^2	grams per square metre
g/m^3	grams per cubic metre
GHG	greenhouse gas
GIS	geographic information system
GJ	gigajoules
GW	gigawatt
ha	hectare
H/W	height-to-width ratio
HCFC	hydrochlorofluorocarbon
HVAC	heating, ventilating and air conditioning
HVCA	Heating and Ventilating Contractor's Association
Hz	hertz
IAQ	indoor air quality
ICF	insulated concrete form
IEA	International Energy Agency
IIR	International Institution of Refrigeration
I/O	indoor/outdoor
I/O	input/output
ISO	International Organization for Standardization
J	joule
K	kelvin
kg	kilogram
kg/m^3	kilograms per cubic metre
KJ	kilojoule
km	kilometre
kWh	kilowatt hours
kWh/kg	kilowatt hours per kilogram
LCC	life cycle cost
L/H	length-to-height ratio
LoopDA	Loop Design Analysis
L/W	length-to-width ratio
m	metre
m^2	square metre
m^3/s	cubic metres per second
Mbps	megabits per second
met	metabolic rate
mg	milligram
mg/m^3	milligrams per cubic metre
MJ	megajoule
mL	millilitre
mm	millimetre
MMVF	man-made vitreous fibre
m/s	metres per second

MW	megawatts
MWh	megawatt hour
n	total sample population size
NADCA	National Air Duct Cleaners Association
NAQS	quarter-sine velocity profile
NO_2	nitrogen dioxide
NO_x	nitrogen oxide
NPL	neutral pressure level
O_3	ozone
OBS	observed
OECD	Organisation for Economic Co-operation and Development
O&M	operation, inspection and maintenance
OVK	Obligatorisk Ventilations Kontroll
OW	open window
Pa	Pascal
PAH	polycyclic aromatic hydrocarbon
PAQ	perceived air quality
Pb	lead
PCM	phase change materials
PDE	partial differential equation
Perl	Practical Extraction and Report Language
PIR	Passive infrared
PMV	predicted mean vote
POW	partially opened window
ppb	parts per billion
PPD	predicted percentage of dissatisfied persons
PIR	passive infrared
ppm	parts per million
PV	photovoltaics
R^2	coefficient of determination
RANS	Reynolds-averaged Navier-Stokes equations
RH	relative humidity
RR	risk ratio
s^{-1}	per second
SBS	sick building syndrome
SET	standard effective temperature
SFP	specific fan power
SI	International System of Units
SMACNA	Sheet Metal and Air Conditioning Contractors National Association
SO_2	sulphur dioxide
SO_x	sulphur oxide
TAB	testing, adjusting and balancing
TARP	Thermal Analysis Research Programme
TRY	test reference year
UK	United Kingdom
μ	micro
UNCHS	United Nations Centre for Human Settlements (Habitat) (*now* UN-Habitat)
US	United States
V	volt
VAV	variable air volume flow rate
VOC	volatile organic compound
W	watt
W/m^2	watts per square metre
WHO	World Health Organization

1

Natural Ventilation in the Urban Environment

Francis Allard and Cristian Ghiaus

Introduction

The invention of boilers and chillers made total indoor climate control technologically possible regardless of outdoor conditions, building architecture and use. Buildings became, as Le Corbusier portended during the 1930s, of the same type: hermetically closed and controlled at a constant temperature in all climates (Mahdavi and Kumar, 1996). This energy-intensive solution is supported by the belief that maintaining constant conditions ensures comfort and satisfaction. By applying this approach, however, the building industry failed in many cases to satisfy the comfort needs of occupants. Different studies claim that as many as 43 per cent of occupants are dissatisfied with heating, ventilating and air conditioning (HVAC), and 56 to 89 per cent of government workers regard HVAC as a problem in Europe and the US (Lomonaco and Miller, 1996; DiLouie, 2002). The tendency for the HVAC industry is, then, to keep indoor temperature at a constant value that will dissatisfy the least number of people without affecting productivity since the cost of salaries is 8 to 13 times the cost of building operation.

The trend towards a greater control of the indoor environment is accentuated by the development of building management systems (BMS). BMS control the indoor temperature within narrow limits in the hope that occupants' complaints will be reduced. To test this hypothesis, let us assume that a female wearing a skirt because of the corporate dress code or current fashion sits at her desk with bare legs. She shares the same thermal comfort zone with a male colleague in a business suit. A comfort evaluation with a predicted mean vote (PMV) index will show that there is no temperature that will satisfy both people and that the offset between preferred temperatures will be 3° Celsius (C). In the absence of individual temperature control, the compromise is to control the building at a constant value based on average clothing (*clo*) and metabolic rate (*met*). If *clo* = 0.7 (average winter/summer) and *met* = 1.2 (office activity), the temperature should be 24.1° C in order to obtain a PMV index of 0. This temperature will dissatisfy both the woman and the man from the previous example (Fountain et al, 1996).

Keeping the indoor temperature at a constant value has a high investment cost and is energy intensive, with implications for resource consumption and environmental impact. Important energy savings can be obtained if the building has a larger range in which it can run freely. This saving can be augmented if ventilation is used for cooling. In fact, field studies show that people accept a larger range of temperature variation in naturally ventilated buildings than in air-conditioned ones (de Dear et al, 1997; Brager and de Dear 1998, 2000).

The European project URBVENT: Natural Ventilation in Urban Areas studied the potential for energy savings when ventilation is used instead of air conditioning and the alteration of this potential by the urban environment. A synthesis of the main findings of this project is given in the following section. First, general aspects about modelling and strategies for natural ventilation in the urban climate are reviewed. Results of the URBVENT project are then presented. A method for estimating the energy savings for cooling by using ventilation is introduced. Although the potential savings may be important, they are affected by the influence of the urban environment through reduced wind velocity, urban heat island, noise and pollution. These changes, which were studied experimentally in the project, are synthesized along with their results in simple algorithms.

The Physics of natural ventilation

Eddy, turbulent and mean description of flow

Airflow is described mathematically through a set of differential equations for mass, momentum and energy conservation based on the solution of the transport equation:

$$\frac{\partial}{\partial t}(\rho\phi) + div(\rho V\phi) = div(\Gamma_\phi grad(\phi)) + S_\phi \qquad (1)$$

where ϕ, Γ_ϕ and S_ϕ (for the k-ϵ model of turbulence) are given in Table 1.1. No general integral of these equations has yet been found and numerical solution for an arbitrary three-dimensional unsteady motion requires the use of a supercomputer.

Turbulent flow is one of the unsolved problems of classical physics. Despite more than 100 years of research, we still lack a complete understanding of turbulent flow. Nevertheless, the principal physical features of turbulent flow, especially with regard to engineering applications, are by now well determined.

Turbulent flow is distorted in patterns of great complexity containing both coarse and fine features. The flow is said to contain *eddies*, regions of swirling flow that, for a time, retain their identities as they drift with the flow, but which ultimately break up into smaller eddies. The velocity field of a turbulent flow can be regarded as the superposition of a large number of eddies of various sizes, the largest being limited by the transverse dimension of the flow and the smallest being those that are rapidly damped out by viscous forces. Mathematical analyses of steady laminar viscous flows show that infinitesimal disturbances to the flow can grow exponentially with time whenever the Reynolds number is sufficiently large. Under these conditions, the flow is unstable and cannot remain steady under practical circumstances because there are always some flow disturbances that may grow spontaneously. The most rapidly growing disturbances are those whose size is comparable to the transverse dimension of the flow. These disturbances grow to form the largest eddies, with a velocity amplitude of generally 10 per cent of the average flow speed. These large eddies are themselves unstable, breaking down into smaller eddies and being replaced by new large eddies that are continually being generated.

The generation and break-up of eddies provides a mechanism for converting the energy of the mean flow into the random energy of molecules by viscous dissipation in the smallest eddies. Compared to laminar flow, turbulent flow is like a short circuit in the flow field; it increases the rate at which energy is lost. As a result, turbulent flow produces higher drag forces and pressure losses than a laminar flow would under the same flow conditions.

When the unsteadiness of the flow is not an overwhelming feature, but rather a small disturbance of the average flow, the velocity field can be expressed as the

Table 1.1 *The dependent variables, effective diffusion coefficients and the source terms in the transport equation*

Equation	ϕ	Γ	S_ϕ
Continuity	1	0	0
U momentum	U	μ_e	$-\frac{\partial p}{\partial x} + \frac{\partial}{\partial x}\left(\mu_e\frac{\partial u}{\partial x}\right) + \frac{\partial}{\partial y}\left(\mu_e\frac{\partial v}{\partial x}\right) + \frac{\partial}{\partial z}\left(\mu_e\frac{\partial w}{\partial x}\right)$
V momentum	V	μ_e	$-\frac{\partial p}{\partial y} + \frac{\partial}{\partial x}\left(\mu_e\frac{\partial u}{\partial y}\right) + \frac{\partial}{\partial y}\left(\mu_e\frac{\partial v}{\partial y}\right) + \frac{\partial}{\partial z}\left(\mu_e\frac{\partial w}{\partial y}\right) - g(\rho - \rho_0)$
W momentum	θ	μ_e	$-\frac{\partial p}{\partial z} + \frac{\partial}{\partial x}\left(\mu_e\frac{\partial u}{\partial z}\right) + \frac{\partial}{\partial y}\left(\mu_e\frac{\partial v}{\partial z}\right) + \frac{\partial}{\partial z}\left(\mu_e\frac{\partial w}{\partial z}\right)$
Temperature	k	Γ_e	q / C_p
Kinetic energy	k	Γ_k	$G_S - \rho\epsilon + G_B$
Dissipation rate	ϵ	Γ_ϵ	$C_1\frac{\epsilon}{k}(G_S + G_B) - C_2\rho\frac{\epsilon^2}{k}$

Source: chapter authors

Table 1.2 *Example of spatial variation of wind pressure coefficients for a rectangular building*

Building level	$C_p(0°)$	Wind pressure coefficients $C_p(90°)$	$C_p(180°)$	$C_p(270°)$
6	0.70	-0.58	-0.36	-0.58
5	0.70	-0.58	-0.36	-0.58
4	0.72	-0.55	-0.35	-0.55
3	0.58	-0.48	-0.34	-0.48
2	0.41	-0.38	-0.29	-0.38
1	0.44	-0.17	-0.28	-0.17

Source: Orme et al (1998)

sum of a mean value, obtained by averaging the velocity in time and a variable component. The variable component has the characteristics of a random noise signal of zero mean. The turbulent kinetic energy, which is a measure of how much kinetic energy has been invested in the random turbulent motion of the flow, amounts, in general, to only a few per cent of the kinetic energy of the time-averaged flow. Even so, the random velocity field produces shear stresses in the flow that are much larger than those that would exist if the flow were laminar. The largest eddies contribute most to the turbulent energy, while the smallest eddies contribute most to energy dissipation.

Wind pressure

The time mean pressure due to wind flow on to or away from a surface is given by:

$$p_w = C_p \rho v^2 / 2 \qquad (2)$$

where C_p is the static pressure coefficient and v [m/s] is the time mean wind speed at a given level, commonly at the height of the building or opening. Pressure coefficients, C_p, are usually measured in wind tunnels or calculated using computational fluid dynamics (CFD) methods. They can have positive or negative values that depend upon building shape and location. According to Walker and Wilson (1994), for an isolated, parallelepipedal building, this variation is:

$$
\begin{aligned}
C_p(\phi) = \frac{1}{2} \Big[& \big(C_p(0°) + C_p(180°) \big) \big(cos^2 \phi \big)^{1/4} \\
& + \big(C_p(0°) - C_p(180°) \big) cos\phi \, \big| cos\phi \big|^{1/4} \\
& + \big(C_p(90°) + C_p(270°) \big) sin^2 \phi \\
& + \big(C_p(90°) - C_p(270°) \big) sin\phi \Big]
\end{aligned}
\qquad (3)
$$

where $C_p(0°)$, $C_p(90°)$, $C_p(180°)$, $C_p(270°)$ are pressure coefficients for four walls making an angle of 0°, 90°, 180° and 270° with the wind. Their values are given in Table 1.2 as a function of wind direction, ϕ. The wind direction and velocity varies in time as a result of wind turbulence and the effects of obstacles, making the pressure coefficients difficult to estimate in the urban environment and for complex-shaped buildings.

Buoyancy pressure

The derivative form of the static pressure is:

$$dp_s = - \rho(z) \cdot g \cdot dz \qquad (4)$$

where the air density depends upon temperature. The gas law, $p = \rho RT$, can be considered in isobaric transformation since the pressure variation, in the order of tens of Pascals, is negligible compared with the atmospheric pressure, $p_0 = 101,325Pa$. It results that:

$$p / R = \rho_0 T_0 = \rho T \approx const. \qquad (5)$$

where subscript 0 designates standard conditions for dry air ($\rho_0 = 1.2929 kg/m^3$, $T_0 = 273.15°K$, $p_0 = 101,325Pa$). Since the pressure differences in ventilation systems is of three to four orders of magnitude lower than p_0, the density variation with temperature may be written as:

$$
\rho = \frac{\rho_0 T_0}{T} = \frac{1.299 \, [kg/m^3] \cdot 273.1 \, [K]}{T \, [K]}
$$

$$
= \frac{352.6 \, [kg \cdot K / m^3]}{T \, [K]} \qquad (6)
$$

For an increase of temperature with height, there will be a corresponding decrease in pressure:

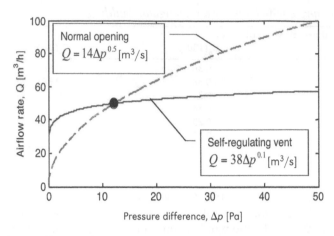

Figure 1.1 *Power law for a normal opening and for a self-regulating vent that provide the same airflow rate at 12Pa*

$$p_s = p_r - \rho_0 g T_0 \int_{z_0}^{z_h} \frac{1}{T(z)} dz \qquad (7)$$

where p_r is the reference pressure at height z_0, and z_h is the coordinate of the height. When the temperature variation on vertical is known, the static pressure may be calculated from equation (7). If the temperature has a linear variation on vertical direction:

$$T(z) = T_{z_0} + k_T z \qquad (8)$$

then the integral (7) becomes:

$$p_s = p_r - \rho_0 g T_0 \left[ln(T_{z_0} + kz) / k \right]_{z=z_0}^{z=h}$$

$$= p_r - \rho_0 g T_0 \left[ln(T_{z_0} + kh) - ln(T_{z_0}) \right] \qquad (9)$$

If the temperature can be considered constant on vertical:

$$T(z) = T_{z_0} \qquad (10)$$

then the integral (7) becomes:

$$p_s = p_r - \rho_0 g \frac{T_0}{T_{z_0}} [z]_{z=0}^{z=h} = p_r - \rho_0 g \frac{T_0}{T_{z_0}} h \qquad (11)$$

The wind velocity and pressure drop across an opening determine the airflow rate through that opening.

Mean flow through openings

The airflow through an opening is derived from the Navier-Stokes equation. By assuming steady, fully developed flow, the pressure drop for a flow between infinite parallel plates or for a flow impinging on a hole, nozzle or orifice in a thin plate is:

$$\Delta p = 0.5 \rho Q^2 / (C_d A)^2 \qquad (12)$$

where Δp is the pressure difference across the opening [Pa], ρ is the air density [kg/m³], Q is the mean airflow rate [m³/s], C_d is the discharged coefficient, a dimensionless number that depends upon the opening geometry and

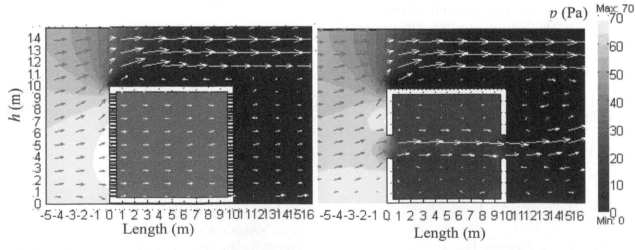

Figure 1.2 *Flow through small and large openings; the area of the openings is equal*

Reynolds number, and A is the cross-sectional area of the opening [m²].

The openings in a building are much less uniform in geometry, and generally the flow is not fully developed. The flow may be described by equation (12) where A is an equivalent area and C_p depends upon opening geometry and the pressure difference. Alternatively, an empirical power law equation is used:

$$\Delta p = (Q / c)^{1/n} \tag{13}$$

where the flow coefficient c [m³/(s·Pa^n)] and the dimensionless flow exponent n are determined from experiments and do not have a physical meaning. For laminar flow, $n = 1$; for turbulent flow, $n = 0.5$; and for transitional flow n is between 0.6 and 0.7. An empirical power law may be used also to model pressure-controlled self-regulating vents, which are inlets that provide relatively constant airflow rates over the pressure range. The low dependence of flow with pressure may be modelled with a small value of the exponent n – for example, $n = 0.1$. A normal opening in a building that would provide the same airflow rate at a design pressure of 12 Pa would have a different behaviour, modelled by the exponent $n = 0.5$ (see Figure 1.1).

Equation (12) and (13) are applicable to small orifices through which the air infiltrates. For larger openings, transforming the wind velocity in pressure through the concept of pressure coefficient is no longer valid.

Flow through small and large openings is very different. Figure 1.2 presents two cases in which the ratio between the opening area and the obstructed area is the same. While in the first case the flow may be estimated by using pressure coefficients, in the second case the inertial forces are very important and the wind pressure loses its meaning.

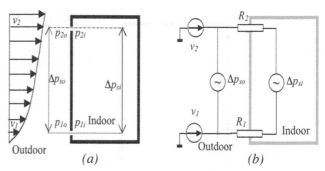

Source: chapter authors

Figure 1.3 *Modelling airflow with lumped parameters: (a) the building; (b) equivalent circuit*

The combined effect of wind and buoyancy

Wind and buoyancy pressures have a combined effect on the flow through openings. For simple cases, empirical models can be applied (Allard, 1998). To solve more complex problems, lumped parameters can be used, as well as an analogy with electrical circuits (see Table 1.3 and Figure 1.3).

By convention, the pressure difference between outdoors and indoors is positive when the building is pressurized relative to outdoors:

$$\Delta p_s = p_i - p_o \tag{14}$$

where p_i and p_o are the indoor and outdoor pressures, respectively. It results from equations (6), (11) and (14) that:

$$\Delta p_s = (p_{ri} - p_{ro}) + \rho_o \left(1 - \frac{T_o}{T_i} \right) gh \tag{15}$$

Table 1.3 *Electrical analogy for modelling with lumped parameters*

Variables	Electrical	SI unit	Fluid	SI unit
Across	Voltage, e	V	Pressure, P	Pa
Through	Current, i	A	Volume flow rate, Q	m³/s
Power	$p \cdot i$	W	$P \cdot Q$	W
Circuit elements		**Symbol**		
Resistance		⊏⊐	$Q = k \cdot (\Delta p)^n$	
Source across	Voltage source	~	Pressure	
Source through	Current source	⊕	Velocity	

Source: chapter authors

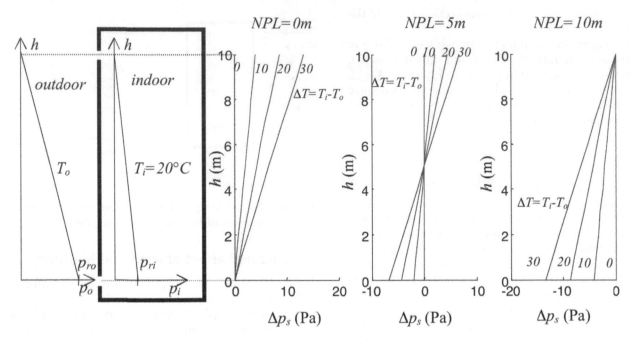

Source: chapter authors

Figure 1.4 *Stack pressure and neutral-level plane*

where indexes i and o correspond to indoor and outdoor, respectively. Wind may induce a positive or a negative pressure inside the building. Consequently, the stack pressure at reference height p_r in equation (11) will be different indoors and outdoors. The level at which the pressure difference between outdoor and indoor pressure is zero is called neutral pressure level (NPL) (ASHRAE, 2001). For a given neutral pressure level, h_{NPL}, the pressure difference between indoors and outdoors is:

$$\Delta p_s = \rho_o \left(1 - \frac{T_o}{T_i} \right) g(h - h_{NPL}) \qquad (16)$$

Figure 1.4 shows the variations of Δp_s with height and temperature difference for different positions of the neutral pressure level.

Natural ventilation strategies

The driving forces for natural ventilation may be used for different strategies: wind variation-induced single-sided ventilation; wind pressure-driven cross-ventilation; and buoyancy pressure-driven stack ventilation. When ventilation is needed for individual rooms, single-sided ventilation, the most localized of all natural ventilation

strategies, may be the only choice available. Cross-ventilation systems provide ventilation to the floor of a building and depend upon building form and the urban environment. Stack ventilation provides ventilation to the building as a whole and depends upon building form and internal layout. Combinations of all these strategies exploit their individual advantages.

Wind variation-induced single-sided ventilation

For the uninitiated, natural ventilation means opening a window to let airflow in a room that is otherwise airtight (see Figure 1.5a). The air flow through the opening is due to wind and buoyancy. The wind has a mean and a fluctuating component that may vary over the opening and produce a 'pumping effect'. When the indoor temperature is higher than outdoors, the buoyancy makes the cold air enter at the lower part and the hot air exit at the upper part of the opening. An empirical model of this complex phenomenon is as follows (de Gids and Phaff, 1982):

$$v_{eff} = (c_1 v_r^2 + c_2 H \cdot \Delta T + c_3)^{1/2} \qquad (17)$$

where c_1 ($c_1 \approx 0.001$) is a dimensionless coefficient depending upon window opening, c_2, c_3 ($c_2 \approx 0.0035$, $c_3 = 0.01$) are constants related to buoyancy and wind, v_r

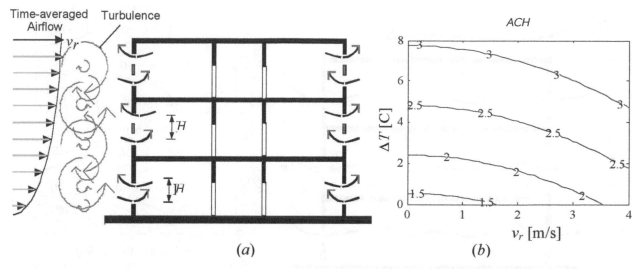

Source: Axley (2001) *(a)*; Cristian Ghiaus *(b)*

Figure 1.5 *(a) Single-sided wind-driven ventilation; (b) air changes per hour calculated for a room of 2.7m in height with windows of 1.5m in height and a window area of one twentieth the floor area*

[m/s] is the mean wind speed for the site, H [m] is the height of the opening, and T [K] is the mean temperature difference between inside and outside.

The flow rate through the opening is:

$$Q = 0.5A_w v_{eff} \qquad (18)$$

where A_w is the effective area of the open window.

Recommendations for single-sided ventilation are: a window area of one twentieth of the floor area, a height of about 1.5m and a maximum room depth of 2.5 times the ceiling height (BRE, 1994).

Let us consider a typical office room with height $h = 2.75$m, window height $H = 1.5$m, and window area one twentieth the flow area, $A_w = A/20$. The volume of this room would be $V = whl = 2.5h \cdot A$. The flow rate through opening would be $Q = 0.5A v_{eff}$. Expressing this flow in air changes per hour (*ACH*), $Q = ACH/3600 \cdot V$, we obtain:

$$ACH = \frac{3600}{V} \cdot 0.5A_w v_{eff} = \frac{3600}{2.5h} \cdot \frac{0.5}{20} \cdot v_{eff} \qquad (19)$$

The dependence of *ACH* as a function of v_{eff} and ΔT is shown in Figure 1.5b. When the difference between indoor and outdoor temperature is low or when the wind velocity is low, the airflow rate is low. Single-sided ventilation is a solution that is not very effective for cooling by ventilation during warm weather periods.

Wind-driven cross-ventilation

Wind airflow over a building tends to induce positive (inward acting) pressures on windward surfaces and negative (outward acting) pressures on leeward surfaces, thereby creating a net pressure difference across the section of the building that drives cross-ventilation airflows. Two-sided or cross-ventilation takes place when air enters the building on one side, sweeps the indoor space and leaves the building on another side (see Figure 1.6a).

The positive windward pressure Δp_{ww} and the negative leeward pressure Δp_{lw} are, in fact, pressure differences from the ambient air pressure of the free-field airflow. While these pressure differences often vary rapidly with time (due to turbulence in the wind airflow) and position (due to the aerodynamic effects of building form), on average, they may be related to a reference time-averaged approach wind velocity, v_r:

$$p_{ww} c_{p-ww} \left(\frac{\rho v_r^2}{2} \right); \quad p_{wl} = C_{p-lw} \left(\frac{\rho v_r^2}{2} \right) \qquad (20)$$

where ρ is the density of the air, $\rho v_r^2/2$ is the kinetic energy per unit volume of the reference wind velocity, and $C_{p-ww} > 0$, $C_{p-lw} < 0$ are the wind pressure coefficients of the windward and leeward surface locations under consideration. The reference wind velocity is most commonly (but not always) taken as the time-averaged

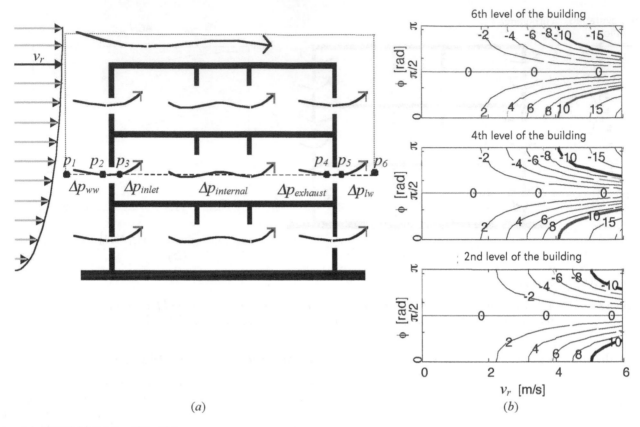

Source: Axley (2001) *(a)*; Cristian Ghiaus *(b)*

Figure 1.6 *Wind-driven cross-ventilation: (a) pressure drops associated with wind-driven cross-ventilation; (b) wind pressure differences for a rectangular, isolated building*

wind velocity at 10m above the building height. The wind pressure difference between the façades:

$$\Delta p_w = p_{ww} - p_{wl} = (C_{p-ww} - C_{p-lw})\frac{\rho v_r^2}{2} \qquad (21)$$

equals the pressure changes along a given cross-ventilation airflow path (see, for example, Figure 1.6a):

$$\Delta p_w = \Delta p_{inlet} + \Delta p_{interval} + \Delta p_{exhaust} \qquad (22)$$

For typical design conditions, the reference wind velocity is around 4m/s, windward wind pressure coefficients are typically around +0.5, leeward wind pressure coefficients are –0.5, and the density of air is approximately 1.2kg/m³. Thus, we may expect the driving wind pressure for cross-ventilation to be approximately 10Pa:

$$\Delta p_w = (C_{p-ww} - C_{p-lw})\frac{\rho U_{ref}^2}{2}$$

$$\approx \left((+0.5) - (-0.5)\right)\frac{1.2\text{kg/m}^3(4\text{m/s})^2}{2} = 9.6\text{Pa}$$

$$(23)$$

A 10Pa driving pressure is small relative to typical fan-driven pressure differences that are of one or two orders of magnitude larger. Thus, to achieve similar ventilation rates, the resistance offered by the natural ventilation system will have to be small relative to ducted mechanical ventilation systems.

This simple natural ventilation scheme suffers from a critical shortcoming: it depends upon wind direction and intensity. As wind direction changes, so do the wind pressure coefficients. Consequently, the driving wind pressure may drop to low values even when windy conditions prevail, causing the natural ventilation airflow rates to drop. When wind speeds drop to low values, the driving wind pressure will again diminish and ventilation airflow will subside regardless of wind direction. For example, Figure 1.6b shows the wind pressure difference calculated for a six-storey building by using equation (3). We can see that the estimated value for the pressure difference, $\Delta p \approx 10$Pa, occurs for a restricted range of wind direction. The variability of wind-induced pressure demands special measures such as self-regulating vents for pressure reduction, wind catchers or a design that makes the building insensitive to wind variations (for example, double façades). The variability of wind-induced pressure puts forward the 'zero-wind' design condition as a critical case, although some limited studies indicate that 'zero-wind' conditions are not only unlikely at many locations, but they may well be short lived (Skaret et al, 1997; Deaves and Lines, 1999; Axley, 2000).

In spite of these shortcomings, wind-driven cross-ventilation has been employed in some recently built non-residential buildings, although its use is uncommon. Examples include the machine shop wing of the Queen's Building of De Montfort University, Leicester, in the UK, designed by Short Ford Associates architects and Max Fordham Associates environmental engineers, and a number of skyscrapers designed by architect Ken Yeang of T. R. Hamzah and Yeang Sdn Bhd in Malaysia.

Even within the time-averaged modelling assumptions, there are significant sources of uncertainty that should be kept in mind. Wind pressure coefficients, C_p, are seldom known with certainty – they vary from position to position over the building envelope, being sensitive to small details of form; they are altered significantly by the shelter offered by other buildings; they vary with wind direction; and they are affected by building porosity. Wind characteristics are generally known with certainty only for regional airports where detailed records are maintained. Consequently, evaluation of the reference wind speed and direction for a given site is always problematic and subject to error. Finally, empirical coefficients associated with flow-resistance models introduce another source of uncertainty, although perhaps not as significant as that due to wind uncertainties.

Buoyancy-driven stack ventilation

Warm air within a building will tend to move up and flow out of upper-level exhausts, while cooler outdoor air will tend to flow in through lower inlets to replace it. For the example shown in Figure 1.7a:

$$\rho_o g \cdot \Delta z - \Delta p_{inlet} - \Delta p_{internal}$$

$$- \rho_i g \cdot \Delta z - \Delta p_{exhaust} = 0 \qquad (24)$$

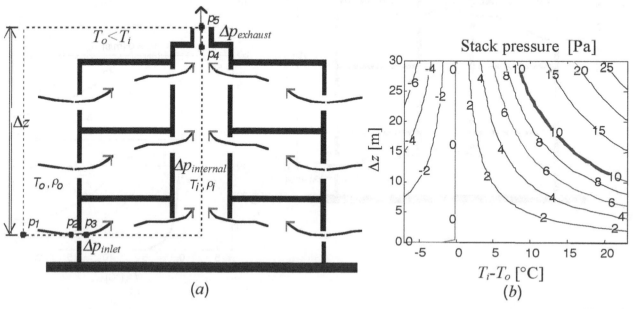

Source: Axley (2001) (a); Cristian Ghiaus (b)

Figure 1.7 *Stack-pressure-driven natural ventilation: (a) pressure drops associated with buoyancy-driven stack ventilation; (b) stack pressure variation as a function of temperature difference and building height*

The stack pressure, $p_s = (\rho_o - \rho_i)g \cdot \Delta z$, equals the pressure losses:

$$\Delta p_s = \Delta p_{inlet} + \Delta p_{internal} + \Delta p_{exhaust} \qquad (25)$$

The driving stack pressure varies with building height, h, and the temperature difference between indoors and outdoors:

$$\Delta p_s = (\rho_o - \rho_i)g \cdot \Delta z = \left(\frac{352.6}{T_o} - \frac{352.6}{T_i} \right) g \cdot \Delta z$$

$$(26)$$

During warm periods, as outdoor temperatures approach indoor air temperatures, the stack pressure differences for all but very tall multi-storey buildings may be expected to be small relative to typical wind-driven pressure differences. Figure 1.7b shows the dependence of stack pressure as a function of temperature difference and height given by equation (26). For a three-storey building of about 10m, the difference between indoor and outdoor temperatures should be about 23° C in order to obtain approximately a 10Pa pressure difference, typical for wind-driven pressure. For an eight-storey building, this temperature difference should be 10° C (see Figure 1.7a). Furthermore, for higher floors, the stack pressure difference available to drive natural airflow will be proportionally smaller. For wintertime air quality control, when large indoor–outdoor temperature differences are usual, buoyancy-driven stack ventilation may be expected to be effective, although differences of air distribution with storey level must be accounted for by proper sizing of inlet vents. But buoyancy-driven stack ventilation alone cannot be expected to be a very effective strategy for cooling. In practice, stack configurations have often achieved acceptable ventilation rates due to the wind forces that also drive flow in stack ventilation systems, thus complicating the system's behaviour. Given that low wind conditions may be unlikely and short lived at most locations, simple buoyancy-driven stack flow is not likely to occur often in practice. Thus, combined wind- plus buoyancy-driven stack ventilation should be considered instead.

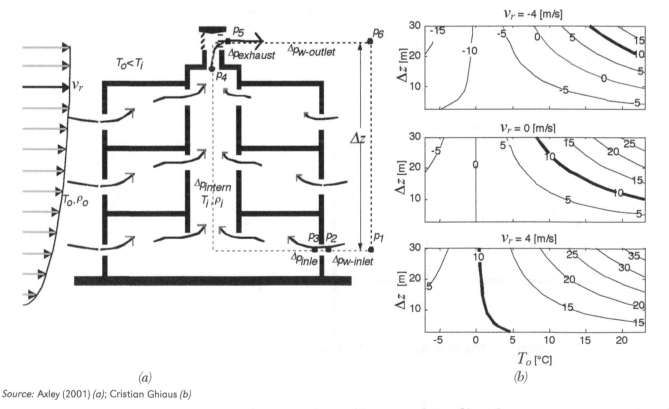

(a)   *(b)*

Source: Axley (2001) *(a)*; Cristian Ghiaus *(b)*

Figure 1.8 *Combined wind- and buoyancy-driven ventilation: (a) pressure drops; (b) total pressure as a function of wind velocity, temperature difference and building height*

Combined wind- and buoyancy-driven ventilation

When properly designed, stack ventilation systems use both wind- and buoyancy-driven pressure differences. For example, let us consider a stack ventilation system under the combined influence of wind and buoyancy differences (see Figure 1.8a). This system is similar to that illustrated in Figure 1.7a but with a stack terminal device added that can respond to the prevailing wind direction to maximize the negative pressure induced by the wind (for example, operable louvres and rotating cowls).

A representative pressure loop – for example, loop p_1–p_2–p_3–p_4–p_5–p_6–p_1, will now include both buoyancy-driven and wind-driven pressure differences that appear as a simple sum:

$$\Delta p_s + \Delta p_w = \Delta p_{inlet} + \Delta p_{exhaust} \tag{27}$$

where:

$$p_s = (\rho_o - \rho_i) g \cdot \Delta z \tag{28}$$

and:

$$\Delta p_w = \left(C_{p-inlet} - C_{p-exhaust} \right) \frac{\rho v_r^2}{2} \tag{29}$$

Pressure loop equations for each of the additional five ventilation loops of Figure 1.8a will assume the same general form, although the values of the various parameters will change. For the pressure loop shown in Figure 1.8a, both the inlet wind pressure coefficient $C_{p-inlet}$ and the exhaust wind pressure coefficient $C_{p-exhaust}$ are likely to be negative as both are on the leeward side of the building. Consequently, the wind-driven pressure difference will act to cause flow in the direction indicated only if the absolute value of the exhaust is greater than that of the inlet. For this reason, driving wind pressure differences for the leeward rooms of stack ventilation systems tend to be smaller than those of the windward rooms. As a result, unless inlet vents are designed accordingly, ventilation rates may be expected to be lower in these rooms and may actually reverse under certain conditions. Figure 1.8b shows the pressure difference obtained by superposing the wind-induced pressure from Figure 1.6b and stack pressure from Figure 1.7b. We can see that when stack ventilation is assisted by wind, the pressure difference may be easily achieved. Self-regulating vents can serve to maintain design-level ventilation rates and thereby mitigate this problem, but cannot inhibit flow reversals or

provide flow when the net driving pressure $\Delta p_s + \Delta p_w$ drops to zero or becomes negative. The stack pressure contribution Δp_s will act to compensate for low or reverse wind pressure; but again this contribution must be expected to be smaller for the upper floors of the building. Consequently, upper leeward rooms tend to experience the lowest driving pressures and, thus, lower ventilation rates (see Figure 1.8b).

Ventilation stacks that extend above nearby roofs, especially when equipped with properly designed stack terminal devices, tend to create negative (suction) pressures that are relatively independent of wind direction. Thus, stack systems serve to overcome the major limitation of simple cross-ventilation systems identified above, while providing similar airflows in the individual rooms of a building. As a result of these advantages, stack ventilation systems – perhaps most often using a central atrium as a shared stack – have become the most popular natural ventilation system used in commercial buildings during recent years, and a number of manufacturers have developed specialized components to serve these systems.

Combinations of fundamental strategies

Most often, the three basic strategies (single-sided, cross- and buoyancy-driven ventilation) are used concurrently in single buildings to handle a variety of ventilation needs, as illustrated in Figure 1.9. A landmark example of such an approach is found in the office buildings of the UK Houses of Parliament in London.

In other buildings that use natural ventilation, the inlets, exhausts and the distribution network are specially designed. One common approach involves the use of in-slab or access-floor distribution of fresh air in order to provide greater control of air distribution across the building section and to temper incoming air to prevent cold draughts (see Figure 1.9). This type of fresh air distribution is similar to displacement ventilation, most commonly implemented mechanically, and offers similar benefits – for example, the use of thermal plumes generated by equipment and by occupants to assist airflow and improved air quality in the occupied zone of rooms.

Solar-assisted ventilation

There is usually little difficulty in providing required airflow rate to a building when wind assists the stack effect (see Figure 1.8b). But since wind speed is reduced in the urban environment, natural ventilation systems in the urban area are usually designed based on buoyancy-driven flow. When buoyancy pressure resulting from the difference between the internal and the external temperature is insufficient, solar-induced ventilation can be an alternative.

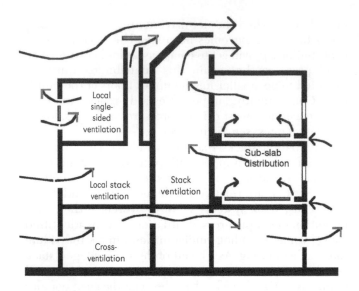

Source: Axley (2001)

Figure 1.9 *Mixed natural ventilation strategies in a single building to satisfy local and global ventilation needs*

The principle is to increase the stack pressure by heating the air in the ventilation stack, resulting in a greater temperature difference than in conventional systems.

The pressure losses for a solar collector are:

$$\Delta p_s = \Delta p_i + \Delta p_d + \Delta p_e \qquad (30)$$

where Δp_i, Δp_d, Δp_e are the inlet, distributed and exit pressure losses, respectively. Depending upon the position of the control damper, Δp_i or Δp_e include the control damper pressure losses. The stack pressure is:

$$\Delta p_s = \rho_0 T_0 [1 / T_e - 1 / T_i] g \Delta z \qquad (31)$$

where T_i is the inlet air temperature of the collector, usually equal to the indoor temperature, and T_e is the exit temperature of the collector (Awbi, 1998):

$$T_e = A / B + (T_i - A / B) exp [- BwH / (\rho_e c_p Q)] \qquad (32)$$

with $A = h_1 T_{w1} + h_2 T_{w2}$, $B = h_1 + h_2$, where h_1, h_2 [W/m²K] are surface heat transfer coefficients for internal surfaces of the collector, T_{w1}, T_{w2} [° C] are surface temperatures of internal surfaces of the collector, w [m] is the collector width, H [m] is the height between inlet and outlet openings, ρ_e [kg/m³] is the air density at exit, c_p

[J/(kg.K)] is the specific heat of air, and Q [m³/s] is the volumetric airflow rate.

The principle of the solar collector may be applied to different types of devices – for example, Trombe walls, double façades, solar chimneys or solar roofs. A Trombe wall is a wall of moderate thickness covered by a pane of glass separated from the wall by a gap of 50mm to 100mm. It may be used for ventilation or heating (see Figure 1.10). A solar chimney is a chimney with a gap of about 200mm placed on the south or south-west façade of the building. Solar roofs are used when solar altitude is large. In this case, a roof has a larger surface area for collecting the solar radiation than a vertical wall or chimney.

Natural ventilation strategies for the urban environment

The use of natural ventilation in the urban environment should take into account the lower wind velocity, as well as noise and pollution. Ventilation systems cannot rely upon low-level inlets since the outdoor air at street levels may be contaminated and inlets will be shielded from winds.

Balanced stack ventilation

A number of ancient Middle Eastern strategies using both roof-level inlets and exhausts – including the traditional Iranian wind towers, or *bagdir*, and the Arabian and Eastern Asian wind catchers, or *malkaf* – are being reconsidered for broader application and technical refinement.

In these *balanced* stack ventilation schemes, air is supplied in a cold stack (that is, with air temperatures maintained close to outdoor conditions through proper insulation of the stack) and exhausted through a warm stack (see Figure 1.11).

Let us consider, for example, the loop through the second level of Figure 1.11. The equation for this pressure loop will be similar in form to the case of combined wind- and buoyancy-driven ventilation:

$$\Delta p_s + \Delta p_w = \Delta p_{inlet} + \Delta p_{internal} + \Delta p_{exhaust} \qquad (33)$$

The stack pressure is determined by the indoor-to-outdoor air density difference and the height difference from the stack exhaust and the floor-level inlet locations, $\Delta p_s = (\rho_o - \rho_i) g \Delta z$, if air temperatures within the cold stack can be maintained close to outdoor levels. Airflow through each floor level will, therefore, be identical to that expected in the simpler single-stack scheme if the airflow resistance of the supply stack (and its inlet and outlet devices) is similar to that provided by the air inlet devices of combined wind and buoyancy ventilation of Figure 1.8.

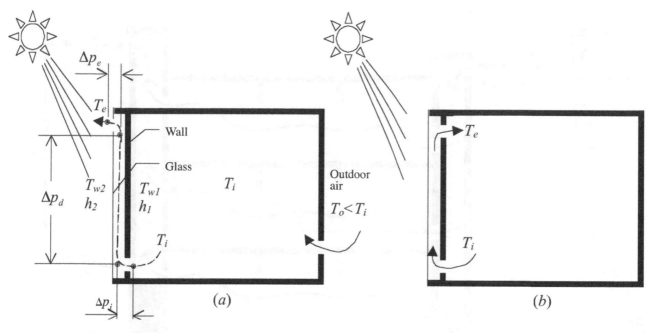

Source: adapted from Awbi (1998)

Figure 1.10 *Solar collector used as: (a) ventilator; (b) heater*

The driving wind pressure is determined by the difference between inlet and exhaust wind pressure coefficients and the kinetic energy content of the approach wind velocity, $\Delta p_w = (C_{p-inlet} - C_{p-exhaust})\rho v_{ref}^2/2$. But, in this case, the high location of the inlet ensures a higher inlet wind pressure and insensitivity to wind direction. This, combined with the potential of a wind-direction-insensitive exhaust stack, makes this scheme particularly attractive for urban environments. Balanced stack systems have been commercially available in the UK for over a century, designed mainly to serve single rooms rather than whole buildings (Axley, 2001). The Windcatcher® natural ventilation systems distributed by Monodraught Ltd in the UK (see Figure 1.12) supplies as high as five air changes per hour under relatively low wind conditions (3m/s) measured 10m above the building (see www.monodraught.co.uk). These systems may also be equipped with coaxial fans to provide mechanical assistance during extreme weather conditions.

In cold conditions, it is possible to achieve ventilation air heat recovery with *top-down* schemes by using coaxial supply.

Passive evaporative cooling

An improvement of the *balanced stack* ventilation system, also based on ancient Middle Eastern and Eastern Asian solutions, consists in adding evaporative cooling to the supply stack. Traditionally, evaporative cooling was achieved through water-filled porous pots within the supply air stream or the use of a pool of water at the base of the supply stack (Santamouris and Asimakopoulos, 1996; Allard, 1998). In more recent developments, water sprayed high into the supply air stream cools the air stream and increases the supply air density, thereby augmenting the buoyancy-induced pressure differences that drive airflow (Bowman et al, 2000).

The loop analysis of the *passive downdraught evaporative cooling* scheme is similar to that of the *balanced stack* scheme; but now the buoyancy effects of the increased moisture content must be accounted for. Let us consider the diagram of such a system shown in Figure 1.13. Two height differences must now be distinguished: z_a – the height above the room inlet location of the moist air column in the supply stack; and z_b – the height of the exhaust above this moist column.

The air density in the moist air supply column, ρ_s, will approach the saturation density corresponding to the outdoor air wet bulb temperature; more specifically, experiments indicate that these supply air conditions will be within 2° C of the wet bulb temperature. Hence, the loop equation describing the (time-averaged) ventilation airflow in this system becomes:

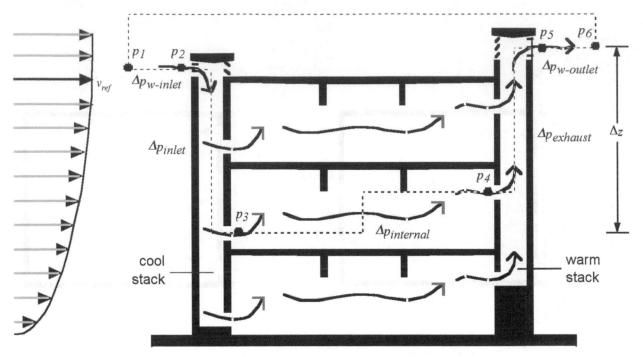

Source: Axley (2001)

Figure 1.11 *Top-down or balanced-stack natural ventilation systems use high-level supply inlets to access less contaminated air and to place both the inlets and the outlets in higher wind velocity exposures*

$$\Delta p_{inlet} + \Delta p_{internal} + \Delta p_{exhaust} = \Delta p_s + \Delta p_w \qquad (34)$$

where:

$$\Delta p_s = [\rho_o z_b + \rho_s z_a - \rho_i(z_a + z_b)]g \qquad (35)$$

$$\Delta p_w = (C_{p-inlet} - C_{p-exhaust})\frac{\rho v_r^2}{2} \qquad (36)$$

For a quantitative measure of the impact of this strategy, let us consider a case similar to the one discussed earlier for wind- and buoyancy-induced natural ventilation, but with a cool moist column height that equals the stack height of 10m (that is, $z_a \approx 0m$ and $z_b \approx 10m$). If the outdoor air has a temperature of 25° C and 20 per cent relative humidty (RH) (that is, with a density of approximately 1.18kg/m³), and is cooled by evaporation to within 2° C of its wet bulb temperature (12.5° C), its dry bulb temperature will drop to 14.5° C, while its density will increase to approximately 1.21kg/m³ and relative humidity to 77 per cent. If internal conditions are kept just within the thermal comfort zone for these outdoor conditions (that is, 28° C and 60 per cent RH), using an appropriate ventilation flow rate given internal gains, then internal air density will be approximately 1.15kg/m³. Consequently, the buoyancy pressure difference that will result will be:

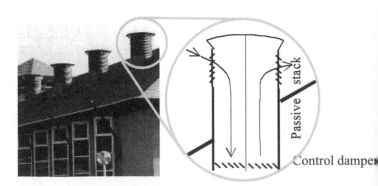

Source: photo reproduced with permission from Monodraught Ltd. Diagram by C. Ghiaus.

Figure 1.12 *Windcatcher for natural ventilation systems*

$$\Delta p_s = \left(1.18\frac{\text{kg}}{\text{m}^3}(0\text{m}) + 1.21\frac{\text{kg}}{\text{m}^3}(10\text{m})\right.$$

$$\left. - 1.15\frac{\text{kg}}{\text{m}^3}(0 + 10\text{m})\right)9.8\frac{\text{m}}{\text{s}^2} = 6.4\text{Pa} \qquad (37)$$

Without the evaporative cooling (that is, with $\Delta z_a \approx 10\text{m}$ and $\Delta z_a \approx 0\text{m}$):

$$\Delta p_s = \left(1.18\frac{\text{kg}}{\text{m}^3}(0\text{m}) + 1.21\frac{\text{kg}}{\text{m}^3}(10\text{m})\right.$$

$$\left. - 1.15\frac{\text{kg}}{\text{m}^3}(10 + 0\text{m})\right)9.8\frac{\text{m}}{\text{s}^2} = 2.9\text{Pa} \qquad (38)$$

Thus, in this example, evaporative cooling more than doubles the buoyancy pressure difference while, at the same time, providing adiabatic cooling.

Double-skin façade

A double-façade construction consists of a normal concrete or glass wall combined with a glass structure outside the actual wall. Double-skin façades offer several advantages. They can act as buffer zones between internal and external conditions, reducing heat loss in winter and heat gain in summer. In combination with ventilation of the space between the two façades, the passive thermal effects can be used to best advantage. Natural ventilation can be drawn from the buffer zone into the building by opening windows in the inner façade. The stack effect of thermal air currents in tall buildings offers advantages over lower buildings. This eliminates potential security and safety problems caused by having windows that open, as well as wind pressure differentials around the building. Double façades can be used for solar-assisted stack ventilation or balanced stack ventilation.

The cooling potential of ventilation

The URBVENT project introduced a method based on the indoor temperature of a free-running building, which, combined with comfort criteria and climate, allows us to assess the energy needed for cooling and the potential for free cooling by ventilation (Ghiaus, 2003). Weather data for this method are available in the public domain from satellite investigations. The method can be applied when buildings similar to existing ones are constructed in a new

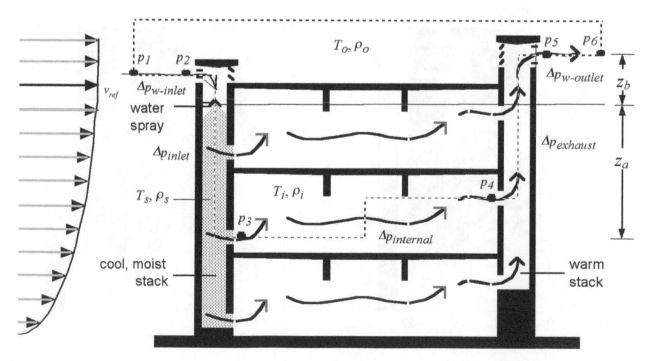

Source: Axley (2001)

Figure 1.13 *Passive downdraught evaporative cooling stack ventilation*

location, when existing buildings are retrofitted or when completely new buildings are designed. The method may be used to interpret the results of building simulation software or field measurements.

Thermal comfort zone

The thermal comfort zone is delimited by a lower, T_{cl}, and an upper comfort limit, T_{cu}. These limits vary with the season or, more precisely, with the mean monthly outdoor temperature. It is argued that the thermal comfort in naturally ventilated buildings has larger seasonal ranges than assumed by International Organization for Standardization (ISO) 7730 and by American Society of Heating, Refrigerating and Air-conditioning Engineers (ASHRAE) 55 standards (de Dear et al, 1997; Brager and de Dear, 1998; Nicol and Humphreys, 2002). On the contrary, field studies have shown that indoor temperature in fully HVAC-controlled buildings have a mean temperature of 23° C, with a standard deviation of 1°–1.5° C and a seasonal shift of 0.5°–1° C, which is narrower than the range of 3° C and the seasonal shift of 3° C required by ASHRAE Standard 55 (Fountain et al, 1996).

Figure 1.15 shows a comparison of comfort zones in fully HVAC-controlled buildings (Fountain et al, 1996), the ASHRAE comfort zone (ASHRAE, 2001) and the

standard for natural ventilation (Brager and de Dear, 2000) applied for European climatic conditions. The approximate parameters of the comfort zones are given in Table 1.4. The summer mean for natural ventilation is considered 25° C, corresponding to mean monthly temperature of 22° C (see Figures 1.14 and 1.15).

Balance temperature and free-running temperature

Energy estimations for HVAC systems are made by using the degree day or the bin methods (ASHRAE, 2001). The concept on which these estimations are based is the balance temperature. The balance temperature for cooling, T_c, is the outdoor temperature for which the building, having a specified indoor temperature, T_{cu}, is in thermal balance. For this temperature, the heat gains (solar and internal) equal the heat losses (ASHRAE, 2001):

$$q_{gain} = K_{tot}(T_b - T_{cu}) \tag{39}$$

where:

- q_{gain} = total heat gains [W];
- K_{tot} = total cooling loss coefficient of the building [W/K];

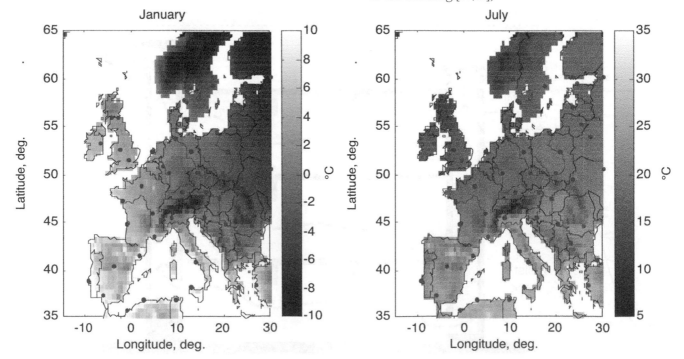

Source: Cristian Ghiaus, based on data from IIASA (2001)

Figure 1.14 *Mean monthly temperature in Europe during January and July*

- T_{cu} = upper limit of comfort temperature [K];
- T_b = balance temperature [K].

The balance point temperature, then, is:

$$T_b = T_{cl} + \frac{q_{gain}}{K_{tot}} \qquad (40)$$

The energy rate needed for cooling is:

$$q_c = \begin{cases} K_{tot}(T_o - T_b), \text{ if } T_o > T_b \\ 0, \text{ if } T_o \leq T_b \end{cases} \qquad (41)$$

where T_o is the outdoor temperature. The energy needed for cooling is:

$$Q_c = \int_{t_{init}}^{t_{fin}} K_{tot}(T_o - T_b)\delta_c dt \qquad (42)$$

where δ_c is the condition for cooling:

$$\delta_c = \begin{cases} 1, \text{ if } T_o > T_b \\ 0, \text{ if } T_o \leq T_b \end{cases} \qquad (43)$$

The total cooling loss coefficient of the building is a function of outdoor temperature and time. The discrete equivalent of the integral (42) is:

$$Q_c = \sum_i \sum_j K_{tot}(i,j)\left[T_o(i,j) - T_b(i,j)\right]\delta_c \Delta t(i) \qquad (44)$$

where summation indexes i and j refer to the time interval (for example, the number of hours) and the bins of the outdoor temperature, T_o, respectively. If the time interval, $\Delta t(i)$, is one hour, the factor:

$$\left[T_o(i,j) - T_b(i,j)\right]\delta_c \Delta t(i) \equiv dh_c \qquad (45)$$

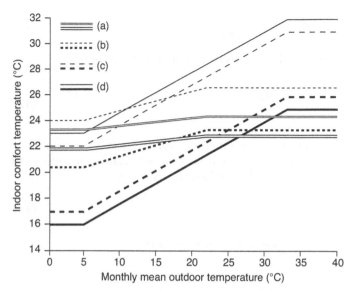

(a) air conditioning; (b) ASHRAE comfort range; (c) natural ventilation, 90 per cent of acceptability limits; (d) natural ventilation, 80 per cent of acceptability limits

Source: Cristian Ghiaus, based on Brager and de Dear (2000) and ASHRAE (2001)

Figure 1.15 *Comfort range in air conditioning and in natural ventilation*

is defined as degree hour for cooling, dh_c [° Ch]. The expression (45) has the disadvantage of using the concept of balance temperature, which implies that the indoor temperature is controlled at a constant value. While this assumption is valid for fully conditioned buildings, it cannot be applicable to buildings with natural ventilation, especially in summer conditions.

We can demonstrate that the degree hours as used in the bin method (ASHRAE, 2001) can have an equivalent expression in terms of the free-running temperature, T_{fr}. The free-running temperature is the indoor temperature when no HVAC system is used; by 'system' we mean heating, air conditioning and ventilation for cooling. From the thermal balance:

Table 1.4 *Comfort zones*

	Mean (° C)	Winter mean (° C)	Summer mean (° C)	Range (° C)	Seasonal shift (° C)
(a) Air-conditioned buildings	23.0	22.5	23.5	1.5	1.0
(b) ASHRAE comfort zone	23.5	22.1	24.9	3.5	2.7
(c) Standard for natural ventilation, 90 per cent of acceptability limits	23.9	19.5	25.0 max 28.3	5.0	8.8
(d) Standard for natural ventilation, 80 per cent of acceptability limits	23.9	19.5	25.0 max 28.3	7.0	8.9

Source: chapter authors, based on Brager and de Dear (2000) and ASHRAE (2001)

$$K_{tot}(T_{fr} - T_o) - q_{gain} = 0 \tag{46}$$

it results that the free-running temperature is:

$$T_{fr} = T_o + \frac{q_{gain}}{K_{tot}} \tag{47}$$

By replacing T_b in equation (45) by the expression (40) and considering the equation (47), we obtain an equivalent expression for degree hour for cooling:

$$dh_c = (T_{fr} - T_{cu})\delta_c \tag{48}$$

where the condition for cooling is:

$$\delta_c = \begin{cases} 1, \text{if } T_{fr} > T_{cu} \\ 0, \text{if not} \end{cases} \tag{49}$$

The cooling need may be satisfied by free cooling or by mechanical cooling. If the outdoor temperature, T_o, is lower than the upper limit of the comfort range, T_{cu}, then free cooling is possible. The condition for free cooling is:

$$\delta_{fr} = \begin{cases} 1, \text{if } T_{fr} > T_{cu} \text{ and } T_o < T_{cu} \\ 0, \text{if not} \end{cases} \tag{50}$$

resulting in the degree hour for free cooling of:

$$dh_{fr} = (T_{fr} - T_{cu})\delta_{fr} \tag{51}$$

If cooling is needed – that is, $T_{fr} > T_{cu}$ – but the outdoor temperature, T_o, is higher than the upper limit of the comfort temperature, T_{cu}, then mechanical cooling is needed. The condition for mechanical cooling is:

$$\delta_{mc} = \begin{cases} 1, \text{if } T_{fr} > T_{cu} \text{ and } T_o \geq T_{cu} \\ 0, \text{if not} \end{cases} \tag{52}$$

resulting in the degree hour for mechanical cooling of:

$$dh_{mc} = (T_{fr} - T_{cu})\delta_{mc} \tag{53}$$

The conditions expressed by the equations (49), (50) and (52) are shown in Figure 1.16. The comfort range, delimited by the lower and the upper limits, T_{cl} and T_{cu}, has a seasonal shift that depends upon the outdoor temperature. The free-running temperature, T_{fr}, may be higher or

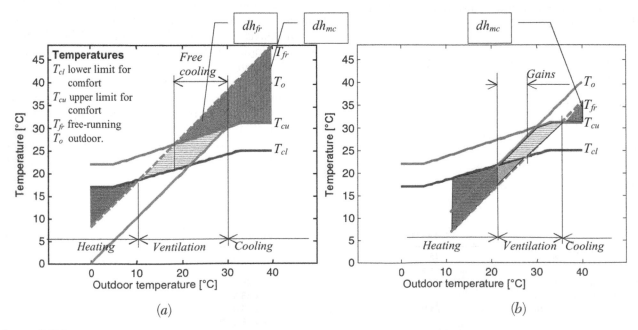

Source: C. Ghiaus

Figure 1.16 *Ranges for heating, ventilation and cooling when the free-running temperature is:*
(a) higher; and (b) lower than the outdoor temperature

lower than the outdoor temperature, T_o. Free cooling is feasible when $T_{fr} > T_o$ (see Figure 1.16a).

By summing the degree hours given by the equations (49), (51) and (53) for bins of outdoor temperature, we obtain the degree-hour distribution as a function of the outdoor temperature:

$$DH_c(T_o) \equiv DH_c(j) = \sum_i [T_{fr}(i,j) - T_{cu}(i,j)]\delta_c \quad (54)$$

$$DH_{fc}(T_o) \equiv DH_{fc}(j) = \sum_i [T_{fr}(i,j) - T_{cu}(i,j)]\delta_{fc} \quad (55)$$

for free cooling,

$$DH_{mc}(T_o) \equiv DH_{mc}(j) = \sum_i [T_{fr}(i,j) - T_{cu}(i,j)]\delta_{mc} \quad (56)$$

and for mechanical cooling. $T_{fr}(T_o)$, $T_{cl}(T_o)$, $T_{cu}(T_o)$ represent the free-running temperature and the comfort temperatures (the lower and the upper limits) that correspond to the bin j centred around T_o; $\sum_{T_{fr}}[\bullet]$ is the sum for all the values in the bin centred around T_o (ASHRAE, 2001).

The integral of degree-hour distribution:

$$DH_c = \sum_j \sum_i [T_{fr}(i,j) - T_{cu}(i,j)]\delta_c \quad (57)$$

for cooling,

$$DH_{fc} = \sum_j \sum_i [T_{fr}(i,j) - T_{cu}(i,j)]\delta_{fc} \quad (58)$$

for free cooling,

$$DH_{mc} = \sum_j \sum_i [T_{fr}(i,j) - T_{cu}(i,j)]\delta_{mc} \quad (59)$$

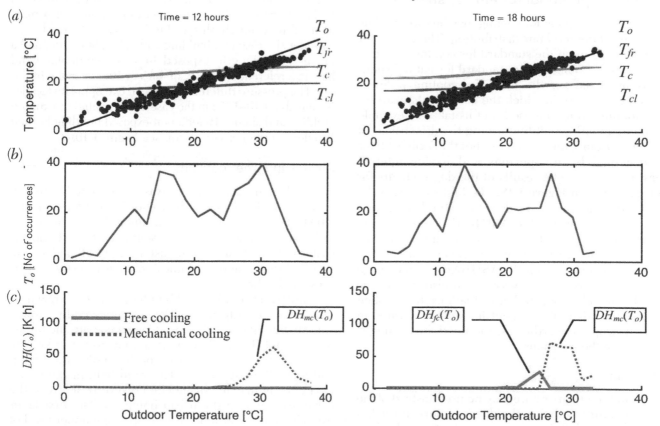

Source: C. Ghiaus

Figure 1.17 *Degree-hour distribution for cooling: (a) free-running temperature; (b) outdoor temperature distribution; (c) degree hour for free cooling and mechanical cooling*

and for mechanical cooling represent the classical definition of degree hours.

Figure 1.17 shows an example of the degree-hour distribution for 12 hours and 18 hours obtained from building simulation. The panels in the first row show the comfort zone, the outdoor temperature and the free-running temperature. The panels in the second row show the frequency distribution of the outdoor temperature in bins of 1° C at 12 hours and 18 hours. The integral of this distribution is the sum of occurrences over a year – that is, 365. The last row shows the degree hours obtained with equations (55) and (56). We may notice that free cooling is not possible at 12 hours because the free-running temperature is lower than the outdoor temperature due to the thermal inertia of the building. The integrals of $DH_{mc}(T_o)$ and of $DH_{fc}(T_o)$ represent the degree hours for mechanical cooling, DH_{mc}, and free cooling, DH_{fc}, respectively. At 18 hours, about 20 per cent of the energy needed for cooling may be saved by using ventilation.

Estimation of potential for energy savings

The energy consumption for cooling may be estimated by the integrals of degree-hour distribution. The energy for cooling depends upon the standard for comfort adopted; since the comfort range in the standard for natural ventilation is much larger, the need for cooling is lower compared to the case in which the comfort standard for air conditioning would be used. Let us take the example of a zone in an office building having the dimensions of 10m long × 7.5m wide × 2.5m in height (72m² of floor area), occupied by ten persons and having internal sources of 30W/m². The results of the degree hours for Europe are given in Figure 1.18. The first row shows the results for the case of the standard for natural ventilation and the second row for the ASHRAE comfort range. Comparison of Figures 1.18a and 1.18c shows approximately the same distribution; however, the energy consumption is twice when the ASHRAE comfort range is used. The pattern of energy savings is also similar (see Figures 1.18b and 1.18d). In both situations, free cooling may save more than 50 per cent for regions located in northern Europe; nevertheless, the need for cooling is much lower in these regions.

Conclusions

The free-running temperature may be used instead of the base temperature to define the degree hours in the bin method. The two definitions are equivalent; but the free-running temperature has the advantage that it can be applied when the indoor temperature varies, which is the case for non-air-conditioned buildings during the summer.

Ventilation has the potential to be used for cooling, especially in Northern Europe. Nevertheless, free cooling alone can cover only a small part of the need for cooling in Southern Europe.

Alteration of ventilation potential in the urban environment

The potential for cooling by ventilation is altered by the urban environment. Lower wind speed, higher temperatures and the need to make the building airtight in order to avoid noise and pollution penetration affect the estimated potential. These factors were experimentally studied in the URBVENT project. The following sections summarize the main results.

Airflow and temperatures in street canyons

In the urban environment, the mean velocity of wind is reduced significantly (about tenfold) and the wind direction is changed. As a result, wind-induced pressure on a building surface is lower. In order to have an approximate idea of how much this reduction amounts to, let us consider the case of a building with a height of 20m and a much larger length, exposed to a perpendicular wind having a reference velocity of 4m/s at 10m over the building. The pressure difference between two opposite façades is then about 10–15Pa in the case of an isolated or exposed building and about 0Pa for a non-exposed building located in a dense urban environment (see Figure 1.19).

Airflow in street canyons

The airflow in street canyons was experimentally studied in the URBVENT project by Georgakis and Santamouris (2003, 2004). The street canyons under investigation had different orientations (north–south; east–west; north-east–south-west; and south-east–north-west) and a wide range of aspect ratio. A synthesis of their investigations is presented below.

Airflow around isolated buildings is better known. It presents a lee eddy and a wake characterized by a lower velocity but higher turbulence, compared with the undisturbed wind. The airflow in urban canyons is less known, especially for lower velocities of the undisturbed wind and for oblique direction. The wind field in the urban environment may be divided into two vertical layers: the urban canopy and the urban boundary. The first layer extends from the ground surface up to the upper level of the buildings; the second is above the buildings (Oke, 1987). The flow in the canopy layer is influenced by the wind in the urban boundary layer, but also depends upon

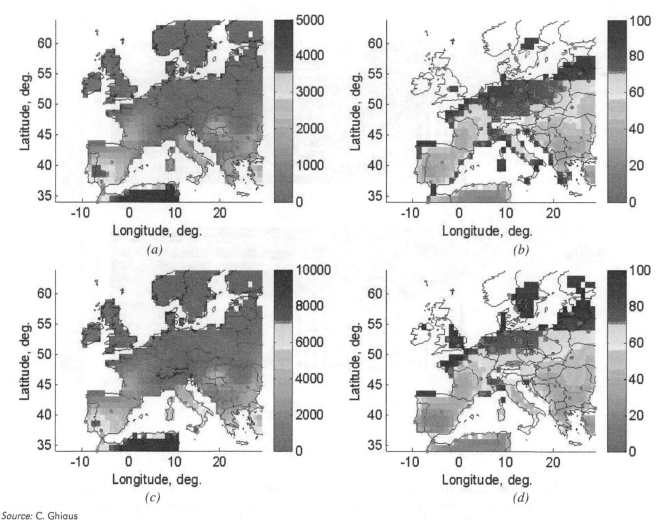

Source: C. Ghiaus

Note: Case of standard for natural ventilation: (a) degree hours for cooling (DH$_{fc}$+DH$_{mc}$); (b) percentage of free cooling. Case of ASHRAE comfort range: (c) degree hours for cooling (DH$_{fc}$+DH$_{mc}$); (d) percentage of free cooling

Figure 1.18 *Energy consumption and savings for an office building as a function of comfort standards.*

the geometry of the buildings and the streets, upon the presence of other obstacle, such as trees, and upon traffic. Generally, the speed in the canopy layer is lower than in the boundary layer.

Wind perpendicular to the canyon axis

The geometry of the street canyons is characterized by H (the mean height of the buildings in the canyon), by W (the canyon width) and by L (the canyon length). Based on these values, the street canyon is characterized by the aspect ratio (AR) H/W and the building by the aspect ratio L/H.

When the predominant direction of the airflow is approximately normal (± 15 degrees) to the long axis of the street canyon, three types of airflow regimes are observed as a function of the building (L/H) and canyon (H/W) geometry (Oke, 1987). When the buildings are well apart, ($H/W > 0.05$), their flow fields do not interact. At closer spacing, the wakes in the canyon are disturbed; the downward flow of the cavity eddy is reinforced. At even greater H/W and density, a stable circulatory vortex is established in the canyon because of the transfer of momentum across the shear layer of roof height. Because high H/W ratios are very common in cities, the skimming airflow regime has attracted considerable attention.

The velocity of the vortex depends upon the undisturbed wind speed. If the undisturbed wind has values higher than 1.5–2m/s and the aspect ratio is $H/W = 1$–1.5, the speed of the vortex increases with the speed of the

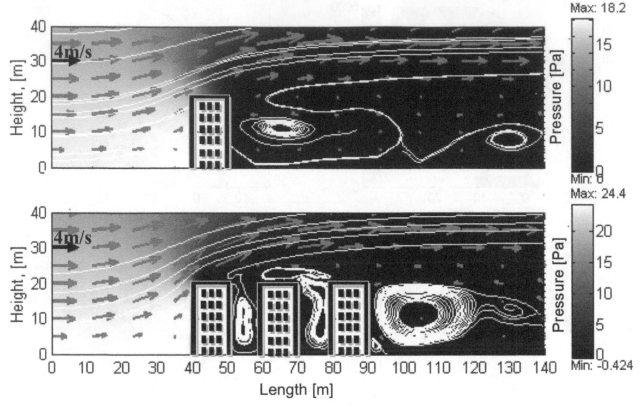

Source: C. Ghiaus

Figure 1.19 *Wind velocity and wind-induced pressure are reduced in the urban environment*

wind (De Paul and Sheih, 1986; Yamartino and Wiegand, 1986; Arnfield and Mills, 1994). If the aspect ratio is higher, a secondary vortex is observed for $H/W = 2$ and even a tertiary for $H/W = 3$ (Hoydysh and Dabbert, 1988; Nakamura and Oke, 1988). Since the lower vortexes are driven by the upper ones, their velocity is 5 to 10 times lower.

For wind speed greater than 5m/s, the relation between the undisturbed wind, u_{out}, and the air velocity in the street canyon, u_{in}, is almost linear:

$$u_{in} = p \cdot u_{out} \tag{60}$$

For $H/W = 1$, the coefficient p has values of between 0.66 and 0.75, air speed, u_{in}, being measured at about 0.06 H and u_{out} at 1.2 H (Nakamura and Oke, 1988).

Wind parallel to the canyon axis
The wind is considered parallel to the canyon axis when the main wind direction is almost along the canyon (± 15 degrees). As in the case of perpendicular winds, the airflow

in the canyon has to be seen as a secondary circulation feature driven by the flow above the roof (Nakamura and Oke, 1988). If the wind speed outside the canyon is below some threshold value (close to 2m/s), the coupling between the upper and the secondary flow is lost and the relation between wind speed above the roof and the air speed inside the canyon is characterized by a considerable scatter (Nakamura and Oke, 1988). For higher wind speeds, the main results and conclusions of the existing studies are that parallel ambient flow generates a mean wind along the canyon axis (Wedding et al, 1977; Nakamura and Oke, 1988), with possible uplift along the canyon walls as airflow is retarded by friction along the building walls and street surface (Nunez and Oke, 1977). This is verified by Arnfield and Mills (1994), who found that for winds that blow along the canyon, the mean vertical canyon velocity is close to zero. Measurements performed in a deep canyon (Santamouris et al, 1999) have also shown an along-canyon flow of the same direction.

Yamartino and Wiegand (1986) reported that the along-canyon wind component, v, in the canyon is directly proportional to the above-roof along-canyon component through a constant of proportionality that is a function of the azimuth of the approaching flow. The same authors found that, at least in a first approximation, $v = U \cdot \cos\theta$, where θ is the incidence angle and U the horizontal wind speed out of the canyon. For wind speeds of up to 5m/s, it was reported that the general relation between the two wind speeds appears to be linear, $v = p \cdot U$ (Nakamura and Oke, 1988). For wind speeds parallel to the canyon axis, and for a symmetric canyon with $H/W = 1$, it was found that p varies between 0.37 and 0.68, air speed being measured at about $0.06 H$ and $1.2 H$, respectively. Low p values are obtained because of the deflection of the flow by a side canyon. Measurements performed in a deep canyon of $H/W = 2.5$ (Santamouris et al, 1999) have not shown any clear threshold value where coupling is lost. For wind speed lower than 4m/s, the correlation between the wind parallel to the canyon and the air velocity along the canyon was not clear. However, statistical analysis has shown that there is a correlation between them.

The mean vertical velocity at the top of the canyon, resulting from mass convergence or divergence in the along-canyon component of flow, w, can be expressed as $w = -H \cdot \partial v/\partial x$, where H is the height of the lower canyon wall, x is the along-canyon coordinate and v is the x-component of motion within the canyon, averaged over time and the canyon cross-section (Arnfield and Mills, 1994). A linear relationship between the wind gradient $\partial v/\partial x$ in the canyon and the wind speed along the canyon was found. According to Arnfield and Mills (1994), the value of $\partial v/\partial x$ varies between -6.8×10^{-2} and 1.7×10^{-2} s^{-1}, while according to Nunez and Oke (1977) $\partial v/\partial x$ varies between -7.1×10^{-2} and 0 s^{-1}.

Wind oblique to the canyon axis

The more common case is when the wind blows at a certain angle relative to the long axis of the canyon. Unfortunately, the existing information on this topic is considerably less compared to perpendicular and along-canyon flows. Existing results are available through limited field experiments, primarily through wind tunnel and numerical calculations. The main results drawn from the existing research have concluded that when the flow above the roof is at some angle of attack to the canyon axis, a spiral vortex is induced along the length of the canyon – a corkscrew type of action (Nakamura and Oke 1988, Santamouris et al, 1999).

Wind tunnel research has also shown that a helical flow pattern develops in the canyon (Dabberdt et al, 1973; Wedding et al, 1977). For intermediate angles of incidence to the canyon long axis, the canyon airflow is the product of both the transverse and parallel components of the ambient wind, where the former drives the canyon vortex and the later determines the along-canyon stretching of the vortex (Yamartino and Wiegand, 1986).

Regarding the wind speed inside the canyon, Lee et al (1994) describe the results of numerical studies in a canyon with $H/W = 1$ and a free stream wind speed equal to 5m/s, flowing at 45 degrees relative to the long axis of the canyon. They report that a vortex is developed inside the canyon whose strength was less than the wind speed above the roof level by about an order of magnitude. Inside the canyon, the maximum across-canyon air speed was 0.6m/s and occurred at the highest part of the canyon. The vortex was centred at the upper middle part of the cavity and, in particular, to about 0.65 of the building height. The maximum wind speeds along canyon were close to 0.8m/s. Much higher along-canyon wind speeds are reported for the downward façade (0.6–0.8m/s) than for the upward façade (0.2m/s). The maximum vertical wind speed inside the canyon was close to 1m/s. Studies have shown that an increase of the ambient wind speed corresponds almost always to an increase of the along-canyon wind speed for both the median and the lower and upper quartiles of the speed (Santamouris et al, 1999).

Experimental values for low velocities of the undisturbed wind

When wind speed outside the canyon is less than 4m/s but greater than 0.5m/s, although the flow inside the street canyon seemed to have chaotic characteristics, extended analysis of the experimental data resulted in two empirical models (Georkakis and Santamouris, 2003). When the direction of the undisturbed wind is along the main axis of the canyon, the values from Table 1.5 can be used. When the direction of the undisturbed wind is perpendicular or oblique to the canyon, the values from Table 1.6 can be used.

Air and surface temperature in urban canyons

The effect of the urban heat island increases with the size of the urban conurbation (see Figure 1.20). But this temperature refers to the undisturbed wind that flows in the urban boundary layer. The temperature distribution in the urban canopy layer is greatly affected by the radiation balance. Solar radiation incident on urban surfaces is absorbed and then transformed to sensible heat. Most of the solar radiation impinges upon roofs and upon the vertical walls of buildings; only a relatively small part reaches ground level. In fact, in street canyons, the air temperature is much lower than in the boundary layer, partially compensating for the heat island effect.

The optical and thermal characteristics of materials used in urban environments, especially the albedo for

Table 1.5 *Values for air speed inside the canyon when the wind blows along the canyon*

Wind speed outside the canyon (U)	Wind speed inside the canyon	Typical wind speed values in the canyon	
		Lowest part	Highest part
0 < U < 1	0.3–0.7m/s	0.3m/s	0.7m/s
1 < U < 2	0.4–1.3m/s	0.4m/s	1.3m/s
2 < U < 3	0.4–1.5m/s	0.4m/s	1.5m/s
3 < U < 4	0.4–2.2m/s	0.4m/s	2.2m/s

Source: Georkakis and Santamouris (2003)

solar radiation and emissivity for long-wave radiation, have a very important impact on the urban energy balance (Santamouris and Georgakis, 2004). The use of high-albedo materials reduces the amount of solar radiation absorbed by building envelopes and urban structures, and keeps their surfaces cooler. Materials with high emissivity are good emitters of long-wave radiation and readily release the energy that has been absorbed as short-wave radiation (Santamouris, 2001).

Georgakis and Santamouris (2003) measured the temperature of the air inside and outside street canyons and the temperatures of wall surfaces. Some representative results are given in the box plots of Figure 1.21. The upper and lower lines of the 'box' are the 25th and 75th percentiles of the sample; the distance between the top and the bottom of the box is the interquartile range. The line in the middle of the box is the sample median. The whiskers extending above and below the box show the extent of the rest of the sample. The plus signs are indications of the data outliers.

The air temperature was 3°–5° C lower inside the canyon than outside. Figure 1.21a shows the median outside air temperature of 30.5° C, while the median of the air temperature inside the canyon was about 26° C for all the heights at which it was measured – that is, 3.5m, 7.5m, 11.5m and 15.5m. Georgakis and Santamouris (2003) explain that the homogeneity of the air temperature inside the canyon is due to the advection. The fact that air temperature outside the canyon is higher than inside can be due to street orientation, which

permits many hours with shadow and a very good airflow inside the canyon due to the large aspect ratio (H/W = 3.3).

The wall surface temperature measurements of Georgakis and Santamouris (2003) show that the maximum simultaneous difference of the two façades was up to 10°–20° C (Figure 1.21b). Comparison of the maximum difference of daily temperatures of the building façades and the surface temperature of the street shows that at street level the temperature was 7.5° C higher than in the lower parts of the canyon.

Noise level and natural ventilation potential in street canyons

High external noise levels are often invoked to justify the use of air conditioning in commercial and residential buildings. Methods of estimating noise levels in urban canyons are necessary if the potential for naturally ventilating buildings is to be assessed. The estimated noise levels can then be compared to the level of noise at which building occupants might be motivated to close windows in order to keep out the noise, which would also compromise the natural ventilation strategy. Studies concerning the noise attenuation in street canyons were carried out by Nicol and Wilson (2003) under the framework of the URBVENT project. A synthesis of their findings is presented as follows.

A series of daytime noise measurements were made in Athens in canyon streets with an aspect ratio (H/W) varying from 1.1 to 5.3. The main purpose of the measure-

Table 1.6 *Values for air speed inside the canyon when the wind blows perpendicular or oblique to the canyon*

Wind speed outside the canyon (*U*)	Wind speed inside the canyon Near the windward façade of the canyon		Wind speed inside the canyon Near the upwind façade
	At the lowest part	At the highest part	
0 < U < 1	0.4m/s	0.7m/s	0.4m/s
1 < U < 2	0.4m/s	1.3m/s	0.4m/s
2 < U < 3	0.6m/s	1.5m/s	0.6m/s
3 < U < 4	0.7m/s	3m/s	0.7m/s

Source: Georkakis and Santamouris (2003)

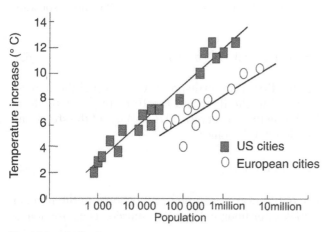

Heat island effect

3° C	London	6° C	Berne
4° C	Essen	7° C	Malmoe, Zurich
5° C	Friebourg	14° C	Paris, Athens
6° C	Gotemborg		

Source: Oke (1982)

Figure 1.20 *Maximum differences in urban and rural temperature for US and European cities*

potential. A simple model of the noise level has been developed using a linear regression analysis of the measured data. The model can be used to predict the fall-off (attenuation) of the noise level with height above street level.

The attenuation is found to be a function of street width and height above the street; but the maximum level of attenuation (at the top of the canyon) is almost entirely a function of the aspect ratio, except in narrow streets. Background noise (L_{90}) suffers less attenuation than foreground noise (L_{10}) with height.

The measured attenuation was compared to results from acoustic simulation. The simulation gives comparable values for attenuation in the canyons. The simulation is used to estimate the effectiveness of balconies for reducing external noise levels in canyon streets. Measurements in a survey throughout Europe are used to estimate the potential for natural ventilation in canyon streets in Southern Europe. The survey and simulations are used to assess the effect of noise on the natural ventilation potential of canyon-type streets, and to suggest limitations to the use of natural ventilation as a function of canyon geometry.

Aspect ratio of street canyons

Canyon-like streets in cities such as Athens vary considerably in the width and height of the buildings that border

ments was to examine the vertical variation in noise in the canyons in order to give advice on natural ventilation

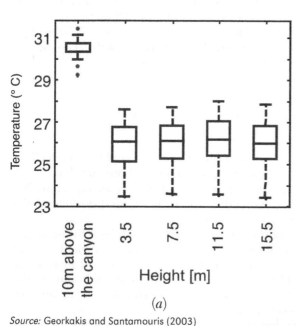

(a)

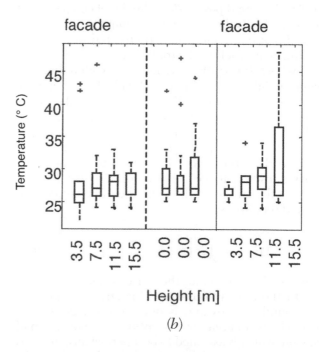

(b)

Source: Georkakis and Santamouris (2003)

Figure 1.21 *Box plots of temperature distribution in a street canyon: (a) vertical distribution in the centre of the canyon; (b) temperature of the walls*

them. While the assumption is that the two sides of the street are of the same height, this is frequently not the case. Nicol and Wilson (2003) have taken the average height of the street façades to give $AR = (H_1 + H_2)/2W$ (where H_1 and H_2 are the respective heights of the sides of the canyon – assumed to be comparable). The façades themselves also vary considerably, some being plain and some with balconies. Most residential streets have balconies; but even some office buildings are constructed with balconies. At ground level the situation can be more complex. The ground floor is often set back with colonnades. Paper stalls and other objects often litter the pavement.

A simple model of noise in canyons

The purpose of Nicol and Wilson's (2003) investigation was to provide a method for estimating the fall-off in noise level with the height in street canyons. The traffic noise, as measured at various locations in the canyons, is a combination of the direct sound and quasi-reverberation in the canyon. The term quasi-reverberation is used to denote a type of reverberation that is not diffuse but consists primarily of flutter echoes between the façades lining the street. Thus, the sound pressure, p, is:

$$p^2 \propto P(dc + rc) \tag{61}$$

where P is the sound power, dc is the direct component of the sound and rc is the reverberant component.

The direct component may be treated in two ways depending upon whether the traffic is considered as a line source (where the traffic stream is considered as the source) or point source (where each vehicle is separately responsible for the noise). For a line source, the direct component, dc, is inversely proportional to the distance from the source; for the point source, the direct component, dc, is inversely proportional to the square of the distance. If the street width is W and the height of the measuring position above the ground is H, assuming the source is in the middle of the road, the distance between source and receiver is:

$$d = \left((W/2)^2 + H^2 \right)^{1/2} \tag{62}$$

For the reverberant sound, the noise is related approximately to the absorption area. This strictly applies to diffuse sound sources and is only approximate in this context. The main area for absorption is the open top of the canyon, which is assumed to be a perfect absorber and whose area per metre of street equals W, the width of the

street. If the absorption coefficient is 0.05, the absorption area is:

$$W_A = W + 0.05W + 2 \cdot 0.05H \tag{63}$$

where the first term corresponds to the top of the canyon, the second to the floor and the third to the two lateral walls. Using the aspect ratio ($AR = W/H$) of the street, the expression (63) becomes:

$$W_A = W(1.05 + 0.1 \cdot AR) \tag{64}$$

The sound power is assumed proportional to the number of vehicles per hour, n. For a line source, its expression is:

$$p^2 = a\frac{n}{d} + b\frac{n}{W_A} + c \tag{65}$$

For a point source, its expression is:

$$p^2 = a\frac{n}{d^2} + b\frac{n}{W_A} + c \tag{66}$$

where a, b and c are constants related to the direct component, the reverberant component and to any background environmental noise entering the canyon, respectively. In general, the contribution of c will be small. Measurements on the roof top of a building in a pedestrian area behind vehicular streets in the centre of Athens gave $L_{Aeq} = 55$dB. In the vehicular streets, few noise levels below $L_{Aeq} = 70$dB were recorded. L_{90} averaged 66dB. The expressions were developed into the form:

$$L_p = 10log_{10}\left(n\left(\frac{a}{d_1} + \frac{b}{W_A} \right) + c \right) \tag{67}$$

where, by the normal definition of sound level in decibels (dB), L_p is the noise level for a sound pressure level p and is equal to $10log_{10}p$ and d_1 is d or d^2 – see equations (65) and (66) – depending upon the assumption about the shape of the noise source.

In equation (67), the value of L_p relates to the height above the canyon floor, H, through the variable d_1. An estimation of the constants a, b and c will enable the change of L_p with H to be determined. The values of the constants a, b and c have been estimated using multiple regression analysis.

Correlation analysis suggested that the line source of sound was a better model for these data. The correlation between measured p^2 and calculated direct noise levels (ignoring the reverberant component) was 0.86 for the linear source assumption, compared to 0.68 for the point source. Further tests showed that there was little change in correlation whether W or W_A were used. The advantage of using W is its simplicity, while W_A would allow the computation of any increase in the absorption coefficient of the façade. However, there is no measured data to evaluate any change of absorption.

Regression analysis gives optimal values for the constants a, b and c for equation (65), which becomes:

$$p^2 = 17.4 \cdot 10^4 . D_2 + 5.34 \cdot 10^4 . RV - 411 * 10^4 \qquad (68)$$

Then,

$$L_{eq} = 10 log_{10} p^2 \qquad (69)$$

where L_{eq} is the noise level at height H above the street; and D_2 is a function of three variables, H, W and n, the number of vehicles (n is assumed to be proportional to the noise generated). Two of these variables (n and W) are also included in RV, together with the aspect ratio, AR, of the canyon. There is a logical problem with a negative value for c since the value of p^2 cannot be negative. This value may be due to a curvature in the relationship which the linear regression cannot take into account. The effect of c on the value of p^2 is, in any case, generally small. Figure 1.23 shows that the measured value is well predicted by the calculated value ($R^2 = 0.75$).

In order to facilitate the visualization, a simplifying assumption has been made that the traffic level is a function of street width. In these data, the correlation between traffic intensity, expressed in number of vehicles per hour, n, and street width W [m] was $R = 0.88$ and the regression relationship (shown in Figure 1.23) was:

$$n = 137W - 306 \qquad (70)$$

Using this simplifying assumption, values of expected noise level at different heights for a particular value of street width w can be calculated. Assuming that the traffic in the canyons follows the relationship shown in equation (70), the expected daytime noise level becomes purely a function of the geometry of the street. Figure 1.22 shows the expected noise levels in Athens at different street widths and heights above the streets. Figure 1.24 shows the implication of this for the natural ventilation potential of office units at a height H above street level.

The results introduced above from the Smart Controls and Thermal Comfort (SCATS) project (McCartney and Nicol, 2002) suggest that the tolerable noise level in European offices is around 60dB. At the same time, the noise attenuation at an open window is accepted as 10–15dB. Thus, an outdoor noise level of 70dB or less is likely to be acceptable. Using special methods and window designs, a further 3–5dB attenuation is possible. For the traffic conditions of Athens, street widths that will give acceptable conditions at different heights above the street are indicated by 'OK'. Street widths that will give unacceptable conditions for buildings with open windows near street level are also indicated by 'Not OK'. Between these two, there are possibilities for acceptable conditions with careful design.

Because the noise measurements reported here were exclusively taken during the day, the implications for natural ventilation potential during the evening or night time can only be estimated, although it should be remembered that in *unoccupied* offices, the outdoor noise level will be irrelevant to night ventilation. The limitations imposed by noise to natural ventilation potential will be important to residences at night. While overall noise levels will almost certainly be lower at night, the reduction will be offset by the greater sensitivity to noise. In addition, the occasional passing vehicle will be almost as loud, though as a point source its attenuation with distance above the street will be according to equation (66) and will be greater than for daytime noise. Notice that the attenuation of the loudest daytime noises (as suggested by the attenuation of the L_{10}) is greater than the attenuation of the L_{eq} for similar reasons.

Configurations in which natural ventilation is possible are indicated (OK), as are those in which it is ruled out (Not OK). Between these two extremes is a region in which there are possibilities for design solutions (based on Figure 1.23).

Attenuation caused by balconies

The drop in noise level close to the building façade caused by the balcony can be estimated from the difference in noise level at 1.5m from the façade compared to that at the level of the façade. This attenuation due to the balcony varies slightly with the street width, but is almost entirely due to the height of the balcony above the street. Figure 1.25 shows the balcony contribution to noise attenuation. The height above the street is signified by the floor. The balcony contribution is just under 2dB at the first floor, and increases slightly on each of the next two floors. On the fourth (top) floor, the contribution rises sharply to over 3dB (presumably because of the relative lack of reflected sound from the canyon walls).

Sound attenuation caused by balconies will mostly be attributable to the depth of the balcony from the building

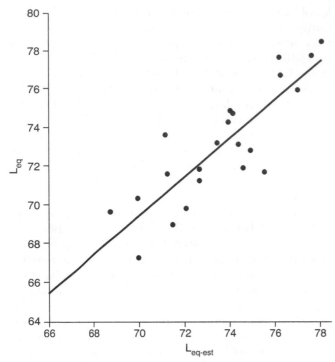

Source: Nicol and Wilson (2003)

Figure 1.22 *Measured L_{eq} (dB) against the predicted $L_{eq\text{-}est}$ (dB); $R^2 = 0.75$*

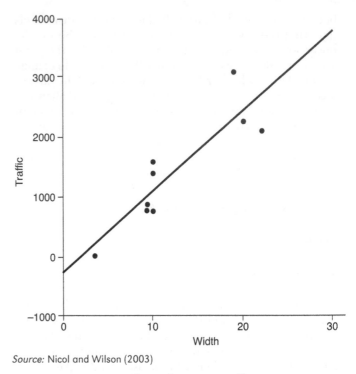

Source: Nicol and Wilson (2003)

Figure 1.23 *Correlation between traffic intensity, n (vehicles per hour) and street width, W (m); $R^2 = 0.78$*

line of the wall and the solidity of the front of the balcony. Balconies that are narrower or that do not have a continuous and solid wall will be likely to give less protection from noise. Much of the noise entering the building through the window will have been reflected from the underside of the balcony on the floor above, particularly on the floors higher above the street where the angle of incidence of the direct noise is higher. In these cases, there may be an advantage to be gained from applying sound-absorbent finishes to the underside of balconies.

Noise in offices

Noise is often used as an argument against natural ventilation and for supporting air conditioning. Noise is thus one of the factors when the potential for natural ventilation is under consideration. The European project SCATS (financed by the European Commission in the Joule Programme) sought to develop control algorithms for both natural ventilation and air-conditioning systems based on the theory of adaptive thermal comfort (McCartney and Nicol, 2002). Instrumentation was built to undertake office surveys in 25 offices in five European countries, amounting to 850 people, on a monthly basis over the year (Solomon et al, 1998). Following research related to the

noise limitations to the use of natural ventilation in buildings in urban areas (Wilson, 1992; Wilson et al, 1993; Dubiel et al, 1996; Nicol et al, 1997), the noise was measured at each work station and a question was answered concerning the noise environment. The response to the acoustic environment was measured on a seven-point scale. The implication of the results is that current noise standards are unnecessarily stringent and that an outdoor L_{Aeq} noise level of some 55–60dB will be accepted. Although there is no direct evidence from Athens, research suggests that in countries such as Greece, where natural ventilation and open windows are common, tolerance to noise is generally greater (Dubiel et al, 1996).

Conclusions

This section described the noise climate of Athens, with its distinctive canyon street formation and single carriageways. The effect on the urban noise climate of different street formations (point-block buildings, wider pavements and tree planting) will generally be to reduce the noise, particularly the reverberant component. Wider streets and dual carriageways will increase the flow and speed of traffic and may have the effect of increasing the noise

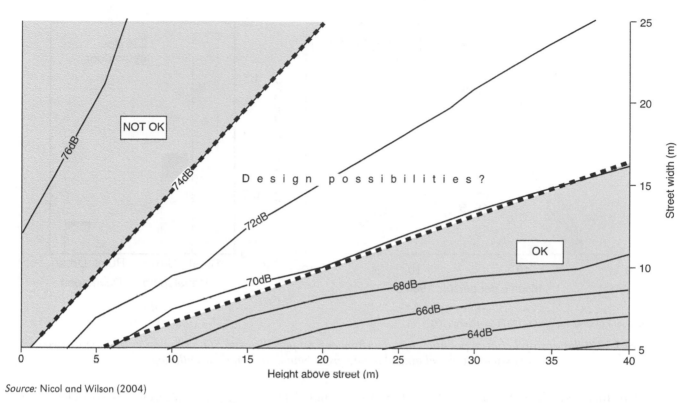

Source: Nicol and Wilson (2004)

Figure 1.24 *Contours of noise level at different heights above the street and at different street widths*

level. Part of the purpose of this study has been to develop a methodology that can be used in other contexts. In different cities (or, indeed, in Athens at different times of the day), the basic components of the noise environment will be the same, although the regression coefficients given in equation (68) could be very different. The correlation between vehicle numbers and street width (see Figure 1.23) may not be so strong in other contexts.

Outdoor–indoor pollutant transfer

Outdoor air pollution is commonly considered as another barrier for using natural ventilation since filters cannot be employed as in mechanical or air-conditioning systems. But two aspects are important in this assessment: the improvement of outdoor air quality with economic development, and the different nature of outdoor and indoor pollutants.

First, economic growth has a tendency to ameliorate outdoor air quality after its initial negative effect (see Figure 1.26a). While material progress is sought, pollution increases with economic growth. But when financial and technological resources are considered sufficient, the cost of pollution is included when evaluating quality of life, and actions to reduce pollution are enforced. It is also

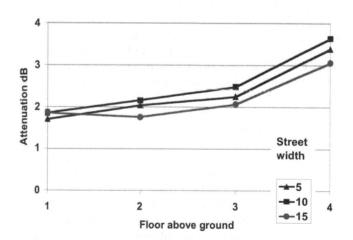

Note: The height of the floors above street level is assumed to be 7m, 10.5m, 14m and 17.5m. Balconies are 1.2m deep and 3m wide, and the values are averaged across the width of the façade.

Source: Nicol and Wilson (2004)

Figure 1.25 *Simulated contribution of balconies to reducing noise level at the surface of the building in five-storey buildings in street canyons*

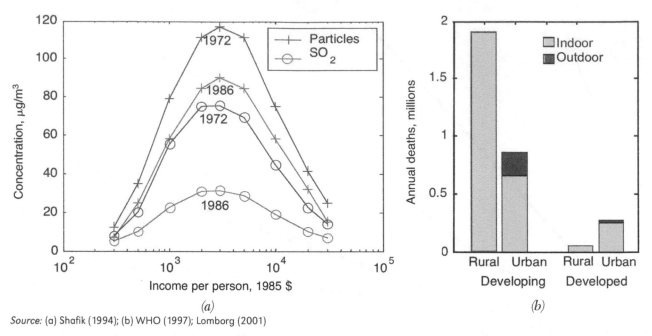

Source: (a) Shafik (1994); (b) WHO (1997); Lomborg (2001)

Figure 1.26 *Relationship between pollution and development: (a) particles and SO₂ pollution in relation to income; (b) estimated global annual deaths from indoor and outdoor pollution*

noticeable that overall outdoor pollution decreases regardless of income.

Second, the type and concentration level of pollutants are different indoors and outdoors. According to the World Health Organization (WHO), the key outdoor pollutants are sulphur dioxide (SO_2), nitrogen dioxide (NO_2), carbon monoxide (CO), ozone (O_3), particulate matter and lead (Pb). Guideline values are given for these pollutants (WHO, 2000). Indoor pollutants include, among others, environmental tobacco smoke, particles (biological and non-biological), volatile organic compounds, nitrogen oxides, lead, radon, carbon monoxide, asbestos and various synthetic chemicals. Indoor air pollution has been associated with a range of health effects, from discomfort and irritation to chronic pathologies and cancers. In an effort to conserve energy, modern building design has favoured tighter structures with lower rates of ventilation (WHO, 2000). The impact of indoor pollution on health is much more important than that of outdoor pollution (see Figure 1.26b). Indoor pollution problems differ in developed and developing countries. In developed countries, indoor pollution is caused by low ventilation rates and the presence of products and materials that emit a large variety of compounds; in developing countries, it is caused by human activity, especially combustion processes.

Epidemiological studies showed associations between health events (such as death and admission in hospitals) and daily average concentrations of particles, ozone, sulphur dioxide, airborne acidity, nitrogen dioxide and carbon monoxide. Although the associations for each of these pollutants were not significant in all studies, taking the body of evidence as a whole, the consistency is striking. For particles and ozone, it has been accepted by many that studies provide no indication of any threshold of effect, and an assumption of linearity was made when defining the exposure–response relationships (WHO, 2000).

Indoor air quality is related to outdoor air pollutant concentration through the rate of air change and reactivity of the pollutant. The façade air tightness, as an intrinsic characteristic of buildings, represents a key factor since it is the main link between indoor and outdoor environments and is an important component of the natural ventilation property of buildings.

Experimental study of the outdoor–indoor pollutant transfer

The key outdoor pollutants (SO_2, NO_2, CO, O_3, suspended particle matter and lead) are usually monitored in large cities. The mean level of sulphur dioxide and lead are equal indoors and outdoors. Ozone and nitrogen dioxide react with building material, resulting in a lower concentration indoors than outdoors when the building is airtight. The particle matter transfer depends upon the

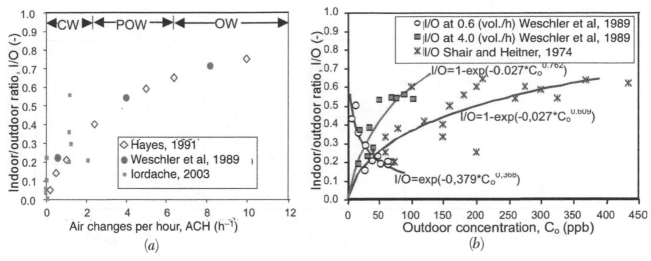

Source: Iordache (2003)

Note: (CW: closed window; POW: partially opened window; OW: open window);

Figure 1.27 *The variation of indoor to outdoor ozone ratio as a function of: (a) air changes per hour (b) outdoor concentration*

particle size. The experimental results show that the ratio between indoor and outdoor concentration (I/O) also depends upon the outdoor concentration of the pollutant.

The indoor–outdoor ratio was studied for ozone, nitrogen dioxide and particle matter in the framework of the URBVENT project and of the French research programme on improving the air quality at local scale (PRIMEQUAL). A literature review shows that the ozone concentration is lower indoors than outdoors and that the ratio increases with the airflow rate (see Figure 1.27a). In the case of closed windows (shown by CW in Figure 1.27a), the transfer is more complex. Our experimental results confirm that this complexity comes from the building façade's air tightness. Other studies showed that the indoor–outdoor ratio also depends upon the outdoor concentration (see Figure 1.27b). These two parameters were considered as explanatory variables in the prediction of the I/O ratio.

Outdoor–indoor transfer mapping

The input–output concentration ratio (I/O) is mapped on outdoor concentration, C_o, and the three main levels of air tightness of the façade: 'airtight', $Q_{4Pa} \approx 0m^3/h$; 'permeable', $Q_{4Pa} \approx 150m^3/h$; and 'very permeable', $Q_{4Pa} \approx 300m^3/h$ (see Figure 1.28). The I/O ratio was determined for closed windows (measurements were taken during the night). Since the room volume was about 150m^3, the maximum air change per hour was approximately 2ACH.

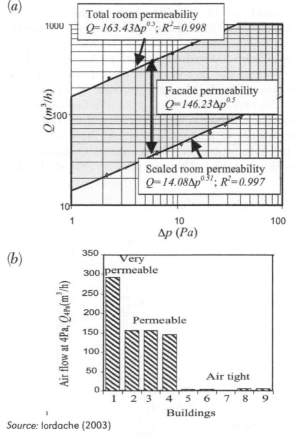

Source: Iordache (2003)

Figure 1.28 *Building classification according to permeability*

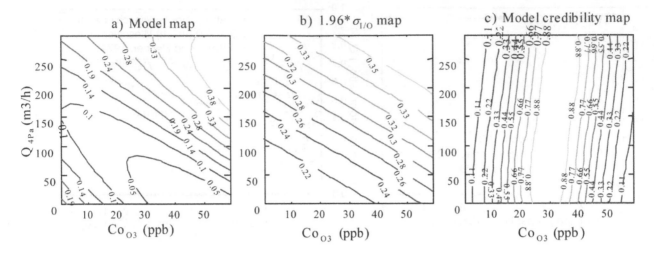

Source: Iordache (2003)

Figure 1.29 *Ozone outdoor–indoor transfer: (a) input/output (I/O) ratio; (b) precision; (c) degree of confidence*

Ozone

The I/O ratio diminishes with the outdoor concentration for the airtight façades and increases for the other two types of façades (see Figure 1.29). Two clusters were found: the first is located in the zone of the airtight façade ($c_{Q4Pa} \approx 5m^3/h$) and the middle-ranged outdoor concentration ($c_{Co} \approx 28ppb$); the second one is placed in the zone of the 'most permeable' façades ($c_{Q4Pa} \approx 292m^3/h$) for middle-ranged outdoor concentration ($c_{Co} \approx 36ppb$). The two peaks of the model are placed in the zone of the 'airtight' façade with low outdoor O_3 concentration and the zone of the 'most permeable' façade with high outdoor O_3 concentration (see Figure 1.29a). The second map (Figure 1.29b)

presents the precision of the model expressed by the dispersion of the points in the database. The map shows that the smallest dispersion of the I/O value is 0.18 while the higher dispersion is 0.38 (Figure 1.29b). The third map presents the credibility of the first two maps (Figure 1.29c). It is higher in the zones where more measurement points were collected – that is, in the proximity of the two clusters. The highest credibility zone ($CR > 0.5$) corresponds to the middle-ranged outdoor O_3 concentrations, between the centres of the two clusters, while the lowest credibility zones ($CR < 0.25$) are for the 'most permeable' façade, with low outdoor concentrations, and the 'airtight' façade, with high outdoor concentrations.

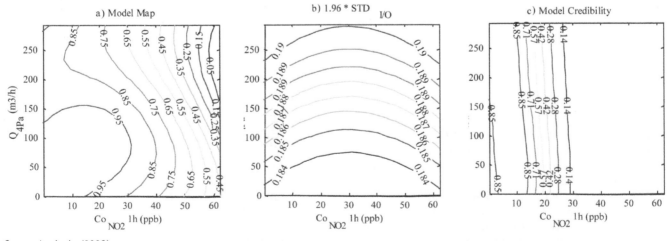

Source: Iordache (2003)

Figure 1.30 *NO_2 outdoor–indoor transfer: (a) I/O ratio; (b) precision; (c) degree of confidence*

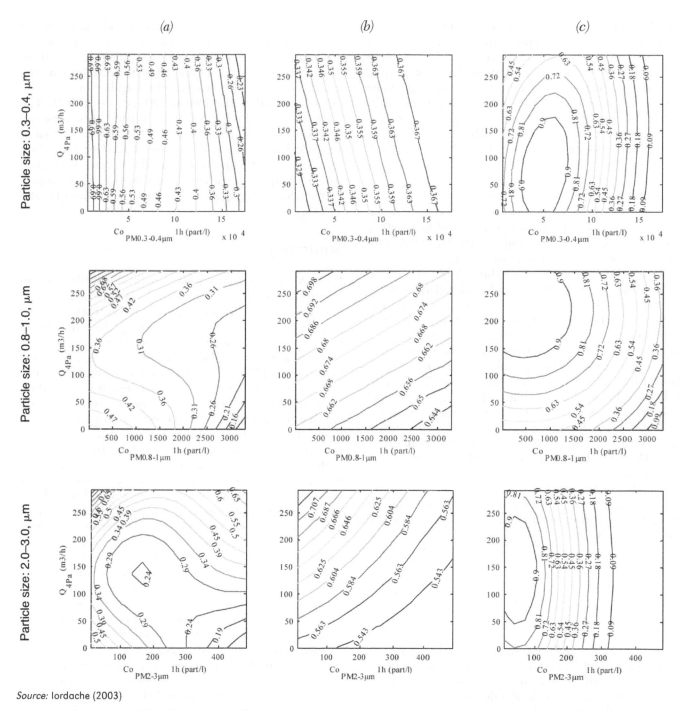

Source: Iordache (2003)

Figure 1.31 *Particle matter outdoor–indoor transfer: (a) I/O ratio; (b) precision; (c) degree of confidence*

Nitrogen dioxide

The same three parameters were calculated for nitrogen dioxide. The I/O ratio diminishes with the outdoor concentration regardless of the façade air tightness. The values of the I/O ratios corresponding to the airtight façades are slightly higher then those corresponding to 'permeable' or 'very permeable' façades (see Figure 1.30a). The model precision is almost the same for the entire domain (Figure 1.30b). The credibility is higher in the zones where more measurements are available – that is, in the proximity of the clusters. Two clusters were found for lower outdoor concentration (Co_{NO2} < 15ppb): one cluster corresponds to the 'airtight' buildings, while the second one corresponds to the 'very permeable' façades. The credibility parameter diminishes with the rise of the outdoor concentration, with values of between 0 and 0.5 for outdoor concentrations higher than 20ppb (see Figure 1.30c).

Particle matter

The same three values are estimated for the penetration indoors of three different-sized intervals of the particle matter: 0.3–0.4μm, 0.8–1μm and 2–3μm. Similar conclusions can be drawn for all three particle sizes.

The I/O ratios diminish with the outdoor concentration regardless of the building façade's air tightness or the size of the particles. For the size interval of 0.3–0.4μm (see Figure 1.31a), the model surface is relatively plane, so the I/O ratio diminishes linearly with the outdoor concentration. For the other two sizes, the model maps present a concavity in the model surfaces for the small values of the outdoor concentration and the 'permeable' façades (see Figure 1.31, middle and bottom). In contrast to the first two size intervals, the model surface of the class 2–3μm presents I/O ratios of 0.65 corresponding to a high outdoor concentration and a 'very permeable' façade. However, the prediction credibility index is very small for that zone.

The dispersion of the I/O ratio presents almost constant values for all outdoor pollution ranges and façade permeabilities. The value of the index characterizing this dispersion is about 0.33 for the first size interval (see Figure 1.31b, top), while it is twofold higher for the last two size intervals (see Figure 1.31b, middle and bottom). The prediction credibility presents the same diminishing trend with the outdoor concentration (see Figure 1.31c).

Conclusions

Outdoor and indoor pollution have a different nature and usually refer to different types of pollutants. After a threshold in wealth is attained, when the financial and technological means are available, outdoor pollution diminishes with economic development.

Under the joint framework of the URBVENT project and the PRIMEQUAL French programme, an experimental study of outdoor–indoor pollution transfer was conducted in nine schools. The pollutants studied were ozone, nitrogen dioxide and 15 sizes of particle matter. Three maps were calculated for every pollutant: the I/O

ratio, the precision of this estimation and the degree of confidence in the I/O ratio. The ratio of indoor–outdoor concentration was determined as a function of airflow through the façade and of the outdoor concentration. The indoor concentration was smaller inside than outside. Ozone presented the lowest I/O ratio (0.1–0.4). The highest I/O ratio for nitrogen dioxide was between approximately 0 and 0.95. The I/O ratio for particle matter depended upon the particle size. The most important variation (0.25–0.70) was measured for particles of small size (0.3–0.4μm); particles of larger size (0.8–3μm) represented a lower, but comparable, variation of the I/O ratio (0.3–0.7).

Summary

When used for free cooling, natural ventilation can replace air-conditioning systems for long periods of time during the year. Consequently, natural ventilation has the potential to save energy for cooling. But the urban environment presents disadvantages in applying natural ventilation: lower wind speeds, higher temperatures due to the effect of the urban heat island, noise and pollution.

The wind in street canyons has much lower values compared to undisturbed wind. When undisturbed wind has values larger than 2–4m/s, a correlation exists between it and the wind in the street canyons. When a 2m/s or stronger wind blows perpendicular to a street canyon, a vortex develops in the canyon. If the wind is parallel to the canyon axis, the vertical velocity in the canyon is very low. Based on experimental data, an empirical model was developed for oblique wind with velocity lower than 4m/s. The temperature measured inside the canyon streets was about 5° C lower than that of the canopy layer, partially compensating for the effect of the urban heat island.

Noise is another barrier to applying natural ventilation. Experimental data permitted the development of a simple model for estimating noise attenuation with height above the street level and the aspect ratio. A monogram, which is useful for decisions in the initial stages of the design, was obtained by making the assumption that traffic intensity (and, consequently, noise level) is dependent upon street width. It was shown that the balconies reduce the noise level by about 2dB on the first floor and by more than 3dB on the fourth floor.

Pollution is also considered to restrict the application of natural ventilation. Two aspects are important when pollution is analysed: first, the type and level of pollution are different outdoors compared with indoors; second, although economic development initially increases pollution levels, when financial and technological resources are available, economic growth induces a reduction in outdoor pollution. Experiments in the URBVENT project and in a related French programme showed that the indoor–outdoor pollutant ratio (I/O) depends upon the façade air tightness and the outdoor concentration. Experiments were conducted for ozone, nitrogen dioxide and particle matter. The most important reduction was noticed for ozone, with an I/O ratio of 0.05 to 0.33 (a higher I/O ratio was measured for a higher outdoor ozone concentration). The I/O ratio for nitrogen dioxide was between 0.05 and 0.95, with lower values for higher outdoor concentration. For particle matter, the I/O ratio was between 0.20 and 0.70, with values depending upon the outdoor concentration and upon the size of the particles.

Acknowledgements

The URBVENT: Natural Ventilation in Urban Areas project was supported by the European Commission in the Fifth Framework Programme for Research and Technological Development. The project on pollutant transfer was supported by the French government in its PRIMEQUAL programme. The authors are very grateful to the participants in these projects: M. Santamouris; C. Georgakis (University of Athens); F. Nicol; M. Wilson (London Metropolitan University); V. Iordache; and P. Blondeau (Université de La Rochelle).

References

Albrecht, F. and Grunow, J. (1935) 'Ein Beitrag zur Frage der vertikalen Luftzirkulation in der Grossstandt', *Metorologie Zeitung*, vol 52, pp103–108

Allard, F. (ed) (1998) *Natural Ventilation in Buildings: A Design Handbook*, James and James, London

Arnfield, A. J. and Mills, G. (1994) 'An analysis of the circulation characteristics and energy budget of a dry, asymmetric, east–west urban canyon. I. Circulation characteristics', *International Journal of Climatology*, vol 14, pp119–134

ASHRAE (American Society of Heating, Refrigerating and Air-conditioning Engineers) (2001) *ASHRAE Handbook Fundamentals*, ASHRAE, Atlanta, GA

ASTM E 779 (1999) *Standard Test Method for Determining Air Leakage Rate by Fan Pressurization*, West Conshohocken, PA

Axley, J. W. (2000) *AIVC TechNote 54: Residential Passive Ventilation Systems: Evaluation and Design*, Air Infiltration and Ventilation Centre, Coventry

Axley, J. W. (2001) *Application of Natural Ventilation for US Commercial Buildings: Climate Suitability and Design Strategies and Methods*, GCR-01-820, National Institute of Standards and Technology, Building and Fire Research Laboratory, Gaithersburg, MD

Awbi, H. (1998) 'Ventilation', *Renewable and Sustainable Energy Review*, vol 2, pp157–188

Bowman, N. T. (2000) 'Passive downdraught evaporative cooling', *Indoor and Built Environment*, vol 9, no 5, pp284–290

Brager, G. and de Dear, R. (1998) 'Thermal adaptation in the built environment: A literature review', *Energy and Buildings*, vol 27, pp83–96

Brager, G. and de Dear, R. (2000) 'A standard for natural ventilation', *ASHRAE Journal*, October, pp21–28

BRE (1994) 'Natural ventilation in non-domestic buildings', *BRE Digest*, no 399, Building Research Establishment, Garston, UK

BS EN 1793-3 (1998) *Road Traffic Noise Reducing Devices: Test Method for Determining the Acoustic Performance, Normalised Traffic Noise Spectrum*, British Standards Institution, London

Chang, P. C., Wang, P. N. and Lin, A. (1971) 'Turbulent diffusion in a city street', in *Proceedings of the Symposium on Air Pollution and Turbulent Diffusion*, 7–10 December 1971, Las Cruces, New Mexico, pp137–144

Dabberdt, W. F., Ludwig, F. L. and Johnson, W. B. (1973) 'Validation and applications of an urban diffusion model for vehicular emissions', *Atmospheric Environment*, vol 7, pp603–618

Deaves, D. M. and Lines, I. G. (1999) 'On persistence of low speed conditions', *Air Infiltration Review*, vol 20, no 1, pp6–8

de Dear, R., Brager, G. and Cooper, D. (1997) *Developing an Adaptive Model of Thermal Comfort and Preference*, Final report, ASHRAE RP-884, American Society of Heating, Refrigerating and Air-Conditioning Engineers, Inc, and Macquarie Research, Ltd, Sydney

de Gids, W. and Phaff, W. H. (1982) 'Ventilation rates and energy consumption due to open windows: A brief overview of research in The Netherlands', *Air Infiltration Review*, vol 4, pp4–5

De Paul, F. T. and Sheih, C. M. (1986) 'Measurements of wind velocities in a street canyon', *Atmospheric Environment*, vol 20, pp445–459

DiLouie, G. (2002) *Personal Control: Boosting Productivity, Energy Savings*, Lighting Controls Association, Rosslyn, VA

Dubiel, J., Wilson, M. and Nicol, F. (1996) 'Decibels and discomfort: An investigation of noise tolerance in offices', in *Proceedings of the Joint CIBSE/ASHRAE Conference, Harrogate, UK*, vol 2, pp184–191

Fountain, M., Brager G. and de Dear, R. (1996) 'Expectations of indoor climate control', *Energy and Buildings*, vol 24, pp179–182

Georgakis, G. and Santamouris, M. (2003) *Urban Environment: Research Report*, URBVENT project, European Commission, Athens

Georgakis, G. and Santamouris, M. (2004) 'On the airflow in urban canyons for ventilation purposes', *International Journal of Ventilation*, vol 3, no 1, pp53–65

Ghiaus, C. (2003) 'Free-running building temperature and HVAC climatic suitability', *Energy and Buildings*, vol 35, no 4, pp405–411

Hayes, S. R. (1991) 'Use of an Indoor Air Quality Model (IAQM) to estimate indoor ozone levels', *Journal of the Air and Waste Management Association*, vol 41, no 2, pp161–170

Hotchkiss, R. S. and Harlow, F. H. (1973) *Air Pollution and Transport in Street Canyons*, US Office of Research and Monitoring, Washington, DC, June

Hoydysh, W. and Dabbert, W. F. (1988) 'Kinematics and dispersion characteristics of flows in asymmetric street canyons', *Atmospheric Environment*, vol 22, no 12, pp2677–2689

IIASA (International Institute for Applied Systems Analysis) (2001) *LUC GIS Database*, Laxenburg, Austria, www.iiasa.ac.at

Iordache, V. (2003) *Etude de l'impact de la pollution atmosphérique sur l'exposition des enfants en milieu scolaire: Recherche de moyens de prédiction et de protection*, PhD thesis, University of La Rochelle, France, pp138–139

Jang, J.-S. R., Sun, C.-T. and Mizutani, E. (1997) *Neuro-Fuzzy and Soft Computing: A Computational Approach to Learning and Machine Intelligence*, Prentice Hall, Upper Saddle River, NJ

Lee, I. Y., Shannon, J. D. and Park, H. M. (1994) 'Evaluation of parameterizations for pollutant transport and dispersion in an urban street canyon using a three-dimensional dynamic flow model', in *Proceedings of the 87th Annual Meeting and Exhibition*, Cincinnati, Ohio, 19–24 June

Lomborg, B. (2001) *The Skeptical Environmentalist*, Cambridge University Press, Cambridge

Lomonaco, C. and Miller, D. (1996) *Environmental Satisfaction, Personal Control and the Positive Correlation to Increased Productivity*, Johnson Controls, Milwaukee, WI, www.johnsoncontrols.com/cg/PersEnv/pe_whitepaper.htm#put_to_work

Mahdavi, A. and Kumar, S. (1996) 'Implications of indoor climate control for comfort, energy and environment', *Energy and Buildings*, vol 24, pp167–177

McCartney, K. and Nicol, F. (2002) 'Developing an adaptive control algorithm for Europe: Results of the SCATS project', *Energy and Buildings*, vol 34, no 6, pp623–635

Nakamura, Y. and Oke, T. R. (1988) 'Wind, temperature and stability conditions in an east–west oriented urban canyon', *Atmospheric Environment*, vol 22, no 12, pp2691–2700

Nicholson, S. E. (1975) 'A pollution model for street-level air', *Atmospheric Environment*, vol 9, pp19–31

Nicol, F. and Humphreys, M. A. (2002) 'Adaptive thermal comfort and sustainable thermal standards for buildings', *Energy and Buildings*, vol 34, no 6, pp563–572

Nicol, F. and Wilson, M. (2003) *Noise in Street Canyons: Research Report*, URBVENT project, European Commission, London Metropolitan University, London

Nicol, F. and Wilson, M. (2004) 'The effect of street dimensions and traffic density on the noise level and natural ventilation potential in urban canyons', *Energy and Buildings*, vol 36, no 5, May, pp423–434

Nicol, F., Wilson, M. and Dubiel, J. (1997) 'Decibels and degrees-interaction between thermal and acoustic interaction in offices', in *Proceedings of the CIBSE National Conference*, London

Nicol, F., Wilson, M. P. and Shelton, J. (2002) 'The effect of street dimensions and traffic density on the noise level at different heights in urban canyons', *Proceedings of the Conference EPIC 2002, Energy Efficient and Healthy Buildings in Sustainable Cities, vol 2*, ENTPE, Lyon, France, pp283–288

Nunez, M. and Oke, T. R. (1977) 'The energy balance of an urban canyon', *Journal of Applied Meteorology*, vol 16, pp11–19

Oke, T. R. (1982) *Overview of Interactions Between Settlements and their Environment*, WMO Experts Meeting on Urban and Building Climatology, WCP-37, World Meteorological Organization (WMO), Geneva

Oke, T. R. (1987) 'Street design and urban canopy layer climate', *Energy and Buildings*, vol 11, pp103–113

Orme, M., Liddament, M. W. and Wilson, A. (1998) *Numerical Data Natural Ventilation Calculations*, AIVC, Coventry

Ribéron, J. (1991) *Guide méthodologique pour la mesure de la perméabilité à l'air des enveloppes de bâtiments, Cahiers du CSTB*, no 2493, Centre Scientifique et Technique du Bâtiment, Paris

Santamouris, M. (2001) *Energy and Climate in the Urban Built Environment*, James and James Science Publishers, London

Santamouris, M. and Asimakopoulos, D. (eds) (1996) *Passive Cooling of Buildings*, James and James, London

Santamouris, M. and Georgakis, G. (2004) 'Energy and indoor climate in urban environments: Recent trends', *Building Services Engineering Research and Technology Building Services Engineering and Technology*, vol 24, no 2, pp69–81

Santamouris, M., Papanikolaou, N., Koronakis, I., Livada, I. and Asimakopoulos, D. N. (1999) 'Thermal and airflow characteristics in a deep pedestrian canyon under hot weather conditions', *Atmospheric Environment*, vol 33, pp4503–4521

Shafik, N. (1994) 'Economic development and environmental quality: An econometric analysis', *Oxford Economic Papers*, vol 46, pp757–773

Shair, F. H. and Heitner, K. L. (1974) 'Theoretical model for relating indoor pollutant concentrations to those outside',

Environmental Science and Technology Journal, vol 8, no 5, p444

Skaret, E., Blom, P. and Brunsell, J. T. (1997) 'Energy recovery possibilities in natural ventilation of office buildings', in *18th AIVC Conference: Ventilation and Cooling*, AIVC, Athens, Greece

Solomon, J., Wilson, M., Wilkins, P. and Jacobs, A. (1998) 'An environmental monitoring system for comfort analysis', in *Proceedings of the EPIC 1998 Conference*, Lyon, France, pp457–462

Walker, I. S. and Wilson, D. J. (1994) 'Practical methods for improving estimates of natural ventilation rates', in *15th AIVC Conference - The Role of Ventilation*, AIVC, Buxton: UK

Wedding, J. B., Lombardi, D. J. and Cermak, J. E. (1977) 'A wind tunnel study of gaseous pollutants in city street canyons', *Journal of Air Pollution Control*, vol 27, pp557–566

Weschler, C. J., Shields, H. C. and Naik, D. V. (1989) 'Indoor ozone exposures', *Journal of Air Pollution Control*, vol 39, pp1562–1568

WHO (World Health Organization) (1997) *Health and Environment in Sustainable Development: Five Years after the Earth Summit*, WHO, www.who.int/environmental_information/Information_resources/htmdocs/execsum.htm

WHO (2000) *Air Quality Guidelines*, WHO, Geneva

Wilson, M. (1992) 'A review of acoustic problems in passive solar design', in *Proceedings of the EuroNoise 1992 Conference*, London, pp901–908

Wilson, M., Nicol, F. and Singh, R. (1993) 'Measurements of background noise levels in naturally ventilated buildings, associated with thermal comfort studies: Initial results', in *Proceedings of the IOA*, vol 15, no 8, pp283–295

Wonnacott, T. H. and Wonnacott, R. J. (1990) *Introductory Statistics for Business and Economics*, fourth edition, John Wiley and Sons, New York, pp142–148

Yamartino, R. J. and Wiegand, G. (1986) 'Development and evaluation of simple models for the flow, turbulence and pollution concentration fields within an urban street canyon', *Atmospheric Environment*, vol 20, pp2137–2156

2

Analytical Methods and Computing Tools for Ventilation

James W. Axley

Introduction

This chapter will provide a critical review of the theories underlying new and emerging methods of ventilation analysis, considering their relation to established methods, their practical applications, their fundamental limitations and the consequent challenges for future development. Consideration will be limited to *macroscopic* methods, with an emphasis placed on multi-zone and sub-zone (zonal) approaches. *Microscopic* methods based on both analytical and computational fluid dynamic (CFD) solutions to governing partial differential continuum equations will only be considered when integrated with *macroscopic* approaches.

At the 16 November 1949 Sessional Meeting of the Institution of Heating and Ventilating Engineers in London, J. B. Dick described experimental investigations of some 20 test houses (Dick, 1949) where tracer gas tests were used to determine ventilation airflow rates Q, and measurements of wind reference mean velocities $\bar{U}_{ref}$, direction θ, façade pressures p_w, and pressure differences across exterior walls Δp were made. While the methods used were not entirely new (see Aynsley et al, 1977, Chapter 1), these comprehensive studies demonstrated that:

- Ventilation airflow rates could be correlated to approach wind velocities using a linear relation of the form:

$$Q = a + b\bar{U}_{ref} \qquad (1)$$

- Façade wind pressures p_w could be correlated to the kinetic energy density of the approach wind through the introduction of a wind pressure coefficient that varied with wind direction $C_p(\theta)$:

$$p_w = C_p(\theta)\frac{1}{2}\rho U_{ref}^2 + p_o \qquad (2)$$

where a and b were empirically determined constants that Dick showed relate, theoretically, to stack- and wind-driven pressures, respectively, ρ is the density of the approach wind, and p_0 is the down-wind ambient atmospheric pressure.

Furthermore, Dick reported that these measurements could be combined with simple theoretical models to indirectly estimate:

- ventilation heat loss rates q_{loss}:

$$q_{loss} = \rho Q c_P \Delta T \qquad (3)$$

- leakage of wall components – assumed by Dick to follow a quadratic relation of the form:

$$\textit{Quadratic model: } \Delta p^e = C^e(Q^e)^2 \qquad (4)$$

where c_p is the specific heat capacity of air, ΔT the inside-to-outside temperature difference, and C^e is a characteristic leakage constant of wall component e.

While Dick did not formally present the details of his theoretical models, in the discussion session following his presentation he 'proceeded to discuss the laws of flow when components were in series or parallel and demon-

BOX 2.1 NOMENCLATURE FOR CHAPTER 2

Notational conventions

v_i variable v associated with zone i; element i of vector $\mathbf{v}$

$\{v\}$ vector $\mathbf{v}$

v^e variable v associated with flow element e

$v_{[k]}$ variable v at iteration k

$\bar{v}$ time average of variable $v(t)$

$\hat{v}$ spatial average of variable $v(x, y)$

$A_{i,j}$ element i, j of matrix $\mathbf{A}$

$[A]$ matrix $\mathbf{A}$

Physical quantities

A cross-sectional area

a empirically determined (constant) parameter

b empirically determined (constant) parameter

b_i, b_{vi} the Bernoulli sum within zone i with and without viscous dissipation, respectively

c_p specific heat capacity at constant pressure

C empirically determined (constant) parameter

C_α concentration of contaminant species α: (mass-α)/(mass-air)

C_d discharge coefficient

C_p wind pressure coefficient

D diameter

D_h hydraulic diameter

e_v viscous loss coefficient

f Fanning friction factor

$\dot{E}_v$ rate of energy dissipated viscously

f^e forward form of an element flow relation

F^e empirical factor to account for relative magnitude of pulsation flows in flow element e

g^e inverse form of an element flow relation

g the acceleration of gravity

I relative turbulence intensity (for velocity, pressure, pressure difference and reference velocity)

K_{tot} total kinetic energy within a control volume

K^e empirical power law coefficient for flow element e

l^e length of flow element e

$\dot{m}^e$ mass flow rate through element e

$\overline{\dot{m}^e}$ mean mass flow rate through element e driven by the mean pressured difference $\overline{\Delta p}^e$

$\overline{\dot{m}^{e\prime}}$ mean mass flow rate through element e driven by the fluctuation pressured difference $\Delta p^{e\prime}$

$m_i^{\&e}$ mass flow rate *into* element e from element port i

M effective molecular weight of dry air

M_α effective molecular weight of contaminant species α

M_i (accumulated) mass of air at node i

N rotation speed

p pressure

p' pressure fluctuation about the time-smoothed mean pressure

p_i reference pressure in zone i

$\hat{p}_i^e$ spatial average pressure at port of flow element e in zone i

p_b buoyancy-induced pressure

Δp_b buoyancy-induced pressure difference

p_w wind-induced pressure

Δp_w wind-induced pressure difference

Δp pressure difference

Δp_{i-j}^e pressure difference across element e from zone i to zone j

q heat transport rate

Q volumetric flow rate

Q^e volumetric flow rate through flow element e

R universal gas constant

R_i empirical constant relating to internal pressures

Re Reynolds number

t time

T temperature

u, v air velocity components in the x and y directions, respectively

u', v' velocity fluctuation about the time-smoothed mean velocity

$u^e(y,z)$ normal air velocity distribution local to port of element e

$\bar{U}_i$ mean air speed at location or node i

$\bar{U}_{ref}$ time-averaged reference wind speed

$\tilde{U}_{ref}$ amplitude of harmonic variation of reference wind speed

V_i volume of zone i

$\dot{W}$ rate of work done

x, y, z spatial coordinates

Φ_{tot} total geo-potential energy within a control volume

μ air viscosity

θ wind direction

ϑ^e size or size parameter of element e

ρ density

ρ air density

$\Im$ factor determined from numerical integration of harmonic approximation of approach wind velocity

Subscripts and superscripts

0	outdoor; outdoor zone
b	buoyancy
d	discharge
eff	effective
in	inflow
$[k]$	iteration index
l	loop index
loc	local
$loss$	loss
L	loop related
out	outflow
p	pressure; at constant pressure
Δp	pressure difference
ref	reference
tot	total
v	velocity
w	wind
α	contaminant species α

Vectors and matrices

$\{\mathbf{f}^e\}$	the system air-mass flow-rate vector contribution of element e
$\{\dot{\mathbf{M}}\}$	the system air-mass flow-rate vector (element $\dot{\mathbf{M}}_i$ is the net air-mass flow rate leaving node i)
$\{\dot{\mathbf{M}}^e\}$	the system air-mass flow-rate vector contribution of element e (i.e. equivalent to $\{\mathbf{f}^e\}$)
$\{\mathbf{p}\}$	the system pressure vector
$\{\hat{\mathbf{p}}^e\}$	the element pressure vector for element e
$\{\hat{\mathbf{p}}_b^e\}$	a 'system-sized' element buoyancy-pressure vector for element e
$\{\hat{\mathbf{p}}_w^e\}$	a 'system-sized' element wind-pressure vector for element e
$[\mathbf{B}^e]$	the Boolean system-to-element pressure-vector transformation matrix for element e
$[\mathbf{J}^e]$	the *element Jacobian* matrix for nodal continuity equations
$[\mathbf{J}]$	the *system Jacobian* matrix for nodal continuity equations
$[\mathbf{J}_L^e]$	the *element loop Jacobian* matrix for loop compatibility equations
$[\mathbf{J}_L]$	the *system loop Jacobian* matrix for loop compatibility equations

strated these laws using a simple house … [and] concluded by showing how the laws could be applied to some typical [design] problems.' This led to a lively discussion where two alternative models were proposed – a power-law model and a 'binomial' model for leakage airflow paths (Dick, 1949):

Power-law model: $\Delta p^e = C^e(Q^e)^n$ \hfill (5)

'Binomial' model: $\Delta p^e = C_1^e Q^e + C_2^e(Q^e)^2$ \hfill (6)

The utility of the 'binomial' model remains in dispute to this day (Etheridge, 1998; Chiu and Etheridge, 2002), although it is now commonly called the 'quadratic' law, and Dick's 'quadratic' model, now most often identified as the *discharge coefficient model*, has remain controversial over the years due to the problematic *discharge coefficient* C_d^e that refuses to be constant. This model, a simplified form of the classic orifice equation, provides a means of estimating airflows given the cross-sectional area of the leakage path A^e:

Discharge Coefficient Model:

$$\Delta p^e = \left(\frac{\rho}{2 A^e C_d^e}\right)^2 (Q^e)^2 \hfill (7)$$

Dick, in this single meeting, managed to:

- present both empirical and theoretical models for ventilation analysis;
- demonstrate their application to the design of ventilation system configuration and sizing system components; and
- estimate the error introduced through uncertain boundary conditions (that is, wind pressure coefficients and approach wind velocity), model parameters (that is, the coefficients and exponent of the component models used) and model laws – not to mention the fact that he also presented a very complete method of tracer gas analysis and its application in a large field study!

The discussion following his presentation ranged broadly over questions of air quality, energy consumption, air distribution within rooms, thermal comfort, smoke control, and the control of natural ventilation rates with self-limiting ventilators, no less. Thus, by the time of this meeting half a century ago, the scope and purposes of ventilation modelling had been largely established.

The methods presented and discussed at this meeting, or consistent extensions of them, remain firmly in place to this day within the broad category of *macroscopic methods* of analysis:

MACROSCOPIC METHODS: analytical methods based on modelling buildings as collections of finite-sized control volumes within which mass, momentum or energy transport behaviour is described in terms of algebraic and/or ordinary differential conservation equations.

Regrettably, many, if not most, of the limiting assumptions associated with the analytical methods discussed in 1949 also remain in place today.

This chapter will critically review new and emerging *macroscopic* methods developed specifically to target these longstanding, limiting assumptions – particularly to address assumptions that:

- Airflow through building openings is solely pressure driven.
- Openings in building envelopes are horizontal and of a constant cross-section.
- Flow resistance provided by rooms is negligible.
- Wind pressure coefficients on sealed buildings are suitable for porous buildings.
- Turbulent wind-driven airflows can be ignored.
- Thermal stratification within rooms has a secondary impact on ventilation.
- Changes in ventilation airflow rates are essentially instantaneous.
- The coupled interactions of heat transfer, moisture transfer and ventilation airflow need not be explicitly considered.
- The conservation of mechanical energy can be (tacitly) ignored.

Discussions of computational methods that now preoccupy ventilation research were conspicuously absent at this seminal 1949 meeting. Two years would have to pass before the UNIVAC I, the first commercially available digital computer, would become available and digital computation would begin to transform scientific and technical analysis across all fields. While the ventilation research community has been slow to make use of digital computation, developments in the past two decades have been very rapid.

Multi-zone methods of airflow analysis – computational generalizations of the *macroscopic methods* discussed in the 1949 meeting – are now commonly applied to investigate 'bulk' airflows in whole-building/ heating, ventilating and air-conditioning (HVAC) systems of arbitrary complexity, while *microscopic methods* of

computational fluid dynamics (CFD) have proven useful to predict the details of airflow (temperature and air quality) distributions within single or well-connected multiple spaces in buildings:

MICROSCOPIC METHODS: analytical methods based on continuum descriptions of mass, momentum and energy transport within discrete physical domains of buildings defined in terms of partial differential conservation equations.

More recently, so-called *sub-zone* (or *zonal*) *methods*, based on extensions of multi-zone methods but utilizing analytical solutions of specific microscopic flow phenomena (for example, jets, plumes and boundary layer flows), have been formulated to provide a computationally less demanding alternative to CFD to predict estimates of airflow details in rooms. Most recently, strategies to *embed* sub-zone and/or CFD methods within multi-zone models are being investigated to meet the pressing need to provide greater within-room detail, while faithfully accounting for whole-building/HVAC system interactions.

Multi-zone and, thus, sub-zone methods have, however, been developed based on the same flawed *macroscopic* principles that were in place at the time of the historic 1949 meeting. Consequently, these computational approaches suffer similar limitations – these will also be considered. In addition, complications resulting from the numerical methods that necessarily must be used to solve multi-zone and sub-zone models of building systems will also be considered and given special emphasis.

More positively, an inverse form of macroscopic methods leads quite naturally to analytical methods that may be used to directly size ventilating system components – these *loop methods* and computational tools based on them will also be considered. Furthermore, computational implementations of both macroscopic and microscopic methods allow the application of stochastic analysis methods that may well be able to characterize the uncertainty in computed results (Rao and Haghighat, 1993; Brohus et al, 2003) – an important subject, but one that must remain beyond the scope of this chapter.

Finally, during the past decade a number of *equation-based simulation environments* have provided an entirely new means of not only implementing multi-zone, sub-zone and even CFD methods, but these tools allow, in principle, coupling of these methods to each other and to other macroscopic and microscopic methods of related coupled heat and mass transfer to effect what has become known as *multi-physics simulation* (Bring, 1992; Lindgren and Sahlin, 1992; COMSOL, 2001; Klein, 2002; LBNL,

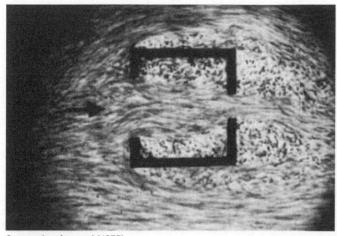 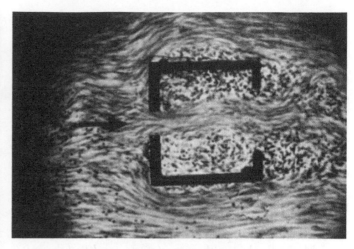

Source: Aynsley et al (1977)

Figure 2.1 *Images of airflow through simple building models recorded in wind tunnel studies by Richard Aynsley*

2002a, 2002b; Modelica Association, 2002). As these general purpose tools appear to be the way of the future, they too will be briefly considered.

Earlier reviews

This chapter will critically review new and emerging methods of ventilation analysis that have become important during the past decade. Yet, of course, much was accomplished in the intervening years between the 1949 meeting discussed above and the end of the 20th century. Fortunately, there are a number of reviews of these earlier approaches to ventilation analysis that the interested reader can turn to.

Aynsley, Melbourne and Vickery, in their excellent early text *Architectural Aerodynamics* (Aynsley et al, 1977), provide a historical review of the earliest developments in the field, a rigorous review of fundamental principles of relevant fluid dynamics, a review of the physical and statistical characteristics of wind, and a chapter on natural ventilation. Importantly, the authors began the longstanding debate relating to the proper selection of the problematic discharge coefficient C_d^e and provided invaluable images of airflow through simple building models (see Figure 2.1) to substantiate their discussion.

This discussion is particularly relevant today because it relates to modelling ventilation in highly porous buildings – the very challenge that has prompted a current re-evaluation of macroscopic methods of airflow analysis in general. Indeed, Aynsley and his colleagues anticipated what is now seen as a central problem of current macroscopic methods:

When openings through buildings are in excess of 20 per cent of the wall area, it becomes increasingly difficult to determine the effective [wind] pressure difference responsible for airflow through the openings from [measured] wind pressure distributions on solid models. (Aynsley et al, 1977)

Significantly, the images published by Aynsley et al (1977) provide compelling evidence that airflow through larger building openings is *not* at all similar to flow through orifices, thus undermining the use of the simplified orifice relation for airflow modelling. Very recent studies reported by Sandberg support this conclusion and place into question the use of (solely) pressure-driven component models for ventilation analysis of porous buildings (Sandberg, 2002). As this particular issue appears to be central to one thread of new developments, this chapter will revisit the theoretical basis of component airflow models in an attempt to add rigour to the discussion.

The energy crisis of the mid 1970s and a number of critical indoor air-quality problems that became apparent by the late 1970s (for example, indoor radon levels, formaldehyde emissions from foam insulation and the dramatic outbreak of Legionnaire's disease in Philadelphia in 1976) revived interest and stimulated research in ventilation analysis. Throughout the following two decades, as improved and new methods were developed, the Air Infiltration and Ventilation Centre (AIVC) published a series of technical reports that served initially to keep the research community abreast of developments in the field, but increasingly became directed to the needs of practising engineers (Liddament and Thompson, 1982;

Liddament, 1983, 1986a, 1986b, 1987, 1991, 1996; Liddament and Allen, 1983; Orme et al, 1998; Orme, 1999). Building on these critical publications and the growing body of literature, much of which was produced for the annual AIVC conferences, a series of handbooks and technical references were also draughted that provide comprehensive reviews of new and established methods, data and guidelines for the application of these methods, worked examples, and increasingly detailed reports of validation studies and related estimates of modelling certainty.

Awbi's (1991) useful text *Ventilation of Buildings* provided an early *second-generation* review of macroscopic modelling fundamentals, with detailed guidance for estimating wind pressures on buildings, and classified available methods as either *empirical, simplified theoretical* or *network* models (Awbi, 1991, Chapter 3, pp60–98). The inclusion of the section on network modelling, although simplified for the purposes of publication, is especially noteworthy as this approach to ventilation analysis had only been recently introduced. Awbi also provides an introduction to the *microscopic* analysis of isothermal and non-isothermal jets to provide tools to model details of airflows within rooms (Awbi, 1991, Chapter 4, pp99–127). Increasingly, jet, plume, boundary-layer and gravity-flow models are being integrated with macroscopic techniques to predict both bulk ventilation and within-room airflows. Awbi's presentation provides a good introduction to the microscopic principles involved and some of the practically useful solutions that are now finding a new use in sub-zone and gravity-flow ventilation analysis.

Santamouris and Dascalaki present more specialized reviews of methods for predicting natural ventilation in buildings and sizing components of natural ventilation systems – first in Dascalaki and Santamouris (1996), and then in more detail in Santamouris and Dascalaki (1998). The earlier publication provides essential details on modelling wind pressures (that is, with equation (2)), with supporting data; a review of simplified and empirical methods to predict ventilation airflow rates; isothermal and non-isothermal flow theory for large openings; and modelling the coupled thermal/airflow interactions in large openings and by boundary-layer pumping. The later publication adds details on semi-empirical models to predict airflow velocities within naturally ventilated buildings. Both publications present elements of the theory underlying multi-zone (network) models, and the later publication touches on sub-zone (zonal) modelling and presents a complete review of CFD modelling theory with some validation results for all three approaches.

Liddament provides a broader, more accessible, review of empirical, simplified, network and CFD methods and an early introduction to the general formulation of the coupled network airflow/thermal analysis problem (Liddament, 1996). Irving and Uys present a comprehensive approach to practical natural ventilation analysis directed at the practitioner, rather than the researcher, and offer an approach to the sizing of components of natural ventilation systems that may be considered to be a precursor to the loop design method that will be considered below (Irving and Uys, 1997).

Finally, two comprehensive reference texts warrant special mention. Etheridge and Sandberg's (1996) *Building Ventilation: Theory and Measurement* presents details of the physical mechanisms and associated theory for flow-through openings, including models for large openings where flows may be bi- or multi-directional, and for microscopic theory and its practical application, including turbulence modelling, boundary-layer flow, jets and gravity flows. The 1998 IEA Annex 26 report *Ventilation of Large Spaces in Buildings* (Heiselberg et al, 1998) provides a complement to Etheridge and Sandberg's text in that it provides additional details on jets, plumes, free convection and gravity flows; provides a presentation of multi-zone (network) models, sub-zone (zonal) models and coupled network airflow/thermal models; presents microscopic CFD methods in detail; and presents some early investigations of *interactive* micro–macro analysis and within-room micro analysis *embedded* in whole-building network analysis – all complementary to the discussion of this chapter.

Naming, rigour and notation

Within the broad field of macroscopic analysis there is a great deal of confusion regarding the theoretical basis of the methods employed and, hence, the naming of them. This is, in part, due to the recent history of the development of these methods and, in part, to different language traditions.

Thus, for example, the first computational implementations of macroscopic methods of whole-building flow analysis were identified as *network* methods (de Gids, 1978; Liddament, 1983; Liddament and Allen, 1983) or *multi-room* methods (Walton, 1982, 1984). For early models limited to isothermal conditions, the dimensions of zones were not relevant and, thus, could be treated as infinitesimal *nodes* connecting discrete airflow paths very much as junctions in electrical *network* models were treated. By the late 1980s, non-isothermal conditions (thus, buoyancy) were considered. Consequently, elevations of airflow paths within zones had to be accounted for and zones could not be thought to be infinitesimal. Hence, some researchers preferred to identify the theory as *multi-zone* (Axley, 1987a; Grot and Axley, 1987; Sahlin and Bring, 1993), *multi-cell* (Feustel and Kendon, 1985; Herrlin, 1985; Etheridge, 1998a) or *multi-room* (Li, 1993) airflow analysis, while others, of course, continued to use the *network* designation (Walton, 1988, 1989). By the late 1990s, duct *network* models were added to *multi-zone* tools (Walton, 1997) that further confused the naming and, thus, the understanding of these methods.

To complicate matters further, attempts to apply macroscopic methods to predict airflow details within zones by subdividing these zones into a relatively small number of control volumes, sometimes called *cells*, were identified as *zonal* methods (Allard et al, 1990; Bouia and Dalicieux, 1991; Inard and Buty, 1991a, 1991b; Li, 1992). To distinguish these methods from the now combined duct *network* and *multi-zone* methods, some researchers identified the latter combined methods as *nodal* or *nodal multi-zone* methods (Lorenzetti, 2002; Mora et al, 2002a), while others preferred to identify these within-zone models as *sub-zonal* methods and the combined whole-building and duct modelling methods as *zonal* methods (Heiselberg et al, 1998)!

MULTI-ZONE VERSUS SUB-ZONE METHODS: within this chapter, methods used to model bulk airflows within whole-building systems (for example, infiltration, exfiltration, inter-room and duct system airflow rates) will be identified as *multi-zone* methods of analysis, and those used to model the details of airflows within individual zones as *sub-zone* methods of analysis.

Flow variable representation

While the notion of a *node* as a specific point within the building airflow system has become central to all macroscopic methods, it is largely misunderstood and must be reconsidered to lay a foundation for a more rigorous approach.

In macroscopic analysis, one limits consideration to mass, momentum and energy transfers into and out of carefully selected control volumes at well-defined inlet or outlet planes. For flow-related transfers, inflow and outflow velocity, pressure, temperature and concentration must be considered. Limiting consideration to a *two-port* control volume (see Figure 2.2), we may most completely represent these flow quantities in terms of their spatial variation (for example, $p_{in}(y, z)$, $v_{in}(y, z)$, $p_{out}(y, z)$, and $v_{out}(y, z)$). Alternatively, we can limit consideration to spatial averages, with some loss of detail, as, for example:

$$\hat{p}_{in} = \frac{\int_{A_{in}} p_{in}(y, z)\,dA}{A_{in}} \; ; \; \hat{p}_{out} = \frac{\int_{A_{out}} p_{out}(y, z)\,dA}{A_{out}} \quad (8a, b)$$

$$\hat{v}_{in} = \frac{\int_{A_{in}} v_{in}(y, z)\,dA}{A_{in}} \; ; \; \hat{v}_{out} = \frac{\int_{A_{out}} v_{out}(y, z)\,dA}{A_{out}} \quad (9a, b)$$

Or, in some instances, it is useful to employ a *bulk* volumetric flow rate Q or mass flow rate $\dot{m}$ defined as:

$$Q_{in} = \int_{A_{in}} v_{in}(y, z)\,dA = A_{in}\hat{v}_{in} \; ;$$

$$Q_{out} = \int_{A_{in}} v_{out}(y, z)\,dA = A_{out}\hat{v}_{out} \quad (10a, b)$$

$$\dot{m}_{in} - \int_{A_{in}} \rho_{in}v_{in}(y, z)\,dA \; ; \; \dot{m}_{out} = \int_{A_{in}} \rho_{out}v_{out}(y, z)\,dA \quad (11a, b)$$

where, ρ_{in} and ρ_{out} are the inlet and outlet air densities, respectively. When these densities are uniform across the section, we obtain the following relations:

Uniform density:

$$\dot{m}_{in} = \rho_{in}\hat{v}_{in}A_{in} = \rho_{in}Q_{in} \; ;$$

$$\dot{m}_{out} = \rho_{out}\hat{v}_{out}A_{out} = \rho_{out}Q_{out} \quad (12a, b)$$

It should be clear that none of these representations of the flow quantities are, in fact, associated with a point in space or *node*! Nevertheless, *pseudo* nodes are often associated with the inflow and outflow planes of control volumes, and pressure variations are (tacitly) assumed to be uniform. For zones that are invariably multi-port control volumes, however, it will be useful to specify *true* nodes within zones of multi-zone models, *cells* of sub-zone models, and simple or multiple junctions in duct networks with which a reference zone pressure (as well as temperature and concentration) is associated.

To distinguish physical variables associated with continua (*microscopic*) descriptions from those using control volume (*macroscopic*) descriptions of flow

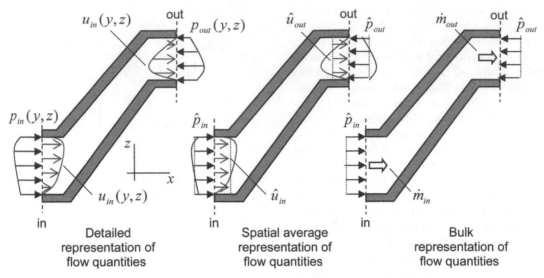

Figure 2.2 *Alternative representations of flow quantities for macroscopic analysis*

phenomena – the latter involving the distinction between zones and discrete flow elements – the notational conventions employed by Lorenzetti will be used (Lorenzetti, 2002). Thus, for example, the pressure at a point within a continuum will be represented simply as p, the reference node pressure associated with a zone i as p_i, and that associated with the port of a flow element e within zone i as p_i^e. In some instances, it will be necessary to make careful distinctions between instantaneous, time-averaged or spatial-averaged quantities. In these instances, the notational convention promoted by Bird will be used (Bird et al, 2002). For pressure, these distinctions are represented by p, $\bar{p}$, and $\hat{p}$, respectively. Finally, as iterative solution methods are often required, the value of a variable at iteration k will be identified with a bracketed subscript notation as, for example, $p_{[k]}$.

When combined, these notational conventions often compromise legibility. Therefore, when the context is clear, a more concise and simplified notation will be used.

Multi-zone nodal methods

A presentation of the conventional approach to multi-zone airflow modelling will set the stage in this section for a discussion of the porous building problem, the conservation of mechanical energy, and the challenge of modelling turbulence-induced effects. The presentation, however, will not be conventional. The intuitively satisfying hydrostatic assumption common to multi-zone theory will be recast as one of a number of possible *field assumptions* to allow a critical evaluation of this important and usually tacit assumption. Furthermore, special care will be taken to consider the possibility of non-uniform pressure and velocity distributions over flow cross-sections.

The conventional approach to airflow analysis

In conventional multi-zone airflow modelling, building systems are idealized as collections of zones linked by discrete airflow paths or duct networks; external wind pressures acting on the building are related to an approach reference wind velocity; temperatures within the zones and duct networks are specified; and specific flow relations are assigned to each of the discrete flow paths and duct network *elements*. Then, equations governing the behaviour of the system may be formed using one of two general approaches – a *mass-flow nodal continuity* approach or a *pressure loop compatibility* approach. These equations are complemented by a *hydrostatic field assumption* for each of the modelled zones in order to achieve closure. The resulting non-linear algebraic *system equations*, defined in terms of zone and duct network pressure variables, are then solved and back-substituted into the *element* equations to determine airflow rates within the system.

While seemingly straightforward in these simple descriptive terms, when the limiting assumptions, physical and mathematical details, and numerical and computational strategies needed to support the conventional approach are presented, the approach takes on its full complexity.

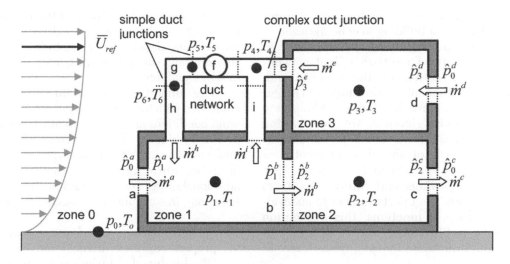

Note: The primary unknown pressure variables p_i are associated with zone *nodes* and duct network *nodes* ● ; additional element pressure variables $\hat{p}_i^e$ are associated with the flow ports of each element indicated above by dotted lines.

Source: chapter author

Figure 2.3 *Representative multi-zone plus duct network model of a hypothetical building system*

Building idealization

A representative integrated multi-zone plus duct network model of a hypothetical building system is illustrated in Figure 2.3. In this model, a single node is associated with the exterior 'zone' (node 0), three nodes are associated with building zones (nodes 1, 2 and 3), and one duct multi-port junction node (node 4) and two simple duct junction nodes (node 5 and 6) are associated with the duct network. Flow between these nodes is modelled using three duct segments (elements e, h and i), a fan (element f), and an elbow fitting (element g) of the duct network and flow elements a, b, c and d of the multi-zone system. Unknown pressures associated with each of the zone nodes (that is, p_1, p_2 and p_3) and the duct network junctions (that is, p_4, p_5 and p_6) will be the primary unknown variables of the building idealization.

Ultimately, we seek to form *system equations* that may be used to determine the element bulk mass airflow rates (that is, $\dot{m}^a$, $\dot{m}^b$, $\dot{m}^c$, ...) resulting from the action of the wind, mechanical system and building temperature distributions (for example, T_1, T_2, T_3, ...). For this building idealization, the system equations may be defined in terms of the seven primary pressure variables – the *system pressure vector*:

System pressure vector:
$$\{\mathbf{p}\} = \{p_0 p_1 p_2 p_3 p_4 p_5 p_6\}^T \tag{13}$$

To proceed to the formulation of the system equations, additional element pressure variables associated with the inflow and outflow ports of each flow element must also be defined. Here, these variables will be collected into an *element pressure vector* as:

$$\{\hat{\mathbf{p}}^a\} = \{\hat{p}_0^a \hat{p}_1^a\}^T , \{\hat{\mathbf{p}}^b\} = \{\hat{p}_1^b \hat{p}_2^b\}^T , ... \tag{14}$$

The element pressure variables, as well as the reference elevations associated with them, are identified by the notational conventions introduced above. Thus, for example, $\hat{p}_i^e$ and z_i^e will represent the spatial average pressure and reference elevation (for example, relative to the zone node i) located in zone i at the port of airflow element e.

The careful reader will note that there are three generic types of control volumes used in the conventional approach – control volumes associated with flow elements, which are currently limited to two-port elements; control volumes associated with zones; and control volumes associated with multi-port duct junctions. The first two types model the primary resistances to flow through the use of a variety of semi-empirical pressure–flow relations – for example, equations (4) to (7) – while the latter models the linking of multiple flow paths, commonly found in building ventilation systems, using either a hydrostatic (for example, for zones) or pressure-equality (for example, for duct junctions) *field assumption*.

This general strategy for idealizing buildings requires some judgement. Any given building may be idealized as a single zone – that is, ignoring all internal resistances to airflow – or by a variety of multi-zone/duct network configurations. The selection of the number and, ultimately, type of flow elements is also subject to judgement. With that said, the approach to building idealization may seem to be otherwise rigorous. Yet, one limiting assumption is invariably and tacitly made: pressure distributions across the ports of the flow elements are commonly assumed to be spatially uniform (that is, $\hat{p}_i^e = p_i^e(y, z)$). For larger openings and external openings subject to oblique winds, duct ports subject to undeveloped flow, and multi-port junctions, this assumption cannot, generally, be justified since velocity distributions are not likely to be uniform and must approach zero at bounding surfaces.

External wind pressures

At surface locations external to a flow element e, wind-induced pressures $\hat{p}_0^e$ at the element port level z_0^e are related to the ambient atmospheric pressure $p_0(z_0^e)$ and the kinetic energy of a reference time-averaged approach wind speed $\bar{U}_{ref}$ through a surface-averaged wind pressure coefficient $\hat{C}_p^e$ as:

$$\hat{p}_0^e = p_0(z_0^e) + \hat{C}_{p2}^{e1}\rho_0\bar{U}_{ref}^2; \quad -1.0 \lesseqgtr \hat{C}_p^e \leqslant 1.0 \tag{15}$$

where ρ_0 is the outdoor air density (usually assumed uniform over the height of the building).

Wind pressure coefficients acting on *sealed* buildings measured in the field and in wind tunnel model studies are available in a number of handbooks (Dascalaki and Santamouris, 1996; Orme et al, 1998; Allard, 1998; Santamouris and Dascalaki, 1998; ASHRAE, 2001; Persily and Ivy, 2001). Increasingly, computational fluid dynamics (CFD) is also used to predict wind pressure coefficients (Murakami, 1993; Holmes and McGowan, 1997; Kurabuchi et al, 2000; Jensen et al, 2002a, 2002b). As surface-averaged wind pressure coefficients vary with wind direction, a number of empirical correlations and computational tools have been developed that account for this variation (Walker and Wilson, 1994; Dascalaki and Santamouris, 1996; Knoll and Phaff, 1996; Knoll et al, 1997; Orme et al, 1998). Finally, the determination of an appropriate reference wind velocity often presents a challenge as wind data is normally available only from local airport weather records where wind conditions may

vary significantly from those at the building site. Again, the handbooks listed here provide guidance for adjusting airport wind data for the given building height, location, surrounding site topography, and shielding effects of nearby buildings.

The uncertainty of wind pressure coefficients combined with the uncertainty of estimating the reference wind conditions are thought to introduce the greater part of the ambiguity of computed results in macroscopic analysis (Bassett, 1990; Fürbringer et al, 1993; Roulet et al, 1996; Fürbringer et al, 1996). Nevertheless, two additional sources of error resulting from limiting assumptions may, in some instance, be equally important:

1   Sealed-building wind pressure coefficients may not properly represent driving forces on porous buildings.
2   Wind-induced turbulence may significantly contribute to ventilation flow rates.

These additional concerns will be considered below.

Element airflow relations

Discrete airflow paths that significantly dissipate flow energy or, conversely, contribute to the flow energy (for example, fans) are modelled with a variety of theoretical, semi-empirical or fully empirical *element relations*. These equations describe the relation between the pressure difference $\Delta\hat{p}^e \equiv \hat{p}_i^e - \hat{p}_j^e$ across the ports of each element e, the element characteristics including key sizing parameters ϑ^e and the air mass flow rate through the element $\dot{m}^e$.

In general functional notation, these component equations can be expressed in either a *forward* or *inverse* form – the latter will be used to form *loop equations* and the former *nodal equations* that govern the behaviour of the ventilation system as a whole:

$$\textit{Forward form:} \quad \dot{m}^e = f^e(\Delta\hat{p}^e, \vartheta^e) \tag{16}$$

$$\textit{Inverse form:} \quad \Delta\hat{p}^e = g^e(\dot{m}^e, \vartheta^e) \tag{17}$$

Libraries of useful element relations are published in the user's manuals of available multi-zone programmes (see, for example, Feustel and Smith, 1997; Dols and Walton, 2000). Two of these will suffice for our purposes here: a linear model and the discharge coefficient model, which is based on the simplified orifice equation presented above in equation (7):

Linear model:

Forward form: $\dot{m}^e = C^e A^e \Delta \hat{p}^e$ (18a)

Inverse form: $\Delta \hat{p}^e = \dfrac{\dot{m}^e}{C^e A^e}$; $\vartheta^e = A^e$ (18b)

Discharge coefficient model:

Forward form: $\dot{m}^e = C_d^e A^e \dfrac{\sqrt{2\rho^e}}{|\Delta \hat{p}^e|^{1/2}} \Delta \hat{p}^e$ (19a)

Inverse form: $\Delta \hat{p}^e = \dfrac{|\dot{m}^e|\dot{m}^e}{2\rho^e (C_d^e A^e)^2}$; $\vartheta_e = A_e$ (19b)

where these equations are written to maintain a sign convention so that a positive air-mass flow rate $\dot{m}^e$ results when $\hat{p}_i^e > \hat{p}_j^e$ and:

- C^e is a constant coefficient associated with element e.
- C_d^e is the *discharge coefficient* which may be expected to vary with flow intensity and element characteristics, but is often assumed practically constant, typically with a value near 0.60.
- ρ^e is the density of air within the flow element.
- A^e is the *free* or unobstructed area of the element's cross-section that will be taken as the component design parameter ϑ_e here.

Anticipating the need to *assemble* the element equations to form and solve the system equations, it is useful to present the flow element equations using matrix notation. To this end, let $\dot{m}_i^e$ be the air mass flow rate *into* element e at port i. Thus, the forward form of the element flow relation may be approximated using a Taylor's series expansion about a current estimate k of the element pressure vector $\{\hat{\mathbf{p}}^e\}_{[k]} = \{\hat{p}_{i[k]}^e \ \hat{p}_{j[k]}^e\}^T$ or, equivalently, the driving pressure difference $\Delta \hat{p}_{[k]}^e = \hat{p}_{i[k]}^e - \hat{p}_{j[k]}^e = [1 - 1]\{\hat{\mathbf{p}}^e\}_{[k]}$ as:

General form:

$$\left\{ \begin{array}{c} \dot{m}_i^e \\ \dot{m}_j^e \end{array} \right\} = f^e(\Delta \hat{p}_{[k]}^e, \vartheta^e) \left\{ \begin{array}{c} 1 \\ 1 \end{array} \right\}$$

$$+ \dfrac{\partial f^e}{\partial \Delta \hat{p}^e}\Big|_{\Delta \hat{p}_{[k]}^e} \begin{bmatrix} 1 & -1 \\ -1 & 1 \end{bmatrix} \left\{ \left\{ \begin{array}{c} \hat{p}_i^e \\ \hat{p}_j^e \end{array} \right\} - \left\{ \begin{array}{c} \hat{p}_i^e \\ \hat{p}_j^e \end{array} \right\}_{[k]} \right\}$$

$$+ O\left\{ \left\| \left\{ \begin{array}{c} \hat{p}_i^e \\ \hat{p}_j^e \end{array} \right\} - \left\{ \begin{array}{c} \hat{p}_i^e \\ \hat{p}_j^e \end{array} \right\}_{[k]} \right\|^2 \right\}$$
     (20a)

The last term on the right is an estimate of the order of error in the approximation; that is, the error diminishes quadratically as $\{\hat{\mathbf{p}}^e\}$ approaches $\{\hat{\mathbf{p}}^e\}_{[k]}$ (Dahlquist and Björck, 1974).

Recognizing the first term on the right-hand side as simply the current estimate of the element air-mass flow rates, we may rewrite this equation in concise notation, dropping the error term, in terms of the *element mass flow-rate vector* $\{\dot{\mathbf{m}}^e\} = \{\dot{m}_i^e \ \dot{m}_j^e\}^T$ as:

$$\{\dot{\mathbf{m}}^e\} \approx \{\dot{\mathbf{m}}^e\}_{[k]} + [\mathbf{J}^e]_{[k]} \left\{ \{\hat{p}^e\} - \{\hat{p}^e\}_{[k]} \right\}$$ (20b)

Here, the matrix $[\mathbf{J}^e]_{[k]}$ is the all-important *element Jacobian* – for the linear and discharge coefficient models it follows directly from the definition implicitly given in equations (20a,b):

Linear model: $[\mathbf{J}^e]_{[k]} = C^e A^e \begin{bmatrix} 1 & -1 \\ -1 & 1 \end{bmatrix}$ (21)

Discharge coefficient model:

$$[\mathbf{J}^e]_{[k]} = C_d^e A^e \dfrac{\sqrt{2\rho^e}}{|\Delta \hat{p}_{[k]}^e|^{1/2}} \begin{bmatrix} 1 & -1 \\ -1 & 1 \end{bmatrix}$$ (22)

Thus, the element Jacobian for the linear model is simply a constant matrix, while that of the discharge coefficient model depends upon the (current estimate of) element pressures – as, indeed, do all other non-linear element models. The Taylor's series expansion – see equations (20a,b) – of the discharge coefficient model simply defines a linear approximation tangent to the actual non-linear model at a current estimate of the driving pressure $\Delta \hat{p}_{[k]}^e$ (see Figure 2.4).

Regrettably, the slope $\partial f^e / \partial \Delta \hat{p}^e$ and, thus, the element Jacobian for the discharge coefficient model becomes unbounded as the (current estimate of) element pressures approach zero:

$$\lim_{\Delta \hat{p}_{[k]}^e \to 0} \left(\dfrac{\partial f^e}{\partial \Delta \hat{p}^e}\Big|_{\Delta \hat{p}_{[k]}^e} = C_d^e A^e \dfrac{\sqrt{2\rho^e}}{|\Delta \hat{p}_{[k]}^e|^{1/2}} \right) = \infty$$ (23)

This flaw, which is common to most other flow element models, leads to system equations that are either *singular* (cannot be solved) or *nearly singular* (leading to slow convergence of iterative solution techniques) whenever the driving pressures in one or more flow elements drop to zero or near-zero conditions, respectively. As the

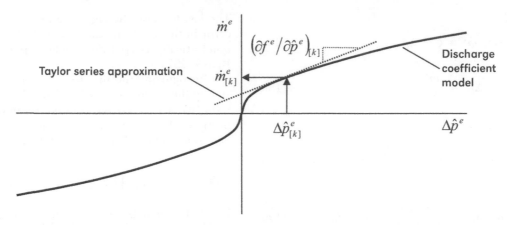

Source: chapter author

Figure 2.4 *The discharge coefficient model and a (linear) Taylor series approximation to it at a current estimate k of the driving pressure* $\Delta p_{[k]}^e$

chances of these conditions arising are relatively great for complex multi-zone building models, in practice airflow elements are invariably defined using a linear model for low Reynolds number flow conditions (for example, $\mathrm{Re}^e < 100$) and a non-linear model for higher Reynolds number flows (Dols and Walton, 2000), where the element Reynolds number Re^e is defined as:

Element Reynolds number:

$$\mathrm{Re}^e = \frac{\rho^e \sqrt{A^e} \hat{v}^e}{\mu^e} = \frac{\sqrt{A^e} \dot{m}^e}{\mu^e A^e} \qquad (24)$$

This strategy is not only numerically necessary, it is physically consistent. Fundamental fluid dynamic theory (for example, the Hagen-Poiseuille equation) demonstrates that a low Reynolds number flow should be expected to be linear (Bird et al, 2002).

Finally, the [1, –1; –1, 1] form of the element Jacobian is symmetric and will be *positive semi-definite* providing $\partial f^e / \partial \Delta \hat{p}^e$ remains positive for all possible values of $\Delta \hat{p}^e$ – a condition that will lead to *symmetric positive-definite* system equations that ensure the existence and uniqueness of solutions and allow for the unrestricted use of specialized, efficient equation-solving methods (Lorenzetti, 2002) (for this reason, fan flow relations that admit negative values of $\partial f^e / \partial \Delta \hat{p}^e$ for certain driving pressures are typically not supported in available multizone programmes).

These element models, and the others that are commonly used in multi-zone modelling, come from the vast body of fluid dynamics literature. Thus, they are based on mature theoretical developments and/or experimental investigations. Nevertheless, their application in

multi-zone analysis is compromised by three key limiting assumptions that:

1 Airflow is solely pressure driven.
2 Dynamic flow effects are not important.
3 All airflow elements may be modelled as two-port elements.

These concerns will be considered below.

Nodal and loop equations

Given a building idealization and design conditions specifying the approach wind velocity and direction, component flow equations may be assembled to form *system equations* that govern the airflow behaviour of the ventilation system as a whole. These system equations may be formulated using one of two fundamental approaches:

1 *Nodal continuity approach* – imposes the *continuity* of mass flow at each modelled node of the building model.
2 *Loop compatibility approach* – imposes *compatibility* of pressure changes as one follows a flow path around a continuous flow *loop* of the given building model.

For the *nodal continuity* approach, we simply demand that the air mass flow rate into each zone i is conserved:

Nodal mass continuity: $\quad -\sum_{\substack{linked \\ elements}} \dot{m}_i^e = \frac{dM_i}{dt} \qquad (25)$

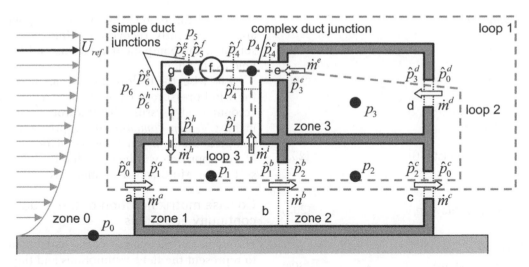

Source: chapter author

Figure 2.5 *Representative integrated multi-zone plus duct network model illustrating three independent loops*

where M_i is the total air mass accumulated and superscript e is permutated through all inflow and outflow elements linked to node i. Commonly, airflow is modelled as a steady phenomena so the accumulation term dM_i/dt is assumed to be zero – two programmes in the CONTAM family of programmes, CONTAMW 2.0 and CONTAM 97, however, allow accumulation (for example, for smoke generation and dispersal studies) (Walton, 1998; Dols, 2001b).

For the *loop compatibility* approach, pressure changes are summed as one proceeds from one flow element to the next around a *flow loop*. The sum of these pressure changes must, of course, add up to zero upon completion of the loop:

Loop pressure compatibility:

$$\sum_{loop\ l} (\hat{p}_i^e - \hat{p}_j^f) = 0 \qquad (26)$$

where i, j, e and f are permuted through all *element* port pressures $\hat{p}_i^e$ as one proceeds around a given loop l. For example, for the flow loop identified as *loop 1* in Figure 2.5 passing through $\hat{p}_0^a, \hat{p}_1^a, \hat{p}_1^b, \hat{p}_2^b, \hat{p}_2^c, \hat{p}_0^c$, back to $\hat{p}_0^a$, one would sum:

$$\sum_{loop\ l} (\hat{p}_i^e - \hat{p}_j^f) = \overbrace{\Delta\hat{p}^a + \Delta\hat{p}^b + \Delta\hat{p}^c}^{\Sigma(\text{element pressure changes})}$$

$$+ \underbrace{(\hat{p}_1^a - \hat{p}_1^b) + (\hat{p}_2^b - \hat{p}_2^c) + (\hat{p}_0^c - \hat{p}_0^a)}_{\Sigma(\text{zone pressure changes})} \qquad (27)$$

Note that the zone node pressures need not be considered in forming the loop equations.

The continuity approach has long been the method of choice for multi-zone building airflow analysis. The compatibility approach, while overlooked by all but a few members of the building simulation community (Wray and Yuill, 1993; Nitta, 1994), provides an analytical approach for designing ventilation system components that will be discussed below and is currently being investigated as a basis of modelling coupled thermal/airflow interactions in buildings at the Hybrid Ventilation Centre in Aalborg, Denmark.

Field assumption closure

Substituting appropriate element equations into the nodal mass continuity relation – see equation (25) – yields a system of n equations, for n nodes, expressed in terms of the m element port pressure variables $\hat{p}_i^e$. As the number of these element port pressure variables will always exceed the number of unspecified node pressures, additional conditions are required for closure. Likewise, in the loop approach, equation (26), additional conditions are needed to establish the zone pressure changes.

Quite reasonably, a *field assumption* is imposed that relates the node pressures p_i to the element port pressures $\hat{p}_i^e$. In conventional multi-zone airflow analysis, two field assumptions are commonly employed – an *equality field assumption* for simple and, possibly, complex duct junctions, and a *hydrostatic field assumption* for building zones and, possibly, complex duct junctions:

Equality field assumption:

$$p(z) = p_i \ (= \text{constant}) \tag{28a}$$

or specifically: $\hat{p}_i^e = p_i$ (28b)

Hydrostatic field assumption:
Non-uniform zone density:

$$p(z) + \int_0^z \rho_i g dz = p_i \ (= \text{constant}) \tag{29a}$$

or specifically: $\hat{p}_i^e + \int_0^{z_i^e} \rho_i g dz = p_i$ (29b)

Uniform zone density:

$$p(z) + \rho_i g z = p_i \ (= \text{constant}) \tag{30a}$$

or specifically: $\hat{p}_i^e + \rho_i g z_i^e = p_i$ (30b)

where z is elevation and z_i^e is the elevation of element port e – both relative to node i.

The use of the *equality field assumption* for simple duct junctions simply presumes that the pressure distribution at the duct section is uniform – a reasonable assumption. For complex duct junctions, the assumption that all connected duct segments have not only uniform but equal pressure distributions may be an acceptable approximation; but it is not generally valid. The hydrostatic field assumption accounts for buoyancy forces due to temperature differences between building zones and the outdoor environment; thus, it plays a central role in modelling building airflow.

In applying the hydrostatic field assumption, one may estimate the density of air using the ideal gas law for dry air:

$$\textit{Dry air density: } \rho = \frac{352.6°K - kg/m^3}{(T_{°C} + 273.15)°K} \tag{31}$$

In ventilation systems involving significant moisture changes, such as those utilizing evaporative cooling to both drive and cool the airflow (Giabaklou and Ballinger, 1996; Santamouris and Asimakopoulos, 1996; Bowman et al, 1997, 2000), however, psychometric methods will have to be employed to provide reasonable estimates of air density (for example, to four significant figures).

By systematically substituting the hydrostatic field assumption for each of the element port pressures $\hat{p}_i^e$ in the nodal mass continuity relation, equation (25), one finally obtains n nodal continuity equations expressed in terms of n node pressures that may then be solved to determine these node pressures. Component flow rates are subsequently recovered using the zone pressures solution, the component relations and the appropriate field assumption.

Likewise, the hydrostatic field assumption along with the wind pressure relation, equation (15), allows one to account for all pressure changes around a given loop when using the *loop compatibility* approach and thereby to complete the loop equations – equation (26). The details of this approach will be considered subsequently in the section on 'Multi-zone loop methods'.

Concise matrix notation of the nodal continuity equations

For mathematical analysis of the nodal theory, it is useful to represent the field assumptions and the mass continuity relations in concise matrix notation. The equality field assumption can be thought to relate an element pressure vector $\{\hat{p}^e\}$ to the system pressure vector $\{p\}$ as:

Equality field assumption: $\{\hat{p}^e\} = [B^e]\{p\}$ (32a)

where $[B^e]$ is a $2 \times n$ Boolean transformation matrix (that is, consisting of ones and zeros) for a two-port element within an n node system idealization. For example, for duct element g of the representative system idealization illustrated in Figures 2.3 and 2.5, this transformation would be written by inspection as:

$$\left\{ \begin{matrix} \hat{p}_5^g \\ \hat{p}_6^g \end{matrix} \right\} = \begin{bmatrix} 0 & 0 & 0 & 0 & 0 & 1 & 0 \\ 0 & 0 & 0 & 0 & 0 & 0 & 1 \end{bmatrix} \left\{ \begin{matrix} p_0 \\ p_1 \\ \vdots \\ p_5 \\ p_6 \end{matrix} \right\} \tag{32b}$$

This Boolean transformation matrix may also be used to transform the vector of element air mass transport rates $\{\dot{m}^e\}$ into a 'system-sized' vector of mass transport rates $\{\dot{M}^e\}$ with the same number of elements as the system pressure vector:

$$\{\dot{M}^e\} = [B^e]^T \{\dot{m}^e\} \tag{33}$$

For the hydrostatic field assumption, a slightly more complex but linear transformation may also be defined:

Hydrostatic field assumption:

$$\{\hat{p}^e\} = [B^e] \Big\{ \{p\} - \{p_b^e\} \Big\} \tag{34a}$$

where $\{\mathbf{p}_b^e\}$ is a 'system-sized' vector of buoyancy-induced pressure terms that is dependent upon zone/node air densities, yet is associated with an element e. For example, for element b of the representative system idealization illustrated in Figures 2.3 and 2.5, the hydrostatic field assumption may be written as:

$$\left\{\begin{matrix}\hat{p}_1^b \\ \hat{p}_2^b\end{matrix}\right\} = \begin{bmatrix} 0 & 1 & 0 & 0 & 0 & 0 & 0 \\ 0 & 0 & 1 & 0 & 0 & 0 & 0 \end{bmatrix}$$

$$\left\{\left\{\begin{matrix}p_0 \\ p_1 \\ p_2 \\ \vdots \\ p_5 \\ p_6\end{matrix}\right\} - \left\{\begin{matrix}0 \\ \rho_1 g z_1^b \\ \rho_2 g z_2^b \\ \vdots \\ 0 \\ 0\end{matrix}\right\}\right\} \qquad (34b)$$

For elements with an exterior surface port, the hydrostatic field assumption must be augmented with the surface wind pressure as:

Exterior field assumption:

$$\{\hat{\mathbf{p}}^e\} = [\mathbf{B}^e]\left\{\{\mathbf{p}\} - \{\hat{\mathbf{p}}_b^e\} + \{\hat{\mathbf{p}}_w^e\}\right\} \qquad (35a)$$

where $\{\hat{\mathbf{p}}_w^e\}$ is a 'system-sized' vector defining a single wind-induced pressure term associated with element e. For example, for element a of the representative system idealization illustrated in Figures 2.3 and 2.5, the exterior field assumption may be written as:

$$\left\{\begin{matrix}\hat{p}_0^a \\ \hat{p}_1^a\end{matrix}\right\} = \begin{bmatrix} 1 & 0 & 0 & 0 & 0 & 0 & 0 \\ 0 & 1 & 0 & 0 & 0 & 0 & 0 \end{bmatrix}$$

$$\left\{\left\{\begin{matrix}p_0 \\ p_1 \\ \vdots \\ p_5 \\ p_6\end{matrix}\right\} - \left\{\begin{matrix}\rho_0 g z_0^a \\ \rho_1 g z_1^a \\ \vdots \\ 0 \\ 0\end{matrix}\right\} + \left\{\begin{matrix}\frac{1}{2}C_p^a \rho_0 U_{ref}^2 \\ 0 \\ \vdots \\ 0 \\ 0\end{matrix}\right\}\right\} \qquad (35b)$$

With these transformations in hand, the Taylor series approximation of a generic element e's flow relation, equation (20b), may then be recast in terms of the system pressure vector $\{\mathbf{p}\}$ as:

$$[\mathbf{B}^e]^T\{\dot{\mathbf{m}}^e\} \approx [\mathbf{B}^e]^T\{\dot{\mathbf{m}}^e\}_{[k]}$$

$$+ [\mathbf{B}^e]^T[\mathbf{J}^e]_{[k]}[\mathbf{B}^e]\left\{\{\mathbf{p}\} - \{\mathbf{p}\}_{[k]}\right\}$$

$$- [\mathbf{B}^e]^T[\mathbf{J}^e]_{[k]}[\mathbf{B}^e]\left\{\{\mathbf{p}_b^e\} - \{\mathbf{p}_b^e\}_{[k]}\right\}$$

$$+ [\mathbf{B}^e]^T[\mathbf{J}^e]_{[k]}[\mathbf{B}^e]\left\{\{\mathbf{p}_w^e\} - \{\mathbf{p}_w^e\}_{[k]}\right\} \qquad (36a)$$

As the buoyancy and wind pressure contributions are fixed for the specified conditions of steady analysis (or for each time step of a dynamic analysis) – that is, $\{\hat{\mathbf{p}}_b^e\} - \{\hat{\mathbf{p}}_b^e\}_{[k]} = \{\mathbf{0}\}$ and $\{\hat{\mathbf{p}}_w^e\} - \{\hat{\mathbf{p}}_w^e\}_{[k]} = \{\mathbf{0}\}$ – this relation simplifies to become:

$$\{\dot{\mathbf{M}}^e\} \approx \{\dot{\mathbf{M}}^e\}_{[k]}$$

$$+ [\mathbf{B}^e]^T[\mathbf{J}^e]_{[k]}[\mathbf{B}^e]\left\{\{\mathbf{p}_w^e\} - \{\mathbf{p}_w^e\}_{[k]}\right\} \qquad (36b)$$

where $\{\dot{\mathbf{M}}^e\}_{[k]}$ is an estimate of element air-mass flow rates based on the current estimate of the element pressures $\{\hat{\mathbf{p}}^e\}_{[k]}$.

Finally, summing all element contributions, we can rewrite the continuity relation, equation (25), for all n nodes/zones simultaneously as:

System mass continuity:

$$-\sum_{e=a,b,c,\dots} \{\dot{\mathbf{M}}^e\} = \frac{d\{\mathbf{M}\}}{dt} = \frac{d\{M_1, M_2, M_3, \dots\}^T}{dt} \qquad (37a)$$

where $\{M_1, M_2, M_3, \dots\}^T$ is the air mass accumulated at each node – again, commonly assumed to be unchanging or $d\{\mathbf{M}\}/dt = 0$. Then, substituting the 'system-sized' Taylor series approximation, equation (36b), we form the Taylor series approximation of the nodal system equations:

$$-\frac{d\{\mathbf{M}\}}{dt} \approx \sum_{e=a,b,c,\dots} \{\dot{\mathbf{M}}^e\}_{[k]}$$

$$+ \sum_{e=a,b,c,\dots} \left[[\mathbf{B}^e]^T[\mathbf{J}^e]_{[k]}[\mathbf{B}^e]\right]\left\{\{\mathbf{p}\} - \{\mathbf{p}\}_{[k]}\right\} \qquad (37b)$$

or:

Nodal mass continuity:

$$- \frac{d\{\mathbf{M}\}}{dt} \approx \{\dot{\mathbf{M}}^e\}_{[k]} + [\mathbf{J}]_{[k]}\left\{\{\mathbf{p}\} - \{\mathbf{p}\}_{[k]}\right\} \qquad (37c)$$

where $\{\dot{\mathbf{M}}\}_{[k]}$ is the current estimate of the net air mass flow out of the nodes/zones and $[\mathbf{J}]_{[k]}$ is the $(n \times n)$ *system Jacobian* matrix, which is directly *assembled* from the *element Jacobian* matrices $[\mathbf{J}^e]_{[k]}$. From equations (37b), (36a) and (20b), it follows that the *assembly* process involves both the assembly of the element Jacobians to form the system Jacobian and the *assembly* of the current estimates of the element mass flow rates to form $\{\dot{\mathbf{M}}\}_{[k]}$:

$$[\mathbf{J}]_{[k]} = \sum_{e=a,b,c,\dots} \left[[\mathbf{B}^e]^T [\mathbf{J}^e]_{[k]} [\mathbf{B}^e] \right] \qquad (38a)$$

$$\{\dot{\mathbf{M}}\}_{[k]} = \sum_{e=a,b,c,\dots} \left\{ [\mathbf{B}^e]^T \{\dot{\mathbf{m}}^e\}_k \right\}$$

$$= \sum_{e=a,b,c,\dots} \left\{ \mathbf{f}^e\left(\{\mathbf{p}\}_{[k]}, \{\mathbf{p}_b^e\}, \{\mathbf{p}_w^e\}, \vartheta^e \right) \right\} \qquad (38b)$$

where, $\{\mathbf{f}^e(\{\mathbf{p}\}_{[k]}, \{\mathbf{p}_b^e\}, \{\mathbf{p}_w^e\}, \vartheta^e)\} = [\mathbf{B}^e]^T f^e(\{\mathbf{p}\}_{[k]}, \{\mathbf{p}_w^e\}, \vartheta^e)$ $\{1\ 1\}^T$, a 'system-sized' current estimate of element e's air mass flow rate, is expressed in terms of element flow relation $f^e(\{\mathbf{p}\}_{[k]}, \{\mathbf{p}_b^e\}, \{\mathbf{p}_w^e\}, \vartheta^e)$, which is functionally equivalent to $f^e(\Delta p_{[k]}^e, \vartheta^e)$, used in equation (20a), but now explicitly accounting for buoyancy and wind effects.

Most importantly, the assembly process maintains the structure of the element Jacobian matrices. Thus, if the element Jacobians $[\mathbf{J}^a]$, $[\mathbf{J}^b]$, $[\mathbf{J}^c]$, ... are symmetric and positive semi-definite (and bounded), then the system Jacobian $[\mathbf{J}]$ will be as well.

Solution of the nodal equations

While the governing non-linear system equations – that is, equation (25) plus the appropriate field assumptions, equations (32), (34) and/or (35) – could be solved, conceivably, by a variety of techniques, most commonly they are solved using the Newton-Raphson or related methods. These methods may be derived from the Taylor's series expansion of the governing nodal equations, equation (37), by recognizing that the right-hand side of these equations only approximately equals the left-hand side. Thus, this equation can be cast into an iterative scheme to provide an improved estimate of the system pressure vector $\{\mathbf{p}\}_{[k+1]}$ from a current estimate $\{\mathbf{p}\}_{[k]}$ (Dahlquist and

Björck, 1974). Specifically, we replace $\{\mathbf{p}\}$ of equation (37c) with $\{\mathbf{p}\}_{[k+1]}$ and substitute equation (38b) defining the 'system-sized' element flow relation $\{\mathbf{f}^e\}$ to obtain:

$$- \frac{d\{\mathbf{M}\}}{dt} \approx \sum_{e=a,b,c,\dots} \left\{ \mathbf{f}^e\left(\{\mathbf{p}\}_{[k]}, \{\mathbf{p}_b^e\}, \{\mathbf{p}_w^e\}, \vartheta^e \right) \right\}$$

$$+ [\mathbf{J}]_{[k]}\left\{ \{\mathbf{p}\}_{[k+1]} - \{\mathbf{p}\}_{[k]} \right\} \qquad (39)$$

Then, with an error estimate defined as $\{\Delta\mathbf{p}\}_{[k+1]} = \{\{\mathbf{p}\}_{[k+1]} - \{\mathbf{p}\}_{[k]}\}$, a simple two-step iterative solution algorithm may be formulated – here, limited to the common steady assumption that $d\{\mathbf{M}\}/dt = \{\mathbf{0}\}$:

Step 1: Solve for error estimate:
$$[\mathbf{J}]_{[k]}\{\Delta\mathbf{p}\}_{[k+1]}$$

$$= -\sum_{e=a,b,c,\dots} \left\{ \mathbf{f}^e\left(\{\mathbf{p}\}_{[k]}, \{\mathbf{p}_b^e\}, \{\mathbf{p}_w^e\}, \vartheta^e \right) \right\} \qquad (40a)$$

Step 2: Update pressure estimate:
$$\{\mathbf{p}\}_{[k+1]} = \{\mathbf{p}\}_{[k]} + \{\Delta\mathbf{p}\}_{[k+1]} \qquad (40b)$$

Thus, with an initial estimate of the system pressure vector, $\{\mathbf{p}\}_{[1]}$, one forms and solves equation (40a) to obtain $\{\Delta\mathbf{p}\}_{[2]}$, which is then substituted into equation (40b) to provide an improved estimate of the system pressure vector $\{\mathbf{p}\}_{[2]}$. This process is repeated until the system pressure estimates converge, $\{\Delta\mathbf{p}\}_{[k+1]} \to \{\mathbf{0}\}$, or, equivalently, until nodal mass flow rates are conserved, $\{\dot{\mathbf{M}}\}_{[k]} = \Sigma\{\mathbf{f}^e\}_{[k]} \to \{\mathbf{0}\}$.

If all flow elements include a linear model (that is, for low flow conditions), the initial estimate $\{\mathbf{p}\}_{[1]}$ may be conveniently determined by assembling the corresponding (linear) element Jacobians and solving $[\mathbf{J}]_{linear}\{\mathbf{p}\}_{[1]} = \{\mathbf{0}\}$ directly. Finally, if one seeks to account for nodal air mass accumulation (that is, $d\{\mathbf{M}\}/dt \neq \{\mathbf{0}\}$), then this two-step iterative procedure would simply be embedded in a time-stepping numerical integration scheme.

As long as the system Jacobian $[\mathbf{J}]_{[k]}$ remains non-singular for each iteration k, the procedure may proceed. If, in addition, the initial estimate of the system pressure vector $\{\mathbf{p}\}_{[1]}$ is sufficiently close to the (unknown) solution $\{\mathbf{p}\}$, the process will converge quadratically (Dahlquist and Björck, 1974). With symmetric and positive semi-definite element Jacobians (that is, with bounded and positive $\delta f^e/\delta\Delta\hat{p}^e$), the assembly process ensures that the system Jacobian will be symmetric and positive definite if one or more of the node pressures is specified (typically,

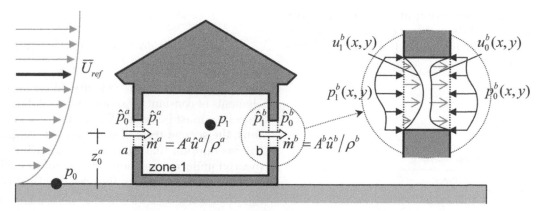

Figure 2.6 *Common single-zone model used to evaluate wind pressure coefficients C_p^e and/or flow element discharge coefficients C_d^e using measured or computational fluid dynamics (CFD)-determined data*

the outdoor zone node pressure p_0 is set to atmospheric pressure, 101,325Pa).

These conditions give rise to unique solutions and allow for the unrestricted use of specialized, efficient equation-solving methods. Even though, with some limitations, these conditions may be relaxed (Lorenzetti, 2002), they are typically enforced.

With the conventional approach to multi-zone building airflow analysis established, we can now move on to new and emerging methods devised to address the shortcomings of this approach and to extend its capabilities.

Porous buildings

To date, multi-zone building airflow analysis has utilized sealed-building wind pressure coefficients to estimate wind forces that drive ventilation; yet Aynsley, working in Australia, has consistently argued that sealed-building wind pressure coefficients may not properly represent these driving forces on porous buildings (Aynsley et al, 1977; Aynsley, 1999). Recent interest in wind-driven natural ventilation for cooling has brought this key problem to the attention of researchers in the Northern Hemisphere (True et al, 2003).

Unfortunately, the evaluation of wind-driving forces acting on buildings has been inextricably linked to the flow element models used to predict flow. Most commonly, single-zone building models with a single windward inflow element and a single leeward outflow element have been considered (see Figure 2.5); wind pressures acting on the building have been modelled with the conventional wind pressure coefficient relation, equation (14); and element flow rates have been modelled

with the discharge coefficient model, equation (19). Applying the loop pressure compatibility relation, equation (26), the familiar wind-driven cross-ventilation model is obtained:

$$\Delta\hat{C}_p\frac{1}{2}\rho\bar{U}_{ref}^2 = \frac{1}{2\rho}\frac{(C_d^aA^a)^2 + (C_d^bA^b)^2}{(C_d^dA^a)^2(C_d^bA^b)^2}\dot{m}^2 \tag{41}$$

where $\Delta\hat{C}_p$ is the algebraic difference of the windward and leeward surface-averaged wind pressure coefficients, $\Delta\hat{C}_p = \Delta\hat{C}_p^a - \Delta\hat{C}_p^b$. Here, the building net air-mass flow rate $\dot{m}$ is, by mass continuity, equal to the mass flow rates through each flow element, $\dot{m} = \dot{m}^a = \dot{m}^b$, and for isothermal conditions, the density ρ is uniform throughout.

When measured or CFD-determined flow rates are compared to those predicted using this standard model, researchers have concluded that sealed wind-pressure coefficients fail to provide reliable estimates of flow rates; and/or flow-dependent discharge coefficients must be used for unidirectional flow in large openings.

Detailed element flow model

The assumption that flow is solely pressure driven, implicit to these modelling decisions, can, however, only be justified if inflow and outflow pressure and velocity distributions are uniform or nearly so. While (near) uniformity of pressure and velocity conditions may be a reasonable assumption for well-developed turbulent flow in ducts, it has not been critically evaluated for *unrestricted* flow in openings such as those found in buildings.

Specifically, the conservation of mechanical energy for the isothermal flow of the outflow element b of Figure 2.6 involves not only pressure 'work', but kinetic and geo-potential energy terms:

$$\overbrace{\left(\int_{A_1} p_1^b u_1^b dA - \int_{A_4} p_0^b u_0^b dA \right)}^{\text{pressure work difference}}$$

$$+ \frac{\rho^b}{2} \overbrace{\left(\int_{A_1} (U_1^b)^2 u_1^b dA - \int_{A_4} (U_0^b)^2 u_0^b dA \right)}^{\text{kinetic energy rate difference}}$$

$$+ \rho^b g \overbrace{\left(\int_{A_1} z_1^b u_1^b dA - \int_{A_4} z_0^b u_0^b dA \right)}^{\text{geo-potential energy rate difference}} - W - \dot{E}_v$$

$$= \frac{d(K_{tot} + \Phi_{tot})}{dt} \qquad (42)$$

Here, $(\rho/2)U_i^{e2}$ is the distribution of the total kinetic energy content of the flow at port i of component e (that is, $U_i^{e2} = u_i^{e2} + v_i^{e2} + w_i^{e2}$ for the component velocities u_i^e, v_i^e and w_i^e).

For completeness, this integral form of the Bernoulli equation includes:

- work done on the airflow $\dot{W}$ (for example, by a fan);
- a viscous dissipation rate $\dot{E}_v$; and
- accounts for the accumulation of kinetic energy K_{tot} and potential energy Φ_{tot} within the flow path – following the example of Bird et al (2002).

Typically, consideration is limited to steady flow conditions and passive resistance, and, for isothermal conditions, the geo-potential term is negligible. Furthermore, on the basis of dimensional arguments, the viscous dissipation term may be related to the kinetic energy density of the flow in the element $(1/2)\rho^b(\hat{u}^b)^2$ and its volumetric flow rate $Q^b = A^b \hat{u}^b$ via an *element loss coefficient* e_v^b as:

$$\dot{E}_v \approx \tfrac{1}{2}\rho(\hat{u}^b)^2 e_v^b(Q_b) = \tfrac{1}{2}\rho A^b(\hat{u}^b)^3 e_v^b \qquad (43)$$

Thus, for flow element b of Figure 2.6, the conservation relation simplifies to:

$$\overbrace{\left(\int_{A_1^b} p_1^b u_1^b dA - \int_{A_0^b} p_0^b u_0^b dA \right)}^{\text{pressure work difference}}$$

$$+ \frac{\rho^b}{2} \overbrace{\left(\int_{A_1^b} (u_1^b)^3 dA - \int_{A_0^b} (u_0^b)^3 dA \right)}^{\text{kinetic energy rate difference}} \approx \tfrac{1}{2}\rho A^b(\hat{u}^b)^3 e_v^b \qquad (44)$$

For (nearly) uniform velocity distributions within elements of constant cross-section, the inflow and outflow velocities must be equal (that is, to conserve mass flow); thus, the second term of equation (44) vanishes. Adding the assumption that the pressure distributions are also (nearly) uniform, equation (44) simplifies further to yield a simple pressure-driven flow relation (that is, with $\hat{u}_1^b = \hat{u}_0^b \equiv \hat{u}^b$ and, by continuity, $\hat{u}_1^b A_1^b = \hat{u}_0^b A_0^b \equiv \hat{u}^b A^b$):

$$(p_1^b - p_0^b) \approx \tfrac{1}{2}\rho(\hat{u}^b)^2 e_v^b \qquad (45)$$

Comparing this result to the discharge coefficient model, the element loss coefficient may be directly related to the element discharge coefficient (that is, for the given assumptions) as:

$$e_v^b = 1 / (C_d^b)^2 \qquad (46)$$

If, however, the pressure and velocity distributions are not (nearly) uniform, then the kinetic energy term may well prove to be significant. As might be expected, for relatively large element cross-sections (that is, for porous buildings), wind-driven pressure differences across the flow elements drop to low values, and yet flow velocities do not; thus, small differences in velocity profiles can play a decisive role in the element flow behaviour.

Total, static and dynamic wind pressure coefficients

To add quantitative scale to this last claim, wind-driven airflows over a small disk, similar to that investigated by Sandberg and his colleagues (Heiselberg, 1999; Jensen et al, 2002b; Sandberg, 2002; True et al, 2003), were studied using direct numerical simulation (DNS) of the Navier-Stokes equations. Computed pressure fields and streamlines for relatively low flow conditions (that is, Re = 1000) are illustrated in Figure 2.7 for three cases:

1   a solid disk of 0.075m in radius and 0.010m in thickness (0 per cent porosity);
2   a disk of 0.075m in radius and 0.010m in thickness with a centred 0.005m radius hole (0.44 per cent porosity); and
3   a disk of 0.075m in radius and 0.010m in thickness with a centred 0.050m radius hole (44 per cent porosity) (note: even at this low Reynolds number a vortex forms and sheds from the leeward edge of the disk; consequently, unsteady flow results were computed).

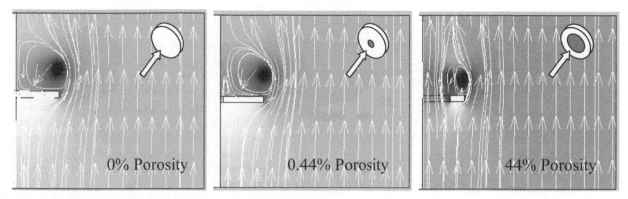

0% Porosity   0.44% Porosity   44% Porosity

Note: Results were computed using finite element solutions of the axysymmetric direct numerical simulation (DNS) of the Navier-Stokes equations with approximately 6500 isoparametric elements in each case adapted to the flow field to achieve grid convergence.

Source: chapter author

Figure 2.7 *Computed pressure fields (grey scale) and flow streamlines for airflow over three disks, each 0.075m in radius and 0.010m thick*

Plots of the corresponding axial distributions of pressure $p(r)$, kinetic energy density $(1/2)\rho u(r)^2$ and their sum acting on the disks are shown in Figure 2.8 – all normalized relative to the approach wind kinetic energy density. The choice of the normalization method used here is significant; it is a generalization of the conventional wind pressure coefficient.

Specifically, it is useful to relate flow conditions to those that would exist for isothermal, inviscid and incompressible flow – conditions that lead to irrotational flow where the Bernoulli sum remains constant throughout the entire flow field (see, for example, Chorin and Marsden, 1993):

Irrotational flow field:

$$p + \frac{1}{2}\rho u^2 + \rho g z = \text{constant} \tag{47a}$$

Or, equivalently, the local Bernoulli sum (*total pressure*) at an opening must be equal to the upstream reference condition (with u_{loc} being the local flow velocity in the opening):

$$p_{loc} + \frac{1}{2}\rho u_{loc}^2 + \rho g z_{loc} = p_0\frac{1}{2}\rho \bar{U}_{ref}^2 + \rho g z_0 \tag{47b}$$

By normalizing the *local total pressure* relative to the approach wind kinetic energy density via algebraic rearrangement of this relation, we obtain:

$$\frac{(p_{loc} - p_0) + \frac{1}{2}\rho u_{loc}^2 + \rho g(z_{loc} - z_0)}{\frac{1}{2}\rho \bar{U}_{ref}^2} = 1.0 \tag{48}$$

On the basis of this result, three specialized wind pressure coefficients may be defined that measure, in a sense, the deviation of the flow from irrotational conditions:

Total wind pressure coefficient:

$$C_{pt} \equiv \frac{(p_{loc} - p_0) + \frac{1}{2}\rho u_{loc}^2 + \rho g(z_{loc} - z_0)}{\frac{1}{2}\rho \bar{U}_{ref}^2} \tag{49}$$

Static wind pressure coefficient:

$$C_{pp} \equiv \frac{(p_{loc} - p_0) + \rho g(z_{loc} - z_0)}{\frac{1}{2}\rho \bar{U}_{ref}^2} \tag{50}$$

Dynamic wind pressure coefficient:

$$C_{pu} \equiv \frac{\frac{1}{2}\rho u_{loc}^2}{\frac{1}{2}\rho \bar{U}_{ref}^2} \tag{51}$$

Note that these coefficients have been defined so that the total wind pressure coefficient is equal to the sum of the other two contributions: $C_{pt} = C_{pp} + C_{pu}$. Furthermore, for a sealed building where the local velocity is $u_{loc} = 0$, the total wind pressure coefficients at any point will simply equal the conventional wind pressure coefficient C_p (although the geo-potential term is often ignored for the latter in the literature, yet is always accounted for in multi-zone airflow analysis programmes). Thus, the total wind pressure coefficient C_{pt} may be thought to be a generalized form of the conventional wind pressure coefficient C_p.

Most importantly, the total wind pressure coefficient C_{pt} accounts for the combined effects of pressure, kinetic

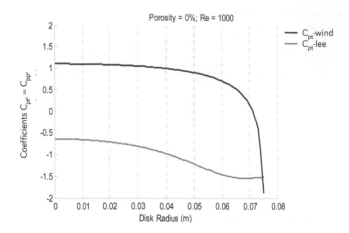

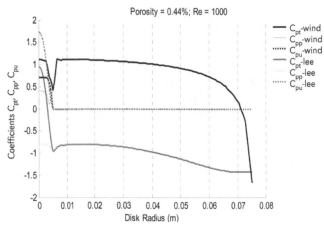

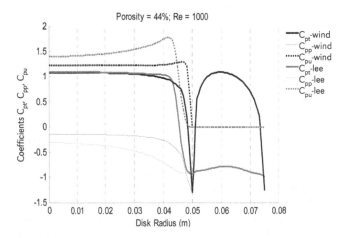

Source: chapter author

Figure 2.8 *Computed windward (black) and leeward (grey) total pressure, static pressure and dynamic pressure distributions normalized relative to the approach wind kinetic energy density for airflow over three disks, each 0.075m in radius and 0.010m thick*

energy density and geo-potential that 'drive' airflow – that is, by equation (42).

With these ideas in mind, we can turn our attention to the results plotted in Figure 2.8. Comparing first the results for the 0.44 per cent porosity disk to the 44 per cent porosity disk, we see that static pressure distributions across the opening in the latter fall to near-zero values. Consequently, the surface-averaged pressure distribution difference $\Delta\hat{C}_{pp}$ for the 0.44 per cent case is relatively large at 1.14, while that of the 44 per cent case is relatively small at 0.272. The surface-averaged dynamic pressure differences, $\Delta\hat{C}_{pu}$, on the other hand, remain similar and negative (!) for both cases: –0.13 for the 0.44 per cent case and –0.12 for the 44 per cent case. Thus, as claimed above, the *relative* importance of the kinetic energy of the flow – the dynamic pressure contribution – increases with increased porosity; yet the conventional macroscopic approach ignores this contribution altogether.

Given the relatively small contribution of the dynamic pressure for the 0.44 per cent porosity case, the surface-averaged total pressure difference $\Delta\hat{C}_{pt}$ over the opening is approximately equal to the surface-averaged static pressure difference $\Delta\hat{C}_{pp}$ (that is, 1.02 compared to 1.14, respectively). Thus, the error resulting from ignoring the dynamic pressure should be expected to be small for this low-porosity case. This is not the case for the 44 per cent porosity case.

Finally, using the computed results to compute the loss coefficient e_v^e for the disk hole using equation (44), e_v^e for the 0.44 per cent porosity disk is 1.5; for the 44 per cent porosity disk, it is 0.0 (that is, within the accuracy of the computations). That is to say, at 44 per cent porosity, the perforated disk provides little resistance to flow as may be expected and is evident from Figure 2.7.

While it is clear that the kinetic energy of the flow in porous buildings must be considered in modelling both the driving wind effects and the element flow relations, only two proposals have been put forward to account for this, and these remain controversial (Kato et al, 1992; Kurabuchi et al, 2004; Ohba et al, 2004). Currently, collaborative and fundamental studies have been initiated by a group of researchers, including the author, that offer the hope of a solution in the near future (Heiselberg, 1999; Jensen et al, 2002b; Sandberg, 2002; True et al, 2003; Kato, 2004). Indeed, the detailed mechanical energy balance and the generalized wind pressure coefficients presented above are being considered by these researchers as a means of improving our understanding and, possibly, of formulating a practical solution for this important problem (Seifert, 2004).

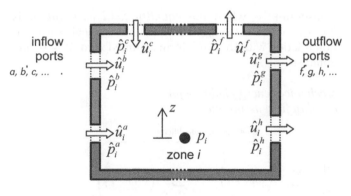

Source: chapter author

Figure 2.9 *A generic multi-port control volume i with associated spatial-averaged port velocities $\hat{u}_i^e$, spatial-averaged port pressures $\hat{p}_i^e$, and elevations z_i^e relative to the zone reference node for each linked flow element*

Multi-port control volumes

To date, all element flow relations used in macroscopic building airflow analysis have been limited to two-port elements. Yet, in duct networks, multi-port junctions are not only commonly used, they are indispensable. Lacking multi-port flow relations, these junctions are normally modelled as diminutive zones using either the equality or hydrostatic field assumptions presented above – that is, assuming that the junction resistance not accounted for may be so small as to be acceptably ignored. While new energy conservation approaches may resolve this problem in the long run (Guffey and Fraser, 1989; Gan and Riffat, 2000), it is useful here to look critically at the consequences of employing these field assumptions as approximations to the airflow within these junctions.

This generic problem actually extends well beyond multi-port duct junctions since building zones are invariably multi-port control volumes as well. Given their size, however, and the typically small air velocities within, modelling flow behaviour within zones with the hydrostatic field assumption has proven so intuitively satisfying that this assumption has been largely unquestioned.

Regrettably, the hydrostatic field assumption produces system equations that generally violate the conservation of mechanical energy. Thus, the conventional approach to multi-zone modelling is flawed at the most fundamental level. Furthermore, initial case studies of this problem clearly indicate that the errors resulting from it may well be significant in practical situations (Axley et al, 2002a, b, c). This section will, therefore, consider the conservation of mechanical energy in multi-

port control volumes and derive the conditions for which the equality and the hydrostatic field assumptions necessarily violate this conservation relation. An alternative, but approximate, (mechanical) energy conserving approach is then presented and briefly evaluated.

Conservation of mechanical energy of multi-port control volumes

The unsteady conservation of mechanical energy for isothermal flow through a generic multi-port control volume (see Figure 2.9) is fundamentally no different from that for a two-port control volume presented in equation (42), although one now must be careful to sum all inflow and outflow port contributions:

$$
\overbrace{\left(\sum_{\substack{e=a,b,c,\dots \\ inflow}} \left(\int_{A_i^e} z_i^e u_i^e dA \right) - \sum_{\substack{e=f,g,h,\dots \\ outflow}} \left(\int_{A_i^e} p_i^e u_i^e dA \right) \right)}^{\text{pressure work difference}}
$$

$$
+ \frac{\rho}{2} \overbrace{\left(\sum_{\substack{e=a,b,c,\dots \\ inflow}} \left(\int_{A_i^e} (U_i^e)^2 u_i^e dA \right) - \sum_{\substack{e=f,g,h,\dots \\ outflow}} \left(\int_{A_i^e} (U_i^e)^2 u_i^e dA \right) \right)}^{\text{kinetic energy rate difference}}
$$

$$
+ \rho g \overbrace{\left(\sum_{\substack{e=a,b,c,\dots \\ inflow}} \left(\int_{A_i^e} z_i^e u_i^e dA \right) - \sum_{\substack{e=f,g,h,\dots \\ outflow}} \left(\int_{A_i^e} z_i^e u_i^e dA \right) \right)}^{\text{geopolitical rate difference}}
$$

$$
+ \dot{W} - \dot{E}_v = \frac{d(K_{tot} + \Phi_{tot})}{dt} \tag{52}
$$

where, port velocities u_i^e here must be considered local velocities perpendicular to the port cross-sections (x_i^e, y^e) and, given that isothermal conditions prevail, the air density ρ in all ports will be identical.

While the integral form of this conservation relation may well be necessary for larger ports (for example, as in the porous building problem), it will be sufficient for our purposes here to consider a simplified form of this conservation relation based on the conventional assumption that pressure and velocity distributions at the ports are (nearly) uniform – that is, $p_i^e(x^e, y^e) \approx \hat{p}_i^e$ and $u_i^e(x^e, y^e) \approx \hat{u}_i^e$, both constant, and $\hat{u}_i^e = \dot{m}^e/(\rho A_i^e) \approx$ by definition:

$$
\overbrace{\left(\sum_{\substack{e=a,\,b,\,c,\,\ldots \\ inflow}} (\hat{p}_i^e \hat{u}_i^e A_j^e) - \sum_{\substack{e=f,\,g,\,h\ldots \\ outflow}} (\hat{p}_i^e \hat{u}_i^e A_j^e) \right)}^{\text{pressure work difference}}
$$

$$
+ \frac{\rho}{2} \overbrace{\left(\sum_{\substack{e=a,\,b,\,c,\,\ldots \\ inflow}} \left((\hat{u}_i^e)^3 A_i^e \right) - \sum_{\substack{e=f,\,g,\,h,\,\ldots \\ outflow}} \left((\hat{u}_i^e)^3 A_i^e \right) \right)}^{\text{kinetic energy rate difference}}
$$

$$
+ \rho g \overbrace{\left(\sum_{\substack{e=a,\,b,\,c,\,\ldots \\ inflow}} (z_i^e \hat{u}_i^e A_i^e) - \sum_{\substack{e=f,\,g,\,h,\,\ldots \\ outflow}} (z_i^e \hat{u}_i^e A_i^e) \right)}^{\text{geo-potential energy rate difference}}
$$

$$
+ \dot{W} - \dot{E}_v = \frac{d(K_{tot} + \Phi_{tot})}{dt} \tag{53}
$$

The *equality field assumption* $\hat{p}_i^e = p_i$ for $e = a, b, c, \ldots f, g, h, \ldots$ only impacts upon the first term; but its imposition causes this term to vanish for steady conditions (that is, by mass conservation):

$$
p_i \left(\sum_{\substack{e=a,\,b,\,c,\,\ldots \\ inflow}} (\hat{u}_i^e A_j^e) - \sum_{\substack{e=f,\,g,\,h,\,\ldots \\ outflow}} (\hat{u}_i^e A_j^e) \right)
$$

$$
= \frac{p_i}{\rho} \left(\sum_{\substack{e=a,\,b,\,c,\,\ldots \\ inflow}} (\rho \hat{u}_i^e A_j^e) - \sum_{\substack{e=f,\,g,\,h,\,\ldots \\ outflow}} (\rho \hat{u}_i^e A_j^e) \right) = 0 \tag{54}
$$

The *hydrostatic field assumption* $\hat{p}_i^e = p_i - \rho g z_i^e$ for $e = a, b, c, \ldots f, g, h, \ldots$ likewise only directly impacts upon the pressure work term, but creates a contribution that is equal to, but of opposite sign to, the geo-potential term:

$$
p_i \left(\sum_{\substack{e=a,\,b,\,c,\,\ldots \\ inflow}} (\hat{u}_i^e A_j^e) - \sum_{\substack{e=f,\,g,\,h,\,\ldots \\ outflow}} (\hat{u}_i^e A_j^e) \right)
$$

$$
- \rho g \left(\sum_{\substack{e=a,\,b,\,c,\,\ldots \\ inflow}} (z_i^e \hat{u}_i^e A_i^e) - \sum_{\substack{e=f,\,g,\,h,\,\ldots \\ outflow}} (z_i^e \hat{u}_i^e A_i^e) \right)
$$

$$
= - \rho g \left(\sum_{\substack{e=a,\,b,\,c,\,\ldots \\ inflow}} (z_i^e \hat{u}_i^e A_i^e) - \sum_{\substack{e=f,\,g,\,h,\,\ldots \\ outflow}} (z_i^e \hat{u}_i^e A_i^e) \right) \tag{55}
$$

Consequently, when the *equality field assumption* is imposed on the *steady* form of the mechanical conservation relation with the work term assumed zero, $\dot{W} = 0$, one obtains:

Mechanical energy balance for equality field assumption:

$$
\frac{\rho}{2} \overbrace{\left(\sum_{\substack{e=a,\,b,\,c,\,\ldots \\ inflow}} \left((\hat{u}_i^e)^3 A_i^e \right) - \sum_{\substack{e=f,\,g,\,h,\,\ldots \\ outflow}} \left((\hat{u}_i^e)^3 A_i^e \right) \right)}^{\text{kinetic energy rate difference}}
$$

$$
+ \rho g \overbrace{\left(\sum_{\substack{e=a,b,c,\ldots \\ inflow}} (z_i^e \hat{u}_i^e A_i^e) - \sum_{\substack{e=f,g,h,\ldots \\ outflow}} (z_i^e \hat{u}_i^e A_i^e) \right)}^{\text{geo-potential energy rate difference}} - \dot{E}_v = 0 \tag{56}
$$

Thus, within the uncertainty of the energy dissipated $\dot{E}_v$, which for building zones may be expected to be relatively small, if not truly negligible, isothermal mechanical energy conservation will only be satisfied when the remaining geo-potential and kinetic energy terms vanish – a condition that will not generally prevail.

When the *hydrostatic field assumption* is applied, under the same limiting conditions, the pressure work and geo-potential terms vanish, but the kinetic energy term remains:

Mechanical energy balance for hydrostatic field assumption:

$$
\frac{\rho}{2} \overbrace{\left(\sum_{\substack{e=a,b,c,\ldots \\ inflow}} \left((\hat{u}_i^e)^3 A_i^e \right) - \sum_{\substack{e=f,g,h,\ldots \\ outflow}} \left((\hat{u}_i^e)^3 A_i^e \right) \right)}^{\text{kinetic energy rate difference}} - \dot{E}_v = 0 \tag{57}
$$

Thus, again mechanical energy will not generally be conserved!

The implications of these conclusions are profound – they imply that conventional multi-zone analysis is flawed at the most fundamental level. Yet, can the error that may result from this flaw be significant? An examination of representative cases demonstrates that this error can, indeed, be significant. For the simple example illustrated in Figure 2.10, flows computed using the conventional

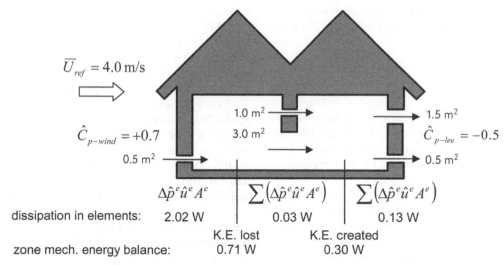

$\overline{U}_{ref} = 4.0 \text{ m/s}$

$\hat{C}_{p-wind} = +0.7$

$\hat{C}_{p-lee} = -0.5$

1.0 m²   1.5 m²

3.0 m²

0.5 m²   0.5 m²

	$\Delta\hat{p}^e\hat{u}^e A^e$	$\sum\left(\Delta\hat{p}^e\hat{u}^e A^e\right)$	$\sum\left(\Delta\hat{p}^e\hat{u}^e A^e\right)$
dissipation in elements:	2.02 W	0.03 W	0.13 W
		K.E. lost	K.E. created
zone mech. energy balance:		0.71 W	0.30 W

Source: Axley et al (2002)

Figure 2.10 *Simple two-zone problem used to evaluate the significance of the mechanical energy error in conventional multi-zone analysis*

hydrostatic field assumption resulted in the destruction of 0.71W of mechanical energy in one zone and the creation of 0.30W in another – both significant in comparison to the mechanical dissipation in the connecting flow paths that ranged from 0.13W to 2.02W (Axley et al, 2002).

Alternative field assumptions

Given the problematic kinetic energy term, it is tempting to impose the irrotational field condition introduced above in the context of generalized wind pressure coefficients – equation (47a) – now, however, adapted to a spatial-averaged form (Axley et al, 2002):

Irrotational field assumption:

$$\hat{p}_i^e + \rho_i g z_i^e + \frac{1}{2}\rho_i(\hat{u}_i^e)^2 = b_i = \text{constant} \quad (58)$$

Here, the sum of pressure, geo-potential and kinetic energy terms are set equal to a zone variable, the *Bernoulli sum* b_i, which is presumed to be constant within the zone.

Solving equation (58) for $\hat{p}_i^e$ and substituting it into the pressure work term of the mechanical energy conservation relation, equation (53), again the pressure work difference will vanish (that is, for steady conditions) and a contribution that is equal to but of opposite sign to the geo-potential difference will result. In addition, a contribution that is equal to but of opposite sign to the kinetic energy difference will also be provided so that all terms except the work term $\dot{W}$, which is zero in the present context, and zone dissipation term $\dot{E}_v$, will vanish, or:

Mechanical energy balance for irrotational field assumption:

$$\dot{E}_v = 0 \quad (59)$$

Thus, within the uncertainty of the power dissipated $\dot{E}_v$, the irrotational field assumption satisfies mechanical energy conservation *in general* – the very condition that the hydrostatic field assumption has commonly been assumed to guarantee but, in fact, does not!

From a fundamental point of view, this result should be expected as irrotational flow results for incompressible, isothermal and inviscid flow (Chorin and Marsden, 1993). This is not to say that the irrotational field assumption provides a means to an 'exact' solution to the multi-zone airflow problem – in general, it does not. Rather, it can be expected to provide an approximate solution to the multi-zone airflow analysis problem that satisfies both the conservation of mass and mechanical energy (that is, in the isothermal case).

In reality, some energy must be dissipated in zones. To capture this aspect of zone behaviour, in the approximate sense, a variation of the irrotational field assumption has been proposed by the author (Axley et al, 2002):

Mechanical energy balance for near irrotational field assumption:

$$\hat{p}_i^e + \rho_i g z_i^e + (1 + e_{vi}^e)\frac{1}{2}\rho_i(\hat{u}_i^e)^2 = b_{vi} = \text{constant} \quad (60)$$

where e_{vi}^e is an element *port dissipation coefficient* and b_{vi} is a modified Bernoulli sum. When substituted into the mechanical energy conservation relation with the accumulation and work terms assumed zero, one obtains:

$$\dot{E}_v = \frac{\rho}{2} \sum_{e=a,b,c,\ldots} \left(e^e_{vi}(\hat{u}^e_i)^3 A^e_i \right) \tag{61}$$

Thus, for this field assumption, zone dissipation is no longer presumed negligible; instead, it is related to the viscous power dissipation associated with all inflow and outflow ports combined (for example, by flow expansion for inflow and contraction for outflow).

These alternative field assumptions will lead directly to alternative formulations of the multi-zone airflow system equations that satisfy both the conservation of air mass flow and the conservation of mechanical energy – in the approximate sense. Yet, as this approach has not been thoroughly investigated, there will be no attempt to develop a concise matrix formulation of the resulting field equations in this section.

Turbulence-induced effects and dynamic analysis

Within the conventional approach to multi-zone airflow analysis, consideration is most commonly limited to mean wind conditions; yet, a number of studies have clearly demonstrated that wind turbulence can also play a significant role in driving building airflows. Recent studies demonstrate that the magnitude of instantaneous airflow rates can easily exceed mean values and may well result in flow reversals at times when the relative magnitude of turbulent driving forces are large (Haghighat et al, 1991; Chorin and Marsden, 1993; Etheridge, 2000a). As a result, mean infiltration air change rates appear to be consistently underestimated by the conventional approach. Results from limited field measurements reveal this underestimation error at approximately 20 per cent (Girault and Spennato, 1999), while those based on theoretical predictions indicate errors of up to 10 per cent (Sirén, 1997) and to as much as 50 per cent (Etheridge, 2000b, 2002) for representative environmental conditions – depending, again, upon the relative magnitude of turbulent driving forces.

Interest in turbulence-induced ventilation dates to at least the mid 1970s (Hill and Kasuda, 1975; Cockcroft and Robertson, 1979; Etheridge and Nolan, 1979); thus, this particular problem is not new. It has been largely ignored in practice due, in part, to the complexity of the problem and, in part, to the absolute scarcity of data needed to model it. Nevertheless, a number of seemingly practical proposals to model turbulence-induced ventilation have appeared (Etheridge and Alexander, 1980; Haghighat et al, 1991; Etheridge and Sandberg, 1996; Sirén, 1997; Etheridge, 1998b, 2000a, 2000b; Girault and Spennato, 1999; Saraiva and Marques da Silva, 1999); but as these models have yet to be critically evaluated, there has been no rush to employ them in practice.

Given the complexity of the problem and the range of modelling approaches considered, only a general introduction will be presented here. The scope and complexity of the problem will be outlined, key elements of dynamic analysis theories to model turbulence-induced ventilation will be presented, and practical methods that may be included in multi-zone airflow analysis will be considered.

Scope and complexity of turbulence-induced ventilation

Turbulent airflow velocities $v(t)$ measured at a point may be represented by the sum of chaotic fluctuations $v'(t)$ about a time-smoothed (moving average) mean value, $\bar{v}(t)$. Similarly, pressures $p(t)$, whole-building pressure differences $\Delta p(t)$ and element pressure differences $\Delta p^e(t)$ induced by these velocities may also be represented by the sum of a time-smoothed mean value and the fluctuation about that mean:

$$v(t) = \bar{v}(t) + v'(t) \; ; \; p(t) = \bar{p}(t) + p'(t) \; ;$$

$$\Delta p(t) = \Delta\bar{p}(t) + \Delta p'(t) \tag{62a, b, c}$$

For all, the relative intensity of turbulent fluctuations may, then, be characterized by the ratio of the standard deviation of the fluctuating component over the mean value:

$$I_v \equiv \frac{\sqrt{(v')^2}}{\bar{v}} \; ; \; I_p \equiv \frac{\sqrt{(p')^2}}{\bar{p}} \; ; \; I_{\Delta p} \equiv \frac{\sqrt{(\Delta p')^2}}{\overline{\Delta p}} \tag{63a, b, c}$$

where I_v is the *velocity*, I_p the *pressure* and $I\Delta p$ the whole-building *pressure difference* relative turbulence intensities. An *element pressure difference* relative turbulence intensity would be defined similarly. The velocity and pressure fluctuations result from the motion of eddies of a variety of scales that manifest turbulent airflow. These scales can be evaluated by nine *integral scales of turbulence* determined statistically from detailed records of the velocity flow field (Haghighat et al, 2000) to provide additional characteristics of the turbulent airflow field.

With these ideas in mind, turbulence-induced ventilation may be thought to fall along a range of possibilities,

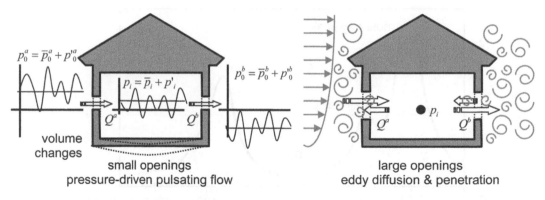

$p_0^a = \overline{p}_0^a + p_0'^a$

$p_i = \overline{p}_i + p'_i$

$p_0^b = \overline{p}_0^b + p_0'^b$

Q^a Q^b

volume
changes

small openings
pressure-driven pulsating flow

p_i

Q^a Q^b

large openings
eddy diffusion & penetration

Source: adapted from Gustén (1989)

Figure 2.11 *The limiting possibilities of turbulence-induced ventilation*

with ventilation airflows driven by pressure fluctuations at one end, for relatively small openings (that is, small relative to the length scales of the ambient turbulence), and by diffusion and/or penetration of turbulent eddies at the other, for relatively large openings (see Figure 2.11).

When envelope openings are similar or large relative to the scale of turbulence, then airflow through these openings will tend to be multi-directional with non-uniform velocity distributions. Under these conditions, macroscopic methods are not well suited to modelling airflows since the usual simple assumptions regarding the *form* of pressure and velocity distributions at flow element ports cannot be made *a priori*. However, microscopic methods of computational fluid dynamics that directly model these eddies appear to offer an approach to modelling these flows faithfully (Jiang and Chen, 2002).

When envelope openings are small relative to the scales of ambient turbulence, airflow will tend to be uni-directional and driven by instantaneous pressure differences across the openings. For these conditions, *pulsating* flows due to fluctuating pressure differences $\Delta p^{e'}$ will be superimposed on the mean flow driven by mean pressure differences Δp^e across each envelope element e. Clearly, the relative importance of the pulsating flow will increase as the relative turbulent intensity $I_{\Delta p}$ increases. Indeed, Etheridge shows (using a non-dimensional parameter equivalent to the inverse of $I_{\Delta p}$) that turbulence-induced effects are likely to be negligibly small when estimating mean flow rates when $I_{\Delta p} < 1.0$, and for practical computations may be ignored with little error when $I_{\Delta p} < 2.5$ for representative cases of single-zone models with two dominant openings (Etheridge, 2000b, 2002).

Beyond this simple but useful observation, the relation between ventilation airflows and the driving

fluctuating pressures is complex and depends upon both the 'the temporal fluctuation and the spatial incoordination' (Haghighat et al, 2000) of the envelope pressure fluctuations. Consequently, to deterministically model the details of airflow within the *time domain*, one would need simultaneous records of pressure for each of the envelope flow elements of a given building. Given the scarcity of wind pressure coefficient data even for the mean flow components, this is clearly out of the question. Consequently, a number of researchers have turned to the frequency domain in the hopes of being able to identify general characteristics of turbulent flow that may lead to practical methods of analysis (Gustén, 1989; Haghighat et al, 1991; Etheridge, 2000a, 2000b).

Given the fluctuating nature of the wind, wind speed and pressure records may be decomposed by Fourier analysis into a sum of harmonic components – that is to say, one may examine wind phenomena in the *frequency domain*. When this is done, some general characteristics of the wind emerge, foremost of which is the spectral distribution of wind power content shown in Figure 2.12 (after Davenport, 1967), which reveals that large-scale, low-frequency turbulence is distinct and separated from small-scale, high-frequency turbulence by a significant spectral gap.

Building airflow system dynamics tend to be relatively rapid. Hence, the dynamic effects of building airflow system responses to the low-frequency portion of the wind spectrum play little role in system response. Consequently, the dynamic response of building airflow systems to these low frequency excitations may be computed using a *quasi-steady* approach based on the conventional multi-zone method presented above (that is, equations (25) or (39) with zone mass accumulation dM_i/dt assumed zero). Dynamic effects may, however, play a

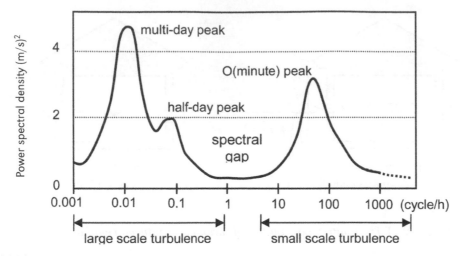

Source: Davenport (1967)

Figure 2.12 *Representative wind power spectral distribution*

significant role when considering the response to the higher frequency portion of the spectrum, depending upon the detailed characteristics of the building airflow system. To capture the full details of system response in these cases, a dynamic approach to analysis is necessary. Likewise, the response of the building airflow system to dynamic excitations driven by the building mechanical system, which also tend to fall in the high-frequency range, may also demand a dynamic approach to analysis.

Elements of dynamic analysis theories

Just what are the dynamic effects of building airflow systems and how may they be modelled? From a fundamental point of view, airflow resists dynamic changes through its *inertia*, viscous dissipation or *damping*, and air mass accumulation or *capacitance* effects.

In the conventional *steady* approach to multi-zone airflow analysis, the air mass flow rate into each zone *i* is presumed to be conserved (that is, from equation (25)) and is simply equal to the sum of the element air-mass flow-rate contributions $\dot{m}_i^e(p_i)$ that are directly dependent upon the unknown zone pressures p_i through either the equilibrium or hydrostatic field assumptions:

Steady approach:

$$-\sum_{\substack{\text{linked} \\ \text{elements}}} \dot{m}_i^e(p_i) = 0 \qquad (64)$$

where, as before, superscript *e* is permutated through all inflow and outflow elements linked to node *i*. Thus, the conventional *quasi steady* approach is defined in terms of a system of (non-linear) algebraic equations.

To account for air mass *capacitance* within zones, we must account for the accumulation of air mass M_i in each zone (that is, equation (25)):

Dynamic capacitance approach:

$$-\sum_{\substack{\text{linked} \\ \text{elements}}} \dot{m}_i^e(p_i) = \frac{dM_i}{dt} \qquad (65)$$

If the air is assumed to behave as an ideal gas, then the accumulation term can be rewritten as below if it is assumed that zone volume V_i, zone pressure p_i and zone temperature T_i all vary with time:

$$-\sum_{\substack{\text{linked} \\ \text{elements}}} \dot{m}_i^e(p_i) = \frac{d(Mp_iV_i / RT_i)}{dt}$$

$$= \frac{Mp_i}{RT_i}\frac{dV_i}{dt} + \frac{MV_i}{RT_i}\frac{dp_i}{dt} - \frac{MV_ip_i}{RT_i^2}\frac{dT_i}{dt} \qquad (66a)$$

where M is the effective molecular weight of air and R is the universal gas constant. If one limits consideration to airflow analysis for specified zone temperatures, then the expansion of the accumulation term becomes simpler:

$$-\sum_{\substack{\text{linked} \\ \text{elements}}} \dot{m}_i^e(p_i) = \frac{d(Mp_iV_i / RT_i)}{dt} \approx \frac{Mp_i}{RT_i}\frac{dV_i}{dt} + \frac{MV_i}{RT_i}\frac{dp_i}{dt}$$

$$(66b)$$

Finally, assuming that the dynamic changes of the zone volume are negligible, $dV_i/dt = 0$, we obtain an accumulation term defined in terms of the zone pressures alone:

Conventional capacitance model:

$$-\sum_{\substack{\text{linked} \\ \text{elements}}} \dot{m}_i^e(p_i) = \frac{d(Mp_iV_i / RT_i)}{dt} \approx \frac{MV_i}{RT_i}\frac{dp_i}{dt} \qquad (66c)$$

This final result defines a system of non-linear ordinary differential equations defined in terms of the unknown zone pressure variables that may be transformed, using one of a number of alternative finite difference schemes, into a time-stepping solution algorithm that demands the solution of a non-linear algebraic problem at each time step. Since it is most commonly used to account for zone *capacitance* in dynamic airflow analysis, it will be identified as the *conventional capacitance model*.

Gustén (1989) argues that the dynamic changes of zone volume may not be negligible and provides some estimates of their likely magnitude. If this is the case, then one would have to solve the dynamic fluid structure interaction problem in parallel with the solution of the dynamic airflow problem – a task that is well beyond the ambitions of any study to date. Likewise, if the temperature rate term of equation (66a) is to be retained, then dynamic airflow analysis would have to be coupled to the building thermal analysis problem – again, a task that has yet to be investigated.

Airflow resists change through its inertia and viscous forces. Both are, in principle, manifest in zones and in flow elements, although in existing dynamic models of airflow analysis both are commonly ignored in zones. One approach to account for these effects follows directly from the unsteady form of the isothermal mechanical energy balance given in equation (42), simplified for the common assumptions of (nearly) constant port pressure and velocity distributions and constant cross-section:

$$\Delta p^e - \tfrac{1}{2}\rho(\hat{u}^e)^2 e_v^e = \frac{1}{\hat{u}^e A^e}\frac{d(K_{tot} + \Phi_{tot})}{dt} \qquad (67)$$

where, K_{tot} and Φ_{tot} are the total kinetic energy and geo-potential energy within the element, respectively. For isothermal conditions, the geo-potential energy term may be dropped and, assuming incompressible flow within the element, the kinetic energy term may be approximated by assuming that the air within the element moves as an incompressible slug at the spatial average element velocity $\hat{u}^e$ – the so-called *slug flow approximation*:

Slug flow approximation:

$$K_{tot} = (\rho l^e A^e)\frac{(\hat{u}^e)^2}{2} \qquad (68)$$

Substituting this relation into the element flow relation, equation (67), and simplifying, we obtain a relatively general dynamic flow element model that accounts for both inertia (that is, the term on the right-hand side below) and viscous damping (that is, the second term on the left-hand side) within a flow element:

$$\Delta p^e - \tfrac{1}{2}\rho(\hat{u}^e)^2 e_v^e = \rho l^e \frac{d\hat{u}^e}{dt} \qquad (69a)$$

Here, l^e is the length of the element along its axis. Alternatively, this dynamic element relation may be written in terms of the element mass flow rate $\dot{m}^e = \rho A^e \hat{u}^e$ as:

Dynamic element model:

$$\frac{e_v^e}{2\rho(A^e)^2}(\dot{m}^e)^2 + \frac{l^e}{A^e}\frac{d\dot{m}^e}{dt} = \Delta p^e \qquad (69b)$$

It may also be rewritten in terms of the element volumetric flow rate $Q^e = \dot{m}^e/\rho$ and the effective element discharge coefficient C_d^e, given $e_v^e \approx 1(C_d^e)^2$ from equation (46), as:

$$(Q^e)^2 + 2A^e(C_d^e)^2 l^e\frac{dQ^e}{dt} = 2(C_d^e A^e)^2\frac{\Delta p^e}{\rho} \qquad (69c)$$

This last relation is identical to the dynamic element relation used by Etheridge if $(C_d^e)^2 l^e$ is taken as the effective length of the element (Etheridge, 2000a, 2000b). Gustén (1989) and, later, Haghighat et al (1991) used a similar model based on a mechanical system analogy that includes an additional term which accounts for air compressibility. Haghighat also linearizes the result to make use of frequency domain response-analysis methods. See Haghighat et al (2000) for a review of these and other approaches used to model pulsation flows driven by wind turbulence.

The dynamic element model, defined in terms of a non-linear ordinary differential equation, is a direct generalization of the inverse form of the discharge coefficient model, equation (19b), a non-linear algebraic equation. Again, as above, this differential equation may be transformed, using finite difference methods, into a series of non-linear algebraic equations (that is, *semi-discrete* element relations) that would be defined at each time step of a time-stepping algorithm.

Finally, to implement a dynamic multi-zone analytical method, one could *assemble* the semi-discrete element (Jacobian) relations within each time step of the mass conservation time-stepping algorithm by applying the appropriate field assumption for each zone (that is, analogously to the procedures defined by equations (37c) and (38a,b)). While this has yet to be implemented for general multi-zone analysis, it has apparently been accomplished for a relatively complex two-zone case with a total of eight flow elements by Sirén (1997) and for a series of single-zone models with two openings by Etheridge (1998b, 2000a, 2000b).

Alternatively, given that the form of the dynamic element model, equation (69b), is an inverse flow form, system equations may be formed by directly summing these element models using a *loop compatibility* approach. Yet, this alternative approach has, apparently, not been investigated.

As one may choose to implement parts or all of these options, a number of different system analysis methods may be formulated. For example:

- *Quasi-dynamic method*: in conventional multi-zone analysis, both zone capacitance dynamics and element inertia are ignored in computing a system's response to time-varying wind and temperature records. Instead, airflow rates are computed for each time increment of a wind (and zone temperatures) record using the steady form of the system equations – equation (40).
- *Semi-dynamic capacitance method*: one may account for zone air mass accumulation alone by assembling only steady-element flow relations at each time step of a time-stepping algorithm based on equation (66c) (Walton, 1998; Dols and Walton, 2000).
- *Semi-dynamic inertia/damping method*: alternatively, at each time step of a time-stepping scheme, one may assemble the semi-discrete dynamic element (Jacobian) relations using the steady form of the node continuity relation to attempt to account for the combined effects of inertia and damping, while ignoring zone capacitance.
- *Full-dynamic method*: finally, by combining the two semi-dynamic methods, one may account for zone capacitance and element flow inertia and damping.

Typically, these complex dynamic models are implemented or, more accurately, prototyped using higher-level programming tools that are sometimes called *simulation environments*. While a variety of these general purpose tools are available, SIMULINK, EES, SIMNON and Dymola, and Modelica® are noteworthy as they have been developed specifically to solve coupled systems of non-linear differential and algebraic equations (DAEs) (Brück et al, 2002; MathWorks, 2002; Modelica Association, 2002; SSPA, 2003; Klein, 2004).

The full dynamic method, as the most complete method, may be expected to provide the most accurate predictions of dynamic system response. Yet, this method and all of the less complete methods have been formulated:

- using the hydrostatic and/or equality field assumptions that have been demonstrated above to systematically violate the fundamental principle of the conservation of mechanical energy in multi-port zones (that is, for isothermal conditions); and
- ignoring zone viscous dissipation and inertia.

Thus, all of these dynamic methods of analysis remained fundamentally flawed! Indeed, errors introduced by use of the conventional field assumptions may well be as significant as the errors that result from ignoring dynamic effects in the first place, although this has yet to be investigated.

Simplified methods of multi-zone dynamic analysis

The detailed deterministic models presented above are generally not suitable for practical multi-zone ventilation analysis, given their complexity and the scarcity of the dynamic wind and/or mechanical system excitation data necessary to apply them. Consequently, simplified semi-empirical models have been developed to fill this need. Two of these simplified approaches are noteworthy. The first, developed by Etheridge, has long been available in the multi-zone tool VENT (Etheridge and Alexander, 1980) and is well documented in section 4.3.1 of *Building Ventilation: Theory and Measurement* (Etheridge and Sandberg, 1996). The second, developed by Sirén (1997), is in some ways similar to the *British Admittance Procedure* developed for dynamic thermal analysis, in that response amplitudes for simple harmonic excitations are used to scale real dynamic excitation data in order to approximate a response analysis.

On close examination, Sirén's approach is broadly similar to Etheridge's. Sirén relates mean element airflows to mean *element* pressure differences that are, through the assumptions made, related to the mean and standard deviation of the approach wind velocity and its fluctuations, respectively. Etheridge relates the mean element airflows to two contributions: one driven by the mean approach wind velocity and the other due to the standard deviation of approach wind velocity fluctuations. However, both approaches ignore inertia, capacitance and compressibility effects and, therefore, present quasi-dynamic models.

The success of these simplified models suggests that these effects may be ignored when predicting pulsation flows for the types of relatively short, small cross-section openings considered by these investigators. That is to say, it appears that pulsation flows may be modelled using these or other similar quasi-steady approaches for infiltration openings. However, Etheridge demonstrates, rather conclusively, that to model turbulence-induced pulsation flow in elongated flow elements (for example, long stacks intended to inhibit flow reversal), inertia effects are likely to be important (Etheridge, 2000a, 2000b). Finally, to come full circle, for openings that are large relative to turbulent eddies, airflow may well be sufficiently complex to be beyond the capabilities of macroscopic methods; thus, the analyst should consider turning to methods of computational fluid dynamics instead.

Multi-zone loop methods

The *node continuity approach* to multi-zone airflow analysis has long been the preferred method; yet, as briefly noted above, another fundamental approach based on *pressure loop compatibility* is available. *Loop methods*, while overlooked by all but a few in the building ventilation community (Li, 1993; Wray and Yuill, 1993; Nitta, 1994), have been used extensively in the piping network analysis community (Jeppson, 1976; Isaacs and Mills, 1980; Wood, 1981; Wood and Rayes, 1981; Demuren and Ideriah, 1986; Wood and Funk, 1993; Savic and Walters, 1996), provide an 'exact' analytical approach to size components of natural and hybrid ventilation systems (Axley, 1998, 1999a, 1999b, 2000a, 2000b, 2001a; Dols and Emmerich, 2003), and offer a number of advantages when compared to the node continuity approach that have not been fully recognized. This section will consider:

- the formation of pressure loop equations for specific *loops* in building airflow systems;
- the formation of whole-building system equations based on these specific loop equations; and
- the use of loop equations to size ventilation system components.

Loop compatibility is, in fact, a general principle that may be applied to the analysis of a number of different classes of systems (for example, hydraulic, electronic and airflow systems) by systematically accounting for changes of a scalar variable (for example, hydraulic head, voltage or pressure) as one traverses a physical *loop* in the system under consideration. While the approach may be applied instantaneously to the dynamic and spatial variation of these scalars, most often time-smoothed spatial averaged values are assumed. Thus, to simplify notation in this section, spatial and time-smoothed averaged variables associated with a flow cross-section i of a flow element e will be represented without the hat and over-bar notation used above. Furthermore, these sections will be identified by consecutive numerical indices rather than by zone and element indices to facilitate discussion.

For example, consider the building model illustrated in Figure 2.13 where four loops are identified that progress from ambient outdoor nodes to a series of flow element cross-sections, then back to the outdoor nodes. For example, loop 1 traverses from node 1 to sections 2, 3, 4, 5, 11, 21 and 23 to nodes 22 and 6, then back to node 1. In all, four flow elements are located along loop 1, including an inlet element a, a door passage element b, a stack duct element e and a stack terminal element j. Thus, time-smooth, spatially-averaged pressures associated with these nodes and cross-sections would then be identified sequentially as p_1, p_2, p_3, p_4, p_5, p_{11}, p_{23}, p_{22}, p_6, then back to p_1. Furthermore, the sequential order establishes a sign convention for airflow direction. Thus, in Figure 2.5 the directed line segments of loop 1 indicate a so-called *forward* loop direction and thereby the positive convention for airflow in elements a, b, e, and i.

Individual pressure loop equations

With the simplified notation established above, the fundamental *loop pressure compatibility* relation of equation (26) becomes:

Loop pressure compatibility:

$$\sum_{loop\ l} (p_j - p_i) = 0 \qquad (70)$$

where indices j and i are permuted sequentially through all nodes and element cross-sections indices of the loop l moving in the *forward* loop direction. If envelope surface pressures are modelled using the conventional wind pressure coefficient relation, equation (14), and hydrostatic assumption, equation (30), then the loop compatibility relation simplifies to:

$$\underbrace{\sum_{loop\ l} (\Delta p^e)}_{\Delta p_{lloss}} = \underbrace{\sum_{loop\ l} C_p^e \frac{1}{2} \rho_o U_{ref}^2}_{\Delta p_{lw}} + \underbrace{\sum_{loop\ l} \rho_{i-j} g(z_i - z_j)}_{\Delta p_{lb}} \qquad (71)$$

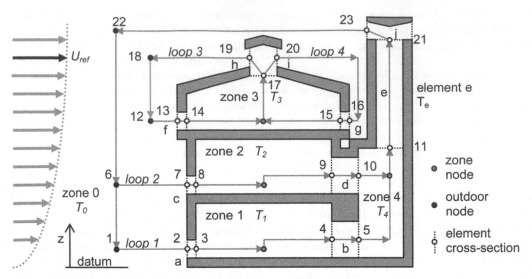

Note: Outdoor nodes are indicated by solid circles and element cross-sections by dotted lines with open circles.

Source: adapted from Irving and Uys (1997)

Figure 2.13 *A representative building section of the Inland Revenue Centre offices designed by Sir Michael Hopkins and Partners*

where ρ_{i-j} is the air density in the zone or element between pressures i and j and ΣC_p^e is the algebraic sum of wind pressure coefficients moving in the forward loop direction. That is to say, increases of pressure induced by the wind while moving in the *forward* loop direction are positive. Thus, the sum of pressure losses in a loop 1 due to flow through elements Δp_{lloss} is simply equal to the sum of the driving wind pressure Δp_{lw} and buoyancy (stack) pressure Δp_{lb} of the loop. It is important to stress, within the assumption of conventional multi-zone airflow analysis, that this general loop equation is 'exact.'

For a specific loop, then, the left-hand side of the general relation would be formed by summing the flow relations that are appropriate to each flow element along the loop. As these relations are generally non-linear relations between element pressure loss Δp^e and air mass flow rate $\dot{m}^e$ (that is, $\Delta p^e = g^e(\dot{m}^e, \vartheta^e)$, equation (17)), it will be useful to form linear approximations to them using, again, a Taylor series expansion about, now, a current estimate k of the element mass flow rate $\dot{m}_{[k]}^e$ and the corresponding estimate of element pressure drop $\Delta p_{[k]}^e$ as:

$$\Delta p^e \approx g^e(\dot{m}_{[k]}^e, \vartheta^e) + \left.\frac{\partial g^e}{\partial \dot{m}^e}\right|_{\dot{m}_{[k]}^e} (\dot{m}^e - \dot{m}_{[k]}^e)$$

$$+ O\left(\left|\dot{m}^e - \dot{m}_{[k]}^e\right|^2\right)$$

$$\approx \Delta p_{[k]}^e + \left.\frac{\partial g^e}{\partial \dot{m}^e}\right|_{\dot{m}_{[k]}^e} (\dot{m}^e - \dot{m}_{[k]}^e)$$

$$+ O\left(\left|\dot{m}^e - \dot{m}_{[k]}^e\right|^2\right) \tag{72a}$$

$$\Delta p^e \approx \Delta p_{[k]}^e + J_{L[k]}^e (\dot{m}^e - \dot{m}_{[k]}^e)$$

$$+ O\left(\left|\dot{m}^e - \dot{m}_{[k]}^e\right|^2\right). \tag{72b}$$

Here the partial derivative term is the current estimate of the *element loop Jacobian* $J_{L[k]}^e$ and the rightmost term indicates that the error in the approximation is proportional to $|\dot{m}^e - \dot{m}_{[k]}^e|^2$.

From the definition implicit to equation (72), we may directly obtain the *element loop Jacobians* for the *linear model*, $\Delta \hat{p}^e = \dot{m}^e/(C^e A^e)$ and the *discharge coefficient model*, $\Delta \hat{p}^e = (\dot{m}^e)^2/(2\rho^e (C_d^e A^e)^2)$, as:

Linear model:

$$J_{L[k]}^e = \frac{1}{(C^e A^e)} \tag{73}$$

Discharge coefficient mode:

$$J^e_{L\,[k]} = \frac{|\,\dot{m}^e_{[k]}\,|}{\rho^e (C^e_d A^e)^2} \tag{74}$$

Thus, unlike the element Jacobian for the forward form, equation (22), the element loop Jacobian for the inverse form of the discharge coefficient model remains bounded (and positive). This will, likewise, be true for the power law (that is, for the usual case of the power law exponent less than or equal to 1.0), the classic duct and the quadratic flow models. Thus, these inverse flow element models will not need to be augmented by bounded forms (for example, linear forms) for near-zero (that is, $\dot{m}^e_{[k]} \to 0$) flow conditions to effect a solution.

Dropping the error term of the Taylor series approximation, then, a specific loop equation may be formed by directly summing all flow element contributions. Thus, for example, the left-hand side of the loop equation for loop 1 of Figure 2.13 would be:

$$\sum_{loop\ 1} (\Delta p^e) \approx \Delta p^\ell_{[k]} + J^a_{L\,[k]}(\dot{m}^a - \dot{m}^e_{[k]})$$

$$+ \Delta p^b_{[k]} + J^b_{L\,[k]}(\dot{m}^b - \dot{m}^e_{[k]}) + \dots \tag{75a}$$

$$\sum_{loop\ 1} (\Delta p^e) \approx \sum_{\substack{loop\ 1 \\ e=a,b,\dots i}} (J^e_{L\,[k]}\dot{m}^e)$$

$$+ \sum_{\substack{loop\ 1 \\ e=a,b,\dots i}} (\Delta p^e_{[k]} - J^e_{L\,[k]}\dot{m}^e_{[k]}) \tag{75b}$$

Finally, generalizing this result and substituting it into the general form of the loop equation, equation (71), one may define an iterative loop equation for a specific loop l:

$$\sum_{loop\ l} (J^e_{L\,[k]}\dot{m}^e_{[k+1]}) = \overbrace{\sum_{loop\ l} C^e_p \frac{1}{2}\rho_o U^2_{ref}}^{\Delta p_{lw}}$$

$$+ \overbrace{\sum_{loop\ l} \rho_{i-j} g(z_i - z_j)}^{\Delta p_{lb}} - \sum_{loop\ l} (\Delta p^e_{[k]} - J^e_{L\,[k]}\dot{m}^e_{[k]}) \tag{76}$$

Here, the left-hand side defines a linear equation in the unknown new estimates of element air-mass flow rates $\dot{m}^e_{[k+1]}$ of loop l, while the right-hand side defines the sum of the driving wind and buoyancy pressures and a correction term $\Delta p_{l[k]}$ (that is, based on the current estimate of element mass flow rates $\dot{m}^e_{[k]}$). When used to form the complete set of *system equations*, one may then apply this relation to determine improved estimates of the element airflow rates $\dot{m}^e_{[k+1]}$ from current estimates $\dot{m}^e_{[k]}$. Repeating this process iteratively until the estimates converge or, equivalently, the pressure correction term $\Delta p_{l[k]}$ vanishes, one may obtain a solution to the multi-zone airflow problem. As in multi-zone nodal methods, linear element equations may be used to determine a first estimate $\dot{m}^e_{[1]}$ to initiate the process.

It will be useful to consider a matrix formulation of equation (76). As an example, the loop equation for loop 1 of Figure 2.13 may be written as:

$$[J^a_{L\,[k]}\ J^b_{L\,[k]}\ 0\ 0\ J^e_{L\,[k]}\ 0\ 0\ 0\ J^j_{L\,[k]}] \begin{Bmatrix} \dot{m}^a_{[k+1]} \\ \dot{m}^b_{[k+1]} \\ \dot{m}^e_{[k+1]} \\ \vdots \\ \dot{m}^j_{[k+1]} \end{Bmatrix}$$

$$= \Delta p_{1w} + \Delta p_{1b} - \Delta p_{1[k]} \tag{77}$$

Graph theory and system equations

The specific loop equation presented in equation (77) for loop 1 of the building idealization of Figure 2.13 defines a single equation with ten unknown airflow rates (that is, $\dot{m}^a_{[k+1]}$, $\dot{m}^b_{[k+1]}$, ... $\dot{m}^j_{[k+1]}$). Thus, to solve for all ten unknowns, we need to assemble nine additional equations. We may, for example, form three additional loop equations for loops 2, 3 and 4, illustrated in Figure 2.13, and augment them with five airflow rate continuity equations for the five zones of the building idealization, plus one for the exterior zone to achieve this objective. But how can we be sure that this set of ten equations will result in a solution, much less an efficient solution? Indeed, we may identify a number of additional loops – some simple, like loop 1, 2, 3, 4, 5, 10, 9, 8, 7, 6, 1 or 12, 13, 14, 15, 16, 20, 19, 18, 12, or others far more convoluted.

The problem of selecting an appropriate set of loops has been one of the major barriers to the use of the pressure loop method for multi-zone analysis (Wray and Yuill, 1993) since the general theory related to this problem is not well known in the building ventilation

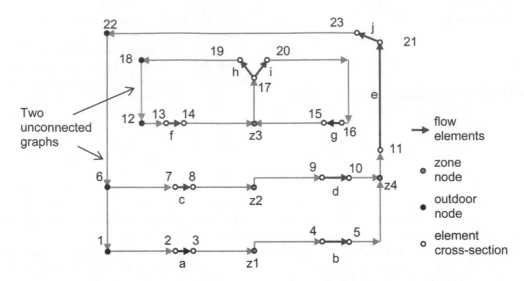

Note: Here zone nodes are labelled z1, z2, z3 and z4, and flow elements are indicated by darker directed line segments.

Source: chapter author

Figure 2.14 *Digraph of the system topology of Figure 2.13*

community and numerical methods to select loops often remain proprietary. Recent work at the Hybrid Ventilation Centre in Aalborg, Denmark, making use of recently published methods to automate loop selection (Savic and Walters, 1996), is, however, removing this critical barrier. To unravel this fundamental problem, it is necessary to introduce some simple ideas from graph theory (Shearer et al, 1971; Diestel, 1991; Walter and Contreras, 1999).

The *topology* of the ventilation system of the building illustrated in Figure 2.13 has been represented by nodes, flow element cross-sections and airflow paths represented by directed line segments. Isolated from the associated control-volume idealization, this topology defines a *graph* of line segments or *directed graph* (*digraph*) of directed line segments and nodes, where, in this context, the flow element cross-sections will be accepted as (pseudo) nodes (see Figure 2.14). The formal study of *graphs* and *digraphs*, the field of *topology*, investigates their mathematical properties – the very properties that relate to the loop equations that interest us here.

A number of key definitions and simple theorems, adapted to the present context, establish the basis for deeper considerations:

- *Path*: a series of alternating nodes and line segments define a (flow) path.
- *Connected graph*: a graph is connected if there is a path between any two nodes of the graph – for example, Figure 2.14 includes two unconnected

graphs. Not surprisingly, the loop equations for each unconnected graph are uncoupled to the others.
- *Node equations*: a node continuity equation can be written for each node of a connected graph. For a graph of n nodes, $(n–1)$ independent continuity equations may be written as the nth equation is simply the sum of the other $(n–1)$ equations.
- *Simple paths and loops*: a simple path does not repeat any node. A *loop* is a *simple path* that begins and is closed by the same node.
- *Tree and spanning tree*: a graph is a *tree* if it is connected and has no loops. A *spanning tree* is the sub-graph of a connected graph that has the same node set and is a tree. Figure 2.15 illustrates the two spanning trees of the corresponding two separate connected graphs of Figure 2.14.
- *Number of line segments*: a tree with n nodes has $(n–1)$ line segments (this may be proved by induction).
- *Number of independent loops*: given a connected graph with m line segments, then $(m–(n–1))$ independent loops exist. The proof is as follows: for the spanning tree we simply replace each of the removed line segments one by one. For each line segment replaced, a loop is created. Thus, we may form only $(m–(n–1))$ independent loop equations.

Consequently, the conditions needed to form the equations that govern the behaviour of the system as a whole – the *system equations* – may be established. For a

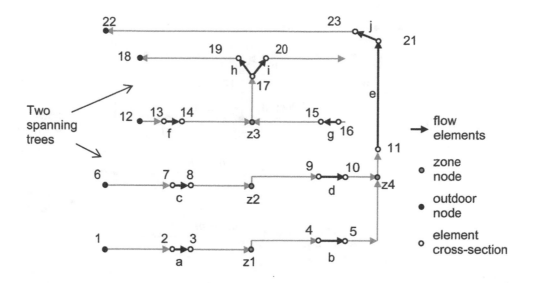

Figure 2.15 *The spanning trees of the two connected graphs of Figure 2.14*

connected system topology of n nodes and m line segments, or, equivalently, m unknown flow rates, we may form $(n–1)$ independent node continuity equations and $(m–(n–1))$ independent loop equations to produce the required set of m independent equations.

Building airflow analysis digraphs are, however, a bit different from other network digraphs. Due to the finite size of building zone control volumes and the field assumptions associated with them, pressure changes within zones are directly defined. Consequently, only the flow elements along line segments introduce flow relations (for example, element loop Jacobians). Furthermore, air mass flow rates along a linear series must be equal by continuity. Thus, we may eliminate a number of intermediate nodes and reduce the number of unknowns to the flow element airflow rates alone. In addition, the contribution of multiple flow elements in a series is simply additive and their airflow rates are equal. Thus, for example, node 21 in Figure 2.15 may also be eliminated. Finally, as outdoor nodes are related by the hydrostatic equation and wind pressure relation alone, rather than flow element relations, from a graph theoretical point of view, they are not independent. With these ideas in mind, the two unconnected digraphs of Figure 2.14 may be greatly simplified as illustrated in Figure 2.16.

For the connected graph of the lower two floors and stack, one independent node and two independent loops remain. As a result, a single node continuity equation and two loop equations may be formed.

Correspondingly, given the serial arrangement of pairs of the flow elements $\dot{m}^a = \dot{m}^b$, $\dot{m}^c = \dot{m}^d$ and $\dot{m}^e = \dot{m}^i$, only three unknown airflow rates need to be determined. Taking the unknown flow rates to be $\dot{m}^a$, $\dot{m}^c$ and $\dot{m}^e$, the system equations may, then, be formed for the lower floors by inspection as:

$$
\begin{bmatrix}
J^a_{L[k]} + J^b_{L[k]} & 0 & J^e_{L[k]} + J^i_{L[k]} \\
0 & J^e_{L[k]} + J^d_{L[k]} & J^e_{L[k]} + J^i_{L[k]} \\
- - - & - - - & - - - \\
1 & 1 & -1
\end{bmatrix}
\begin{Bmatrix}
\dot{m}^a_{[k+1]} \\
\dot{m}^b_{[k+1]} \\
\dot{m}^e_{[k+1]}
\end{Bmatrix}
$$

$$
= \begin{Bmatrix}
\Delta p_{1w} + \Delta p_{1b} - \Delta p_{1[k]} \\
\Delta p_{2w} + \Delta p_{2b} - \Delta p_{2[k]} \\
- - - - - - \\
0
\end{Bmatrix}
\tag{78a}
$$

These equations are partitioned with the loop equations above the partition line and the node continuity equation below. In the general case, the matrix of coefficients associated with the continuity equations will simply be an array of 1s, 0s and −1s. For concise matrix notation, the array above the partition line above will be identified as (the current estimate of) the *system loop Jacobian* $[\mathbf{J}_{L[k]}]$, the vector of element mass flow rates will be identified as $\{\dot{\mathbf{m}}_{[k]}\}$, and the loop pressure vectors as the *system* wind pressure vector $\{\Delta\mathbf{p}_w\}$, buoyancy pressure vector $\{\Delta\mathbf{p}_b\}$ and (current estimate of the) pressure correction term $\{\Delta\mathbf{p}_{[k]}\}$.

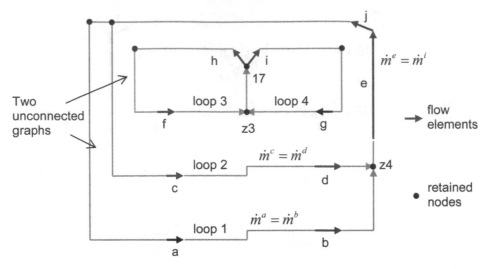

Source: chapter author

Figure 2.16 *Simplified digraph of the ventilation system topology of the building illustrated in Figure 2.13*

This set of equations may then be solved iteratively by forming the current system loop Jacobian $[\mathbf{J}_{L[k]}]$ (that is, for the current estimate the element mass flow rates $\{\dot{\mathbf{m}}_{[k]}\}$) and the pressure correction term $\{\Delta\mathbf{p}_{[k]}\}$, and solving for the new estimate of the element mass flow rates $\{\dot{\mathbf{m}}_{[k+1]}\}$. Then, using the solution, one may update the system Jacobian and pressure correction term and repeat the process. This basic Newton-Raphson solution method may be improved by a number of refinements (Lorenzetti and Sohn, 2000; Lorenzetti, 2002); but for our purposes here it will be sufficient.

For the simple example used here, the process of identifying the spanning tree and thereby a set of independent loops for each connected graph could be done by inspection. This is, however, not generally possible in practice. Consequently, automated topological analysis algorithms have been developed to identify acceptable sets of independent loops (Savic and Walters, 1996) that are now being applied to multi-zone problems.

The system equations that result from the loop method presented above have no apparent structure that may be used to evaluate questions relating to the existence and uniqueness of the solution. One may, however, go one step further using graph theoretic principles to answer these questions with two additional definitions and a related theorem:

- *Planar graph*: a graph that can be drawn so that all line segments lie on a plane without crossing each others is known as a *planar graph*. By inspection, the two connected graphs of Figure 2.16 are planar graphs.

- *Mesh of a planar graph*: the *mesh* of a planar graph are the enclosed 'windows' of the planar graph. The number of meshes of a given planar graph is equal to the number of independent loops (again this is evident by inspection for Figure 2.16).

Thus, one may define a so-called *mesh flow* for each mesh so that mass continuity is defined in terms of these *mesh flows* – for example, as illustrated in Figure 2.17. When this is done, with all mesh flows circulating in the same sense, the system equations not only become especially compact, they prove to be symmetric (Shearer et al, 1971):

$$\begin{bmatrix} J^a_{L[k]} + J^b_{L[k]} + J^e_{L[k]} + J^i_{L[k]} & J^e_{L[k]} + J^i_{L[k]} \\ J^e_{L[k]} + J^i_{L[k]} & J^c_{L[k]} + J^d_{L[k]} + J^e_{L[k]} + J^i_{L[k]} \end{bmatrix}$$

$$\begin{Bmatrix} \dot{m}_{1[k+1]} \\ \dot{m}_{2[k+1]} \end{Bmatrix} = \begin{Bmatrix} \Delta p_{1w} + \Delta p_{1b} - \Delta p_{1[k]} \\ \Delta p_{2w} + \Delta p_{2b} - \Delta p_{2[k]} \end{Bmatrix} \qquad (79)$$

or, in concise notation:

$$[\mathbf{J}_{L[k]}]\{\dot{\mathbf{m}}_{[k+1]}\} = \{\Delta\mathbf{p}_w\} + \{\Delta\mathbf{p}_b\} + \{\Delta\mathbf{p}_{[k]}\} \qquad (80)$$

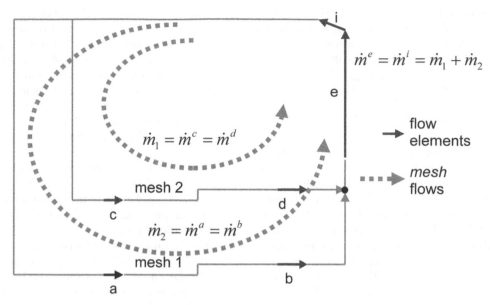

$$\dot{m}^e = \dot{m}^i = \dot{m}_1 + \dot{m}_2$$

$$\dot{m}_1 = \dot{m}^c = \dot{m}^d$$

flow
elements

mesh
flows

mesh 2

$$\dot{m}_2 = \dot{m}^a = \dot{m}^b$$

mesh 1

Source: chapter author

Figure 2.17 *Minimal digraph of the ventilation system topology of the building illustrated in Figure 2.13, now defined in terms of two mesh flows $\dot{m}_1$ and $\dot{m}_2$*

Given that the element Jacobians $J^e_{[k]}$, $e = a, b, c, ...$ remain bounded and positive for the discharge coefficient, power-law, classic duct, quadratic and linear element models, assemblages of these common elements will lead to positive definite *system Jacobians* $[\mathbf{J}_{[k]}]$ and, thus, the assurance that a solution to the airflow problem exists and is unique. If fan element models are included that also maintain positive-definiteness (that is, given that sums of element Jacobians are involved, negative contributions are acceptable as long as the sums remain positive), then existence and uniqueness remain assured.

Finally, to account for the dynamic variation of zone air mass M_i, each of the zone nodes must be retained and nodal continuity must be explicitly accounted for. Thus, in general, the loop system equations for zone mass accumulation will have the following form:

$$\begin{bmatrix} J_{L[k]} \\ -- \\ B_L \end{bmatrix} \{\dot{\mathbf{m}}_{[k+1]}\} = \begin{Bmatrix} \{\Delta\mathbf{p}_w\} + \{\Delta\mathbf{p}_b\} + \{\Delta\mathbf{p}_{[k]}\} \\ --------- \\ \dfrac{d\{\mathbf{M}\}}{dt} \end{Bmatrix}$$

(81)

where, a *loop Boolean* matrix $\{\mathbf{B}_L\}$ accounts for the nodal continuity and the nodal air mass accumulation term

$d\{\mathbf{M}\}/dt$ is defined in terms of the nodal mass vector $\{\mathbf{M}\} = \{M_1 M_2 ...\}^T$ defined as before. Note that this form accounts for nodal mass accumulation but is not a full dynamic formulation of the airflow problem as discussed in the earlier section on 'Turbulence-induced effects and dynamic analysis'.

The loop design method

The *loop design method* is a procedure for sizing airflow components of natural and hybrid ventilation systems based on the pressure loop equations presented above – see equation (71). It allows direct sizing of a variety of airflow components, unambiguous consideration of the combined interaction of buoyancy, wind and mechanically induced pressures, and allows control and operational strategies, as well as other non-technical design constraints, to be included in the design process.

With a preliminary proposal for the building form and ventilation system configuration and topology in hand, and design conditions and requirements specified, driving design wind pressures Δp_{lw} and stack pressures Δp_{lb} for each loop l of the ventilation system may be directly (numerically) evaluated. Furthermore, airflow rates through each of the ventilation system components $\dot{m}^e$ will be determined by the specified design requirements. These may then be directly substituted into appropriate element flow relations $\Delta p^e = g^e(\dot{m}^e, \vartheta^e)$ for each of the ventilation system components. Consequently, a number of loop equations equal to the number of independent loops of the system topology may be formed in terms of the unknown component size parameters ϑ^e:

$$\underbrace{\sum_{loop\ l} (g^e(\vartheta\,{}^{\vartheta}))}_{\substack{component \\ losses}} = \underbrace{\Delta p_{lw}}_{\substack{numerical\ value\ of \\ wind\ pressure}} + \underbrace{\Delta p_{lb}}_{\substack{numerical\ value\ of \\ stack\ pressure}} \qquad (82)$$

The loop equations so formed define combinations of component size parameters ϑ^e that will satisfy the specified design conditions and design requirements. In this sense, then, they define all feasible design solutions. Furthermore, with properly defined flow relations, the loop equations will be hyperbolic in form and, thus, bounded by limiting asymptotes that determine minimal acceptable sizes for each of the components of the ventilation system – information that will prove invaluable to the system designer in the search for acceptable component sizes.

With these ideas in mind, a general procedure for the design of natural and hybrid ventilation systems may be outlined that involves eight distinct steps:

1   *Lay out global geometry, topology and loops*: lay out the global geometry and topology (that is, the flow component types and connectivity) of the proposed natural or hybrid ventilation system and systematically identify all relevant loops. This is typically done in a building sectional drawing.
2   *Identify pressure nodes*: identify an ambient pressure node and additional (pseudo) pressure nodes at entry and exit ports of each flow component along each loop.
3   *Establish design conditions*: establish surface node wind pressure coefficients, design outdoor temperature, wind speed and direction, and desired interior temperature conditions (for a given building idealization and longer-term weather record, design conditions may be evaluated in terms of the time variation of the driving pressures; this will be discussed below).
4   *Establish the design requirements*: establish the required design ventilation rate for each inlet and, by continuity, determine the objective design flow rates required for each flow component.
5   *Form the pressure loop equations*: form the loop equations for each loop selected in step 1 (it is useful here to keep the stack and wind pressure contributions separate so that *with-wind* and *without-wind* operational strategies can be readily evaluated).

6   *Evaluate asymptotic limits*: determine the minimum feasible sizes for each of the flow components by evaluating the asymptotic limits of each component design parameter for each of the loop equations. Note: for components common to a number of loops, one of the loop equations (that is, typically the loop with the minimal driving pressures) will establish the governing minimum (again, it is useful to compute these limits for the *with-wind* and *without-wind* cases separately).
7   *Apply design constraints*: develop and apply a sufficient number of technical or non-technical design rules or constraints to transform the *under-determined* design problem defined by each loop equation into a *determined* problem.
8   *Devise an operational strategy*: develop an appropriate operational strategy to accommodate the regulation of the ventilation system for variations in design conditions (for example, for the *with-wind* and *without-wind* cases).

Each pressure loop equation will be defined in terms of the unknown design (size) parameters ϑ^e of the components along that loop. Thus, the loop equations will generally be *under-determined* and, therefore, will not have a unique solution. Instead, many feasible solutions to the design problem may be formulated. Consequently, the designer must impose design rules, constraints and operational strategies to transform each loop equation into a *determined* problem in order to determine a final 'design solution' – that is, one of the many feasible solutions possible.

Defining design rules, design constraints and operational strategies may seem to be the most elusive part of this methodology. It is important to emphasize that this is not a flaw in the methodology, but is intrinsic to the design of natural ventilation systems – systems that inevitably may be designed in a number of different ways. The methodology offers and supports the designer's need to introduce practical design considerations (for example, off-the-shelf component sizes, an architectural constraint that all windows be of certain sizes, etc.) that necessarily constrain design in the real world. In a more complex application, these design rules could define, for instance, objective functions to be minimized in search of a design solution using optimization procedures.

Given that the design wind pressure Δp_{lw} and stack pressure Δp_{lb} terms of the loop equations – that is, the right-hand side of equation (82) – are independent of component sizes, they may be evaluated *a priori*. Furthermore, if the given steady formulation of the loop equations is assumed to accurately model dynamic variations of building airflows – the common *quasi-dynamic* assumption of current macroscopic theory – then the dynamic variation of the stack and wind pressure terms may also be evaluated *a priori*, given outdoor wind and temperature time histories. If this is done, design conditions may be established using a statistical evaluation of these driving pressures – an approach that must be expected to provide a more reliable basis for the specification of design conditions.

For the loop design method to be practically useful, the designer must have a relatively large and varied *library* of component relations to properly account for the variety of ventilation system components that may be used in practice. As a macroscopic approach, this library may include inverse forms of the element flow relations commonly used in multi-zone nodal analysis that are published in a number of handbooks and programme user's manuals (Liddament, 1986a, 1986b; Pelletret and Keilholz, 1997; Orme et al, 1998; Orme, 1999; Haas, 2000; ASHRAE, 2001; Persily and Ivy, 2001; Dols et al, 2000; Dols, 2001a, 2001b).

A full understanding of the loop design method demands consideration of detailed worked examples – examples that cannot reasonably be considered here. Thus, the reader is directed to earlier publications that present details of the application of the loop design method to a series of residential design problems of increasing complexity (Axley, 1998, 1999a, 1999b, 2000a) and to larger building design problems, with some limited consideration given to hybrid system components (Axley, 2000b, 2001a; Axley et al, 2002a, 2002b; Ghiaus et al, 2003). For examples of using a problem-specific statistical analysis of driving stack and wind pressures to establish environmental design conditions, the reader is directed to Axley (2000a, 2001a).

Since the loop design method employs the same theoretical base and diagrammatic conventions used by multi-zone airflow analysis programmes, it may also be implemented within the interface of these programmes. LoopDA (Loop Design Analysis) is such an implementation (Dols and Emmerich, 2003) – it provides an implementation within the CONTAMW interface that is directly integrated with the CONTAMW simulation engine. LoopDA supports the *loop design* of system components and subsequent *analysis* of system performance. Another similar but more ambitious tool development project has recently been reported by Mansouri (Mansouri, 2003; Mansouri and Allard, 2003). In this project, the loop design method was implemented within the SPARK simulation environment (LBNL, 2002a, 2002b) to create a preliminary design-sizing tool, and design development performance evaluation was realized with an integrated multi-zone airflow and thermal analysis tool based on the integration of COMIS (Haas, 2000) and TRNSYS (Klein et al, 1988). Using these tools, a number of representative natural ventilation systems were designed and their long-term performance was evaluated.

Sub-zone and embedded modelling methods

Conventional multi-zone methods were developed to predict bulk airflow rates through the discrete airflow paths and/or duct networks that link zones of a given multi-zone building idealization. As conceived, therefore, they were not developed to predict the details of airflow within building zones. This is not to say that conventional multi-zone methods are based on the assumption that the zones are well mixed, as is often claimed – they are based on the far less restrictive hydrostatic field assumption that admits the possibility of non-uniform density distributions within modelled zones. Nevertheless, conventional multi-zone methods provide no more intrazonal detail than estimates of zone airflow velocities $\hat{u}_i^e$ at the ports of each element e linked to a given zone i of the building idealization as $\hat{u}_i^e = \dot{m}_i^e/(\rho^e A^e)$.

Yet, the building design and research community has long recognized the central importance of zone airflow details when the health, safety and comfort of occupants or the efficacy of some manufacturing operations must be considered. Computational fluid dynamic (CFD) models can provide the needed detail; but due to computational limits, they cannot yet be applied to studies of complex whole-building systems, they are often limited to steady driving conditions, and they demand personnel and computational resources well beyond that of current multi-zone modelling tools. Consequently, a number of proposals have been put forward over the years to extend available multi-zone models to provide at least an approximate evaluation of the details of room airflows for whole-building simulation studies without the computational and personnel overhead associated with CFD modelling approaches. In this section, two closely related approaches to this general problem will be considered – (isolated) *sub-zone* and *embedded* modelling methods (see Figure 2.18).

Sub-zone methods

Sub-zone or *zonal* models, as they are often called, have been formulated by isolating zones of building systems and subdividing them into a relatively small number of discrete control volumes or *cells*. Early sub-zone models

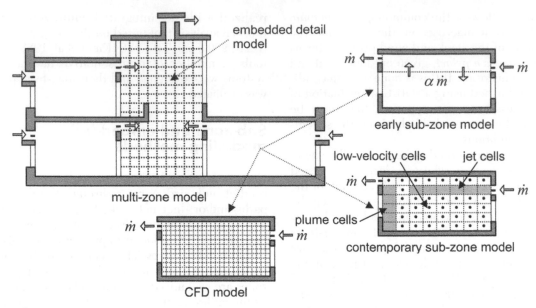

Figure 2.18 *Modelling of flow details within zones using either isolated or embedded sub-zone or computational fluid dynamics (CFD) methods*

then adopted one of a number of empirical relations between the zone supply and intrazonal flow rates to model displacement ventilation, short circuiting, bypass or recirculating airflow within the zone (Waters and Brouns, 1991). For example, for the common short-circuiting airflow model shown in Figure 2.18, a fraction α of the zone supply airflow rate $\dot{m}$ is presumed to mix with air within the zone and the rest is assumed to flow directly to the zone exhaust. Using these empirical models, researchers could then investigate energy or mass transport processes dependent upon the intrazonal airflows.

More recent sub-zone models have taken deterministic approaches based on extensions of multi-zone methods. Early on, pressures within each cell i of the sub-zone model were assumed to vary hydrostatically relative to a reference cell node pressure p_i, and airflow between adjacent cells $\dot{m}_{i,j}$ was modelled by applying an infinitesimal form of the discharge coefficient flow model to cell boundaries (Allard et al, 1990; Bouia and Dalicieux, 1991). However, the researchers involved in this early work quickly realized that cells falling within the influence of plumes, jets and thermal boundary layers had to be treated differently (Inard and Buty, 1991a, 1991b).

Contemporary sub-zone models apply different cell flow models to different sets of cells depending upon the context (see Figure 2.19). Airflow rates from those cells thought to be in the influence of jets, plumes or boundary layers are *specified* using one of the several well-established jet flow relations (for example, see Awbi (1991) and Etheridge and Sandberg (1996)), plume relations or boundary-layer flow relations (for example, see Heiselberg et al, 1998). Finally, for cells not falling within the influence of jets, plumes or boundary layers – that is to say, for lower-velocity flow regimes within the zone – cell-to-cell airflow is commonly modelled using, again, an infinitesimal form of the discharge coefficient model.

With these modelling tactics in hand, *system* equations governing the airflow of the sub-zone idealization may be formed as in conventional multi-zone analysis using either a nodal mass continuity or loop compatibility approach with, now, the hydrostatic field assumption applied to the low-velocity cells (Axley, 2001b). The low-velocity cell-to-cell flow relations are assembled to form the system equations needed to predict the unknown low-velocity flows given the boundary conditions imposed on the low-velocity regimes by the jet, plume and boundary-layer analytical solutions. While conceptually straightforward, the number of possible boundary conditions to be considered can be daunting. Haghighat and colleagues enumerated a total of eight different generic flow-boundary conditions between plumes, jets and boundary layer and the low-velocity flow regimes (Haghighat et al, 1999, 2001; Lin et al, 1999). Furthermore, the assignment of cell-type must be based on anticipated flow conditions and, thus, requires considerable judgement. Nevertheless, Wurtz and his colleagues

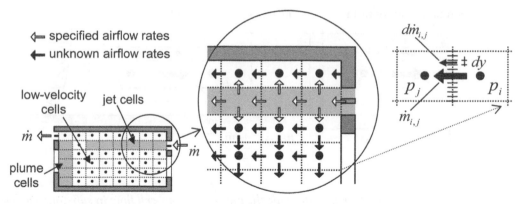

Note: Cell-to-cell flow relations in low-velocity regions are based on infinitesimal discharge coefficient flow models.

Source: chapter author

Figure 2.19 *Contemporary sub-zone (zonal) models specify airflow rates within jets, plumes and boundary layers, and solve for airflow rates in low-velocity regions*

have managed to automate this process successfully (Musy et al, 1999; Wurtz et al, 2001, 2003; Musy et al, 2002).

The computational advantage gained by this sub-zone approach lies in the fact that the systems of algebraic equations produced are relatively small and far easier to solve than those resulting from CFD methods. Consequently, longer-term (quasi) dynamic studies not currently possible using CFD methods are commonly undertaken with sub-zone methods. Furthermore, these sub-zone models may be directly and readily integrated with sub-zone thermal, moisture and contaminant transport models to account for the multiple physical phenomena that invariably impact upon aspects of the health, safety, comfort and/or efficacy being studied (Rodriguez et al, 1993; Haghighat et al, 2001; Wurtz et al, 2001, 2003).

When jets, plumes and boundary-layer flow areas are properly identified, these sub-zone models provide estimates of the details of flow that can be in good agreement with both measured results and predictions provided by detailed CFD analysis (Haghighat et al, 1999, 2001; Lin et al, 1999; Wurtz et al, 1999, 2001, 2003; Mora et al, 2002a). When low-velocity regimes dominate the flow, however, the commonly used low-velocity pressure-driven cell-flow relations cannot be expected to capture details of airflows where inertia, momentum transport and/or shear stresses play an important role (for example, in regions of recirculation) (Rodriguez et al, 1993).

Nevertheless, as the approach ensures the conservation of bulk airflows and provides what may be a representative, rather than faithful, characterization of mixing within the zone, computed results may still prove

to be useful. As these results are invariably coupled to other thermal and mass transport relations, the relative importance of errors in the details of flow magnitude and structure may prove to be insignificant when these other phenomena are the focus of study.

Low-velocity cell-to-cell flow relations

From a macroscopic point of view, the subdivision of zones into cells and the application of established jet, plume and boundary-layer solutions to sets of these cells are fundamentally sound strategies. The mass continuity or loop compatibility principles used to assemble the system equations properly account for mass conservation, but, as in conventional multi-zone analysis, tacitly ignore the conservation of mechanical energy at the cell level (for example, see the earlier section on 'Multi-port control volumes'). The impact of this fundamental flaw has, however, yet to be investigated.

Perhaps more significantly, the commonly used low-velocity cell-to-cell flow relations are based on assumptions that have little physical justification and fail, by orders of magnitude, to model the resistance offered to airflow in simple flow regimes. As sub-zone models have been typically used in isolation of larger whole-building models, their ability to properly account for resistance has not been an issue. Thus, this error has gone unnoticed. As investigators attempt to *embed* sub-zone models within larger multi-zone building models, this error may well become problematic. In any event, an understanding of this error and consideration of strategies for correcting it should be expected to result in improved sub-zone modelling methods.

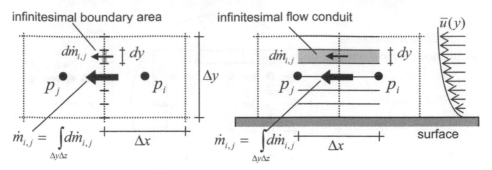

Figure 2.20 *Conventional and surface-drag idealizations of cell-to-cell flow in sub-zone models*

The commonly used low-velocity cell-to-cell flow models have been based on methods employed for large openings (Etheridge and Sandberg, 1996; Allard, 1998) – hence, the large opening models share similar shortcomings. In this approach, all flow dissipation is assumed to occur at the boundary between cells (see Figure 2.20), and a power-law relation is assumed to govern the differential mass flow $d\dot{m}_{i,j}$ through infinitesimal areas of the boundary dA as:

$$d\dot{m}_{i,j} = \rho C(\Delta p_{i,j})^n dA \tag{83}$$

where C is an empirical 'permeability' constant, analogous to the orifice discharge coefficient.

Expressions relating the cell-to-cell air-mass flow rate $\dot{m}_{i,j}$ to the corresponding nodal pressure difference $\Delta p_{i,j} = p_i - p_j$ may then be derived by integrating equation (83) over the cell boundary area. For isothermal conditions and a rectangular subdivision of the zone with cell dimensions $(\Delta x, \Delta y, \Delta z)$, one obtains:

$$CBPL\ model:\ \Delta p_{i,j} = \left(\frac{\dot{m}_{i,j}}{\rho C \Delta y \Delta z}\right)^{1/n} \tag{84}$$

where the cell-to-cell air-mass flow rate $\dot{m}_{i,j}$ here, corresponds to flow along the x direction passing through the $\Delta y \Delta z$ boundary area between cells i and j.

This conventional boundary power law (CBPL) model implicitly assumes that viscous dissipation occurs only at the imaginary boundaries between cells – the boundary perpendicular to the flow. When applied to airflow through a room, the total pressure drop along any given stream tube will depend linearly upon the number of cell boundaries crossing that stream tube. Thus, when using this conventional approach, finer cell subdivisions will not

lead to convergence, but instead simply increase the apparent resistance offered by the room – clearly an unrealistic outcome.

An alternative *surface-drag* approach that accounts for dissipation due to shear transfer to adjacent surfaces has been proposed (Axley, 2000c, 2001b). In this approach, a momentum balance on infinitesimal flow conduits linking adjacent cells is used to relate the gradient of mean flow shear stresses $\bar{\tau}_{xy}$ to the nominal cell-to-cell pressure difference $\Delta p_{i,j}$ (see Figure 2.20). A family of different but related cell-to-cell flow models has been developed based on this approach. Of these, a three dimensional (3D) variant based on Newton's approximation $\bar{\tau}_{xy} = -\mu(d\bar{u}/dy)$ and an assumed quarter-sine velocity profile (NAQS) that is appropriate for laminar flow provides a useful example here:

$$NAQS\ model:\ \Delta p_{i,j} = \frac{\mu\ \pi^2 \Delta x}{2\rho\ \tau^2 \Delta y \Delta z}\dot{m}_{i,j} \tag{85}$$

When applied to modelling flow in a duct segment – a test case devised to unambiguously evaluate the ability of the models to predict flow resistance – the NAQS model is observed to be in good agreement with the exact solution for laminar flow, and another complex surface-drag model based on a mixing length shear stress model and assumed power-law velocity profile – the MLPL model – approximates the classic Darcy Weisbach equation for turbulent flow (see Figure 2.21). In comparison, the conventional CBPL model fails by two orders of magnitude to predict the correct pressure drop in the duct segment for all flow rates when four cell subdivisions are used along the length of the duct segment. Finer cell divisions will simply exacerbate this error of the CBPL model.

Here, the significant result of the surface-drag approach is that the computed results are grid convergent

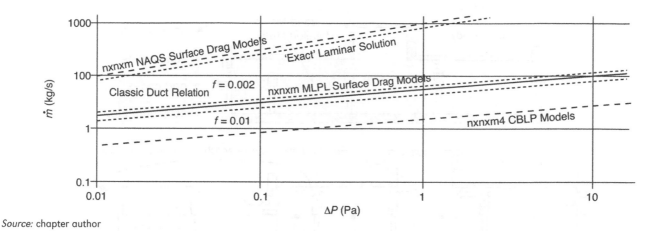

Figure 2.21 *Comparison of modelled duct flow behaviour using the conventional boundary power law (CBPL) and two surface-drag models (the NAQS and the MLPL models) with the classic Darcy-Weisbach equation for turbulent flow and the 'exact' Hagen-Poiseuille solution for laminar flow*

due to the direct dependency of pressure drop $\Delta p_{i,j}$ on the corresponding grid subdivision Δx – a feature not shared by the conventional CBPL approach.

Another approach that also provides a grid-convergent, pressure-driven cell-to-cell flow model may be developed directly from a finite difference approximation of Darcy's law:

Darcy's law model:

$$\Delta p_{i,j} \approx \left(\frac{\mu\,\Delta x}{2\rho\,k\Delta y\Delta z}\right)\dot{m}_{i,j} \; ; \; \Delta p_{j,k} \approx \left(\frac{\mu\,\Delta x}{2\rho\,k\Delta y\Delta z}\right)\dot{m}_{j,k}$$

$$(86a,\ b)$$

While Darcy's law is normally used to model flow in porous media, from a more general perspective it models flow in domains where flow is driven solely by pressure differences. Thus, to the extent that low-velocity airflows in sub-zone idealizations are driven solely by pressure differences, it is reasonable to use Darcy's law to model these flows. Again, these relations define a cell-flow model that is directly related to the cell subdivision dimension Δx and therefore will be grid convergent.

It must be emphasized, however, that the finite-difference approximation used here to draw comparisons with the other low-velocity cell-flow models presented above is relatively primitive. In practice, one should turn, instead, to higher-order finite-difference or finite-element approximations to obtain accurate results in a computationally efficient manner.

In sub-zone modelling, the low-velocity cell-flow

relations are the 'glue' that allows one to patch together reliable jet, plume and boundary-layer solutions to effect an approximate and computationally straightforward analysis of the details of airflow within zones. These flow relations are thus critical to the eventual success of sub-zone approaches. This subsection has shown that the commonly used low-velocity cell-flow relations based on boundary power-law models are critically flawed, since they will not lead to grid-convergent solutions and must be expected to misrepresent zone resistance by orders of magnitude. The approaches presented based on surface-drag momentum considerations and Darcy's law correct the grid-convergence flaw and may accurately model zone resistance in some cases; but they, too, are limited by the assumption that flow is solely pressure driven. As inertia, convective transport of momentum and/or viscous shear transport may be expected to play a role in even low-velocity flow regimes, the problem of modelling low-velocity flows in sub-zone idealizations remains unresolved, thus warranting additional study. As one alternative, coarse-grid Reynolds-averaged Navier-Stokes (RANS) CFD solutions using the standard k-ϵ turbulence model appear to hold some promise (Mora et al, 2002b, 2003).

Embedded detailed analysis

Established methods of computational fluid dynamics and the more recent and less mature methods of sub-zone analysis have been applied to isolated problems of detailed airflow analysis within rooms; yet room airflows do not occur in isolation – they result from and interact with airflows from the larger whole-building systems in

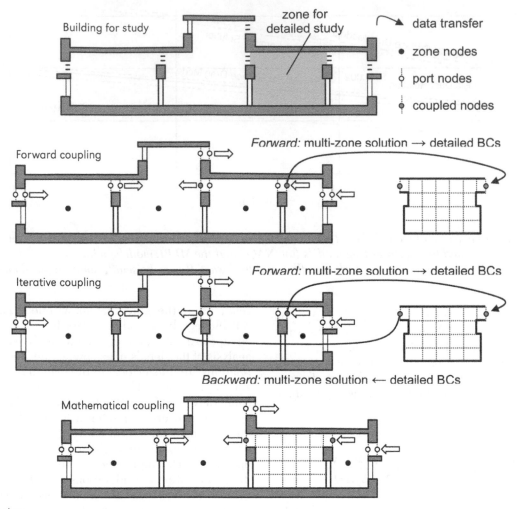

Source: chapter author

Figure 2.22 *Coupling strategies for embedded detailed analysis – the upper two are ad hoc computational strategies and the bottom one is a mathematically rigorous but practically difficult-to-achieve strategy*

which they are embedded. Consequently, current interest has turned to the problem of embedding detailed sub-models within larger whole-building multi-zone models to more faithfully model airflow details within zones and, importantly, to model them over extended time periods. In the rush to combine these modelling capabilities, however, few have paused to consider the problem of embedded analysis from a fundamental rather than applied perspective.

A number of approaches to this general problem have been considered (Mora et al, 2003), all involving the communication of boundary conditions and computed results to and from the multi-zone model to the detailed model. Currently, three general strategies may be distinguished (see Figure 2.22):

1   *Forward computational coupling*: in this approach, a multi-zone model of the building is first used to estimate airflows and pressures within the building system. Then the solution results computed for the interface between the whole-building and detailed model – the *coupled nodes* – are used to establish the boundary conditions of the detailed model, which is solved. Li and Holmberg (1993) used this approach to directly establish inflow boundary conditions for a series of 15 detailed CFD analyses corresponding to the 15 rooms of the three-storey building studied.

2   *Iterative computational coupling*: this approach is initiated as in forward coupling; but the detailed solution results are then passed back to the multi-zone

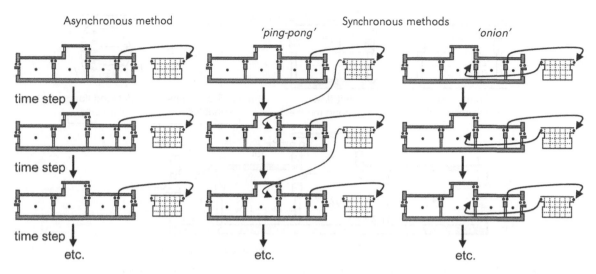

Asynchronous method   Synchronous methods

'ping-pong'   'onion'

time step

time step

time step

etc.   etc.   etc.

Source: adapted from Mora et al (2003)

Figure 2.23 *Three* ad hoc *computational schemes for dynamic embedded detailed analysis*

model to establish a new set of boundary conditions for the multi-zone model. The multi-zone model is solved again and the process is repeated iteratively with the hope of achieving convergence. Schaelin et al (1993) considered this approach conceptually and applied it by coupling a CFD model of two upper rooms of a two-storey, three-room building and the nearby outdoor environment with a multi-zone model of the remaining first-floor room. In their attempt to classify the types of data exchanges needed, Schaelin et al (1993) established the fact that the determination of data transfer protocol is problem dependent and complicated by the need to define well-posed analytical problems for both the detailed and multi-zone models at each step of the iterative process.

3   *Mathematical coupling*: in principle, it is possible to form a coupled system of equations for both the multi-zone and detailed models combined that then can be solved using one of a number of well-established non-linear solution techniques (for example, by the Newton-Raphson method or its variants). While this option may not yet be practical for embedded CFD models, it has been implemented with embedded sub-zone models within the SPARK simulation environment by Mora et al (2003) (see also LBNL, 2002a, 2002b).

These coupling strategies may then be placed in a time-marching scheme to provide a means of analysing the dynamic response of the embedded zone to wind, stack and mechanical excitations acting on the whole-building system. After Mora et al (2003), three types of coupled time-marching schemes can be considered – asynchronous and synchronous computational schemes and time integration of the mathematically coupled system alternative. One asynchronous and two synchronous computational schemes – commonly identified as the *ping-pong* and *onion* methods – are illustrated schematically in Figure 2.23.

These dynamic analysis schemes, while intuitively satisfying, must be recognized to be *ad hoc* strategies in that there is no theoretical assurance that they will provide accurate or even convergent results in a time-efficient manner. In contrast, a number of numerical methods exist for the time integration of the mathematically coupled system of non-linear equations that can provide this assurance – at least, in a conditional sense (see, for example, Dahlquist and Björck, 1974; Gerald and Wheatley, 1990; Press et al, 1992; Stoer and Bulirsch, 1993; Borse, 1997; Sowell and Haves, 1999). The current preoccupation with these *ad hoc* schemes is thus misguided – research resources would be better spent pursuing strategies based on one or more of the well-established and thoroughly studied numerical methods.

While not immediately obvious, there are a number of complications relating to the coupling of boundary conditions to computed results that may compromise these approaches and to demands that must be placed on the physical consistency of the coupled models that have also been largely ignored.

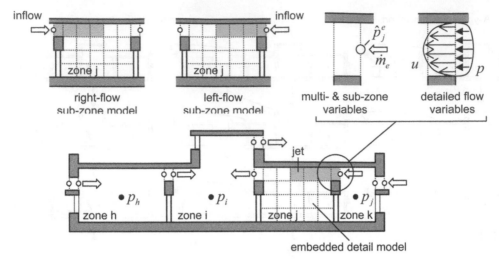

Note: Solid circles represent (multi-zone) zone reference pressures, open circles element port spatial-average pressures, and open arrows bulk mass flow rates.

Source: chapter author

Figure 2.24 *A hypothetical but representative case of a detailed model embedded in a multi-zone idealization of a four-zone building*

Coupling boundary conditions and results

Figure 2.24 presents a diagram of a hypothetical but representative case of a detailed model embedded in a multi-zone idealization of a four-zone building. Here, the embedded detail model may be considered to be either a sub-zone or CFD idealization. Multi-zone analysis, using either the nodal continuity or the loop compatibility approach, would be used to determine the bulk air-mass flow rates and the spatial average pressures at the coupled nodes of the embedded system – for example, $\dot{m}^e$ and $\hat{p}^e_j$ illustrated in Figure 2.24. For a fully embedded detail model, these multi-zone results would then be used to establish the boundary conditions at the coupled nodes of the embedded detail model (a partially embedded detailed model – for example, for a zone including an exterior wall – may have some boundary conditions established independently of the multi-zone results and, depending upon the direction of airflow, may have only a forward-coupled relation to the multi-zone model). Conversely, computed results from a detailed analysis could be used to establish new boundary conditions for the multi-zone model if either an iterative computational or mathematically coupled method is to be pursued. While conceptually straightforward, the actual association of computed results with boundary conditions may be complicated by the direction of the conversion (that is, from multi-zone to detailed model or vice versa), the direction of airflow and the type of detailed model used.

Embedded sub-zone models present the least problematic alternative since their boundary conditions and computed results may also be defined in terms of bulk air-mass flow rates and spatial average pressures so that the association of multi-zone and embedded model variables is one to one. For deterministic sub-zone models (that is, with all cell flow relations defined in terms of pressure-driven flow relations), the sub-zone cell-flow relations may be directly assembled within the multi-zone model to form mathematically coupled systems of equations that govern the combined multi-zone and embedded sub-zone models (Mora et al, 2003; Ren and Stewart, 2003; Stewart and Ren, 2003).

For sub-zone models with inflow conditions modelled by analytical jet equations, one must, however, distinguish inflow boundary conditions from outflow boundary conditions. Consequently, when flow reversals are possible, separate sub-zone models would have to be formulated for each jet-flow possibility as jet cells must be distinguished from low-velocity cells *a priori* – as, for example, in the right-flow and left-flow models shown in Figure 2.24. With each possible sub-zone model, inflow jet-port mass flow rates would most reasonably be associated with the corresponding multi-zone mass flow rates, and multi-zone spatial-averaged pressures would be associated with outflow ports. Furthermore, transitional low-flow conditions that may not be well modelled by a jet-flow region may demand additional sub-zone models.

In addition, as the outflow results from the sub-zone model will, in general, add 'internal' boundary conditions to the original multi-zone model, topologically distinct multi-zone models would need to be defined for each of the possible sub-zone models if an iterative computational coupling strategy is to be pursued. For example, for the simple problem illustrated in Figure 2.19, one would need five topologically distinct models for:

1 an initial multi-zone whole-building idealization with boundary conditions established at the three envelope surface elements;
2 a right-flow sub-zone model;
3 a corresponding modified multi-zone model with an internal boundary condition at the right-flow detailed room exhaust;
4 a left-flow sub-zone model; and
5 a corresponding modified multi-zone model with an internal boundary condition at the left-flow detailed room exhaust.

In general, then, the number of sub-zone/multi-zone model possibilities could not only become unacceptably large, but the numerical complications of solving the resulting discontinuous non-linear problem could prove impossibly challenging. Nevertheless, by limiting consideration to problems with unidirectional flow, Mora and his colleagues have initiated studies to make progress in this direction (Mora et al, 2002a, 2003).

Coupling CFD detailed models to a multi-zone whole-building simulation shares similar problems since distinct inflow and outflow boundaries must be identified to proceed with the detailed analysis. In addition, the conversion of multi-zone results to detailed CFD boundary conditions is inherently indeterminate and dependent upon the specific type of detailed CFD model used. In contrast, the conversion of CFD results to multi-zone boundary conditions follows directly from the definition of the spatial-averaged values used in multi-zone analysis – see equations (10), (11) and (12).

For detailed models based on the direct numerical simulation (DNS) of the Navier-Stokes equations, the coupled boundary conditions may be defined in terms of *distributions* of:

• airflow velocity $u(y, z)$ or pressure $p(y, z)$ (*Dirichlet* conditions);
• spatial derivatives of the velocity (*Neumann* conditions); or
• the combination of both (*Robin* condition).

Multi-zone analysis provides only (time-smoothed) spatial-averaged values, yielding no information regarding the spatial variation of velocity. Thus, the conversion of multi-zone results to DNS boundary conditions is indeterminate and can only be made via simplifying assumptions.

For detailed models based on the Reynolds-averaged Navier-Stokes equations (RANS) using, for example, the k-ϵ two-equation turbulence model, Dirichlet boundary conditions would be defined in terms of time-averaged velocity $\bar{u}(y, z)$, pressure $\bar{p}(y, z)$, turbulent kinetic energy $k(y, z)$, and dissipation rate of turbulence energy $\epsilon(y, z)$ distributions. Needless to say, multi-zone analysis provides no information relating to $k(y, z)$ and $\epsilon(y, z)$; thus, an added dimension of indeterminacy results.

Some investigators have, apparently, assumed time-averaged velocity distributions to be uniform for detailed CFD inflow boundaries:

Uniform velocity assumption:

$$\bar{u}(y, z)\big|_{CFD} = \hat{u}\big|_{multi-zone} \qquad (87)$$

Yet, this assumption will violate the zero-slip boundary condition most reasonably applied to surfaces adjacent to the inflow boundary. This will not only introduce an unrealistic singularity at the edges of the boundary, but may compromise convergence of the detailed analysis.

Alternatively, a parabolic velocity distribution, for laminar flow conditions, or a power-law distribution, for turbulent conditions, could be assumed to avoid the singularity problem and, perhaps, more faithfully model actual inflow conditions (Bird et al, 2002).

The conversion of the (time-smoothed) spatial-averaged pressure computed from multi-zone analysis to an (outflow) pressure distribution for detailed CFD analysis is less problematic and a uniform field assumption would seem acceptable:

Uniform pressure assumption:

$$\bar{p}(y, z)\big|_{CFD} = \hat{p}\big|_{multi-zone} \qquad (88)$$

The few reports of attempts to embed detailed CFD models within multi-zone whole-building idealizations provide little specific information regarding coupling boundary conditions and computed results. Yet it is clear that this coupling necessarily introduces a level of empiricism that can only be evaluated by trial-and-error validation. Nevertheless, beyond the uncertainty that results, additional barriers to embedded analysis may also be related to the physical consistency and completeness of the detailed and multi-zone approaches used in embedded analysis.

Physical consistency and completeness

As established above, conventional multi-zone theory:

- ignores the resistance to flow that may be offered by zones;
- systematically violates the conservation of mechanical energy; and
- ignores dynamic flow effects associated with capacitance, inertia and damping.

Conventional sub-zone models implicitly include zone resistance, but violate mechanical energy conservation and also ignore dynamic effects. CFD models based on the Navier-Stokes equations, on the other hand, are complete models in that they account for all flow phenomena, including, in the current context, zone resistance, energy conservation and dynamic effects. Ideally, embedded and multi-zone models should not only be physically consistent and complete if they are to be iteratively coupled, they should also be accurate. If not, attempts to iteratively couple them may be expected to encounter solution-convergence difficulties and/or to produce erroneous results.

When coupling embedded detailed models that include zone resistance, the detailed model will, in effect, present a zone resistance to the multi-zone model that may, if significant, alter the magnitude or even the direction of the whole-building airflows. Unfortunately, the conventional approach to sub-zone modelling is now known to produce large errors in zone resistance, and the effective resistance offered by candidate CFD methods

has yet to be critically evaluated.

Ironically, if zone resistance is truly negligible, then there is no reason to iteratively couple detailed models with multi-zone models since the latter must be expected to alone provide accurate results. In this case, multi-zone results may simply be used directly in a forward-coupling strategy to set the boundary conditions for detailed analysis. Thus, detailed analysis can be conducted as a post-processing operation and the difficulties of iteratively coupled embedded analysis can be avoided altogether.

If zone resistance is not negligible, one could attempt to extend multi-zone analysis to properly account for zone resistance (see the earlier section on 'Multi-port control volumes') and again avoid the complexities of iteratively coupled embedded analysis. A recent study reported by Chen and Wang (2004) involving the analysis of buoyancy-driven airflows in nine different building stair towers by both CFD and a multi-zone approach provides some initial support for this conclusion. Chen and Wang (2004) showed in these studies that both CFD and multi-zone approaches provided nearly identical bulk airflow results when (stairwell) zone resistances were accurately accounted for (see Figure 2.25). However, additional multi-zone analyses using 'phantom zones' that provided no zone resistance provided flow results that agreed within 11 per cent with those computed by CFD – suggesting that zone resistance was, in fact, negligible in all but two of the nine cases studied. Significantly, these two cases corresponded to 5- and 11-storey stair towers with closed treads – that is, the two cases for which stairwell (zone) resistance should be expected to be greatest.

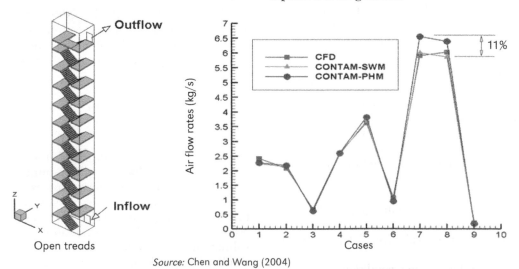

Source: Chen and Wang (2004)

Figure 2.25 *Computed airflow rates in a stair tower by computational fluid dynamics (CFD) (square markers), multi-zone analysis using an empirical flow model for stairwells that accounts for stairwell (zone) resistance (triangular markers), and multi-zone analysis using 'phantom zones' that ignores zone resistance (circular markers)*

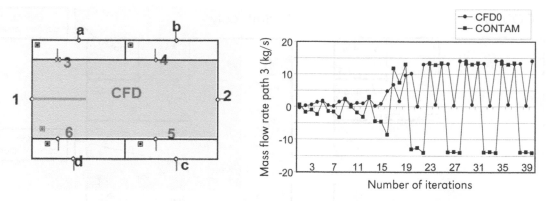

Source: Chen and Wang (2004)

Figure 2.26 *A diagram of a CFD0-detailed model embedded within a CONTAM multi-zone model of a five-zone building and computed results for airflow through door at path 3*

Problems associated with the mechanical energy flaw of multi-zone models follow a similar logic. When coupled with CFD models that do properly account for energy conservation, one might expect a non-convergent flip-flopping of computed results since the multi-zone model allows energy conservation violations, while the CFD model does not. Again, results recently reported by Chen and Wang (2004) appear to exhibit this behaviour. In an embedded analysis of a five-zone building consisting of a central hall flanked by four offices (see Figure 2.26), Chen iteratively coupled a CFD0 model (Chen and Xu, 1998) of the central hall to a CONTAM multi-zone model (Dols and Walton, 2000) of the four flanking offices. Computed results of the airflow through one of the doorways (path 3) linking the central hall to the offices, for example, not only diverged with each iteration, but changed direction back and forth (see Figure 2.26).

While it appears that no project has yet been reported that investigates the dynamic response of an embedded detailed model, the current lack of dynamic terms in multi-zone airflow analysis theory may again result in convergence problems when iteratively coupled to CFD-detailed models that do account for dynamic effects.

Two lessons can be learned from this discussion and the few reported cases that support it:

1 Embedded detailed and multi-zone models need not be iteratively coupled if both models are complete, physically consistent and accurate in their own domains. When this is the case, a forward coupling from multi-zone analysis to embedded detailed analysis should be sufficient. When this is not the case, solution convergence problems may well result.

2 Conventional multi-zone models are currently incomplete in that zone resistance and dynamic effects are ignored and mechanical energy is not conserved. Consequently, when iteratively coupled to detailed CFD models, convergence problems may result unless zone resistance is truly negligible and the detailed CFD model properly models this resistance, and any energy conservation violation also proves to be negligible. Ironically, if these two conditions are satisfied, iterative coupling would not seem to be needed.

Thus, there is a clear need to investigate:

- the effective zone resistance produced by both sub-zone and CFD modelling techniques (that is, analogous to the studies reported above for sub-zone models based on the conventional boundary power-law cell-flow relations);
- the importance of mechanical energy conservation errors inherent in multi-zone modelling; and
- available strategies to include dynamic effects in multi-zone modelling.

Sub-zone multi-physics and computational tools

This section has approached sub-zone and embedded detail methods as methods devised solely for airflow analysis; yet, invariably these methods have been developed as a means of investigating the complexities of multiple physical phenomena, or *multi-physics*. Beyond the somewhat incidental consideration of thermal comfort and its relation to detailed room airflows, sub-zone and

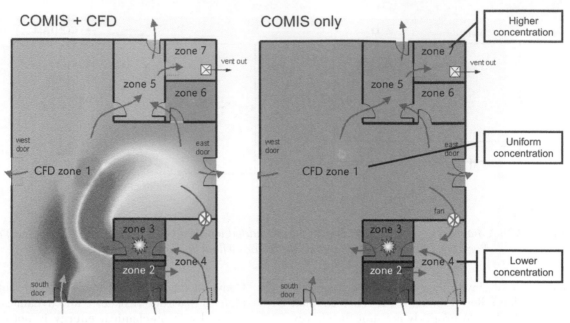

Source: Lorenzetti et al (2003)

Figure 2.27 *Comparison of a contaminant release in zone 3 of a of seven-zone building modelled by both an embedded CFD model within a COMIS multi-zone model and by a COMIS multi-zone model alone*

embedded detail methods have been more often directed to a variety of problems associated with building thermal and contaminant dispersal analysis – for example, problems associated with thermal comfort and its relation to detailed room air temperatures; smoke movement; moisture transport; the transport of volatile organic compounds emitted and adsorbed by building materials; and, most recently and regrettably, the spread of chemical and/or biological contaminants introduced in buildings with malicious intent.

The consideration of the detailed modelling methods used to account for these various areas of multi-physics is well beyond the scope of this chapter – although the subsequent section will treat them broadly at the multi-zone level. Consequently, the discussion will be limited to one representative example and a broad discussion of the computational tools that have proven essential to these multi-physics investigations.

Figure 2.27 presents the results of an embedded analytical study completed by Lorenzetti et al (2003). In this study, the investigators iteratively coupled a CFD model to a COMIS multi-zone model to compute bulk airflows between zones of a seven-zone model and detailed zone airflows for one of the seven zones – the large irregularly shaped zone identified as zone 1 in Figure 2.27. Using the results from the embedded airflow analysis exercise, they

then iteratively coupled CFD and COMIS contaminant dispersal models to predict time histories of averaged contaminant zone concentrations in the macroscopic zones and detailed concentrations in the microscopic zone.

Figure 2.27 compares contaminant concentrations due to a release in zone 3 computed using an embedded CFD model within a COMIS multi-zone model to that using a COMIS multi-zone model alone. The embedded analysis provides results that appear to realistically account for the detailed flow conditions within zone (1) – that is, the dilution of the contaminant due to fresh air inflow at the south door and the high concentration plume flowing into the room from the release zone. It also appears to more realistically account for the macroscopic down-stream dispersal into zones 5, 6 and 7.

While the utility of obtaining both improved multi-zone analytical results and greater detail for selected zones is clearly very promising, it must be noted that these results were not obtained without difficulty. In this particular case, coupling the multi-zone model with a two-dimensional (2D) CFD model of the larger room converged without difficultly. However, when a three-dimensional (3D) CFD model was used, convergence was complicated by flow reversals.

To couple existing CFD and multi-zone programmes, one must devise a computational strategy to pass data back and forth between the programmes. Lorenzetti et al (2003) used the programming and scripting language Perl (Practical Extraction and Report Language) (Wall, 2004) to launch the coupled programmes, extract key results from the output files created by these programmes and prepare input files to be read by the programmes to effect the coupling (Lorenzetti et al, 2003). That is to say, data was passed back and forth via files written to a computer's hard disk drive. In order to effect a more rapid exchange of data, Chen and Wang (2004) modified the source codes of both the CFD programme (CFD0) and the multi-zone programme (CONTAMW) so that they could exchange data via a *coupled data interface* maintained in high speed memory of the computer's central processing unit (CPU).

As noted above, coupling detailed sub-zone and multi-zone models may be achieved directly by assembling both subsystems of non-linear algebraic equations to form one combined system of equations that may then be solved using standard (for example, Newton-Raphson) methods. Indeed, the author of this chapter used the multi-zone programme CONTAM96, without modification, to implement sub-zone analysis using both the conventional boundary power-law and surface-drag cell-flow models to investigate their behaviour – an inelegant but possible strategy (Axley, 2000d, 2001b). Stewart and Ren (2003) modified the COMIS multi-zone programme to create a new programme, COwZ, to more elegantly and efficiently add sub-zone modelling capabilities to a multi-zone programme (Ren and Stewart, 2003; Stewart and Ren, 2003). Finally, Mora et al (2003) implemented coupled sub-zone and multi-zone models within the very general and thus flexible equation-based SPARK simulation environment that, in addition, allowed Mora to add additional multi-physics capabilities (LBNL, 2002a, 2002b; Mora et al, 2003).

Multi-zone multi-physics analysis

The subject of this chapter has focused principally on a discussion of multi-zone methods of building airflow analysis. Yet, from their inception these methods have invariably been developed to investigate the complexities of the multiple physical phenomena (*multi-physics*) associated with building airflows. For example, Walton (1982, 1984) developed an early multi-zone airflow modelling capability in the AIRMOV subroutines of the Thermal Analysis Research Programme (TARP) (Walton, 1982, 1984) that established the basis of the AIRNET programme (Walton, 1988), which was, in turn, later integrated within the multi-zone contaminant dispersal analysis programmes CONTAM86 and CONTAM87

(Axley, 1987b, 1988a; Grot and Axley, 1987) to create the current CONTAM family of integrated airflow and contaminant dispersal analysis programmes (Grot, 1990; Grot and Axley, 1990; Walton, 1994; Walton, 1997; Dols and Walton, 2000; Dols et al, 2000). Furthermore, the architecture of the CONTAM86 programme itself was based on an earlier building thermal analysis programme DTAM1 (Axley, 1988b). While the CONTAM programme family did not (originally) attempt to iteratively couple these three broad areas of analysis – *thermal, airflow* and *contaminant dispersal* analysis – the need to couple these areas was immediately identified (Axley, 1987a) since building airflows are generally sensitive to air density variations in buildings that depend upon air temperature and composition (that is, contaminant concentration); conversely, thermal and concentration transport dynamics are driven in large part by building airflows.

Over the years, a large variety of building multi-physics problems linked to building airflows have been modelled. The vast majority of these problems have fallen within the same three broad areas of analysis. Consequently, in addition to multi-zone airflow analysis, multi-zone methods have been developed for thermal and contaminant dispersal analysis. Indeed multi-zone methods of building thermal analysis pre-date those of airflow and contaminant dispersal analysis by at least two decades.

Multi-physics modelling assumptions

Multi-zone methods of thermal, airflow and contaminant dispersal analysis approach their respective domains in a very similar manner (see Figure 2.28):

- *Control volumes*: buildings are idealized as collections of zone- and transport link-control volumes. For thermal analysis, the transport links include conduction/convection, radiation, and advection energy, while for contaminant dispersal analysis they include adsorption/desorption, chemical and advection mass transport links.
- *Zone state variables*: zone state variables of pressure p_i, temperature T_i and species α concentration $C_{\alpha,i}$ are associated with specific locations or *nodes* within each zone i for airflow, thermal and contaminant dispersal analysis, respectively. As before, intermediate nodes are associated with links (for example, T_i^e, $C_{\alpha,i}^e$ and p_i^e associated with the port of element e within zone i).
- *Link transport relations*: transport relations are specified for each of the respective transport links e that relate transport rates of air mass $\dot{m}^e$, heat q^e, enthalpy $\dot{h}$ or species mass $\dot{m}_\alpha^e$ to the appropriate zone state

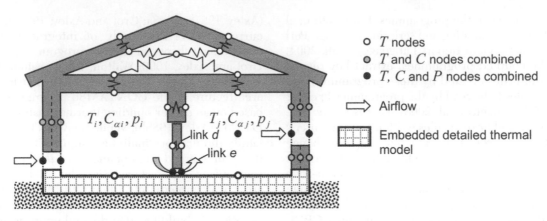

Source: chapter author

Figure 2.28 *Hypothetical multi-zone idealization for combined thermal, airflow and contaminant dispersal analysis*

variables. In general, these transport relations may be linear or non-linear; but the important advection transport relations for thermal and contaminant dispersal analysis – that directly couple airflow to thermal and contaminant dispersal behaviour – are particularly simple. For example, for the enthalpy $\dot{h}^{\&}$ and species mass $\dot{m}_\alpha^e$ transport rates, these would be:

$$\dot{h} = c_p \dot{m}^e T_i^e \; ; \text{ for air mass flow } from \text{ zone } i \qquad (89)$$

$$\dot{m}_\alpha^e = \dot{m}^e C_{\alpha,i}^e \; ; \text{ for air mass flow } from \text{ zone } i \qquad (90)$$

where, c_p is the specific heat capacity of air and species mass concentration $C_{\alpha,i}^e$ is expressed as a mass fraction.

Although not immediately obvious, these advection relations lead to asymmetric contributions to the system equations for thermal and contaminant dispersal analysis (Axley, 1989), while a simple thermal resistance, as might be used to model the rate of heat transfer q^d through the door link d of Figure 2.28, would lead to a symmetric contribution:

$$q^d = \frac{A^d}{R^d}(T_i^d - T_j^d) \; ;$$

for heat flow positive from zone i to zone j. (91)

Here, A^d is the area available for heat transfer and R^d is the resistance of the link e.

Additional link transport relations for thermal analysis, hygrothermal analysis and thermal system

control are concisely presented in the recent publications of Mendes (Mendes et al, 2001a, 2001b, 2002, 2003; Oliveira et al, 2003), and for contaminant dispersal analysis, in the user's manuals of the popular programme families COMIS and CONTAM and associated publications (Feustel, 1990; Feustel and Raynor-Hoosen, 1990; Feustel and Smith, 1992; Walton, 1994, 1997; Axley, 1995; Feustel and Smith, 1997; Pelletret and Keilholz, 1997; Dols and Walton, 2000; Dols et al, 2000; Haas, 2000).

- *Capacitance*: for dynamic formulations, air mass, thermal or species mass capacitance may be associated with control volumes. Here, these will be identified by variables M_i for the air mass in zone i, which provides 'capacitance' for both airflow and contaminant dispersal analysis, M_{Tj} for the thermal mass associated with thermally massive control volumes, and M_{Cj} for the species sorption capacity associated with some building materials, such as the ubiquitous gypsum wallboard products that can play a central role in contaminant dispersal analysis. M_{Tj} and M_{Cj} would be associated, typically, with finite difference discretizations of walls (for example, as indicated diagrammatically by the grey nodes in Figure 2.28).

For the discussion of this subsection, it will be useful to represent zone air mass in terms of the product of zone air density ρ_i and zone volume V_i (assumed constant here):

$$M_i = \rho_i V_i \qquad (92)$$

• *Zone field assumptions*: as for airflow analysis, the finite size of zones and the general indeterminacy of temperature and concentration fields within them demand that simplifying field assumptions be made to effect closure. The representative transport relations presented above, equations (89), (90) and (91), have been written in terms of local temperatures and concentrations at the intermediate link nodes, thus allowing non-uniform temperature and concentration field assumptions to be considered. Nevertheless, most commonly – indeed, almost universally – zones are assumed *well mixed* for multizone thermal and contaminant dispersal analysis (hence the interest in embedded analysis discussed in the earlier section on 'Embedded detailed analysis'). For the present discussion, we will accept the *well-mixed zone field* assumption for these two areas of analysis and move on.

Multi-physics system equations

With these modelling assumptions made, link equations may then be assembled to form system equations that govern, independently, the thermal, airflow and contaminant dispersal behaviour of the building modelled. Methods to form and solve the system equations for airflow analysis have been considered in detail above. Similar methods may be used for both thermal and contaminant dispersal analysis, although these problem areas often lead to computationally more benign linear, rather than non-linear, systems of equations (Axley, 1988b, 1989). Nevertheless, a discussion of these methods for thermal and contaminant dispersal analysis is beyond the scope of this chapter. Instead, we will turn directly to these system equations.

From the previous sections, system equations for airflow analysis have been derived using both nodal continuity and loop compatibility approaches that are presented here – with the zone air-mass accumulation terms expressed in terms of zone density:

*Nodal (mass conservation) continuity form
(from equation (39a)):*

$$[\mathbf{J}]_{[k]}\{\mathbf{p}\} + [diag(V_i)]\frac{d\{\rho\}}{dt}$$

$$= -\sum\{\mathbf{f}^e(\{\mathbf{p}\}_{[k]}, \{\mathbf{p}_b^e\}, \{\mathbf{p}_w^e\})\} - [\mathbf{J}]_{[k]}\{\mathbf{p}\}_{[k]}$$

(93a)

$$\{\dot{\mathbf{m}}^e\} = \{\mathbf{f}^e(\{\mathbf{p}\}, \{\mathbf{p}_b^e\}, \{\mathbf{p}_w^e\})\}$$

(93b)

Loop compatibility form:

$$\begin{bmatrix} \mathbf{J}_{L[k]} \\ -- \\ \mathbf{B}_L \end{bmatrix}\{\dot{\mathbf{m}}\} = \begin{Bmatrix} \{\Delta\mathbf{p}_w\} + \{\Delta\mathbf{p}_b\} + \{\Delta\mathbf{p}_{[k]}\} \\ ------ \\ [diag(V_i)]\frac{d[\rho]}{dt} \end{Bmatrix}$$

(94)

Here, zone air densities have been collected together in a density vector $\{\mathbf{\rho}\} = \{\rho_1\rho_2...\}$ and the $[k+1]$ iteration index has been dropped from the left-hand-side dependent variable to simplify the discussion.

Note that when using the conventional nodal continuity approach, one must first form the system equations, equation (93a), and solve (iteratively) for the system pressure vector $\{\mathbf{p}\}$. Then, using this solution, one must back-substitute into the element flow relations, equation (93b), to obtain the individual element (link) mass flow rates $\{\dot{\mathbf{m}}^e\}$. Using the loop compatibility approach, on the other hand, one forms the *loop* system equations, equation (94), and solves directly for all the element mass flow rates $\{\dot{\mathbf{m}}\}$.

The system equations for thermal analysis are invariably based on a nodal continuity approach, although now for energy rather than mass conservation. These equations assume the form of a system of (often linear) ordinary differential equations as:

Nodal (energy conservation) continuity form:

Thermal analysis:

$$[\mathbf{K}]\{\mathbf{T}\} + [\mathbf{M}_T]\frac{d\{\mathbf{T}\}}{dt} = \{\mathbf{E}_{Tpas}\} + \{\mathbf{E}_{Tact}\}$$

(95)

where $\{\mathbf{T}\} = \{T_1 T_2...\}^T$ is the collection of node temperatures identified as the system temperature vector, $[\mathbf{K}]$ is the system *conductance matrix* assembled from link relations such as those given in equations (89) and (91) above, $[\mathbf{M}_T]$ is the thermal mass matrix, and $\{\mathbf{E}_T\}$ is the thermal excitation due to passive sources (for example, solar, outdoor air temperature and internal gains due to occupants) and active sources (for example, mechanical heating and cooling systems).

Two special analytical cases bound the limits of application of these system equations – ideal mechanical control and free-floating of indoor temperatures. In a building with ideal mechanical control of indoor temperatures, $d\{\mathbf{T}\}/dt$ will be zero (or possibly specified), $\{\mathbf{T}\}$ will be specified, $\{\mathbf{E}_{Tpas}\}$ will be determined *a priori* from climatic and occupancy data, and $[\mathbf{K}]$ may be formed given a complete physical description of the building if flow element mass flow rates are known. Thus, the analyst may use equation (95) to solve for the mechanical excitation $\{\mathbf{E}_{Tact}\}$ needed to *actively* control indoor temperatures – a procedure commonly called *loads analysis*. For these conditions, multi-zone airflow analysis may be directly used to determine these heat transport rates as $\{\mathbf{T}\}$; thus, zone air densities $\{\boldsymbol{\rho}\}$ will be specified. That is to say, airflow analysis is only *forward* coupled to thermal analysis for ideal mechanical control of indoor air temperatures.

When mechanical control is not active – that is, when $\{\mathbf{E}_{Tact}(t)\}=\{\mathbf{0}\}$ – indoor temperatures will float freely and zone air densities will vary with time. As a result, building airflows will vary with time as well. Under these circumstances, airflow analysis is forwardly coupled through the element air-mass flow rates to thermal response, while thermal analysis is *backward* coupled to airflow response through zone air densities which depend upon temperature.

Actual building thermal behaviour invariably falls between these two extremes, floating freely during moderate temperature periods and in the *dead-band* range of building thermal control devices. Thus, beyond the central need for iteratively coupled thermal/airflow analysis for passively controlled (for example, naturally ventilated) buildings, there is a need for iteratively coupled analysis for more accurate analysis of mechanically conditioned building operation.

Multi-zone contaminant dispersal analysis is also governed by systems of ordinary differential equations – equations that are typically linear for dispersal of *passive* (that is, non-interactive) contaminants or non-linear when certain chemical or sorption transport processes are important. Again, these equations are most commonly formed using a nodal continuity approach, now considering the conservation of contaminant mass flow rate. For a single contaminant α, the governing system equations assume the form:

Nodal (species mass conservation) continuity:

Contaminant dispersal analysis:

$$[\dot{\mathbf{M}}]\{\mathbf{C}_\alpha\} + [diag(\rho_i V_i)]\frac{d\{\mathbf{C}_\alpha\}}{dt} = \{\mathbf{E}_{\alpha\,env}\} + \{\mathbf{G}_\alpha\}$$

(96)

Here, $\{\mathbf{C}_\alpha\} = \{C_{\alpha 1}C_{\alpha 2}...\}^T$ is the system concentration vector (that is, the collection of zone concentrations for species α), $[\dot{\mathbf{M}}]$ is the system *flow matrix* assembled from link relations such as those given in equation (90) using air mass flow rates $\dot{m}^e$ specified or determined from multi-zone airflow analysis, $[\mathbf{M}] = [diag(\rho_i V_i)]$ is the zone air

mass matrix, $\{\mathbf{E}_{\alpha env}\}$ is the *environmental excitation* (that is, the mass flow rate of contaminant due to infiltration), and $\{\mathbf{G}_\alpha\} = \{G_{\alpha 1}G_{\alpha 2}...\}$ is the mass generation rate of the contaminant within each zone.

Again, two special analytical cases bound the limits of application of these system equations – trace and non-trace contaminant dispersal analysis. As for trace analysis, the dispersal of the contaminant has a negligible impact upon zone densities (and, thus, building airflows), while for non-trace analysis the opposite is true. Thus, trace analysis is inherently a forward-coupled problem, while non-trace analysis requires iterative coupling in both forward and backward directions.

Here, a 'contaminant' could be any substance that is not normally part of the dry air composition – water moisture, smoke and those contaminants commonly falling within the scope of indoor air quality concerns such as volatile organic and inorganic gases and a variety of aerosols. Indeed, the dispersal of moisture is often of primary importance in iteratively coupled contaminant dispersal/airflow analysis since it may well exist at concentration levels that significantly affect variations in air density.

Computational coupling strategies

Intuitive considerations of the forward and backward causal interactions of thermal, airflow and contaminant dispersal behaviour, like those presented above, have led directly to the development of computational strategies to couple existing multi-zone programmes to effect multi-physics analysis. While intuitively satisfying, these computational coupling strategies are, in fact, *ad hoc* procedures that generally lack theoretical justification. Nevertheless, these procedures will be considered in detail in this section. The following section will turn to more recent rigorous, but difficult, approaches based on mathematically coupling all relevant system equations.

From the perspective of this section, the principal coupled relations between airflow, thermal and contaminant dispersal analysis can be diagrammed as illustrated in Figure 2.29:

- *Forward relations*: given zone air densities ρ_i, multi-zone airflow analysis can be used to predict element mass flow rates $\dot{m}_i^e$.
- *Backward relations*: multi-zone thermal and contaminant dispersal analysis can be used to transform these element mass flow rates $\dot{m}_i^e$ into zone temperatures T_i and concentrations C_i that then determine zone densities given the equation of state.
- *Equation of state.*

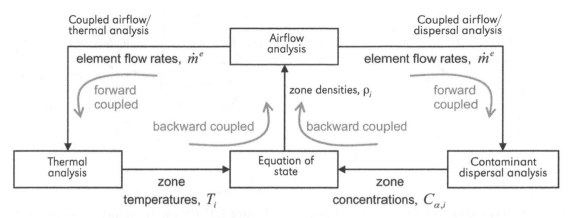

Figure 2.29 *Coupling possibilities of multi-zone airflow analysis with multi-zone thermal and contaminant dispersal analysis*

Thus, for ideal mechanical thermal control and/or trace contaminant dispersal, multi-zone airflow analysis may be forward coupled to multi-zone thermal and contaminant dispersal analysis, respectively, to complete airflow/thermal/contaminant dispersal multi-physics analysis. For free-floating thermal response (or other circumstances short of ideal mechanical control) or non-trace dispersal analysis, multi-zone airflow analysis must, however, be iteratively coupled through both forward and backward coupling relations via an equation of state. While inherently complex, these coupled relations can be directly defined in terms of algebraic relations of the state variables shared by all three areas of analysis.

For example, for a single gaseous contaminant, the equation of state may be reasonably based on Dalton's law for gas mixtures that leads to the following equation for zone density:

Gas phase equation of state:

$$\rho_i = \left(\frac{M_\alpha M_{air}}{(1 - C_a)M_\alpha + C_a M_{air}}\right)\left(\frac{p_i}{RT_i}\right) \quad (97)$$

where M_α and M_{air} are the (effective) molecular weights of contaminant α and dry air, respectively, and R is the universal gas constant. Note that for trace dispersal $(1–C_\alpha) \approx 1$ and $C_\alpha \approx 0$; thus, this relation simplifies to be the equation of state for dry air alone.

Using the system equations enumerated above, the forward and backward coupling strategies may be diagrammed more specifically as illustrated in Figure 2.30 for airflow/thermal analysis couplings, and in Figure 2.31 for airflow/contaminant dispersal analysis. For the

former, starting with an initial estimate of zone densities, one may form and solve the nodal continuity airflow equations for zone pressures. This solution is then used to determine element mass flow rates, which may be used to form the thermal analysis conductance matrix so that the thermal problem can be solved (that is, at the current time step). When backward coupling is required (for example, the dotted lines in Figure 2.30), one would then use the zone temperature solution and the equation of state to compute zone densities, and the process may be repeated.

This schematic accounts for dynamic variations of the airflow buoyancy vector $\{p_b^e\}$ and zone density $d\{\rho\}/dt$, yet ignores other coupled relations to zone density that appear in the flow-element equations. While these additional coupled relations may be included computationally with little effort, the primary physical coupling must be expected to be via the buoyancy vector $\{p_b^e\}$ – that is, the multi-zone equivalent of the Boussinesq assumption commonly used in CFD. Significantly, few current implementations of the iteratively coupled airflow/thermal problem include the coupling of the zone density $d\{\rho\}/dt$ term, which may conceivably play a significant role in fire analysis. The research version of the CONTAM family of programmes, CONTAM97, developed by Walton (1998) and the computational coupling of the S3PAS and COMIS programmes reported by Rodriguez and Allard (1992) both, however, include this coupled term.

The schematic illustrated in Figure 2.31 for coupling multi-zone airflow and contaminant dispersal analysis is much the same as that discussed for coupling thermal analysis, but now element air-mass flow rates from the

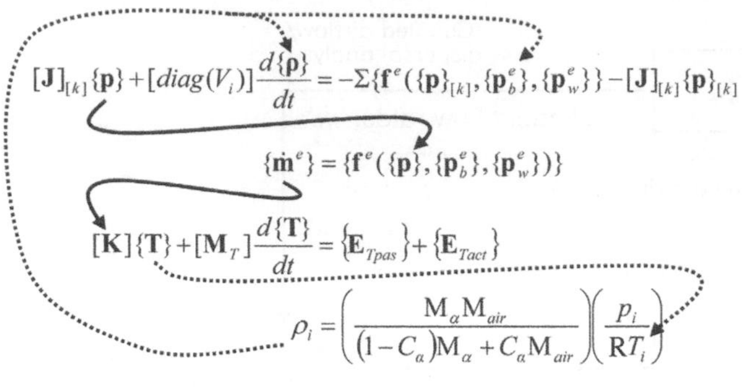

$$[\mathbf{J}]_{[k]}\{\mathbf{p}\} + [diag(V_i)]\frac{d\{\boldsymbol{\rho}\}}{dt} = -\Sigma\{\mathbf{f}^e(\{\mathbf{p}\}_{[k]},\{\mathbf{p}_b^e\},\{\mathbf{p}_w^e\})\} - [\mathbf{J}]_{[k]}\{\mathbf{p}\}_{[k]}$$

$$\{\dot{\mathbf{m}}^e\} = \{\mathbf{f}^e(\{\mathbf{p}\},\{\mathbf{p}_b^e\},\{\mathbf{p}_w^e\})\}$$

$$[\mathbf{K}]\{\mathbf{T}\} + [\mathbf{M}_T]\frac{d\{\mathbf{T}\}}{dt} = \{\mathbf{E}_{Tpas}\} + \{\mathbf{E}_{Tact}\}$$

$$\rho_i = \left(\frac{M_\alpha M_{air}}{(1-C_\alpha)M_\alpha + C_\alpha M_{air}}\right)\left(\frac{p_i}{RT_i}\right)$$

Source: chapter author

Figure 2.30 *Forward coupling (solid line) and backward coupling (dotted line) of conventional (nodal continuity) multi-zone airflow analysis and thermal analysis (for a single time step of a dynamic analysis)*

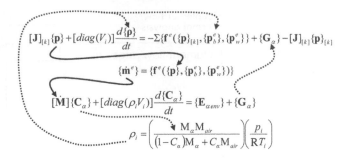

$$[\mathbf{J}]_{[k]}\{\mathbf{p}\} + [diag(V_i)]\frac{d\{\boldsymbol{\rho}\}}{dt} = -\Sigma\{\mathbf{f}^e(\{\mathbf{p}\}_{[k]},\{\mathbf{p}_b^e\},\{\mathbf{p}_w^e\})\} + \{\mathbf{G}_\alpha\} - [\mathbf{J}]_{[k]}\{\mathbf{p}\}_{[k]}$$

$$\{\dot{\mathbf{m}}^e\} = \{\mathbf{f}^e(\{\mathbf{p}\},\{\mathbf{p}_b^e\},\{\mathbf{p}_w^e\})\}$$

$$[\mathbf{M}]\{\mathbf{C}_\alpha\} + [diag(\rho_i V_i)]\frac{d\{\mathbf{C}_\alpha\}}{dt} = \{\mathbf{E}_{\alpha env}\} + \{\mathbf{G}_\alpha\}$$

$$\rho_i = \left(\frac{M_\alpha M_{air}}{(1-C_\alpha)M_\alpha + C_\alpha M_{air}}\right)\left(\frac{p_i}{RT_i}\right)$$

Source: chapter author

Figure 2.31 *Forward coupling (solid line) and backward coupling (dotted line) of conventional (nodal continuity) multi-zone airflow analysis and contaminant dispersal analysis (for a single time step of a dynamic analysis)*

(nodal) airflow analysis are used to form the flow matrix $[\dot{\mathbf{M}}]$ for dispersal analysis and the zone concentration solution is passed to the equation of state for backward coupling (principally) to the buoyancy term. However, for non-trace dispersal analysis, especially smoke dispersal analysis, the generation of 'contaminant' $\{\mathbf{G}_\alpha\}$ must be included in the airflow nodal mass conservation relations, as indicated in Figure 2.31.

The coupling of the loop compatibility multi-zone airflow analysis method to both thermal and contaminant dispersal analysis would appear to be more straightforward since back-substitution to determine element air-mass flow rates is not required. Although this strategy has apparently been overlooked in the past, it is currently being investigated by Rasmus Jensen in his doctoral studies at the Hybrid Ventilation Centre of the University of Aalborg, Denmark. A possible schematic for coupling the loop compatibility airflow to the thermal equations is illustrated in Figure 2.32 – coupling strategies for dispersal analysis will follow a similar logic.

Dynamic computational coupling strategies

Multi-zone multi-physics analysis is often pursued to investigate the seasonal or annual response of a building system to environmental and use excitations; thus, dynamic response analysis is often a central objective. With this objective in mind, a variety of dynamic computational strategies may be formulated using the forward and backward coupling strategies presented above. Using the diagram illustrated in Figure 2.29 as a diagrammatic building block, a number of dynamic coupling strategies may be formulated (see Figures 2.33 and 2.34).

Forward coupling (from airflow analysis to thermal and dispersal analysis) may be employed in an asynchronous or synchronous manner (see Figure 2.33 left), where airflow analysis is simply used to solve a series of airflow problems in time and the element mass flow rate results are used to complete a 'loads analysis' type of thermal analysis and/or a trace contaminant dispersal analysis. It appears that the 'load analysis' option has been overlooked by the building research community, while the quasi-dynamic trace dispersal analysis is, of course, the basis of many multi-zone indoor air-quality analysis tools such as the CONTAM or COMIS families of programmes (Feustel and Raynor-Hoosen, 1990; Walton, 1994, 1997; Feustel and Smith, 1997; Pelletret and Keilholz, 1997; Dols and Walton, 2000; Dols et al, 2000; Haas, 2000).

When configured synchronously – moving from airflow analysis to thermal and/or dispersal analysis at one time step, then using the latter results to compute zone densities for airflow analysis at the next time step – one may simulate the iteratively coupled interactions that govern free-floating thermal and/or non-trace dispersal analysis. This particular synchronous scheme, as applied solely to airflow/thermal iterative coupling, was introduced and identified by Hensen as a 'ping-pong' approach (Hensen, 1990; Hensen and Clarke, 1990). When applied simultaneously to thermal and contaminant dispersal analysis, one must imagine the (airflow) player standing between two 'ping-pong' tables competing simultaneously against two opponents – a vivid but not particularly rigorous metaphor.

A scheme falling between the schemes illustrated in Figure 2.33 involving a synchronous 'ping-pong' time integration coupling of airflow/thermal analysis, yet an

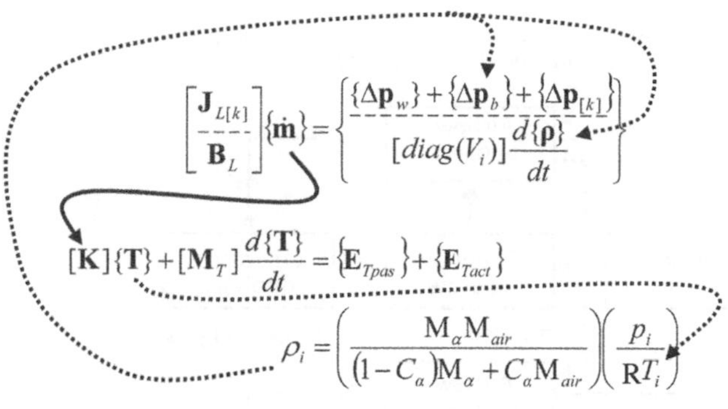

$$\begin{bmatrix} \mathbf{J}_{L[k]} \\ \hline \mathbf{B}_{L} \end{bmatrix} \{\dot{\mathbf{m}}\} = \left\{ \frac{\{\Delta \mathbf{p}_{w}\} + \{\Delta \mathbf{p}_{b}\} + \{\Delta \mathbf{p}_{[k]}\}}{[diag(V_{i})]\dfrac{d\{\boldsymbol{\rho}\}}{dt}} \right\}$$

$$[\mathbf{K}]\{\mathbf{T}\} + [\mathbf{M}_{T}]\frac{d\{\mathbf{T}\}}{dt} = \{\mathbf{E}_{Tpas}\} + \{\mathbf{E}_{Tact}\}$$

$$\rho_{i} = \left(\frac{M_{a}M_{air}}{(1 - C_{a})M_{a} + C_{a}M_{air}} \right) \left(\frac{p_{i}}{RT_{i}} \right)$$

Source: chapter author

Figure 2.32 *Forward coupling (solid line) and backward coupling (dotted line) of conventional (loop compatibility) multi-zone airflow analysis and thermal analysis (for a single time step of a dynamic analysis)*

asynchronous 'forward' coupling of airflow-to-contaminant dispersal analysis (see Table 2.1), is apparently the basis of the TFCD programme developed by Klobut and colleagues (1991).

Figure 2.34 illustrates two additional methods. The leftmost figure presents a dynamic strategy in which both thermal and dispersal analysis are iteratively coupled with airflow analysis within each time step until convergence of computed results is obtained – an approach also introduced and identified by Hensen as the 'onion' method

when applied solely to coupling airflow/thermal analysis (Hensen, 1990; Hensen and Clarke, 1990). The rightmost figure illustrates one of six possible mixed strategies where thermal analysis is iteratively coupled to airflow analysis ('onionSync'), while contaminant dispersal analysis is forward coupled to airflow analysis ('forwardAsync') within each time step.

The 'onionSync–onionSync' method of Figure 2.34 is supported by the research version of CONTAM, CONTAM97 (Walton, 1998) and the S3PAS and COMIS tool reported by Rodriguez (Rodriguez and Allard, 1992), but has not been studied extensively. The 'onionSync–forwardAsync' method is also supported by CONTAM97 and appears to be employed by a number of multi-zone tools based on coupling the COMIS airflow/trace-dispersal analysis programme to the TRNSYS thermal analysis programmes (Dorer, 1998a, 1998b; Haas, 2000; Albrecht et al, 2002; Seifert et al, 2002; Weber et al, 2002), although some of these tools may be based on the simpler 'pin-pongSync–forwardAsync' variant instead. For additional reviews of these and other investigations of *ad hoc* multi-zone computational coupling strategies, the reader may wish to consult the general reviews published by Orme (1999), Heiselberg et al (1998) and Kendrick (1993).

In spite of some problems of solution stability and convergence (see, for example, Axley, 2001a; Axley et al, 2002a, 2002b), the results of applications of these compu-

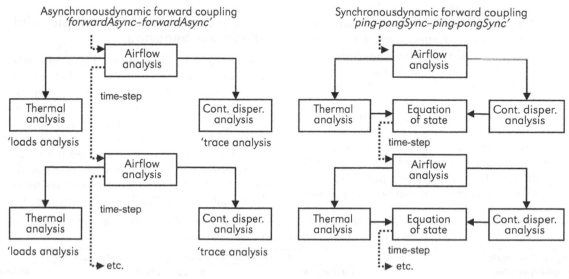

Note: Solid lines indicate data transfer within one time step, while dotted lines indicate data transfer from time step to time step.

Source: chapter author

Figure 2.33 *Asynchronous and synchronous dynamic coupling strategies based on forward coupling of multi-zone airflow analysis to thermal and contaminant dispersal analysis*

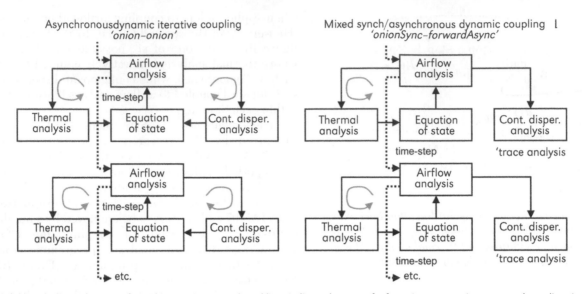

Asynchronousdynamic iterative coupling
'onion-onion'

Mixed synch/asynchronous dynamic coupling │
'onionSync-forwardAsync'

Note: Solid lines indicate data transfer within one time step, dotted lines indicate data transfer from time step to time step, and grey lines indicate an iterative loop.

Source: chapter author

Figure 2.34 *Two synchronous dynamic coupling strategies based on forward and backward coupling of multi-zone airflow analysis to thermal and contaminant dispersal analysis*

tational coupling strategies offer great promise for natural ventilation, smoke spread, moisture transport and air quality analysis. Yet, the message is becoming rather clear – the proliferation of a large number of *ad hoc* computational coupling strategies without critical and rigorous evaluation has left the field in a state of confusion – notwithstanding the apparent successes of individual applications. Even within the classification framework presented above, a total of nine different approaches to the problem of computationally coupling multi-zone airflow analysis to thermal and dispersal analysis exist (see Table 2.1). If we further consider variants of these strategies based on those that account for zone air-mass accumulation in the airflow analysis equations and those that do not, an additional eight approaches may be identified that bring the total to 17 possible coupling strategies. Yet, not one of these strategies approaches the problem

by the method of choice of numerical analysts – that is, using a Newton-Raphson iteration scheme based on the (full) Jacobian of the mathematically coupled combined system of equations for all three analytical domains!

Mathematical coupling, multiple solutions and chaotic behaviour

In principle, the airflow, thermal and contaminant dispersal analysis approaches may be coupled mathematically to form one combined system of (non-linear) differential and algebraic equations (DAEs) (Axley and Grot, 1989, 1990). In comparison to embedded CFD analysis, where coupling of interface boundary conditions is inherently indeterminate, the coupled variables here share a one-to-one relation across all three areas of analysis. Thus, forming a combined system of equations is tractable, albeit challenging. While challenging, a mathematical

Table 2.1 *Matrix of nine possible strategies for computationally coupling multi-zone airflow analysis to thermal and contaminant dispersal analysis*

Thermal analysis coupling strategy	Contaminant dispersal analysis coupling strategy		
	forwardAsync	ping-pongSync	onionSync
forwardAsync	Figure 2.33 left	✓	✓
ping-pongSync	✓	Figure 2.33 right	✓
onionSync	Figure 2.34 right	✓	Figure 2.34 left

Source: chapter author

approach to the problem leverages the vast numerical methods literature to advantage and enables (mathematical) analysis of the inherent stability of the governing coupled-system equations and the convergence characteristics of the methods used to solve them.

There are currently two paths to mathematically couple the multi-zone airflow, thermal and contaminant dispersal analysis equations – as well as any additional multi-physics or control equations that an analyst might wish to consider. The obvious, but generally impractically difficult, approach is to combine all governing equations formally using conventional mathematical operations – possibly aided by symbolic math processors such as MATHEMATICA®, Maple, MathCAD or MatLAB (MathWorks, 2000; Wolfram Research, 2004; MapleSoft, 2004a, 2004b). An alternative and increasingly important approach is provided by *equation-based simulation environments*.

Before considering these two alternatives in detail, it should be noted that some analysts in the building research community have the expertise and, thus, the command of advanced numerical methods to devise computational strategies that are tailored to solve the specific multi-zone multi-physics tasks considered without actually resorting to full mathematical coupling of the combined system equations. Woloszyn and her colleagues provide one such example where they subdivided the combined system equations into so-called *blocks* (that is, to effect a more manageable solution strategy), and investigated the condition number of the combined Jacobians of selected block types to devise a block partitioning strategy that would lead to an efficient solution of the system equations (Woloszyn et al, 2000a, 2000b). In simple terms, Woloszyn and her colleagues explicitly applied the numerical solution strategies that are implicitly provided by the numerically more advanced simulation environments – for example, the SPARK and IDA simulation environments (Bring, 1992; Lindgren and Sahlin, 1992; Sahlin, 1993; LBNL, 2002a, 2002b).

Formal coupling of the multi-physics equations: Multiple solutions and chaotic behaviour

Yuguo Li and collaborators in Australia, Sweden and Denmark have formally assembled and solved the coupled airflow/thermal analysis system equations for a series of one- and two-zone problems of increasing complexity to investigate the general behaviour of the resulting non-linear equations under a variety of thermal and airflow conditions (Li and Delsante, 1998; Delsante and Li, 1999; Li et al, 2000, 2001; Chen and Li, 2002; Li, 2002; Heiselberg, 2004). They have shown that under certain specific conditions when buoyancy forces tend to oppose wind or mechanical forces:

- Multiple solutions to the (steady) system equations are possible, of which more than one of the multiple solutions may be stable.
- Flow may abruptly switch from buoyancy-dominated flow, characterized by stratification, to wind-dominated flow, characterized by well-mixed conditions that may lead to catastrophic transformations of smoke layers in fires.
- Under dynamic variations of thermal and wind conditions, airflows may exhibit hysteresis – that is to say, they may exhibit, for example, different flow intensities and directions for increasing buoyancy conditions from those for decreasing buoyancy conditions.

Small-scale experimental studies using water tunnels to simulate wind-driven effects and salt for buoyancy effects completed by these researchers and, independently, by researchers at the University of Cambridge (Hunt and Linden, 2000; Hunt et al, 2000, 2001) have confirmed the theoretical work demonstrating both close correspondence to theory and the presence of the multiple stable flows predicted by the analytical results. The analytical and experimental results were also validated by detailed CFD studies which further demonstrated that the observed hysteretic behaviour is sensitive to the assumed initial conditions used for the CFD analysis.

Furthermore, Lorenzetti (2002) demonstrates that the key finding that multiple solutions may exist for the simple one- and two-zone cases considered extends, in general, to multi-zone coupled thermal/airflow theory based on the nodal continuity approach, while Nitta (1996, 1997) demonstrates this also to be true for the multi-zone theory based on the loop compatibility approach.

The implications of these findings are profound – solution uniqueness cannot, in general, be assumed for multi-zone coupled airflow/thermal analysis and, by extension, for multi-zone multi-physics analysis, in general. In addition, due to the possibility of multiple stable solutions, and the inherent numerical characteristics of the Newton-Raphson method used to solve the non-linear coupled airflow/thermal problem, computation may – again under certain specific conditions – degenerate to produce not simply divergent, but chaotic solution results (Nitta, 1999, 2002; Li et al, 2001; Li, 2002). Significantly, Li et al (2001) conclude:

Unfortunately, there seems to be no general approach to distinguish the true steady-state solution [for the steady form of the coupled airflow/thermal problem] from the spurious [chaotic] solutions in practical simulations.

While it is well known that even the Navier-Stokes equations can admit multiple solutions, and multiple airflow patterns can and do exist in practical situations (see Zhang et al, 2000, for a particularly compelling example), the prospect that a multi-zone analysis tool may, at any time during a longer-term simulation, produce unrealistic or completely spurious results clearly undermines the utility of multi-zone analysis as a practical tool.

Nevertheless, Li's sobering conclusion may be overly pessimistic, as these studies did not include dynamic effects in the airflow theory considered. Thus, there is some hope that if a fully dynamic formulation of the airflow theory is coupled to the (dynamic) thermal theory, then computed results may avoid the problematic numerical chaotic behaviour, while being able to capture the hysteretic behaviour observed in the experimental and CFD studies reported.

Equation-based simulation environments and advanced computational tools

In concise terms, *equation-based simulation environments* allow the analyst to define the link transport, system conservation and excitation relations for a given coupled (or uncoupled) problem mathematically. The simulation environment then automatically forms the system equations (for example, via forming the system Jacobian matrix) and solves them using state-of-the-art numerical methods that are often beyond the expertise and, thus, the command of the analyst. The analyst is therefore able to focus on the elemental physics, using more or less conventional mathematical notation, without having to be overly concerned with the methods used to solve the posed problem:

> *The modeller can forget entirely about the way equations are solved, he/she is merely required to* formulate *the problem, not to* solve *it. This enables domain experts to concentrate on their specialty and not have double duty as numerical analysts.* (Sahlin, 2000)

The more advanced simulation environments solve the fully coupled problem – or a numerical approximation of it – so that the question of coupling strategy need not be considered. If the general methods used to achieve this fully coupled solution are reliable, then the search for effective coupling strategies, which have preoccupied

multi-physics analysis in the building research community, need no longer be pursued. Three of the limited number of simulation environments have been developed specifically with the needs of the building research and design community in mind. Thus, these simulation environments are likely to provide an acceptable, if not optimal, level of efficiency and reliability. These programmes include the EES programme, which is directed to *ad hoc* and more modest modelling tasks (Klein, 2004), and the SPARK and IDA programmes, which are directed to the development of simulation tools (Bring, 1992; LBNL, 2002a).

Other simulation environments that have been developed as general tools, or tools for other specific application domains, have also been applied to building thermal, airflow and contaminant dispersal analysis. These programmes include SIMNON, a programme designed to solve ordinary differential and difference equations that typically govern 'dynamical systems' (SSPA, 2003), SIMULINK a 'toolbox' within the MATLAB math processing environment based on a (signal processing) block diagram, rather than a control volume approach to system simulation (MathWorks, 2002), Dymola, a general purpose simulation environment directed towards engineering systems analysis (Brück et al, 2002), Modelica®, a simulation environment originally developed for robotic, automotive and aerospace applications, and even system dynamics programmes originally directed to business management needs, such as Vensim® (Ventana Systems, 2004).

Furthermore, simulation environments invariably allow users to define link transport or subsystem equations as *objects* that can be interconnected as necessary to define the topology of the system being studied. That is to say, these environments often support an *object-oriented* approach to systems analysis so that users can create *libraries* of objects that may be reassembled in different system configurations (topologies) as needed. As these objects may be readily exchanged with other users, object-oriented simulation environments allow largely independent component development efforts to be readily integrated to create complex component-based simulation tools. Consequently, research efforts focusing on specific objectives can be undertaken by an individual; yet the product of this individual work may be combined with that of others to realize a larger tool development objective (see, for example, the products of the IDA community of programme developers (Grot and Axley, 1987; Kolsaker 1991; Vuolle and Sahlin, 2000) and the work being done at the University of La Rochelle using SPARK under the direction of Wurtz (Wurtz et al, 2001; Mora et al, 2002a; Musy et al, 2002).

To facilitate the use of the object-oriented approach, simulation environment programmes typically include interfaces that allow the analyst to lay out system topologies graphically (for example, the SIMULINK, Modelica® or Vensim® user interfaces) or have separate graphic user interface programmes to achieve this same objective (Lindgren and Sahlin, 1992; Sahlin, 1993; LBNL, 2002b).

As promising as simulation environments are, they are currently directed at dynamic systems governed by (non-linear) ordinary differential and algebraic equations and therefore cannot support continua thermal, airflow or contaminant transport analysis governed by partial differential equations. Within the continua analysis fields, the evolution of programme capabilities is somewhat similar, although equation-based simulation environments have been very rare. Recently however, FEMLAB, an equation-based partial differential equation (PDE) simulation environment, has emerged (COMSOL, 2003). FEMLAB may be used to solve a broad range of PDEs, optionally defined mathematically by the user, in 1D, 2D and/or 3D Cartesian or axysymmetric coordinate spatial domains using general finite-element solution procedures and optimized solver routines. Additionally, FEMLAB offers the capability of integrated analysis within the MATLAB and, importantly, SIMULINK environments. Consequently, the integrated use of the FEMLAB and SIMULINK environments supports equation-based simulation analysis of not only multi-physics coupled microscopic analysis, but embedded microscopic/macroscopic analysis, as well. A demonstration of the potential of this unusual capability has been reported by van Schijndel (2003) who applied these tools to coupled thermal/airflow, thermal bridge and moisture transport problems.

Summary

By the time of the Sessional Meeting of the Institution of Heating and Ventilating Engineers in London more than half a century ago (Dick, 1949), the broad purposes of building ventilation analysis were clear. It could address questions relating to air quality, moisture transport, energy consumption, room air distribution, thermal comfort and smoke movement – all governed by buoyancy-, wind- and mechanically driven airflows. It could serve to improve our understanding of these phenomena, to predict system behaviour related to these phenomena and to design systems to control these phenomena for the health, safety and welfare of building occupants.

While many elements of the governing physics of building ventilation were appreciated at the time, they were deemed so complex that the potential of building ventilation analysis was thought to be limited:

> The process of air change in a room is produced by a complex pattern of forces due to the impact of wind and to temperature differences between internal and external air columns. These forces act on the network of air resistances (formed by the flues, cracks around windows and doors, etc. in the house), and thus the magnitude and direction of the air flow through the individual resistances is determined not only by the acting [building] pressures, but also by the magnitudes of the resistances elsewhere in the network... It is thus extremely difficult to predict on theoretical grounds the air-change rate in a room of a house. (Dick, 1949)

Ironically, this conceptual model of flow *resistances* configured in a *network* acted upon by wind and buoyancy *forces* became the basis of present-day multi-zone ventilation analysis methods that have effectively provided the predictive capability that was thought to be impossibly difficult 50 years ago.

Unfortunately, these same methods are now critically limited by the theoretical assumptions put into place by the time of the 1949 sessional meeting – assumptions that airflow was solely pressure driven, that the flow resistance of rooms was negligible, that sealed-building wind pressure coefficients, mean wind pressures and mean room temperatures would be sufficient for building airflow analysis, and that the complexities of the dynamic effects of inertia and capacitance, the coupled interactions of heat and mass transfer, and even the need to conserve not only mass and thermal energy but also mechanical energy could be ignored.

This chapter has reviewed the physical theory, mathematical and numerical methods and computational strategies that have, together, been combined to devise multi-zone methods of building ventilation analysis – methods to analyse the very *networks* that Dick thought to be 'extremely difficult', if not impossible, to analyse in 1949. This review has been organized to systematically consider these several assumptions that now critically limit the field.

The conventional and popular multi-zone *nodal continuity* and the less popular *loop compatibility* approaches to multi-zone ventilation analysis have been presented in a concise matrix notation:

- to place a central emphasis on the *hydrostatic field assumption* that is commonly applied to zones to effect closure of the modelling theory and to demonstrate that other field assumptions may be considered as well;
- to provide a mathematically rigorous formulation of the theory so that numerical questions of solution existence and uniqueness may be evaluated;
- to provide a complete description of the commonly used Newton-Raphson solution algorithm so that conditions that lead to computationally efficient, stable and convergent solutions may be identified; and
- so that strategies to couple methods of building airflow analysis to thermal and contaminant dispersal analysis can be considered in detail.

In an attempt to set the stage for a more rigorous discussion of the shortcomings of existing theory, the theory has been formulated so that continual descriptions of variables are distinguished from those that are, in fact, spatial or temporal averages used to describe bulk airflow quantities.

Three problematic areas of multi-zone ventilation analysis are then addressed:

1   the problem of modelling the behaviour of porous buildings;
2   the fundamental flaw in existing theory relating to the conservation of mechanical energy; and
3   the challenge of modelling dynamic effects in building airflows.

It is shown that multi-zone airflow analysis will, in general, violate the conservation of mechanical energy when pressure variations within zones are modelled with the commonly used *hydrostatic field assumption*. Furthermore, the error that will result is likely to be significant. Thus, conventional multi-zone airflow analysis is flawed at the most fundamental level. Furthermore, theoretical and computational results are presented that demonstrate that both improved wind pressure coefficients and models of airflow resistances which properly account for mechanical energy transformations are needed to improve the modelling of airflows in porous buildings.

Multi-zone ventilation analysis is most commonly approached as a quasi-steady phenomenon under dynamic variations of buoyancy, wind and mechanical forces – that is, assuming that airflow dynamics are so rapid as to be essentially instantaneous. Yet, wind turbulence is known to increase infiltration airflows well above those due to mean airflow conditions, and dynamic flow phenomena such as back-draughting, pulsation and turbulence penetration are observed aspects of building airflows in the field. Consequently, in the interests of improved accuracy and fidelity, a number of proposals have been put forward over the years to model these dynamic effects. Both theoretical and semi-empirical approaches to the problem have been reviewed and strategies for including the former in multi-zone ventilation analysis have been outlined. However, this chapter stops short of presenting a fully dynamic formulation of the multi-zone theory since this has not yet been pursued in the field.

The nodal continuity approach to multi-zone airflow analysis has been the method of choice for the past two decades. Yet, another approach, the so-called *loop compatibility* or simply *loop* method, has been available and occasionally implemented. While there are a number of apparent advantages of this approach, the challenge to form the loop equations has been a key barrier to its use. Simple graph theoretical principles reviewed in this chapter should help to remove this barrier.

Fortuitously, the inverse form of the multi-zone loop method may be used as the basis of a preliminary design method for sizing natural and hybrid ventilation systems components. This method – the *loop design method* – allows the designer to select feasible combinations of component sizes that satisfy specified design conditions and design criteria. The basis and an outline of this 'exact' approach have been briefly considered in this chapter.

Multi-zone methods of airflow analysis can effectively, if not also accurately, predict bulk airflow rates through the discrete airflow paths that link zones of a building. Yet, these methods provide little information about the details of airflows within the zones. Consequently, a number of studies have turned to the problem of modelling both the details of airflows within rooms (*detailed analysis*) and the complexities of the multiple physical phenomena (*multiphysics*) associated with them. Both established methods of computational fluid dynamics (CFD) and the more recent and less mature methods of *sub-zone* (or *zonal*) analysis have been applied to isolated problems of detailed airflow analysis within rooms. However, room airflows do not occur in isolation – they result from, and interact with, airflows within the larger whole-building systems in which they are embedded. This has led some investigators to embed detailed models within larger whole-building models, hoping to more faithfully model these detailed airflows and, importantly, to model them over extended time periods. Consequently, this chapter also reviewed the current state of sub-zone modelling techniques and strategies to embed these and CFD models within multi-zone models of whole-building systems.

From their inception, multi-zone methods of airflow analysis were developed to investigate the complexities of the multiple physical phenomena (*multi-physics*) associated with building airflows. The coupled relation of these airflows to building thermal and contaminant dispersal behaviour is, by now, familiar to all – airflows depend upon the distribution of air densities which, in turn, depend upon mass and thermal transport processes that, to come full circle, depend upon airflows. Of the number of multi-physics problems that could be addressed, the coupled interaction of building airflows, thermal response and contaminant dispersal has received the greatest attention since multi-zone methods that share similar modelling assumptions are available in each of these areas. The last section of this chapter, therefore, reviewed some of the work completed during the past two decades to couple these three areas of analysis.

Modelling assumptions and system equations for multi-zone thermal and contaminant dispersal analysis were outlined and a range of computational coupling strategies for both steady state and dynamic analysis were identified. Notwithstanding the apparent successes of some of these efforts to *computationally* couple these areas of multi-zone analysis, the proliferation of an unreasonably large number of *ad hoc* computational coupling strategies without critical evaluation has left the field in a state of confusion. Fortunately, the alternative and rigorous strategy of mathematically coupling these three areas of analysis is possible, thus effectively side-stepping the problematic question of selecting a computational coupling strategy altogether. Currently, there are two paths to mathematically couple the multi-zone airflow: thermal and dispersal analysis equations – traditional mathematical manipulation and equation-based *simulation environments*. These and the problems that may result from them then have been reviewed in this chapter.

During the past 50 years, advances in computational hardware and software have fostered the development of numerical methods and computational strategies that now allow building analysts to routinely model large and complex multi-zone systems thought to be impossibly difficult to analyse beforehand. A type of computational tyranny has, however, accompanied these advances – building analysts have had to master these very numerical methods and computational strategies (and to find the time to implement them) to be truly effective. This has left little time to consider the more fundamental issues of their domain. Consequently, the fundamental principles of ventilation analysis – and, importantly, the limiting assumptions underlying these principles – have changed little in the past half century. This chapter has attempted to critically review the theory of ventilation analysis in order to highlight these limiting assumptions. Today, it appears that the very promising *simulation environments* will free members of the ventilation community from the tyranny of computational advances so that once again the community can turn to the central task of the field – to advance the state of the theory.

References

Albrecht, T., Gritzki, R., Grundman, R., Perschk, A., Richter, W., Rösler, M. and Seifert, J. (2002) *Evaluation of a Coupled Calculation for a Hybrid Ventilated Building From a Practical and Scientific Point of View* (P. Heiselberg, ed), Hybrid Ventilation Centre, Aalborg University, Aalborg, Denmark

Allard, F. (ed) (1998) *Natural Ventilation in Buildings: A Design Handbook*, James and James Ltd, London

Allard, F., Inard, C. and Simoneau, J.-P. (1990) 'Phénomènes convectifs intérieurs dans les cellules d'habitation – Approaches expérimentales et numériques', *Revue Générale de Thermique*, vol 340, pp216–225

ASHRAE (American Society of Heating, Refrigerating and Air-conditioning Engineers) (2001) *ASHRAE Handbook – Fundamentals*, ASHRAE, Atlanta, GA

Awbi, H. B. (1991) *Ventilation of Buildings*, E. and F. N. Spon, London

Axley, J. W. (1987a) *Indoor Air Quality Modeling Phase II Report*, US DOC, NBS, Gaithersburg, MD

Axley, J. W. (1987b) 'The NBS Multi-Zone IAQ Model', in *Proceedings of the Pacific Northwest International Section of the APCA*, PNWIS APCA, Seattle, WA.

Axley, J. W. (1988a) *Progress Toward a General Analytical Method for Predicting Indoor Air Pollution in Buildings: Indoor Air Quality Modeling Phase III Report*, US DOC, NBS, Gaithersburg, MD

Axley, J. W. (1988b) *DTAM1: A Discrete Thermal Analysis Method for Building Energy Simulation: Part I Linear Thermal Systems with DTAM1 Users Manual*, US DOC, NIST, Gaithersburg, MD

Axley, J. W. (1989) 'Multi-zone dispersal analysis by element assembly', *Building and Environment*, vol. 24, no 2, pp113–130

Axley, J. W. (1995) *New Mass Transport Elements and Components for the Next Generation NIST IAQ Model*, NIST, Gaithersburg, MD

Axley, J. (1998) 'Introduction to the design of natural ventilation systems using loop equations', in *19th AIVC Conference, Ventilation Technologies in Urban Areas*, AIVC, Oslo, Norway

Axley, J. W. (1999a) 'Passive ventilation for residential air quality control', *ASHRAE Transactions*, vol 105, Part 2, pp864–876

Axley, J. W. (1999b) 'Natural ventilation design using loop equations', in *Indoor Air 99*, ISIAQ and AIVC, Edinburgh

Axley, J. W. (2000a) *AIVC TechNote 54: Residential Passive Ventilation Systems – Evaluation and Design*, Air Infiltration and Ventilation Centre, Coventry, UK

Axley, J. W. (2000b) 'Design and simulation of natural ventilation systems using loop equations', in *Healthy Buildings 2000*, FiSIAQ and ISIAQ, Espoo, Finland

Axley, J. W. (2000c) 'Surface-drag flow relations for zonal modeling', Presented to CGE/LBL-Workshop, Thermal and Airflow Simulation in Buildings, Conférence de Grande Écoles and Lawrence Berkeley National Laboratory, Berkeley, CA

Axley, J. W. (2000d) 'Zonal models using loop equations and surface drag cell-to-cell flow relations', in *RoomVent 2000, Seventh International Conference on Air Distribution in Rooms*, University of Reading, UK

Axley, J. W. (2001a) *Application of Natural Ventilation for US Commercial Buildings: Climate Suitability, Design Strategies and Methods, Modeling Studies*, NIST, Gaithersburg, MD

Axley, J. W. (2001b) 'Surface-drag flow relations for zonal modeling', *Building and Environment*, vol 36, no 7, pp843–850

Axley, J. and Grot, R. (1989) 'The coupled airflow and thermal analysis problem in building airflow system simulation', in *ASHRAE Symposium on Calculation of Interzonal Heat and Mass Transport in Buildings*, ASHRAE, Vancouver, BC, Canada

Axley, J. and Grot, R. (1990) 'Coupled airflow and thermal analysis for building system simulation by element assembly techniques', in *RoomVent 1990: Engineering Aero and Thermodynamics of Ventilated Rooms, Second International Conference*, NORSK VVS, Norwegian Association of Heating, Ventilating and Sanitary Engineers, Oslo, Norway

Axley, J., Wurtz, E. and Mora, L. (2002a) 'Macroscopic airflow analysis and the conservation of kinetic energy', in *Room Vent 2002: Air Distribution in Rooms, Eighth International Conference*, Copenhagen, Denmark

Axley, J., Emmerich, S. J. and Walton, G. (2002b) 'Modeling the performance of a naturally ventilated commercial building with a multizone coupled thermal/airflow simulation tool', *ASHRAE Transactions*, vol 108, part 2, pp1260–1275

Axley, J., Emmerich, S. J., Dols, S. and Walton, G. (2002c) 'An approach to the design of natural and hybrid ventilation systems for cooling buildings', in *Indoor Air 2002*, ISIAQ, Monterey, CA

Aynsley, R. M (1999) 'Unresolved issues in natural ventilation for thermal comfort', in *HybVent Forum 1999*, Sydney, Australia

Aynsley, R. M., Melbourne, W. and Vickery, B. J. (1977) *Architectural Aerodynamics*, Applied Science Publishers, London

Bassett, M. (1990) *Technical Note AIVC 27: Infiltration and Leakage Paths in Single Family Houses – A Multizone Infiltration Case Study*, Air Infiltration and Ventilation Centre, Coventry, UK

Bird, R. B., Stewart, W. E. and Lightfoot, E. N. (2002) *Transport Phenomena*, second edition, John Wiley and Sons, New York

Borse, G. J. (1997) *Numerical Methods with MATLAB: A Resource for Scientists and Engineers*, PWS Publishing Company, Boston

Bouia, H. and Dalicieux, P. (1991) 'Simplified modeling of air movements inside dwelling room', in *Building Simulation 1991*, IBPSA (International Building Performance Simulation Association), Sophia-Antipolis, Nice, France

Bowman, N. T., Lomas, K., Cook, M., Eppel, H., Ford, B., Hewitt, M., Cucinella, M., Francis, E., Rodriguez, E., Gonzales, R., Alvarez, S., Galata, A., Lanarde, P. and Belarbi, R. (1997) 'Application of passive downdraught evaporative cooling (PDEC) to non-domestic buildings', *Renewable Energy*, vol 10, no 2/3, pp191–196

Bowman, N. T., Eppel, H., Lomas, K. J., Robinson, D. and Cook, M. J. (2000) 'Passive downdraught evaporative cooling', *Indoor and Built Environment*, vol 9, no 5, pp284–290

Bring, A. (1992) *IDA SOLVER: User's Documentation*, Department of Building Services Engineering, KTH (Royal Institute of Technology), Stockholm, Sweden

Brohus, H., Frier, C. and Heiselberg, P. (2003) 'Measurements of hybrid ventilation performance in an office building', *International Journal of Ventilation*, vol 1 (Hybrid Ventilation Special Edition), pp77–88

Brück, D., Elmqvist, H., Mattson, S. E. and Olsson, H. (2002) 'Dymola for multi-engineering modeling and simulation', in *Second International Modelica Conference*, Modelica Association, Oberpfaffenhofen, Germany

Chen, Q. Y. and Wang, L. (2004) *Coupling of Multizone Program CONTAM with Simplified CFD Program CFD0-C*, School of Mechanical Engineering, Purdue University, West Lafayette, IN

Chen, Q. and Xu, W. (1998) 'A zero-equation turbulence model for indoor air simulation', *Energy and Buildings*, vol 28, pp137–144

Chen, Z. and Li, Y. (2002) 'Flow bifurcations of buoyancy driven natural ventilation in a single-zone building', in *RoomVent 2002, Eighth International Conference on Air Distribution in Rooms*, Technical University of Denmark and Danvak, Copenhagen, Denmark

Chiu, Y. H. and Etheridge, D. W. (2002) 'Calculations and notes on the quadratic and power law equations for modelling infiltration', *International Journal of Ventilation*, vol 1, no 1, pp65–77

Chorin, A. J. and Marsden, J. E. (1993) *A Mathematical Introduction to Fluid Mechanics*, Springer-Verlag, New York

Cockcroft, J. P. and Robertson, P. (1979) 'Ventilation of an enclosure through a single opening', *Building and Environment*, vol 11, pp29–35

COMSOL (2001) *FEMLAB Reference Manual*, COMSOL, Inc, Burlington, MA

COMSOL (2003) *FEMLAB 3.0 User's Guide*, COMSOL, Inc, Burlington, MA

Dahlquist, G. and Björck, A. (1974) *Numerical Methods*, Prentice-Hall, Inc, Englewood Cliffs, New Jersey

Dascalaki, E. and Santamouris, M. (1996) 'Natural ventilation', in Santamouris, M. and Asimakopoulos, D. (eds) *Passive Cooling of Buildings*, James & James, London, Chapter 9, pp220–306

Davenport, A. G. (1967) *The Dependence of Wind Loads on Meteorological Factors*, International Research Seminar on Wind Effects on Buildings and Structures, University of Toronto Press, Ottawa

de Gids, W. F. (1978) 'Calculation method for the natural ventilation of buildings', *Verwarming Vent*, no 7, pp552–564

Delsante, A. and Li, Y. (1999) 'Natural ventilation induced by combined wind and thermal forces in a two-zone building', in *HybVent Forum 1999*, Sydney, Australia

Demuren, A. O. and Ideriah, J. K. (1986) 'Pipe network analysis by partial pivoting method', *Journal of Hydraulic Engineering*, vol 112, no 5

Dick, J. B. (1949) 'Experimental studies in natural ventilation of houses', *Journal of the Institution of Heating and Ventilating Engineers*, December, pp420–466

Diestel, R. (1991) *Graph Theory*, Springer-Verlag, New York

Dols, W. S. (2001a) *NIST Multizone Modeling Website*, NIST, Gaithersburg, MD

Dols, W. S. (2001b) 'A tool for modeling airflow and contaminant transport', *ASHRAE Journal*, vol 43, no 3, pp35–44

Dols, W. S. and Emmerich, S. J. (2003) *LoopDA – Natural Ventilation Design and Analysis Software*, US DOC NIST, Gaithersburg, MD

Dols, W. S. and Walton, G. N. (2000) *CONTAMW 2.0 User Manual: Multizone Airflow and Contaminant Transport Analysis Software*, US NIST, Gaithersburg, MD

Dols, W. S., Walton, G. N. and Denton, K. R. (2000) *CONTAMW 1.0 User Manual: Multizone Airflow and Contaminant Transport Analysis Software*, US NIST, Gaithersburg, MD

Dorer, V. (1998a) *Thermal Summer Condition Evaluation of a Large Naturally Ventilated Test Laboratory Hall at EMPA*, EMPA

Dorer, V. (1998b) *Air Quality and Thermal Comfort in a Naturally Ventilated School Building with Glazed Double Façade*, EMPA

Etheridge, D. W. (1988) 'Modeling of air infiltration in single- and multi-cell buildings', in *Energy and Buildings*, Elsevier Sequoia, The Netherlands

Etheridge, D. W (1998a) 'A note on crack flow equations for ventilation flow modeling', *Building and Environment*, vol 33, no 5, pp325–328

Etheridge, D. W. (1998b) 'Unsteady flow effects due to fluctuating wind pressures in natural ventilation design – mean flow rates', *Building and Environment*, vol 35, no 2, pp111–133

Etheridge, D. W. (2000a) 'Unsteady flow effects due to fluctuating wind pressures in natural ventilation design – instantaneous flow rates', *Building and Environment*, vol 35, no 4, pp321–337

Etheridge, D. W. (2000b) 'Unsteady flow effects due to fluctuating wind pressures in natural ventilation design – mean flow rates', *Building and Environment*, vol 35, no 2, pp111–133

Etheridge, D. W. (2002) 'Nondimensional methods for natural ventilation design', *Building and Environment*, vol 37, no 11, pp1057–1072

Etheridge, D. W. and Alexander, D. K. (1980) 'The British Gas multi-cell model for calculating ventilation', *ASHRAE Transactions*, vol 86, no 2, pp808–821

Etheridge, D. W. and Nolan, J. A. (1979) 'Ventilation measurements at model scale in a turbulent flow', *Building and Environment*, vol 14, pp65–68

Etheridge, D. W. and Sandberg, M. (1996) *Building Ventilation: Theory and Measurement*, John Wiley and Sons, Chichester, UK

Feustel, H. E. (1990) 'The COMIS air flow model: A tool for multizone applications', in *Indoor Air 1990, Fifth International Conference on Indoor Air Quality and Climate*, Canada Mortgage and Housing Corporation, Toronto, Canada

Feustel, H. E. and Kendon, V. M. (1985) 'Infiltration models for multicellular structures – A literature review', in *Energy and Buildings*, Elsevier Sequoia, The Netherlands

Feustel, H. E. and Raynor-Hoosen, A. (eds) (1990) *COMIS – Fundamentals*, Lawrence Berkeley National Laboratory, Berkeley, CA

Feustel, H. E. and Smith, B. V. (eds) (1992) *COMIS 1.2 User Guide*, Lawrence Berkeley National Laboratory, Berkeley, CA

Feustel, H. E. and Smith, B. V. (eds) (1997) *COMIS 3.0 – User's Guide*, Lawrence Berkeley National Laboratory: Berkeley, CA

Fürbringer, J.-M., Dorer, V., Huck, F. and Weber, A. (1993) 'Air flow simulation of the LESO Building, including a comparison with measurements and sensitivity analysis', in *Indoor Air 1993, Fifth International Conference on Indoor Air Quality and Climate*, Helsinki University of Technology, Helsinki, Finland

Fürbringer, J.-M., Roulet, C. A. and Borchiellini, R. (eds) (1996) *Annex 23: Multizone Air Flow Modeling: Evaluation of COMIS: Appendices*, vol 1/2, Swiss Federal Institute of Technology, Institute of Building Technology, Lausanne, Switzerland

Gan, G. and Riffat, S. B. (2000) 'Numerical determination of energy losses at duct junctions', *Applied Energy*, vol 67, pp331–340

Gerald, C. F. and Wheatley, P. O. (1990) *Applied Numerical Analysis*, fourth edition, Addison-Wesley Publications, Reading, MA

Ghiaus, C., Allard, F. and Axley, J. (2003) 'Natural ventilation in an urban context', in Santamouris, M. (ed) *Solar Thermal Technologies for Buildings: The State of the Art*, James & James, London, Chapter 6

Giabaklou, Z. and Ballinger, J. A. (1996) 'A passive evaporative cooling system by natural ventilation', *Building and Environment*, vol 31, no 6, pp503–507

Girault, P. and Spennato, B. (1999) 'The impact of wind turbulence on the precision of a numerical modeling study', *Indoor Air 1999*, ISIAQ and AIVC, Edinburgh

Grot, R. A. (1990) *Users Manual – NBSAVIS – CONTAM88*, NIST, Gaithersburg, MD

Grot, R. and Axley, J. (1987) 'The development of models for the prediction of indoor air quality in buildings', in *Proceedings of the Eighth AIVC Conference: Ventilation Technology Research and Application*, AIVC, Überlingen, Federal Republic of Germany

Grot, R. A. and Axley, J. W. (1990) 'Structure of models for the prediction of air flow and contaminant dispersal in buildings', in *11th AIVC Conference on Ventilation System Performance*, Air Infiltration and Ventilation Centre, Belgirate, Lake Maggiore, Italy

Guffey, S. E. and Fraser, D. A. (1989) 'A power balance model for converging and diverging flow junctions', *ASHRAE Transactions*, vol 95, Part 2, pp1661–1669

Gustén, J. (1989) *Wind Pressure on Low-Rise Buildings: An Air Infiltration Analysis Based on Full-scale Measurements*, Division of Structural Design, Chalmers University of Technology, Gothenburg, Sweden

Haas, A. (2000) *COMIS 3.1*, EMPA – Swiss Federal Laboratories for Materials Testing and Research, Zurich

Haghighat, F., Brohus, H. and Rao, J. (2000) 'Modelling air infiltration due to wind fluctuations: A review', *Building and Environment*, vol 35, no 5, pp377–385

Haghighat, F., Li, Y. and Megri, A. C. (2001) 'Development and validation of a zonal model – POMA', *Building and Environment*, vol 36, no 9, pp1039–1047

Haghighat, F., Lin, Y. and Megri, A. C. (1999) 'Zonal model: A simplified multiflow element model', in *HybVent Forum 1999*, Sydney, Australia

Haghighat, F., Rao J. and Fazio, P. (1991) 'The influence of turbulent wind on air change rates – a modelling approach', *Building and Environment*, vol 26, no 2, pp95–109

Heiselberg, P. (1999) 'The hybrid ventilation process – theoretical and experimental work', *Air Infiltration Review*, vol 21, no 1, pp1–4

Heiselberg, P., Murakami, S. and Roulet, C.-A. (eds) (1998) *Ventilation of Large Spaces in Buildings – Analysis and Prediction Techniques*, IEA Annex 26: Energy Efficient Ventilation of Large Enclosures, Aalborg University, Aalborg, Denmark

Heiselberg, P. (2004) 'Experimental and CFD evidence of multiple solutions in a naturally ventilated building', *Indoor Air*, vol 14, no 1, pp43–54

Hensen, J. L. M. (1990) *ESPmfs, A Building and Plant Mass Flow Network Solver*, FAGO, Eindhoven University of Technology, Netherlands

Hensen, J. L. M. and Clarke, J. A. (1990) 'A fluid flow network solver for integrated building and plant energy simulation', in *Third International Conference on System Simulation in Buildings*, Liège, Belgium

Herrlin, M. K. (1985) 'MOVECOMP: A static multicell airflow model', *ASHRAE Transactions*, vol 91, Part 2

Hill, J. and Kasuda, T. (1975) 'Dynamic characteristics of air infiltration', *ASHRAE Transactions*, vol 81, no 1, pp168–185

Holmes, M. J. and McGowan, S. (1997) 'Simulation of a ccomplex wind and buoyancy driven building', in *Building Simulation 1997, Fifth International IBPSA Conference*, IBPSA, Prague

Hunt, G. R., Cooper, P. and Linden, P. F. (2001) 'Thermal stratification produced by plumes and jets in enclosed spaces', *Building and Environment*, vol 36, pp871–882

Hunt, G. R., Cooper, P. and Linden, P. F. (2000) 'Thermal stratification produced by plumes and jets in enclosed spaces', in *RoomVent 2000: Ventilation for Health and Sustainable Environment*, University of Reading, Elsevier Press

Hunt, G. R. and Linden, P. F. (2000) 'Multiple steady airflows and hysteresis when wind opposes buoyancy', *Air Infiltration Review*, vol 21, no 2, pp1–3

Inard, C. and Buty, D. (1991a) 'Simulation of thermal coupling between a radiator and room with zonal models', in *12th AIVC Conference on Air Movement and Ventilation Control within Buildings*, Ottawa, Canada, Air Infiltration and Ventilation Centre, Coventry, UK

Inard, C. and Buty, D. (1991b) 'Simulation of thermal coupling between a radiator and room with zonal models', in *Building Simulation 1991*, IBPSA (International Building Performance Simulation Association), Sophia-Antipolis, Nice, France

Irving, S. and Uys, E. (1997) *CIBSE Applications Manual: Natural Ventilation in Non-domestic Buildings*, CIBSE, London

Isaacs, L. T. and Mills, K. G. (1980) 'Linear theory methods for pipe network analysis', in *Journal of the Hydraulics Division*, ASCE

Jensen, J. P., Heiselberg, P. and Nielsen, P. V. (2002a) *Numerical Simulation of Airflow through Large Openings*, International Energy Agency – Energy Conservation in Buildings and Community Systems – Annex 35: Hybrid Ventilation in New and Retrofitted Office Buildings

Jensen, J. P., Heiselberg, P. and Nielsen, P. V. (2002b) 'Numerical simulation of airflow through large openings', in *RoomVent, Eighth International Conference on Air Distribution in Rooms*, Technical University of Denmark and Danvak, Copenhagen, Denmark

Jeppson, R. W. (1976) *Analysis of Flow in Pipe Networks*, Ann Arbor Science, Ann Arbor, MI

Jiang, Y. and Chen, Q. (2002) 'Effect of fluctuating wind direction on cross natural ventilation in buildings from large eddy simulation', *Building and Environment*, vol 37, no 4, pp379–386

Kato, S. (2004) 'Flow network model based on power balance as applied to cross-ventilation', *The International Journal of Ventilation*, vol 2, no 4, pp395–408

Kato, S., Murakami, S., Mochida, A., Akabayashi, S. and Tominaga, Y. (1992) 'Velocity-pressure field of cross ventilation with open windows analyzed by wind tunnel and numerical simulation', *International Journal of Wind Engineering and Industrial Aerodynamics*, vol 41–44, pp2575–2586

Kendrick, J. (1993) *AIVC Technical Note 40: An Overview of Combined Modelling of Heat Transport and Air Movement*, AIVC, Coventry, UK

Klein, S. A (2002) *E. E. S. – Engineering Equation Solver*, F-Chart Software, Madison, WI

Klein, S. A. (2004) *E. E. S. – Engineering Equation Solver*, F-Chart Software, Madison, WI

Klein, S. A. et al (1988) *TRNSYS – A Transient System Simulation Program, Version 12.2*, University of Wisconsin, Madison

Klobut, K., Tuomaala, P., Siren, K. and Seppanen, O. (1991) 'Simultaneous calculation of airflows, temperatures and contaminant concentrations in multi-zone buildings', in *12th AIVC Conference: Air Movement and Ventilation Control Within Buildings*, Ottawa, Canada, Air Infiltration and Ventilation Centre, Coventry, UK

Knoll, B. and Phaff, J. C (1996) *Two Unique New Tools for the Prediction of Wind Effects on Ventilation, Building Construction, and Indoor Climate*, TNO Building Construction and Research, Delft, The Netherlands

Knoll, B., Phaff, J. C. and de Gids, W. F. (1997) *Pressure Simulation Program*, TNO Building Construction and Research, Delft, The Netherlands

Kolsaker, K. (1991) 'An NMF-based component library for fire simulation', in *Building Simulation 1991*, IBPSA (International Building Performance Simulation Association), Sophia-Antipolis, Nice, France

Kurabuchi, T., Ohba, M., Arashiguchi, A. and Iwabuchi, T. (2000) 'Numerical study of airflow structure of a cross-ventilated model building', in *RoomVent: Ventilation for Health and Sustainable Environment*, University of Reading and Elsevier Science Publications

Kurabuchi, T., Ohba, M., Endo, T., Akamine, Y. and Nakayama, F. (2004) 'Local dynamic similarity model of cross-ventilation: Part 1 – theoretical framework', *The International Journal of Ventilation*, vol 2, no 4, pp371–381

LBNL (2002a) *LBNL and ASAI ASA, SPARK 1.0.2 Reference Manual – Simulation Problem Analysis and Research Kernel*, Lawrence Berkeley National Laboratory, Berkeley, CA

LBNL (2002b) *LBNL and ASAI ASA, VISUALSPARK 1.0.2 User's Guide, Simulation Problem Analysis and Research Kernel*, Lawrence Berkeley National Laboratory, Berkeley, CA

Li, Y. (1992) *Simulation of Flow and Heat Transfer in Ventilated Rooms*, Royal Institute of Technology, Stockholm, Sweden

Li, Y. (1993) 'Predictions of indoor air quality in multi-room buildings', in *Indoor Air 1993: Fifth International Conference on Indoor Air Quality and Climate*, Helsinki University of Technology, Helsinki, Finland

Li, Y. (2002) 'Spurious numerical solutions in coupled natural ventilation and thermal analysis', *International Journal of Ventilation*, vol 1, no 1, pp1–12

Li, Y. and Delsante, A. (1998) 'On natural ventilation of a building with two openings', in *19th AIVC Conference: Ventilation Technologies in Urban Areas*, AIVC, Oslo, Norway

Li, Y. and Holmberg, S. (1993) 'General flow and thermal-boundary conditions in indoor air flow simulation', *Indoor Air 1993: Fifth International Conference on Indoor Air Quality and Climate*, Helsinki University of Technology, Helsinki, Finland

Li, Y., Delsante, A, Chen, Z. and Sandberg, M. (2000) 'Some examples of solution multiplicity in natural ventilation', in *RoomVent 2000*, University of Reading and Elsevier Science Publication

Li, Y., Delsante, A., Chen, Z., Sandberg, M., Anderen, A., Bjerre, M. and Heiselberg, P. (2001) 'Some examples of solution multiplicity in natural ventilation', *Building and Environment*, vol 36, pp851–888

Liddament, M. (1983) 'The Air Infiltration Center's Program of Model Validation', *ASHRAE Transactions*

Liddament, M. (1986a) *Air Infiltration Calculation Techniques – An Applications Guide*, Air Infiltration and Ventilation Centre, Coventry, UK

Liddament, M. (1986b) 'Air infiltration and ventilation calculation techniques', *Air Infiltration Review*, vol 8, no 1, pp6–7

Liddament, M. (1987) *Technical Note AIVC 21: A Review and Bibliography of Ventilation Effectiveness – Definitions, Measurement, Design and Calculation*, Air Infiltration and Ventilation Centre, Coventry, UK

Liddament, M. (1991) *Technical Note AIVC 33: A Review of Building Air Flow Simulation*, Air Infiltration and Ventilation Centre, Coventry, UK

Liddament, M. (1996) *A Guide to Energy Efficient Ventilation*, AIVC, Coventry, UK

Liddament, M. and Allen, C. (1983) *Technical Note AIVC 11: The Validation and Comparison of Mathematical Models of Air Infiltration*, Air Infiltration and Ventilation Centre, Coventry, UK

Liddament, M. and Thompson, C. (1982) *Technical Note AIC 9: Mathematical Models of Air Infiltration: A Brief Review and Bibliography*, The Air Infiltration Centre, Bracknell, UK

Lin, Y., Megri, A. C. and Haghighat, F. (1999) 'Zonal models – A new a generation of combined airflow and thermal model', in *Indoor Air 1999: The Eighth International Conference on Indoor Air Quality and Climate*, Building Research Establishment Ltd, Edinburgh

Lindgren, M. and Sahlin, P. (1992) *IDA Modeller: A User's and Programmer's Guide*, Swedish Institute of Applied Mathematics, Göteborg, Sweden

Lorenzetti, D. M. (2002) 'Computational aspects of nodal multizone airflow systems', *Buildings and Environment Journal*, vol 37, pp1083–1090

Lorenzetti, D. M. and Sohn, M. D. (2000) 'Improving the speed and robustness of the COMIS Solver', in *RoomVent 2000*, University of Reading, Elsevier Science Publication

Lorenzetti, D., Jayaraman, B., Gadgil, A., Hong, S. and Chiczewski, T. (2003) 'Results from a coupled CFD plus multizone simulation', Presented to NIST Workshop on Non-uniform Zones in Multizone Network Models, Indoor Environment Department, Lawrence Berkeley National Laboratory, Berkeley, CA

Mansouri, Y. (2003) 'Conception des enveloppes de bâtiments pour le renouvellement d'air par ventilation naturelle en climats tempérés: Proposition d'une méthodologie de conception', in *Mécanique Thermique et Génie Civil*, Ecole Polytechnique de l'Université de Nantes, Nantes, France

Mansouri, Y. and Allard, F. (2003) 'Methods and methodological tools for the elaboration of natural ventilation strategy', in *Healthy Building 2003: Seventh International Conference of Energy-Efficient Healthy Buildings*, Singapore

MapleSoft (2004a) *Maple-9*

MathSoft (2004b) *MathCAD11*

MathWorks (2000) *MATLAB: The Language of Technical Computing – Using MATLAB Version 6*, The MathWorks, Natick, MA

MathWorks (2002) *SIMULINK: Model-Based and System-Based Design – Using SIMULINK Version 5*, The MathWorks, Natick, MA

Mendes, N., Lamberts, R. and Philippi, P. C. (2001a) 'The Umidus 2.1 program and the prediction of hygrothermal performance of porous building elements', in *CLIMA 2000: Seventh REHVA World Congress – Indoor Environment Technology: Towards a Global Approach*, AICARR (Italian Association of Air Conditioning, Heating and Refrigerating Engineers), Napoli, Italy

Mendes, N., Oliveira, G. H. C. Araújo, H.X. and Conceição, R.I. (2001b) 'Building thermal performance analysis by using MATLB/SIMULINK', in *Building Simulation 01*, IBPSA, Rio de Janeiro, Brazil

Mendes, N., Oliveira, G. H. C. and Araújo, H. X. (2002) 'The use of Matlab/Simulink to evaluate building heating processes', in *ESDA 2002: Sixth Biennial Conference on Engineering Systems Design and Analysis*, ASME, Istanbul, Turkey

Mendes, N., Oliveira, G. H. C., Araújo, H. X. and Coelho, L.S. (2003) 'A MATLAB-based simulation tool for building thermal performance analysis', *IBPSA News*, vol 13, no 2, pp54–61

Modelica Association (2002) *Modelica® – A Unified Object-Oriented Language for Physical Systems Modeling: Language Specification Version 2.0*, Modelica Association

Mora, L. (2003) 'Prédiction des performances thermo-aérauliques des bâtiments par association de modèles de différents niveaux de finesse au sein d'un environnement orienté objet', in *Spécialité Génie Civil*, Université de La Rochelle, La Rochelle, France

Mora, L., Gadgil, A. J. and Wurtz, E. (2003) 'Comparing zonal and CFD model predictions of isothermal indoor airflows to experimental data', *Indoor Air*, vol 13, pp77–85

Mora, L., Wurtz, E. and Axley, J. W. (2002a) 'Prediction of indoor environmental quality in multizone buildings using zonal models', in *RoomVent 2002: Air Distribution in Rooms – Eighth International Conference*, Copenhagen, Denmark

Mora, L., Gadgil, A. J., Wurtz, E. and Inard, C. (2002b) 'Comparing zonal and CFD model predictions of indoor airflows under mixed convection conditions to experimental data', in *EPIC Conference: Third European Conference on Energy Performance and Indoor Climate in Buildings*, Lyon, France

Murakami, S. (1993) 'Comparison of various turbulence models applied to a bluff body', *Journal of Wind Engineering and Industrial Aerodynamics*, vols 46 and 47, pp21–36

Musy, M., Wurtz, E., Winklemann, F. and Allard, F. (1999) *Generation of a Zonal Model to Simulate Natural Convection in a Room with Radiative/Convective Heater*, Draft report

Musy, M., Winklemann, F., Wurtz, E. and Sergent, A. (2002) 'Automatically generated zonal models for building air flow simulation: Principles and applications', *Building and Environment*, vol 37, no 8–9, pp873–881

Nitta, K. (1994) *Calculation Method of Multi-Room Ventilation: Memoirs of the Faculty of Engineering and Design*, vol 42, Kyoto Institute of Technology, Kyoto

Nitta, K. (1996) 'Study on the variety of theoretical solutions of ventilation network', *Journal of Architecture, Planning and Environmental Engineering – Transactions of AIJ*, vol 480, pp31–38

Nitta, K. (1997) 'Analytical study on a variety of forms of multi-room ventilation', in *International Symposium on Building and Urban Environment Engineering 1997*, Tianjin, China

Nitta, K. (1999) 'Variety modes and chaos in smoke ventilation by ceiling chamber system! Unexpected end of formula', in *Sixth International IBPSA Conference (BS 1999)*, IBPSA, Kyoto, Japan

Nitta, K. (2002) *Variety Modes and Chaos in Natural Ventilation or Smoke Venting Systems*, Draft report

Ohba, M., Kurabuchi, T., Endo, T., Akamine, Y., Kamata, M. and Kurahashi, A. (2004) 'Local dynamic similarity model of cross-ventilation: Part 2 – Application of local dynamic similarity model', *The International Journal of Ventilation*, vol 2, no 4, pp383–393

Oliveira, G. H. C., Coelho, L. S., Mendes, N. and Araújo, H. X. (2003) 'A MATLAB-based simulation tool for building thermal performance analysis', *IBPSA News*, vol 13, no 2, pp54–61

Orme, M. (1999) *AIVC Technical Note 51: Applicable Models for Air Infiltration and Ventilation Calculations*, AIVC, Coventry, UK

Orme, M., Liddament, M. W. and Wilson, A. (1998) *Numerical Data for Air Infiltration and Natural Ventilation Calculations*, AIVC, Coventry, UK

Pelletret, R. Y. and Keilholz, W. P. (1997) 'COMIS 3.0 – a new simulation environment for multizone air flow and pollutant transport modeling', in *Building Simulation 1997: Fifth International IBPSA Conference*, IBPSA, Prague

Persily, A. K. and Ivy, E. M. (2001) *Input Data for Multizone Airflow and IAQ Analysis*, NIST, Gaithersburg, MD

Press, W. H., Teukolsky, S. A., Vetterling, W. T. and Flannery, B. P. (1992) *Numerical Recipes in C: The Art of Scientific Computing*, second edition, Cambridge University Press, Cambridge

Rao, J. and Haghighat, F. (1993) 'A procedure for sensitivity analysis of airflow in multi-zone buildings', *Building and Environment*, vol 28, no 1, pp53–62

Ren, Z. and Stewart, J. (2003) 'Simulating air flow and temperature distribution inside buildings using a modified version of COMIS with sub-zonal divisions', *Energy and Buildings*, vol 35, pp257–271

Rodriguez, E. A. and Allard, F. (1992) 'Coupling COMIS airflow model with other tranfer phenomena', *Energy and Buildings*, vol 18, pp147–157

Rodriguez, E. A., Alvarez, S. and Cáceres, I. (1993) 'Prediction of indoor temperature and air flow patterns by means of simplified zonal models', in *ISES Solar World Conference*, Hungarian Energy Society, Budapest

Roulet, C.-A., Fürbringer, J.-M. and Borchiellini, R. (1996) 'Evaluation of the multizone air flow simulation code COMIS', in *Roomvent 1996: Fifth International Conference on Air Distribution in Rooms*, Yokohama, Japan

Sahlin, P. (1993) *IDA Modeller: A Man-Model Interface for Building Simulation*, Department of Building Services Engineering, Royal Institute of Technology, Stockholm, Sweden

Sahlin, P. (2000) *The Methods of 2020 for Building Envelope and HVAC Systems Simulation – Will the Present Tools Survive?*, EQUA Simulation Technology Group AB, Sweden

Sahlin, P. and Bring, A. (1993) *The IDA Multizone Air Exchange Application*, Swedish Institute of Applied Mathematics and Department of Building Services Engineering, Royal Institute of Technology, Stockholm, Sweden

Sandberg, M. (2002) 'Airflow through large openings – a catchment problem?', in *RoomVent 2002: Eighth International Conference on Air Distribution in Rooms*, The Technical University of Denmark and Danvak, Copenhagen, Denmark

Santamouris, M. and Asimakopoulos, D. (eds) (1996) *Passive Cooling of Buildings*, James & James, London

Santamouris, M. and Dascalaki, E. (1998) 'Prediction methods', in Allard, F. (ed) *Natural Ventilation in Buildings: A Design Handbook*, James & James, London, Chapter 3, pp63–157

Saraiva, J. G. and Marques da Silva, F. (1999) 'Atmospheric turbulence influence on natural ventilation air change rates', in *Indoor Air 99*, ISIAQ and AIVC, Edinburgh

Savic, D. and Walters, G. A. (1996) 'Integration of a model for hydraulic analysis of water distribution networks with an evolution program for pressure regulation', *Microcomputers in Civil Engineering*, vol 11, pp87–97

Schaelin, A., Dorer, V., van der Maas, J. and Moser, A. (1993) 'Improvement of multizone model predictions by detailed flow path values from CFD calculations', *ASHRAE Transactions*, vol 99, Part 2

Seifert, J. (2002) *Coupled Air Flow and Building Simulation for a Hybrid Ventilated Educational Building*, Dresden University of Technology, Dresden, Germany

Seifert, J. (2004) 'Wind-driven cross ventilation in buildings with large openings', Draft paper to be submitted to *Journal of Industrial and Wind Engineering*

Shearer, J. L., Murphy, A. T. and Richardson, H. H. (1971) *Introduction to System Dynamics*, Addison-Wesley Publications, Reading, MA

Sirén, K. (1997) 'A modification of the power-law equation to account for large-scale wind turbulence', in *18th AIVC Conference: Ventilation and Cooling*, AIVC, Athens, Greece

Sowell, E. F. and Haves, P. (1999) *Numerical Performance of the SPARK Graph-Theoretic Simulation Program*, LBNL, Berkeley, CA

SSPA (2003) *SIMNON Introduction and Tutorial*, SSPA Sweden AB Foundation, Chalmers University of Technology, Göteborg

Stewart, J. and Ren, Z. (2003) 'Prediction of indoor gaseous pollutant dispersion by nesting sub-zones within a multi-zone model', *Building and Environment*, vol 38, pp635–643

Stoer, J. and Bulirsch, R. (1993) *Introduction to Numerical Analysis*, second edition, Springer-Verlag, New York

True, J. J., Sandberg, M., Heiselberg, P. and Nielsen, P. V. (2003) 'Wind driven cross-flow analysed as a catchment problem and as a pressure driven flow', *International Journal of Ventilation*, vol 1, Hybrid Ventilation Special Edition, pp88–102

van Schijndel, A. W. M. (2003) 'Modeling and solving building physics problems with FemLab', *Building and Environment*, vol 38, no 2, pp319–327

Ventana Systems (2004) *Vensim 5.3*, Ventana Systems Inc, Harvard, MA

Vuolle, M. and Sahlin, P. (2000) 'IDA indoor climate and energy – a new generation simulation tool', in *Healthy Building 2000*, SIT Indoor Air Information Oy, Helsinki, Finland

Walker, I. S. and Wilson, D. J. (1994) 'Practical methods for improving estimates of natural ventilation rates', in *15th AIVC Conference: The Role of Ventilation*, AIVC, Buxton, UK

Wall, L. (2004) *Perl 5.8.0 Documentation*, programmer: Larry Wall, www.perl.com

Walter, G. G. and Contreras, M. (1999) 'Compartmental modeling with networks', in Bellomo, N. (ed) *Modeling and Simulation in Science, Engineering and Technology*, Birkhauser, Boston

Walton, G. N. (1982) 'Airflow and multiroom thermal analysis', *ASHRAE Transactions*, vol 88, Part 2

Walton, G. N. (1984) 'A computer algorithm for predicting infiltration and interroom airflows', *ASHRAE Transactions*, AT-84-11, no 3

Walton, G. N. (1988) *AIRNET: A Computer Program for Building Airflow Network Modeling*, NIST, Gaithersburg, MD

Walton, G. N. (1989) 'Airflow network models for element-based building airflow modeling', in *ASHRAE Symposium on Calculation of Interzonal Heat and Mass Transport in Buildings*, ASHRAE, Vancouver, BC

Walton, G. N. (1994) *CONTAM93 User Manual*, National Institute of Standards and Technology (NIST), Gaithersburg, MD

Walton, G. N. (1997) *CONTAM96 User Manual*, NIST, Gaithersburg, MD

Walton, G. N. (1998) *Notes on Simultaneous Heat and Mass Transfer*, Draught report, NIST, Gaithersburg, MD

Waters, J. R. and Brouns, C. E. (1991) 'Ventilation effectiveness – the AIVC guide', in *12th AIVC Conference: Air Movement and Ventilation Control Within Buildings*, Ottawa, Canada, Air Infiltration and Ventilation Centre, Coventry, UK

Weber, A., Koschenz, M., Holst, S., Hiller, M. and Welfonder, T. (2002) 'TRNFLOW: Integration of COMIS into TRNSYS TYPE 56', in *AIVC-EPIC Conference Proceedings*

Wolfram Research (2004) *MATHEMATICA-5*, Wolfram Research, Champaign, IL

Woloszyn, M., Duta, A., Rusaoven, G. and Hubert, J. L. (2000a) 'Combined moisture, air and heat transport modelling methods for integration in building simulation codes', in *RoomVent 2000: 7th International Conference on Air Distribution in Rooms*, University of Reading, UK, and Elsevier Science Publications

Woloszyn, M., Rusaouen, G., Rooux, J.-J. and Daguse, T. (2000b) 'Adapting block method to solve moist air flow model', *Mathematics and Computers in Simulation*, vol 53, no 4–6, pp423–428

Wood, D. J. (1981) *Algorithms for Pipe Network Analysis and Their Reliability*, Research Report No 127, University of Kentucky, Water Resources Research Institute, Lexington.

Wood, D. J. and Funk, J. E. (1993) 'Hydraulic analysis of water distribution systems', in Cabrera, E. and Martinez, F. (eds) *Water Supply Systems: State of the Art and Future Trends*, Computational Mechanics Publications, pp43–68

Wood, D. J. and Rayes, A. G. (1981) 'Reliability of algorithms for pipe network analysis', in *Journal of the Hydraulics Division*, ASCE

Wray, C. P. and Yuill, G. K. (1993) 'An evaluation of algorithms for analyzing smoke control systems', *ASHRAE Transactions*, vol 99, Part 1, pp160–174

Wurtz, E., Nataf, J.-M. and Winkelmann, F. (1999) 'Two- and three-dimensional natural and mixed convection simulation using modular zonal models in buildings', *International Journal of Heat and Mass Transfer*, vol 42, pp923–940

Wurtz, E., Deque, F., Musy, M. and Mora, L. (2001) 'A thermal and airflow analysis tool using simplified models based on the zonal method', in *CLIMA 2000: 7th REHVA World Congress – Indoor Environment Technology: Towards a Global Approach*, AICARR (Italian Association of Air Conditioning, Heating and Refrigerating Engineers), Napoli, Italy

Wurtz, E., Deque, F., Mora, L., Bozonnet, E. and Trompezinsky, S. (2003) 'SIM_ZONAL: A software to evaluate the risk of discomfort: Coupling with an energy engine, comparison with CFD codes and experimental measurements', in *Building Simulation 2003: For Better Building Design*, Technische Universiteit Eindhoven, International Building Performance Simulation Association (IBPSA)

Zhang, G., Morsing, S., Bjerg, B. and Svidt, K. (2000) 'A study on the characteristics of airflow in a full scale room with a slot wall inlet beneath the ceiling', in *RoomVent 2000: Ventilation for Health and Sustainable Environment*, University of Reading, UK, and Elsevier Science Publications

3

Ductwork, Hygiene and Energy

François Rémi Carrié and Pertti Pasanen

Introduction

Ductwork is a vital element of many ventilation systems, whether natural or mechanical. In fact, a duct system is the most common answer to the need for a ventilation system that involves the transport of ventilation air independently of room air motion. There are four major reasons for this need (Malmström, 2002):

1   to ease the control of airflow rates within the building;
2   to enclose polluted exhaust or extract air;
3   to keep supply air clean and conditioned; and
4   to make heat recovery from exhaust air possible.

For this, air ducts are commonly used in residential and non-residential buildings.

Health concerns are of central importance in duct systems since they are directly related to the primary function of a ventilation system – that is, to provide fresh and clean air to occupants. However, there are many other boundary conditions that must be dealt with, and Malmström (2002) proposes to classify those in four categories: energy use, minimization of cost, fire safety and acoustics.

Because of the potential health implications, ductwork system performance is an active field of research. During the last decade, issues of duct hygiene and energy use have been highlighted, probably encouraged by the increasing international concern for sustainable development.

In fact, Nordic countries had identified the need for research and development in setting up a framework for designing, installing and commissioning ductwork systems since the late 1950s – see, for instance, the AMA framework (VVS AMA98, 1998) – and their approach included health and energy issues. However, this concern broke through at the international level through pioneering studies that looked at the significance and the effects of dust deposition, microbial growth and deficiencies such as poor insulation or air tightness. The general consensus that can be drawn from these investigations is that ductwork systems merit further attention on health and energy issues, and that significant efforts are needed for improving commissioning and maintenance procedures.

The purpose of this chapter is to review the recent major contributions in these areas, and to identify future challenges.

Duct hygiene and health aspects

Intrinsic sources and contaminants

Ventilation systems and ductwork, as their largest component, are purpose built to deliver sufficient amounts of fresh air to occupied spaces. However, due to several reasons, the objective is not always achieved and ventilation systems may act as sources of emissions. Heating, ventilating and air-conditioning (HVAC) systems may have their intrinsic sources originating from the manufacture, transportation and storage of components; or the debris inside a ductwork may originate from a building construction site; or some dust and debris may accumulate during the operation period of the system. Poor design and maintenance procedures may increase the dust accumulation rate in the system. Typical contaminants and their sources are presented in Table 3.1.

Table 3.1 *Intrinsic emission sources and contaminants in heating, ventilating and air-conditioning (HVAC) systems*

Type of contaminant	Examples of sources	References
Volatile organic compounds (VOCs) and odours	Oils used in manufacturing processes of ducts and components, seals, adhesives and caulks	Girman et al (1993); Leovic et al (1993); Fransson (1996); Pasanen et al (1995); Morrison and Hodgson (1996); Asikainen et al (2002)
Construction residues and soil contaminants	Improper covering during transportation and storage; deterioration of coatings; cutting dust with a side grinder	Robertson (1988); Rothenberger et al (1989); Pasanen et al (1992); Fransson (1996); Luoma and Kolari (2002)
Lubricating oils; grease	Fan, motors and bearings in air stream	Levin and Moschandreas (1990); Pasanen et al (1995)

Source: Pasanen (1998)

Sources of volatile organic compounds (VOCs) in new installations

HVAC ducts and joint components are usually made of galvanized sheet metal that needs corrosion protection to avoid oxidation of the zinc surfaces during storage. The metal is protected either by mineral oil-based products, or with chromic acid treatment, or with organic passivation to obtain raw material with a low oil content. The chromic acid-treated material is mostly used in the manufacture of the spiral seamed ducts; but the components that need more machining or sharpening also require external lubrication to decrease friction between the tools and sheet metal. The typical lubrication products developed for machine tools usually leave a thin layer of oil residue that evaporates slowly from the surface of the component. The evaporation rate depends upon the lubrication product used, and it is more significant during the first few months after installing the components. Vegetable oil-based products are also used; but they do not necessarily promote the hygiene of the HVAC system due to oil oxidation. They also offer some nutrients for micro-organisms (Pasanen et al, 1995). If oil-based lubricants are used in the production of spiral seam ducts, the oil residue flows to the bottom of the duct where it forms a sticky residue covering approximately 10 per cent of the duct surface in smaller duct diameters.

Sealants, adhesives, caulks and gaskets used in HVAC systems are usually made of synthetic organic polymers, which typically contain latex acrylic, butadiene rubber, neoprene rubber, butyl rubber, silicone and polyurethane (Leovic et al, 1993). These materials may emit VOCs, including solvents, unreacted monomers, reaction by-products and additives – for example, plasticizers, antioxidants and stabilizers (Morrison and Hodgson, 1996). As for oil residues, the emission rates from these kinds of soft materials normally decrease within a few months to less than half of the original level and therefore are a concern only in new and renovated buildings. To speed up the decrease of the emissions from the surfaces of the ventilation system, it is recommended to keep the ventilation of the building working constantly without breaks during the first year.

Usually, contamination by oil residues in older HVAC installations is low and the existence of emissions is very difficult to prove because the similar hydrocarbons are present in the outdoor air. In some studies, the HVAC system has proven to be an occasional source of VOCs (Mølhave and Thorsen, 1991; Sundell et al, 1993); but the particular sources are not identified.

Construction residues and soil debris (new installations)

Dust concentrations are often elevated in building construction sites, where concentrations as high as $2.9mg/m^3$ have been measured (Luoma, 2000). This leads to dust deposition onto every horizontal surface, and if open-end air ducts are present, the dusty air is driven with airflow into the ducts. The debris may be composed of construction dust, sand, pieces of concrete (Finke and Fitzner, 1993), wood shavings, and even lunch packets and drink cans (Robertson, 1988). Brick tiles have been found in air ducts, where they have been used to restrict airflow when balancing flows (Pasanen et al, 1992). Sheet metal filings formed during cutting of the ducts make a narrow dust stripe on the bottom of ducts near cleaning hatches, junctions and duct endings. Such debris increases the roughness of a duct surface and accelerates dust accumulation (Wallin, 1994).

The amount of accumulated dust in recently installed air ducts depends a lot upon working practices during the building and installation work. On the components made and stored in factories with a proper quality control, the average amount of accumulated dust is low: $0.1g/m^2$ (Luoma, 2000). The components will gain most of their dust load during the installation phase (storage at the building site and installation process). In a study reported by Luoma (2000), two ductwork systems were built according to the cleanliness classification procedure

Table 3.2 *Contaminants in HVAC systems and their sources*

Type of contaminant	Typical example of source	References
Dust	Outdoor air pollution; low efficiency of filter or frame leakage; construction material; reversed airflow	Nielsen et al (1990); Downing and Bayer (1991); Mølhave and Thorsen (1991); Krzyzanowski (1992); Pasanen et al (1992); Pejtersen et al (1992); Kjaer and Nielsen (1993); Wallin (1994); Pasanen (1995)
Fibres	Fibre shedding; fibre release from new filters	Esmen et al (1980); Gamboa et al (1988); Morey and Shattuck (1989); Samini (1990); Schumate and Wilhelm (1991); Price and Crumb (1992); Jacob et al (1993); Christensson and Krants (1994)
Bacteria	Humidifier; cooling tower; moist insulation material; condenser drip pan; drain pan	Ager and Tickner (1983); McJilton et al (1990); Morey and Feeley (1990); Nielsen et al (1990); Morey and Williams (1991); Hugenholtz and Fuerst (1992); Morey (1992); Gilmour et al (1995); Hung et al (1995); Maus et al (1996); Neumeister et al (1996); Sugawara (1996)
Fungi	Moist and dirty insulation; plenum; duct surface; cooling coil; drain pan; moistened air filters; bird droppings	Ager and Tickner (1983); Bernstein et al (1983); Pellikka et al (1986); Morey and Williams (1990, 1991); Sverdrup and Nyman (1990); Ahearn et al (1992); Morey (1992); Martiny et al (1994); Hung et al (1995); Kemp et al (1995a, 1995b); Neumeister et al (1996); Nielsen et al (1990); Chang et al (1996); Foarde et al (1996a, 1996b); Iida et al (1996); Möritz (1996); Parat et al (1996); Pejtersen (1996a); Sugawara (1996)
Volatile organic compounds (VOCs), odorous compounds and polycyclic aromatic hydrocarbons (PAHs)	Filters, sound adsorbers and insulation material; deposited dust	Fanger et al (1988); Rothenberger et al (1989); Bluyssen (1990); Downing and Bayer (1991); Mølhave and Thorsen (1991); Kjaerboe and Strindehag (1993); Kjaer and Nielsen (1993); Sundell et al (1993); Teijonsalo et al (1993); Schleibinger et al (1995); Bluyssen et al (1996); Morrison and Hodgson (1996); Pejtersen (1996a, 1996b)
Cleaning agents and biocides	Use of biocides, disinfectants and deodorizers	Burge et al (1989); Downing and Bayer (1991); Luoma et al (1993); Pasanen et al (1997)
Boiler additives	Use of anti-corrosives; biocides; slimicides; pH control neutralizers	National Research Council (1983); Morey and Shattuck (1989); Halas (1991a, 1991b)

Source: Pasanen (1998)

established by FiSIAQ (2001) in order to obtain a 'clean' system, while no specific procedure or criteria were set for a third system. The averages of dust accumulations were 0.6g/m² and 0.7g/m² in the two first buildings, whereas dust accumulation was as high as 3.1g/m² in the third case. The power of requirements for cleanliness is also supported by data presented by Holopainen et al (2002c), where the mean amount of accumulated dust was 0.9g/m² in 139 buildings where specific requirements were set. This index reached 2.3g/m² in 44 buildings that did not have special requirements on cleanliness. Note also that in Finland, before any national guidance to build clean ventilation systems was set, the amount of accumulated dust in new supply air-duct installations was about 5g/m² in buildings aged less than one year (Pasanen, 1996).

Emission characteristics of the components

Dust and debris in heating, ventilation and air-conditioning (HVAC) surfaces

During the normal use of the ventilation system, some particulate contaminants are driven into the system either from internal or external sources, which increases the load of organic and inorganic dust layer. Typical contamination problems in HVAC systems are presented in Table 3.2.

Dust in air ducts

Table 3.3 summarizes dust accumulation results from various studies. Note that the mass of accumulated dust on the duct surfaces is determined by using different methods, which makes it difficult to compare the results. Note also that in many studies, ventilation systems have not been cleaned regularly; therefore, the range in the amounts of accumulated dust is wide even within the same study.

Table 3.3 *Amount of accumulated dust on supply air-duct surfaces in various studies*

Type of building	n	Age	Amount of dust accumulation		Annual deposition	Method[†]	Reference
			Mean	Range	g/m² per		
		Years	g/m²	g/m²	year		
School; office	13	3–29	6.8	1.1–50.9	0.7	Vac. met. blade	Nielsen et al (1990)
Residential	33	0–45	0.2	limit of detection–2.7	< 0.1	Vac. National Air Duct Cleaners Association (NADCA)	Ager and Tickner (1983)
School; office; public building	14	3–34	13.2	1.2–158	1	Vac. pl. blade	Pasanen (1995)
Residential	23	2–16	1.2	0.2–3.9	n/a	Vac. brushing	Kalliokoski et al (1995)
Not reported	5	19–37	2.6	1.9–3.0	0.2–0.3	Tape	Fransson (1996)
Not reported	4	22–32	7.5	n/a	n/a	Wiping	Ito et al (1996)
Office; public building	8	0–1	5.1	0.2–8.4	n/a	Vac. pl. blade	Pasanen (1996)
Public building	13	3–30	18.8	4–131	1.1	Wiping with sol.	Küchen (1998)
Public building	17	3–30	7.0	0.2–82	0.5	Wiping	Küchen (1998)
School; office; day care centre; public building	9[‡]	New	0.9	0.4–2.9	n/a	Vac. pl. blade	Holopainen et al (2002b)
	9	New	2.3	1.2–4.9			

Notes:

† vac. = vacuum sampling and gravimetric analysis; met. blade = sample loosened with blade; pl. blade = sample loosened with plastic blade; tape = sample taken to sticky tape; wiping = sample wiped with a pre-weighed cloth; wiping with sol. = sample wiped with a cloth immersed with solvent

‡ First line refers to systems build according to cleanliness classification (FiSIAQ, 2001).

n/a = not available

Sources: specified in Reference column

The deposition rate of airborne dust particles is affected by many mechanisms. The particle size, shape and density (aerodynamic size) and the properties of airflow are the most important factors that determine the settling velocity onto duct surfaces. The main mechanisms are gravitational sedimentation; inertial impaction; Brownian diffusion; and turbulent diffusion. The large particles ($d_p > 1\mu m$) are primarily affected by the first two mechanisms that transport particles to the bottom of the duct and onto the surfaces close to a sharp radius of curvature of the streamlines (bends, diffusers, dampers, etc.). At higher air velocities, the inertial impaction is a dominant mechanism in particle deposition. The deposition of smaller particles is controlled by the two latter mechanisms and deposition increases on rough surfaces, especially with high air velocities (Wallin, 1994). As a layer of particles settles on the surface, it will increase the accumulation rate of the particles on the surface.

The main particle deposition occurs in the air intake chamber/duct and on the air filters. Furthermore, the accumulation rate has a negative correlation to the horizontal distance from the air handling unit (AHU); but the correlation is not as clear for the vertical ducts if the AHU is located on top of the building (Pasanen, 1995). The effect of filtering efficiency is also seen in the annual accumulation of dust, both in the field (Pasanen, 1995) and in the theoretical calculations (Pasanen et al, 1992). Roughly, these studies show that the dust deposition rate in ductwork is reduced by a factor of ten if efficient F8 filters are used instead of G3 filters (see Figure 3.1). This is very significant for the cleaning interval. Efficient air filters are the most effective way of diminishing particle deposition on heat exchangers, sound attenuators, ducts and their components (Pasanen et al, 1992; Fransson, 1996).

Fibres

Some non-mineral fibres found in air ducts belong to the normal part of dust derived from indoor sources, such as clothing, furniture and paper constituents (Rothenberger et al, 1989; Jaffrey, 1990). Some man-made vitreous fibres (MMVFs) found in air ducts are derived from insulation materials installed in other parts of the building (Van der Wal et al, 1989). However, a common fibre contamination of air ducts is caused by damaged sound attenuators in air handling units. For example, gradual deterioration of a duct liner has been considered the most problematic contaminant in a multi-storey building (Samini, 1990). Forces from airflow may overstrain the components in the

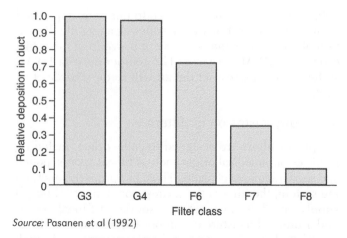

Source: Pasanen et al (1992)

Figure 3.1 *Relative dust accumulation rate with different filtering efficiencies*

ductwork during use. In an AHU, a loose airflow steering blade chafed the surface, perforated the aluminium-coated sound attenuation surface and released a remarkable quantity of fibres from insulation material into the air supply stream. Occupants subsequently suffered from irritation symptoms, and the source was eventually found.

Overall, the highest concentrations of fibres (around one fibre per cubic centimetre – f/cm^3) have been measured during the handling (for example, installing) of insulation or acoustic material containing MMVF (Jaffrey, 1990; Jacob et al, 1993). The fibre release from new thermal or acoustic insulation materials has been studied in test facilities. Glass wool insulation exposed to air velocities of 12–25m/s released fibres in concentrations of $0.5 \times 10^{-3} f/cm^3$ (Gamboa et al, 1988). Air velocity strongly affects the release of fibres from surfaces. At velocities of 7–8m/s and 5m/s, fibre emissions from acoustical glass and rock wool were two and three orders of magnitude lower.

Fibre emissions from glass fibre filters have also been discussed as a potential health hazard in indoor environments. The results of 42 filters tested at nominal air velocities (0.079m/s and 0.131m/s) showed that fibre emissions decreased rapidly during the first 15 minutes of use, after which the rate of decrease slowed (Esmen et al, 1980). Similarly low emissions from EU7 glass fibre filters have also been reported by Schumate and Wilhelm (1991) and Christensson and Krants (1994). Glass fibre released from duct liners, duct boards (Gamboa et al, 1988) and recently installed fibre filters did not have a significant impact on indoor air fibre concentrations according to several studies (Esmen et al, 1980; Schumate and Wilhelm 1991; Christensson and Krants, 1994).

Bacteria

In HVAC systems, problems associated with bacteria mainly focus on humidifiers (McJilton et al, 1990; Morey and Feeley, 1990) and other water reservoirs. Contaminated moist or wet insulation may act as a reservoir for bacteria where concentrations of bacteria (for example, *Bacillus* and *Corynebacterium*) have ranged from 10^4 to 10^6 colony-forming units per square centimetre (CFU/cm^2) (Morey and Williams, 1991). Elevated bacterial concentrations (104 CFU/cm^2) have also been found on condensate drip pan surfaces during the cooling season; lower bacterial concentrations were found on cooling coil surfaces (Hung et al, 1995). More serious problems occurred in a building where people suffered from symptoms of humidifier fever. Here, water-damaged furnishings and debris from the HVAC system contained *Thermoactinomycetes* and a protozoan, *Acanthamoeba polyphaga* (Morey and Feeley, 1990).

Bacterial concentrations of 50–5000CFU/g have been found under dry conditions in dust collected from air ducts of non-problem buildings (n = 13) (Nielsen et al, 1990). Concentrations were slightly higher in dust in systems that recirculated air, indicating the existence of a bacterial source in these buildings (Nielsen et al, 1990). Higher levels of bacteria (10^4–$10^5 CFU/g$) were found in dust collected from HVAC systems in Japanese office buildings (Sugawara, 1996). Bacterial concentrations are probably related to the height of an air intake – for example, high bacterial levels in air ducts have been found in systems with a low (2m) height of air intakes (Laatikainen, 1993). Surface densities of bacteria ($10^3 CFU/cm^2$) have also been detected on heat exchangers (Nyman and Sandström, 1991).

Air filters usually effectively collect microbial particles. The number of heterotrophic bacteria in air filters of public buildings ranges typically from 10^3–$10^4 CFU/g$, and the highest counts elevate up to $10^5 CFU/g$ (see, for example, Martikainen et al, 1990). However, it is reported that air filters are not a favourable growth media for bacteria (Martiny et al, 1994; Maus et al, 1996).

Fungal growth in dust and in moist insulation materials

The surface density of viable fungal spores in the dust found in HVAC systems has varied from detection limit to $10^6 CFU/m^2$. The most abundant species recognized in the dust are *Penicillium*, *Cladosporium*, *Aspergillus*, *Aureobasidium*, *Chaetomium* and *Alternaria*, all of which may originate from the outdoor environment (Sverdrup and Nyman, 1990; Valbjørn et al, 1990; Kalliokoski et al,

1995; Sugawara, 1996). In non-problem systems, the levels of fungal spores on the supply duct surfaces are usually low – that is, a few colony-forming units per 100cm^2.

In several case studies, fungal spore concentrations have been analysed in samples that have usually been taken after a problem in a HVAC system is suspected. Fast-growing *Penicillium* species have been identified on surfaces of wet and dirty insulation with total counts of 0.12–1.1 × 10^5 CFU/cm^2 (see Morey and Williams, 1990, 1991). Similar levels of *Cladosporium* species have been determined on surfaces of wetted insulation material (Morey, 1992).

Handling of contaminated material greatly affects the airborne concentration of fungal spores. In one case, concentrations of *Cladosporium*, *Epicoccum* and *Penicillium* were 5000 CFU/g in the insulation of an induction unit. When the system operated, the concentration of airborne *Cladosporium* in a room where complaints were reported was only 50 CFU/m^3. However, the concentration increased dramatically up to 22,000 CFU/m^3 when contaminated insulation material was removed and the ventilation was simultaneously on (Morey and Williams, 1991). In another case, dirty insulation material in an HVAC system was suggested to be a reservoir or, in addition to outdoor air, a secondary source for *Cladosporium* spores (Morey and Williams, 1991). Other studies also have revealed high counts and proportions of *Cladosporium* on a heavily contaminated HVAC system (Ahearn et al, 1991), and *Cladosporium* and *Penicillium* on surfaces of drain pans (Iida et al, 1996).

Heavy fungal contamination has also deteriorated the quality of supply air with noxious odours in the HVAC system of a new building. Although the insulation material appeared dry, the surface was covered with a dense mycelial web of *Penicillium* and *Talaromyces* spp. A xerophilic *Eurotium herbariorum* had also colonized the plastic facing of the glass fibre material (Ahearn et al, 1992). This case demonstrates that xerophilic fungi may grow in conditions without condensed water if the relative humidity (RH) of air is high enough.

According to laboratory studies, the amount of dust significantly affects the fungal growth rate on the glass fibre insulation material and especially on the sheet metal surfaces. A realistic low level of 5–10g/m^2 of dust on porous insulation hastened fungal growth at 97 per cent of RH within six weeks, but did not on galvanized sheet metal. Such a support of the growth could not be seen at lower relative humidities on either material. When the soiling level was ten times higher, the growth was also seen at lower humidities of 85 per cent within four to five weeks, and even faster at higher humidities (Chang et al,

1996). The material used for five to ten years supported growth, also indicating that, besides cleaning, the limitation of moisture is the best way of preventing microbial growth on HVAC surfaces. The temperature in HVAC systems does not restrict the growth factor (Foarde et al, 1996a, 1996b).

Micro-organisms on air filters

Supply air filters are designed to collect dust, including particles of biological origin such as fungal spores, pollen and seed fragments of plants. Thus, in a properly functioning system, the filter is the dirtiest component. Although supply air filters may act as a source of fungal spores under unusual moisture conditions (Bernstein et al, 1983; Kemp et al, 1995b), or when fungi have grown through the filters (Elixmann et al, 1987), most studies have shown that air filters decrease the airborne concentrations of spores (Reponen et al, 1989; Pasanen et al, 1990b; Kuehn et al, 1996; Möritz, 1996; Neumeister et al, 1996; Parat et al, 1996; Pejtersen, 1996b). A used glass fibre filter in field conditions exposed to 40 per cent or 80 per cent RH for 18 weeks did not release viable fungal spores (Pejtersen, 1996b). Even higher humidity (90 per cent) did not support growth on a polymer filter in the air stream, but a slight growth was observed on a glass fibre filter after a year. The increase of artificial nutrients on the filter showed the potential for rapid growth (Kemp et al, 1995b). Besides limited moisture, the lack of nutrients limits fungal growth on air filters, and the airflow seems to decrease the viability and growth of micro-organisms on the filters in field conditions.

The potential for microbial growth on dusty air filters has been tested in many laboratory studies (Ruden and Botzenart, 1974, cited in Möritz, 1996; Pasanen et al, 1990a; Pasanen et al, 1991b; Kemp et al, 1995a, 1995b; Möritz, 1996). Tests have demonstrated that, in undisturbed conditions at relatively high humidity, the growth of micro-organisms takes place without any external inoculation of spores or nutrients. When temperature decreases below the optimum temperature (20–25° C), the growth rate of micro-organisms decreases and more water activity is needed to support growth (Pasanen et al, 1991a). At sufficiently high relative humidity (> 90 per cent), fungal growth may expand through the filter medium within a few weeks (Kemp et al, 1995a). Ecological competition for living space also occurs on filters. Fast-growing species such as *Penicillium* spp will easily replace the original abundant species, such as *Cladosporium*, *Aspergillus* and *Aureobasidium*, present in the outdoor air (Pasanen et al, 1991b).

Despite continuous settling of fungal spores and bacteria on air filters, counts of viable spores do not

Table 3.4 *Average proportions of odour pollution loads from building materials, ventilation systems and the occupants in different building types*

Building category	Pollution load						Reference
	HVAC system		Materials		Occupants		
	olf/m²	%	olf/m²	%	olf/m²	%	
Offices	0.25	55	0.12	27	0.08	18	Fanger et al (1988)
Assembly halls	0.28	39	0.32	45	0.11	16	Fanger et al (1988)
Day care centres	0.32	42	0.07	9	0.38	49	Thorstensen et al (1990)
Schools	0.20	39	0.11	22	0.20	39	Pejtersen et al (1991)

Note: The pollution load in olf/m² is calculated to correspond to the floor area of a ventilated space. 'olf' is a unit for measuring air pollution in terms of odours perceived by humans, developed by Fanger et al (1988).

Source: Pasanen (1998)

increase continuously with loading in normal use. Martiny et al (1994) determined a survival time of three days for micro-organisms in field conditions. The study was continued by Neumeister et al (1996), who revealed that survival times differ with species, and a common outdoor air fungal species, *Cladosporium* spp, was the most abundant. Laboratory tests have shown that micro-organisms that produce spores for proliferation or to survive are able to survive very well; however, bacteria with lower resistance to dry conditions (*Micrococcus luteus* and *Escherichia coli*) lose their viability from 5 per cent up to 70 per cent within an hour, and 99–99.9 per cent in five days' exposure to airflow (Maus et al, 1996).

The existence of micro-organisms on air filters can be concluded as follows. In operation, air filters collect a number of viable spores and bacteria, as well as particles of inorganic and organic origin that compose a suitable medium for microbial growth if physical conditions and water are available. The airflow through the filter decreases the viability of the micro-organisms. Varying the airflow within the day causes stress for micro-organisms; during the shut-off period of a ventilation system (nights and weekends), the spores will germinate and grow. During the day when the system is functioning, the micro-environment becomes disadvantageous and the sprouting growth is destroyed (Möritz, 1996). Considering these facts, the existence of micro-organisms on air filters is a relatively common contamination of HVAC systems, and identification of other sources is more important if the HVAC system is suspected as a source of micro-organisms.

Odour evaluations

In a pioneering study by Fanger et al (1988), ventilation systems contributed to an average of 42 per cent of the total sensory load in office buildings, based on evaluations by an untrained panel. However, individual odour sources were not evaluated. The same method was used in investigations of offices (Fanger et al, 1988; Pejtersen et al, 1990; Bluyssen et al, 1996), assembly halls (Fanger et al, 1988), schools (Thorstensen et al, 1990), and day care centres (Pejtersen et al, 1991). These studies point out that ventilation systems often contribute a major part of the total sensory pollution load.

The distribution of odour pollution from major sources in buildings with different uses from three studies is presented in Table 3.4. Occupants are not necessarily the most important pollution sources, an assumption upon which most ventilation standards are based. Rather, emissions from building materials and the ventilation system are the main odour sources, although considerable variation is seen.

Odours in ventilation systems have been studied in detail by Pejtersen et al (1989). Unclean supply air filters, rotary heat exchangers and humidifiers were the most significant pollution sources in HVAC systems. These results have been supported by Finke and Fitzner (1993) and Ishikawa et al (1996). Laboratory evaluations have indicated significant odour emissions from uncleaned new and old air ducts, while emissions from cleaned new air ducts were low (Torkki and Seppänen, 1996). Odour emissions from cleaned components of AHU were lower than before cleaning; thus, cleaning of AHU improves the perceived air quality. However, dirty components such as uncleaned ductwork may cancel improvements (Björkroth et al, 1997a, 1997b).

Odour emissions from supply air filters used for a few months to a year have been reported to increase the perceived air pollution load (Bluyssen, 1990; Hujanen et al, 1991; Finke and Fitzner, 1993; Pejtersen, 1996b). The strength of the emissions appears to depend upon the amount of accumulated dust (Bluyssen, 1990; Pejtersen, 1996b) and composition of the accumulated dust (Hujanen et al, 1991). Pejtersen et al (1996b) did not find effects of the airflow on odour emissions and there was no

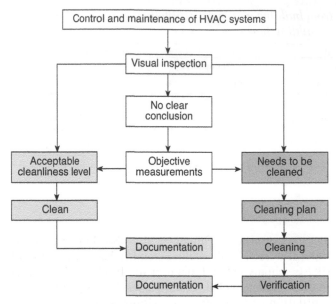

Source: Holopainen et al (2003a)

Figure 3.2 *A flow chart of the phases included in the inspection of cleanliness*

difference between filters used at 40 per cent or 80 per cent RH. At the moment, no relationship between the odour perception and total concentration of volatile organic compounds has been found.

Evaluation of the cleanliness of HVAC systems

Inspection of functionality and cleanliness of HVAC systems is a part of proper maintenance and cleanliness that is all the more important since health and comfort demands are set at high levels. Basically, the inspection of cleanliness may arise in two instances:

1   to check if the HVAC system is dirty and needs to be cleaned (exceeding a 'trigger' level); and
2   to evaluate cleanliness after the cleaning work (effectiveness of cleaning).

In the first case, the verification methods must be so that:

• both landlords and tenants will accept the result of the inspection; and
• the result of inspection gives reliable grounds for ordering cleaning work.

In the second case, the verification methods are needed by the cleaning company for quality control of their work, and also by the building owner as a customer in order to

verify the cleaning result. A flow chart of an evaluation procedure to detect the need for cleaning is presented in Figure 3.2.

Visual inspection

Different methods are applied to evaluate the cleanliness of HVAC systems. Objectivity and efficiencies of the methods vary. Visual inspection is a fast method of evaluating cleanliness, and it is usually good enough to detect, for example, microbial growth on water reservoirs or deposits from major malfunction of filtration. The visual inspection may be assisted with technical devices such as cameras, endoscopes and remote control cameras with video footage (Loyd, 1997). In some references, the visual inspection developed is more easily repeated by using special forms and notebooks, which make the inspection more systematic (Lavoie and Lazure, 1994; HVCA, 1998b). The visual inspection method is sometimes considered too subjective to give reliable results. If the visual inspection is done systematically, and especially if reference scales are used, the objectivity of the method will improve (Holopainen et al, 2002a; Asikainen et al, 2003; Holopainen et al, 2003a, b).

According to the procedure used in Finland, the inspection is started with a systematic visual inspection in which the dustiness of air ducts is compared to the reference scale of six photos in which the dustiness is also expressed as numerical values in g/m^2. The inspector estimates the amount in a numerical scale so that the result can be compared with the limit values. In cases where the system is clearly clean (for example, the average amount of dust is less than $0.5g/m^2$), or when the ductwork is clearly dirty (for example, the average amount of dust is more than $5g/m^2$), visual inspection by a trained person is a sufficient method to evaluate the need for cleaning. If the building owner and contractor do not agree whether the system should be cleaned or not, the objective methods (described later) should be used for evaluating cleanliness.

In a Canadian guide on the prevention of microbial growth in ventilation systems, the checklist includes components from outdoor air intakes, different sites in an AHU, and both supply and exhaust air ducts with peripheral units. The cleanliness is classified in a scale from 1 to 4 in which 1 refers to very clean and 4 signifies some stage of reduction in airflow (Lavoie and Lazure, 1994).

Gravimetrical methods

Most of the quantitative methods to verify cleanliness of ventilation systems are based on the measurement of the mass of the dust and debris deposited on a known surface area. In the *filter sampling method*, dust vacuumed on a filter is weighed either without filter housing, or weighed together with the housing. In the latter method, the measurement may be biased by the dust deposited on the walls of the filter housing. Several methods have been developed to loosen the dust from the surface. The dust may be loosened with the aid of a metal blade (Nielsen et al, 1990), which may also take some metallic zinc particles from rough surfaces (Fransson, 1996). Although soft plastic scrapers are not so effective for the tightly fastened dust, they are recommended because they leave the metal surface untouched (Pasanen, 1995). The mouthpiece of a filter cassette is also used for loosening the dust from a rejected area with a 100cm² flexible template. In one standardized method, the dust is loosened by air stream in a narrow slit between the template and filter holder. The method takes the loose particles to the sample (NADCA, 2003), and thus the result of the method describes the amount of potential particles that can be driven into the air stream, rather than the total dust deposition if the dust is tightly fastened on the surface. This method was developed to verify cleanliness after the cleaning.

The sample may be taken without vacuuming with the *wiping method* on the filter or filter-like cloth (Ito et al, 1996; Kumagai et al, 1997). The use of a solvent may augment the loosening of the dust from the surface, which makes the method very effective, especially for greasy solids (Fitzner et al, 1999; Müller et al, 1999).

A *sticky tape* is also used to collect deposited dust particles from the surface (Fransson, 1996, Holopainen et al, 2002a). The tape is weighed before and after collection of the dust, and the difference of the mass is used in calculating the dust density on the surface. The shape and dimensions of the tape keep the sampling area constant. The method is rapid if the balance is used in the field. According to preliminary studies with different tapes, the moisture and hygroscopicity of the tape material affects the reliability of the method. On very dusty surfaces, the collection capacity of the tape with dusty surfaces does not yield objective values. However, tests on surfaces with recently deposited dust revealed that the recovery is good from surfaces with dust accumulation levels lower than 4g/m² (Pasanen, 1999).

Sampling site

Sampling of the site and size of the surface is selected and determined variously in different studies. The National Air Duct Cleaners Association (NADCA) standard method determines a constant 100cm² area, and the distance from the surface is determined by the

thickness (1mm) of the template. In most studies, the sample is collected from the duct bottom because most of the accumulation occurs at the bottom surface of the duct. In circular air ducts, the definition of the bottom is not so clear; therefore, the diameter of the duct affects the broadness of the accumulation area. In Finnish studies (Pasanen et al, 1992; Lahtivuori, 1996), the sample area is determined so that one of the lower-quarter sectors from the lowest line of the duct to the widest line of the duct is chosen as the sampling area. This means that different sizes of templates are needed for different duct diameters.

Comparing the methods

Loosening efficiencies of the evaluation methods vary substantially (see Table 3.5) (Fransson, 1996; Fitzner et al, 1999; Holopainen et al, 2002a). The collection recoveries (efficiencies) have been determined only for cloth wiping and NADCA standard methods. For the cloth wiping method, average recoveries of dust varied from 87 to 95 per cent (Ito et al, 1996). The recoveries depend upon the surface to be sampled. Those of the NADCA method tested using typical surfaces found in HVAC systems were 70 per cent on galvanized sheet metal, 40 per cent on duct liner and 16 per cent on fibre board surfaces (Anon, 1995). Fitzner et al (1999) have determined efficiency factors by comparing the other methods based on the vacuuming technique to their cloth-wiping method with solvent (see Table 3.5). The use of propanol as a solvent increased the efficiency of the collection of the cloth even more than scraping with a blade, which was ranked as the most efficient method in the Fransson (1996) study.

The performance of dust collection efficiency of gravimetrical dust tape and filter methods has been compared in laboratory conditions with dust collected from used ventilation filters. In the dust density range used, 0.3–15g/m², the gravimetrical methods have given equal results in the sub-range of 0.3–4g/m². At higher levels, the dust tape did not have enough capacity to take all the dust from the surface. Thus, the tape method could be used for purposes where the amount of dust is low or the method is used only for detection if the amount of dust exceeds the set trigger value (Pasanen, 1999).

Methods for sampling microbial contaminants

The amount of microbial contamination is usually determined with the *cultivation method* that also enables the identification of the genera and species of micro-organisms. The determination of the fungal spore and bacteria counts can be done *from the dust sample* collected as described previously. The only restriction is

Table 3.5 *Relative efficiencies of the sampling methods developed for the solid deposits in the HVAC systems*

Method	Standard or note	Efficiency of the method
Wiping with cloth	Solvent	1
Filter with vacuum	Scraping with blade	0.9
Wiping with cloth	Japanese Air Duct Cleaners Association (JADCA)	0.5
Tape	Gravimetric	0.35
Filter with vacuum	With brush	0.15
Filter with vacuum	Scraping with filter holder	0.1
Filter with vacuum	National Air Duct Cleaners Association (NADCA)/ Heating and Ventilating Contractor's Association (HVCA)	0.02

Source: Fitzner et al (1999)

that the mass of the sample must be large enough, at least 100mg, for reliable determination. In this method, the sample is mixed and shaken up in a known volume of dilution water from which it is plated on suitable nutrient agar for bacteria and fungi.

Direct counting of spores or microbial cells with the aid of microscopy is usually impossible because of the high density of dust particle with various light-reflecting properties in the samples. The cultivation method is also used for *water samples* from humidifiers or other water reservoirs. The *insulation and other soft material samples* can be treated as dust samples.

The surface sample can be collected with the *wiping method,* in which a known area is wiped crosswise with a cotton wool stick wetted in sterile dilution water. The sample is cultivated as the dust sample. Both methods give results in *colony-forming units per square meter* (CFU/m^2) if the dust sample is collected from a known area. From flat surfaces, the sample may also be taken with a *contact method.* Whichever method is used for microbial analysis, attention should be paid to preventing the contamination of samples during sampling and during treatment of the sampling instruments. Bacteria, in particular, may originate from the person who takes the samples.

Conclusions

In many countries, an increasing interest in ventilation system cleanliness exists in order to maintain high indoor air quality. This offers many kinds of business challenges, such as duct cleaning work, as well as opportunities for industries that manufacture the cleaning devices. Objective methods to evaluate the cleanliness of HVAC systems are needed by building contractors and landlords in order to inspect if the HVAC systems need to be cleaned; such methods must also be used by the cleaning companies during their quality control.

Surfaces in HVAC systems carry many kinds of debris. Solid particles originate from the original building construction or from outdoor air during the normal use of the building. Contaminants may also contain living cells or microbial colonies supported by exceptional moisture sources, as well as oil residues from the time of manufacturing the components. All of these different groups of contaminants need methods of their own. A list of them is displayed in Table 3.6. Note that different methods proposed have different efficiencies (see Table 3.5).

Energy performance

The international context of the past decade, emphasizing energy conservation as a key element of any governmental energy policy, has highlighted concern for the energy performance of duct systems. The industry, practitioners, designers, research institutes and policy-makers have contributed to significant progress, whether in learning, implementing or developing new concepts and products to improve the energy performance of duct systems. In fact, these systems account for a large fraction of the energy use in a building, and a significant body of literature exists that shows that there are great energy-saving opportunities in this field. These are linked to various duct design issues mentioned in Table 3.7 and discussed below. Table 3.7 contains the major metrics related to the issues of concern, as well as key references.

Low pressure drops: Layout and sizing

The duct system layout has a major influence on pressure drop and, therefore, on the fan energy needed to transport the air through the ductwork. Ideally, the duct system layout should be integrated within a building design optimization scheme to account for issues such as space demand, energy use, acoustics and fire safety.

Table 3.6 *Summary of the methods used for evaluating the cleanliness of HVAC systems*

Method	Units	Note
Visual inspection		
Non-systematic inspection		No scaling for cleanliness; subjective
Systematic		Grades for the cleanliness; semi-objective
		Optical and electrical devices may be used to store the views and records
Quantitative methods for dust		
Filter sampling	(g/m^2)	Most common; repeatable
Cloth wiping	(g/m^2)	Effective when used with solvent
Tape sampling	(g/m^2)	Suitable for low levels ($< 4g/m^2$)
Quantitative methods for micro-organisms		
Cultivation of dust sample	(CFU/g)	Identification of the cultivable species
Cultivation of liquid sample	(CFU/ml)	Identification of the cultivable species
Cultivation of swab sample	(CFU/m^2)	Identification of the cultivable species
Cultivation of contact sample	(CFU/m^2)	Identification of the cultivable species
Counting of spores in dust sample	(number per gram)	Gives total spore count; needs a specific separation technique

Source: chapter authors

Unfortunately, there are no such methods available today and the duct design is often limited to a sub-optimization scheme with boundary conditions constrained by the building design.

However, provisions must be taken at the early stages of the building design to bring some consistency to this kind of approach. Key questions that must be addressed include the number and type of systems that will be installed; the location of the fans and air handling units; the location of air intakes and exhausts; the location of the shafts; and the space demand for the ducts and fan rooms.

Sizing is another important step that will greatly influence the pressure drop in the system. There are several sizing methods (see ASHRAE, 2001, Chapter 34; Malmström, 2002) to guide the designer at this point, both in the details of the layout and in product selection.

The *specific fan power* (SFP) is a metric related to this issue. It is a measure of the pressure losses p_{tot} in the system, generally expressed in $kW/(m^3/s)$:

Table 3.7 *Issues and associated metrics regarding the energy performance of duct systems*

Issue(s)	Related metric(s)	Symbol	References[*]
Low pressure drop: layout and sizing	Specific fan power	SFP	Jagemar (1994)
Air tightness	Effective leakage area	ELA_{ref}	ASHRAE (2001)
Air tightness	Leakage class	K or C_L	Eurovent 2/2 (1996); ASHRAE (2001)
Air tightness	Leakage flow ratio	q_{vl}/q_v	Delp et al (1998)
Thermal insulation	Thermal resistance or transmittance	R or U	ASHRAE (2001)
Thermal insulation	Conduction effectiveness	ϵ_s	Delp et al (1998b)
Thermal insulation and air tightness	Energy delivery efficiency	ϵ	Delp et al (1998b)
Thermal insulation and air tightness	Normalized airflow rate (airflow per unit duct surface area)	q_V/A	ASHRAE (2001)
Thermal insulation and air tightness	Duct-loss power ratio	DLPR	Xu et al (2003)
Fouling	Fouling time	τ_{foul}	Siegel et al (2002)
Heat recovery	Sensible effectiveness	ϵ_T	Irving (1994)
Exergy consumption	Exergetic efficiency	η	Shukuya and Hammache (2002) Franconi (1998)

Note: * not exhaustive

Source: chapter authors

Table 3.8 *Total efficiencies for different types of fans according to VVS AMA 98*

Centrifugal fan; backward curved blades	65%
Centrifugal fan; forward curved blades	50%
Centrifugal fan; straight curved blades	50%
Vane-axial fan with guide vanes (ducted)	60%
Unducted axial fan	55%

Source: VVS AMA 98 (1998)

$$SFP = \frac{P}{q_v} \qquad (1)$$

where P is the sum of the fan power demands (W) and q_V is the airflow rate through the building (m³/s) (Jagemar, 1994). This index can be used for individual fans. Physically, it represents the fan power necessary to distribute airflow of 1m³/s through the system.

According to Jagemar (1996), measured values of SFP for individual fans were typically in the range of 1–2kW/(m³/s) at that time. The TIP-Vent source book (Wouters et al, 2001) mentions values as low as 0.5kW/(m³/s) for very good system designs. One reference value adopted in the French building code (RT 2000) is 0.9kW/(m³/s).

Reference values adopted in the Allmän Material och Arbetsbeskrivning (AMA) guidelines for the total fan efficiencies are listed in Table 3.8. Measured values of the total fan efficiency for HVAC applications are in the range 35–80 per cent, depending upon fan type and size (Jagemar, 1996). Total efficiencies higher than 70 per cent are rare. With normal fan efficiencies (50–65 per cent), the total pressure drop in the system connected to the fan must be lower than 500–650Pa to draw less power than 1kW (m³/s). Considering a pressure drop at the air handling unit of 200–300Pa, and at the air terminal device of 40–100Pa, the total pressure loss in the duct system should be lower than 200–300Pa.

Ductwork air tightness

Duct leakage is detrimental to energy efficiency, comfort effectiveness, indoor air quality and, sometimes, even to health. A ductwork air-tightness limit should be required to:

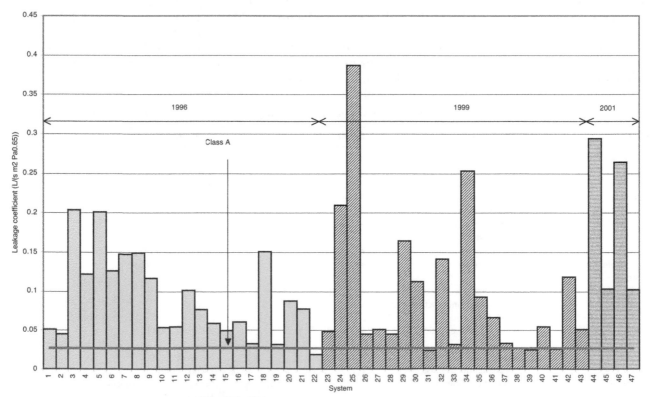

Note: Measurement campaigns were performed in 1996, 1999 and 2001.

Source: Carrié et al (1999); Carrié and Barhoun (2002)

Figure 3.3 *Measured leakage coefficients normalized per duct surface area of 47 duct systems in France*

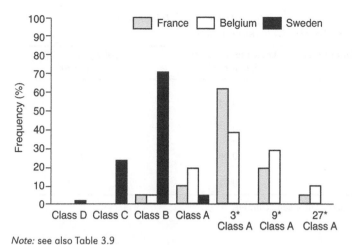

Note: see also Table 3.9

Source: Carrié et al (1999)

Figure 3.4 *Distribution of Eurovent leakage classes measured in Belgium, France and Sweden*

- minimize the cost and the energy penalty due to an oversized or inefficient plant;
- ease the flow-balancing process;
- have control over the leakage noise; and
- limit the infiltration or exfiltration to unconditioned spaces (with potentially large effects on energy use, power demand, indoor air quality and comfort effectiveness).

There is a substantial body of research on air distribution system leakage that includes both detailed field characterizations and energy analyses (Cummings et al, 1990; Proctor and Perrick, 1992; Davis and Roberson, 1993; Babawale et al, 1993; Modera, 1993; Parker et al, 1993; Jump and Modera, 1994; Walker et al, 1998; Carrié et al, 1999; Carrié et al, 2000; Fisk et al, 2000; Carrié and Barhoun, 2002). Published material on these subjects indicate that duct leakage can be very significant (see Figure 3.3) and that duct system inefficiencies account for a significant fraction of the space-conditioning energy use.

The general consensus that can be drawn from these studies is that there are two major ways to waste energy through duct leakage:

1 The fan has to work harder. The airflow passing through the fan is directly affected by duct leakage. In order to meet the required airflow rates at the air terminal devices, the fan must be sized and operated at detrimental conditions for energy use. If the fan power is scaled approximately with the third power of the airflow rate for an existing duct system, a leakage flow rate of 6 per cent should imply a fan power demand increase of 20 per cent ($=1/(1-0.06)^3 - 1$).

2 There may be net thermal losses when the ducts pass through unconditioned spaces. Supply make-up air leaking out to unconditioned spaces is simply lost, along with the energy that was used to condition that air. Insufficient heat recovery or recycling may also result from duct leakage in extract and return ducts (Pittomvils et al, 1996).

Duct leakage may also affect the ventilation rates of a building and, therefore, ventilation energy losses. This question has been investigated by Babawale et al (1993) and Modera (1993).

The benefits of tight air ducts have been identified in Sweden since 1960, with the first requirements stipulated in the 1960 version of AMA. Today's stringent requirements and control procedures in this country explain the striking difference between the ductwork air tightness measured in Belgium, France and Sweden (see Figure 3.4).

Inadequate or missing sealing media, worn tapes, poor workmanship around duct take-offs and fittings, ill-fitted components and physical damage during inspection and maintenance work are frequent deficiencies mentioned in a number of reports (Sherman and Walker, 1998; Walker et al, 1998; Carrié et al, 1999). This situation has motivated research and development on retrofitting techniques – for example, the aerosol duct-sealing research and development programme at the Lawrence Berkeley National Laboratory in Berkeley, California (see www.ducts.lbl.gov).

There are three major metrics that are used for evaluating the air tightness of a duct system: the *effective leakage area*, the *leakage class* and the *leakage flow ratio*.

The *effective leakage area (ELA)* concept is commonly employed to characterize the leakiness of a building envelope (see Chapter 4). For duct leakage applications, the discharge coefficient is usually set to 1 and the reference pressure should be close to the ductwork operating pressure (Modera, 1993).

The Eurovent *leakage classes* are based on maximum values of the leakage coefficient per square metre of duct surface area (K in L/(s·m²·Pa$^{0.65}$)), setting the flow exponent arbitrarily to 0.65:

$$K = \frac{q_{vl}}{A\Delta p_{ref}^{0.65}} \qquad (2)$$

where A is the duct surface area (m); Δp_{ref} is the reference pressure at which the tightness test is performed (Pa); and q_{vl} is the leakage volume flow rate (m³/s).

Table 3.9 *Eurovent 2/2 and corresponding American Society of Heating, Refrigerating and Air-conditioning Engineers (ASHRAE) leakage classes*

Eurovent 2/2 leakage class[*] $l/(s \cdot m^2 \cdot Pa^{0.65})$	ASHRAE leakage class (in SI units) $ml/(s \cdot m^2 \cdot Pa^{0.65})$	Leakage at 100Pa l/s per m²	Leakage at 400Pa l/s per m²
Class A: $K < K_A = 0.027$	27	0.54	1.33
Class B: $K < K_B = 0.009$	9	0.18	0.44
Class C: $K < K_C = 0.003$	3	0.06	0.15
Class D: $K < K_D = 0.001$	1	0.02	0.05

Note: * Note that leakage class D is not defined in Eurovent 2/2 but is used in some European countries.

Source: chapter authors

Note that the flow exponent, set to 0.65, actually varies considerably (see, for instance, Carrié et al, 1999). According to HVCA (1983), this value is consistent with Swedish tests on a variety of constructions.

Because of uncertainties in determining the flow exponent, the choice of the reference pressure can considerably affect the resulting leakage class or calculated leakage flow. This is illustrated in Figure 3.5, where a class C compliant system for pressures greater than 50Pa has a leakage flow rate greater than permitted for pressures lower than 50Pa. This is due to the slope, directly derived from the flow exponent, equal in that case to 0.55.

The ASHRAE *leakage classes* are based on the leakage flow in cubic feet per minute (cfm) per 100ft² of duct surface area at 1 inch of water, generally termed C_L. Its definition differs in SI units since 2001. It is simply 1000 times the leakage coefficient K defined above.

The leakage classes, like the ELA, can be measured with a fan-pressurization technique (Eurovent 2/2, 1996).

It is often recommended that the *leakage flow ratio* (that is, the ratio of the leakage airflow rate to the airflow rate transported in the ductwork) does not exceed 6 per cent. In typical good-quality systems, according to HVCA (1998a), it is generally accepted that this leakage flow ratio corresponds to leakage class A. Note that for a duct system that transports 10l/s per m² of duct surface area at 100Pa, the leakage flow ratio would be of 5.4 per cent for a class A system based on equation (2). However, this statement is difficult to generalize given that the ratio of flow rate to the duct surface area and to the pressure drops varies widely. In any case, the fan power implications of leakage flow ratios of 6 per cent, combined with potential heat losses in HVAC systems, suggest higher demands.

In practice, the leakage flow rate is very difficult to measure directly. Whether at the air handling unit, in a duct or at a grille, the measurement of the airflow is very delicate and requires experienced technicians. McWilliams (2003) reviews the measurement techniques that can be used and gives information on the accuracy of the results. Perhaps the most delicate task concerns the measurement at the air terminal devices in low-pressure systems since the presence of a flow measurement device such as a standard flow hood can seriously affect the pressure distribution in the system. The active flow hood method developed in conjunction with TNO (the Netherlands Organisation for Applied Scientific Research, in Delft) during the mid 1980s and discussed, for instance, by Liddament (1996) or Walker et al (2001), appears interesting in this respect since it aims at making the flow hood 'transparent' by reproducing the same boundary conditions 'seen' by the air terminal device with and without the flow hood.

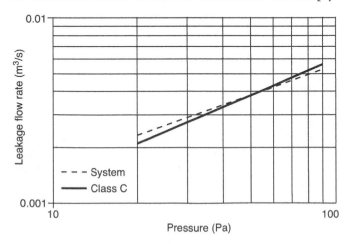

Note: The duct surface area is 100m²; the flow exponent is 0.55; and the leakage flow at 50Pa and above matches class C.

Source: chapter authors

Figure 3.5 *Leakage flow rate versus pressure*

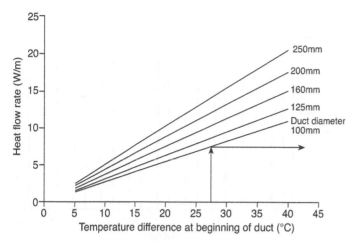

Note: Airflow rate = 0.1m²/s; 60mm-thick insulation

Source: chapter authors

Figure 3.6 *Heat flow rate and temperature loss per unit length of duct for different duct diameters*

Another method to measure the leakage flow rate is the DeltaQ test (Walker et al, 2002). The method is based on measuring changes in the leakage airflow rate through the ductwork and the building envelope as the pressure across the duct leaks changes. The pressure changes are created by a blower door (see Chapter 4) and the ductwork system's fan (on or off). The key advantage of this method is that it directly determines, with good accuracy and repeatability, the leakage airflow rates to unconditioned spaces at operating conditions.

Thermal insulation

Ductwork insulation is of paramount importance for energy conservation when a thermodynamic function is combined with the system. Energy losses associated with insufficient insulation are commonly called conduction losses. Building codes in several countries include thermal insulation requirements to limit those losses. Requirements may be expressed in terms of a minimum *thermal resistance*. However, performance loss in terms of watts (W) per metre or °C per metre of duct length can be easily evaluated with standard heat transfer equations (see Figure 3.6). The designer should evaluate the need for higher levels of insulation based upon the significance of those energy losses.

The significance of conduction losses has been investigated by several researchers (Babawale et al, 1993; Davis and Roberson, 1993; Modera, 1993; Parker et al, 1993; Delp et al, 1998a, 1998b; Fisk et al, 2000; Xu et al, 2002). One interesting concept discussed in detail by Delp et al

(1998c) regards the *conduction effectiveness*, $\epsilon_{s,i}$, which is defined as the ratio of the delivered capacity at an air terminal device (ATD) (index i) to the capacity available at the air handling unit, neglecting duct leakage:

$$\epsilon_{s,i}(t) = \frac{\text{Energy flux at ATD } i}{\begin{array}{c}\text{Energy flux available for ATD } i \text{ downstream}\\ \text{of the heat exchanger, no leaks}\end{array}}$$

$$= \frac{q_{m,i}(h_{s,i} - h_{amb})}{q_{m,i}(h_{e,s} - h_{amb})} \tag{3}$$

where $h_{e,s}$, $h_{s,i}$, and h_{amb} are the specific enthalpy of the air downstream of the heat exchanger, at the ATD, and in the room served (J/kg); $q_{m,i}$ is the mass airflow rate at ATD i (kg/s).

Delp et al (1998c) report conduction effectiveness in the range of 80–95 per cent. This metric can be used to evaluate the overall delivery efficiency of a system.

Thermal insulation and air tightness

The concept of *energy delivery efficiency* has been used by several authors (for example, Delp and et al, 1998a, 1998b) as one means of characterizing the energy efficiency of heating or cooling duct systems. The underlying principle is to rate the useful energy (that is, the energy actually delivered at ATD s) in comparison with the energy that is put into the system. The concept is detailed here for a system that includes one heat exchanger.

The instantaneous energy delivery efficiency, $\epsilon_s(t)$, is defined as the ratio of the energy flux delivered to the ATDs to the energy flux available downstream of the heat exchanger (Delp et al, 1998a):

$$\epsilon_s(t) = \frac{\text{Energy flux at ATD } s}{\begin{array}{c}\text{Energy flux available downstream}\\ \text{of the heat exchanger}\end{array}} \tag{4}$$

Therefore:

$$\epsilon_s(t) = \sum_i \left(\frac{q_{m,i}}{q_{m,fan}}\right)\epsilon_{s,i}(t) \tag{5}$$

where $q_{m,fan}$ is the fan mass airflow rate (kg/s).

On the extract side, the efficiency of the system is defined as:

$$\epsilon_r(t) = \frac{\text{Minimum energy flux to condition indoor air}}{\text{Actual energy flux at the heat exchanger}}$$

$$(6)$$

The overall efficiency of the system is:

$$\epsilon(t) = \frac{\text{Energy flux delivered at ATD } s}{\text{Actual energy flux at the heat exchanger}} \quad (7)$$

Note that $\epsilon(t) = \epsilon_s(t)\epsilon_r(t)$ as the minimum energy flux necessary to condition the indoor air is equal to the energy flux available downstream of the heat exchanger. Energy flux available at the heat exchanger:

$$= q_{m,fan}(h_{e,s} - h_{amb}) \ . \quad (8)$$

The instantaneous delivery efficiency concept is often extended to the more pragmatic average delivery efficiency simply by integrating the energy fluxes over time.

This approach is very interesting to quantify energy losses in an air distribution system. The delivery efficiency can be evaluated at the design stage or in the field. However, there are two major drawbacks:

- The efficiency varies with the operating conditions and climate.
- *In situ* evaluation of the delivery efficiency requires the accurate measurement of the airflow rates at the ATDs, which is very delicate and time consuming in practice.

Delp et al (1998a) detail the measurement methods and results in ten small commercial buildings. They report average delivery efficiencies in the range of 50–80 per cent, with an average of 65 per cent and one value below 30 per cent (see Figure 3.7). Today, there is a significant body of literature on this subject regarding US buildings that consistently show low-delivery efficiencies. In Europe, the only data we found has been published by Carrié and Barhoun (2002), who report values for individual branches in the range of 65–80 per cent and some values lower than 40 per cent (see Figure 3.8).

Similar concepts of first law energy efficiencies – that is, energy output to energy input – have been discussed by several authors (for example, Reddy et al, 1994; Franconi, 1998). The variations with the energy delivery concept discussed above concern the inclusion of other energy use, including, for instance, fan energy or reheat coil loads.

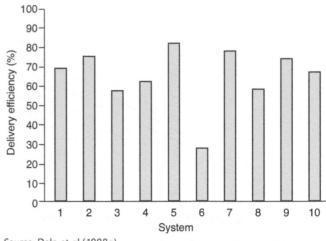

Source: Delp et al (1998a)

Figure 3.7 *Delivery efficiencies in small service-sector buildings in California*

The *normalized airflow rate* (the system airflow rate normalized by duct surface area) is another metric that can be useful in relation to the air tightness and thermal insulation issues. Indeed, leakage and conduction losses are likely to increase with increasing duct surface area. Although there is no direct relationship between the delivered airflow rate and a system's total surface area, ASHRAE (2001, Chapter 34) provides typical values of total fan flow divided by duct surface area of 10–25l/s per m^2.

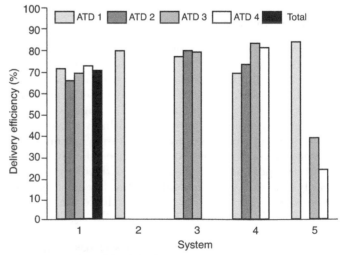

Source: Carrié and Barhoun (2002)

Figure 3.8 *Average delivery efficiencies of individual supply branches in five systems of service-sector buildings in France*

Finally, the *duct-loss power ratio* (DLPR), defined as the ratio of the fan power for an airtight, perfectly insulated duct system ($P_{ideal\ ductwork}$) to the fan power for the duct system of interest (P_{actual}), is a measure of the effect of distribution losses on the fan energy use:

$$DLPR = \frac{P_{ideal\ ductwork}}{P_{actual}} \quad (9)$$

Although interesting, this metric is seldom used, probably because its *in situ* evaluation is very delicate (Xu et al, 2003).

Heat recovery

Ventilation heat recovery consists of transferring some exhaust air stream energy to fulfil a specific task within the building, such as preconditioning of fresh air. There is an array of indices that characterize the efficiency of these devices, described in detail by Irving (1994). Typical *sensible effectiveness* of air-to-air heat recovery units range from about 50 per cent up to about 80 per cent (see Table 3.10). Water-loop heat exchangers have relatively low efficiencies (40–60 per cent).

While heat recovery can be successfully implemented, there are some hidden losses that can seriously affect the energy benefits of such systems (see Irving, 1994; Pittomvils et al, 1996; Carrié et al, 2000):

- The fan (electric) energy use is increased (there are two fans and increased pressure drop due to the heat exchanger).
- The system must not be short circuited; in particular, the building construction needs to be fairly airtight.
- The conduction and convective losses in the supply and extract ducting (for example, due to poor air tightness or poor insulation) must be limited.

If the climate is mild, and considering the hidden losses discussed above, a balanced ventilation system may consume more primary energy than can be recovered by air-to-air heat recovery. The first-order calculation of the heating energy savings, as well as the increased fan (electric) energy use due to heat recovery units, is straightforward and is based on degree days, specific fan power and airflow rate (see, for example, Liddament, 1996, Chapter 12). Made at early stages of design, these calculations can help the designer to decide between different options. They are also useful to understand and decide upon the constraints that must be put on fan power demand and building air tightness, as well as ductwork air tightness and insulation, to achieve the expected

Table 3.10 *Minimum sensible effectiveness as defined in VVS AMA 98*

Type of heat exchanger	Class A (%)	Class B (%)
Rotary air-to-air heat exchanger	70	80
Fixed-plate cross-flow heat exchanger	50	60
Fixed-pipe heat exchanger[*]	50	60
Heat exchanger with two-phase medium[**]	45	55

Notes: [*] Heat exchanger where one of the airstreams passes through the inside of the pipes and the other on the outside of the pipes.
[**] The heat-pipe heat exchanger.

Source: VVS AMA 98 (1998)

energy savings. Note, however, that beyond the energy savings, the quality of the air distribution is the deciding factor for the choice of system.

Fouling

Evidence shows that duct systems and HVAC equipment can be seriously fouled, which can increase the system's pressure drop considerably (Wallin, 1994). Besides indoor air-quality issues resulting from this fact (see for instance, Siegel, 2002), this unwanted increased pressure drop may result in deficiencies, such as insufficient airflow rates, and augmented fan and air-conditioner energy use.

As for coil fouling, Siegel et al (2002) defined the *fouling time* as the time it takes for the pressure drop of a cooling coil to double at constant airflow rate. They found fouling times of about 7.5 years for typical residential system coils in the US and a corresponding efficiency degradation of the unit of 2–4 per cent.

Exergy consumption

Exergy is a measure of the quality of the energy: it is the part of energy that can disperse, as opposed to anergy, which is the part that is already dispersed. Exergy can be consumed in a system due to irreversibilities in the thermodynamic processes, which result in entropy generation (Shukuya and Hammache, 2002).

Several *exergetic efficiencies* can be defined to rate the quality of thermodynamic processes involved in a duct system (Franconi, 1998; Shukuya and Hammache, 2002). In their space heating examples, Shukuya and Hammache (2002) show that the space heating exergy loads are 6–7 per cent of the chemical exergy input to the boiler. Furthermore, they note that the chemical exergies required by the fans are comparable to the space heating exergy loads. Once again, this stresses the importance of careful duct design to minimize fan energy use, which is actually pure exergy.

Maintenance

Need for maintenance

Proper functioning of HVAC systems requires planned and regular maintenance. A preventive maintenance programme should include inspection and maintenance of the following components: outdoor air intake louvres; damper controls; air filters and air cleaning devices; heat exchangers; drip pans; heating and cooling coils; fan belts; humidification equipment and controls; fire dampers; and air distribution systems (Axelrad, 1995; Gallo, 1999). According to many studies, interviews of practitioners and inspections in the field have shown that documented operation and maintenance programmes are rare, and the main activity is usually focused on inspection or replacement of filters or checking belts (see, for example, McGuinness, 1993). Interviews of randomly selected maintenance personnel from 50 Finnish office buildings showed that design and operation documents were missing in 40 per cent of the buildings. In many cases, the maintenance personnel had not received training, and therefore had incorrect information concerning the operation and design of the HVAC system (Tuomaala and Paananen, 1991). The lack of proper maintenance manuals was also observed in a Swedish study summarized by Lindström in 1998 (cited in Carrié et al, 1999).

The lack of maintenance, or faults due to other reasons, is troubling due to the fact that 40–90 per cent of the systems, depending upon the purpose of building, were accepted in the Swedish performance checks (Månsson, 1999). The lowest acceptance rates (40 per cent) were for hospitals and offices; the systems in dwellings performed better (70 per cent) and those in schools and nurseries the best. The most common fault was inadequate airflow rate. In a Finnish study focused on the performance of control devices, the accepted fraction of more than 2000 HVAC systems was reported to be 33 per cent. The detailed analysis showed that many of the faults and failures were connected to poor tuning of the control system, calibration errors of sensors, and bugs in the application software (Laitila and Kosonen, 1995).

Many national guidelines and standards order maintenance as a preventive action to keep the ventilation system functioning in the designed way to deliver sufficient amounts of clean air, conditioned to selected temperatures and humidity, to occupied spaces (see, for example, Boverket, 1994; Gallo, 1999; Nathanson, 1999; HVCA TR/17, 1998b; FiSIAQ, 2001; D2, 2003). With a proper maintenance programme, the systems are more likely to maintain comfortable and good indoor air quality.

Maintenance actions and indoor air quality (IAQ)

Contamination of the ventilation system can become far reaching and an economical burden. The following consequences are listed in Loyd (1992):

- health risks for building occupants and health outcomes by exposure to microbial agents;
- dust entering from ventilation systems and reaching indoor air and surfaces;
- increase of pressure drop that reduces airflow rates in the system and increases the need of energy;
- decrease of fire safety due to clogging of the fire dampers;
- increased risks of fires if kitchen extracts are greasy and fatty at high temperatures;
- damage to sensitive surfaces of furniture, decor and arts; and
- infection risks in hospitals.

All maintenance actions that restrict airflow rate and air distribution to design values improve the pleasantness of the indoor climate. These include maintenance and checking of control devices; changing of filters; cleaning and disinfection of humidifiers; and cleaning of the remaining surfaces of ductwork if needed. One should remember that weather conditions may also have a serious impact (for example, snow and rain in a gale) upon properly designed and installed HVAC equipment, possibly causing moisture problems in the system for short periods.

Filter maintenance

Unless properly and regularly maintained, filters will not function properly. For short periods of time, the particle concentrations may be higher downstream of the filter than upstream when the fan is turned on (Krzyzanowski, 1992). Conversely, Kuehn et al (1996) showed no effect of fan cycling on particle release from loaded filters in laboratory tests. Consequently, the release of the particles from air filters seemed to be negligible in that study.

Pressure drop has been used as a criterion for filter replacement. Unfortunately, filter maintenance is often done improperly, and leakage between frames and filters is common (Pasanen et al, 1990b, 1991b; Ottney, 1993; Hagström et al, 1996). With leaks, the pressure drop may not attain the value given by the manufacturer for filter replacement. Therefore, additional criteria using visual inspection and a certain time interval to change the filters have been used. However, visual inspection may be very subjective and its use has lead to increased particle concentrations in occupied spaces due to insufficient maintenance (Black and Bayer, 1988; Wallin, 1994). Basing filter replacement intervals upon routine monitoring of suspended particle concentration is unreliable because of local and vertical variation (Pasanen et al, 1991b). In 90 per cent of HVAC systems, however, filter replacement intervals were based on the fixed time interval, and dates of the last and the next filter replacement were available from 71 per cent of maintenance personnel (Pasanen et al, 1990b).

It has been suggested that filters should not be replaced when they look dirty because the efficiency of a filter increases with loading. However, replacement of flat panel type filters has been recommended when the downstream side looks dirty (FiSIAQ, 2001). The most reasonable time for a filter replacement is just before heating or cooling seasons (Ottney, 1993). Washing or cleaning of filters is not recommended due to a decrease in filtering efficiency.

Malfunctions of filters enable coarse particles to penetrate the HVAC system and to accumulate on its inner surfaces (Nielsen et al, 1990; Laatikainen et al, 1991; Valbjørn et al, 1990; Wallin, 1994; Fransson, 1996; Lahtivuori, 1996). Some particles will be transferred into occupied space. Dust in HVAC systems downstream of the main filter shows that filtration is inadequate (Ottney, 1993) or that the debris originated during building construction.

Cleaning and IAQ

Cleaning actions have improved the perceived indoor air quality, especially if the systems have been running for a long time without cleaning. In a study by Holopainen et al (1999), the occupants of the buildings perceived the IAQ better after than before cleaning. In a building with a higher number of occupants, the result was statistically significant. However, contradictory results were obtained with a trained sensory panel who evaluated that the odour emissions were higher after cleaning than before (Holopainen et al, 1999). Note that complaints regarding the dustiness of surfaces also decreased after cleaning. A study with a more detailed analysis of the influence of cleaning actions on the perceptions of air quality in 15 office buildings revealed improvements in almost all parameters of the work environment. The most significant improvements were achieved in 'stuffy, bad air', 'dry air' and 'dust and dirt' (Kolari et al, 2003). The frequency of workplace-related symptoms (itching and irritation of eyes, irritated, stuffy or runny nose and a hoarse, dry throat) also decreased after air ducts were cleaned. However, the IAQ measurements in these non-problem buildings also showed low levels of airborne dust, micro-organisms and volatile organic compounds before the cleaning of air ducts; thus, a notable decrease in these parameters could not achieved. The ventilation rate was already high before the cleaning – on average as high as 23l/s per person; after cleaning, it was, on average, 27l/s per person (Kolari et al, 2003).

Accessibility of the system

Many building codes state that air distribution systems will be designed, manufactured and installed in a manner such that cleaning of component surfaces is possible (see, for example, D2, 2003; EN 12097–R5, 2003). This means that the system should have enough openings in places that are easy to access. Access doors should be easy to open and there should be enough space around the doors so that cleaning work is possible and unrestricted. The minimum dimensions of the access openings for different diameters of ducts are listed in the European standard. Access doors are required on both sides to maintain and properly clean duct-mounted components. These comprise damper blades, heating and cooling coils, sound attenuators with internal pods or baffles, filter sections, in-duct fans, heat recovery devices, airflow control devices and air-tuning vanes. The standard presents schematic pictures describing how the access openings should be placed in the ductwork (EN 12097–R5, 2003).

HVAC system cleaning

Several field studies have shown that dust and debris decrease due to cleaning, which definitely supports the fire safety of the buildings, especially if the debris is combustible material. However, in many cases the cleaning result is not satisfactory if the demanded cleanliness level is as high as approved for other indoor surfaces (Kalliokoski et al, 1995; Kolari, 2002; Holopainen et al, 2003a, b).

Dry and wet cleaning methods are applied to clean HVAC systems. The method is usually selected according to the required cleaning result, space and shape, and component of the surface to be cleaned; of course, the selected method should avoid any deterioration of the surface. The commonly used dry cleaning methods are mechanical brushing, vacuuming and compressed air cleaning (HVCA, 1998b; Luoma et al, 1993; NADCA, 1995; Loyd, 1997). Mechanical brushing is an effective way of cleaning circular and rectangular air ducts. The power for rotating the brushes is from an electrical motor, or a pneumatic motor may rotate the brushes. Efficient brushing in larger air ducts is obtained with a hydraulic motor. In all methods, the dust should be removed from the surface and then transported out from the ventilation

Table 3.11 *Cleaning techniques for various components according to the Building Services Research and Information Association*

Component	Cleaning method*
Air handling unit	ABCDJK
Small ducting, below 400mm x 600mm	ACFGHIJK
Small circular ducting, ø < 800mm	CFGHIJK
Large ducting, above 400mm x 600mm; ø > 800mm	ACGHIJK
Extract grille and baffle	ABDK
Supply diffuser and baffle	ABDK
Flexible ducting	DFGHIK
Air-turning vane or flow straightener	ABCHK
Volume-control damper	ABCHK
Fire damper	ABCHK
Silencer or attenuator	ABCGHIK
Plenum (ceiling and floor)	ABCHK
Mixing box (dual duct)	ABCGHIK
Reheater; chiller coil	ADEGH
Extract fan	ABCDEGHK
Induction unit	ABCDGHIK
Sensor	ABCGHI
Kitchen extract fan, ducting and canopy	ABCDL
Grease filter	D
Air filters; recommend to change	(ABC)

Notes:

A = manual vacuuming
B = hand wash/wipe
C = hand wipe in clean environment
D = steam washing
E = chemical spray
F = mechanical brushing
G = air jetting
H = high-volume air blast
I = sectional extraction
J = sectional blocking
K = sealing or encapsulation
L = hand scrape

Source: Loyd (1997)

system. The transportation is usually carried out with airflow along the vacuum cleaner hose or along the under-pressurized ductwork. If the air ducts are used for extraction, they are vacuumed with a special vacuum unit containing effective filters to remove particles before the air is exhausted. If the vacuum unit is placed indoors, it should be equipped with high-efficiency bag filters so that the particles are not transported to the indoor air (NADCA, 1995). For efficient removal and transportation of the loosened dust from the duct, the velocity of extraction air should be at least 15m/s. To get sufficient velocity, usually only a part of the duct line is vacuumed. The section is plugged, for example, with a piece of foam. The cleaning direction is normally the same as the airflow direction in the ducts. Clients should be made aware of the cleaning methods to be used before cleaning occurs.

Cleaning of the air handling unit with dampers, filter banks, coils, fans, humidifiers and sound attenuators usually needs a lot of hand work. Duct surfaces in the air handling unit and its vicinity, containing atmospheric deposits, could be vacuumed by a hand *vacuum cleaner*. In some cases, wet *hand-washing or wiping* with non-corrosive cleaning agents is a suitable means of loosening the debris from the surface. The coils usually collect dust on their laminae, which should be removed dry with vacuuming or, more effectively, with *compressed air jets*. If the dust gets wet and no mechanical force can be applied, the dust will take shape along the surface and the particles will become much more difficult to loosen after the surface dries. Table 3.11 shows possible cleaning techniques for components in ventilation systems. For some components, there are several possible cleaning methods.

Some components of the ventilation systems require chemicals and cleaning agents to get clean. The cleaning and disinfection chemicals should be selected so that they are non-corrosive; they must be safe for users and for building occupants. Concerning the selection of disinfection agents, one should identify the micro-organisms that are to be fought. Some of the agents are effective for fungi, some for bacteria, and some work for both. Disinfection agents do not work on dirty surfaces; therefore, the surface should always be cleaned before disinfection.

Performance checks

In 1992, the Swedish parliament and government introduced a regulation on compulsory inspection of ventilation systems. The objective was to evaluate and improve the quality of ventilation systems through regular field diagnostics. To this end, the so-called OVK procedure (Obligatorisk Ventilations Kontroll) was set up (Boverket, 1994). It involves regular checks of ventilation installations (see Table 3.12) on items including documentation (availability and accuracy of operation/maintenance instruction manual); ductwork (air intake, duct joints and air tightness, dirty ducts, and recirculation of exhaust air); fan rooms; air handling units (for example, sound/vibration, accessibility and filters); feedback control; room air distribution; measurements; and miscellaneous observations. The defects observed during the inspection must be fixed either by the next inspection or (in the event of a severe problem) in a shorter time, after which another inspection will be performed.

In 1998, B. Lindström from Boverket, Sweden, presented the analysis of more than 8000 reports at the SAVE-DUCT international seminar (see Carrié et al, 1999). It showed that only 37 per cent of the systems had been approved. The distribution of approved systems for the different types of buildings was as follows:

Table 3.12 *Requirements for performance checks of ventilation systems in Sweden*

Buildings	Last date for first inspections of existing building	Inspection intervals	Inspector qualifications class
Day care centres, schools, healthcare centres, etc.	31 December 1993	Two years	K
Blocks of flats, office buildings, etc., with balanced ventilation	31 December 1994	Three years	K
Blocks of flats, office buildings, etc., with mechanical exhaust ventilation	31 December 1995	Six years	N
Blocks of flats, office buildings, etc., with natural ventilation	31 December 1995	Nine years	N
One- and two-family houses with balanced ventilation	31 December 1995	Nine years*	N

Notes: Class N qualified inspectors can investigate only simple installations; class K qualified inspectors can investigate all types of installations.

* Since 1999, only the first inspection of an installation brought into use.

Source: Boverket (1994)

- apartment buildings: 25 per cent;
- office buildings: 43 per cent;
- schools: 37 per cent;
- day care centres: 51 per cent;
- healthcare centres: 32 per cent.

The most common defaults are summarized in Table 3.13. One could argue that this picture of the existing installations is not as good as one would like; however, Sweden remains the only country that has an accurate status of existing ventilation installations and of the progress that can be made in this area. Furthermore, note that the non-compliant systems have to be fixed; therefore, globally, the situation is likely to improve over time as the OVK procedure is applied.

Porrez and Wouters (2001) and Barles et al (2001) have tested the OVK approach in three Belgian and five French buildings, respectively. Their conclusions converge on the potential of similar pragmatic procedures for stimulating better design and maintenance, thereby improving air quality, comfort and energy management.

Table 3.13 *Most common defaults found during an inspection of Swedish ventilation systems*

Wrong airflow rate	61%
Missing maintenance manuals	48%
Deposits in fans	40%
Deposits in ducts	37%
Defects in fans	30%
Control and guidance equipment	27%
Deposits in filters	25%
Defects in supply and exhaust devices	23%
Deposits in supply air devices	22%
Defects in filters	20%

Source: based on the analysis of more than 8000 reports in 1998

At present, the OVK procedure is the most advanced and practical performance-oriented approach for ventilation system checking. Note, however, that although there exist other commonly used procedures for testing, adjusting and balancing (TAB) or control, the only published statistical analyses we found are based on controls performed for the French government (Carrié and Garin, 2004). These controls are part of building code compliance checks performed by the Centre d'Etudes Techniques de l'Equipement (CETE) network in about 250 new building projects (multi-family buildings and grouped individual houses) per year on a number of aspects, including fire safety, accessibility and acoustics. Regarding ventilation, the control is limited to the checking of ventilation equipment (for example, the type of ATD) and extract airflow rates. The results displayed in Figure 3.9 for year 2000 show the considerable efforts needed.

The National Building Code of Finland, D2, gives regulations for performance checks of ventilation systems for new or renovated buildings. The code demands that commissioning should include the following checks: checking or measuring of air tightness; inspection of the cleanliness of the system; adjusting and measuring the airflow rates; and measuring the specific electrical power of the system (D2, 2003). The systems are classified in three tightness classes and the air tightness is commonly measured according to the standard SFS 3542.

The measurements are more demanding the higher the tightness class is. In the lower tightness demands, if the system is built by using air tightness classified components (class C or better), the tightness measurements are not necessary for the whole system; spot checks are adequate. The inspection of installations without measuring the tightness of the systems is sufficient for systems that serve only one apartment or restricted part of the building. The tightness of whole systems should be measured if toxic or corrosive agents are transported in the ventilation ducts.

The building code orders that the cleanliness of HVAC systems should be inspected; if necessary, cleaning must occur before balancing and measuring airflows. Airflow rates have to be adjusted to fulfil the designed values, and measured values should be documented and added to the commissioning documents of the building. The ventilation system should be clean before the building is taken in to use.

Besides the national building code, a voluntary guideline, *Classification of Indoor Climate 2000* (FiSIAQ, 2001), is widely accepted and used both in new buildings and in the renovation of buildings. The guideline gives two cleanliness classification categories, P1 and P2, which can be applied both in new and existing buildings. In cleanliness category P1, the ductwork should be built from cleanliness classified components and the amount of dust should not exceed $1g/m^2$; the corresponding value for category P2 is $2.5g/m^2$. According to the guideline, the supply air ductwork will be inspected at least every five years. The dirtiness of the inner surface of the ducts can be determined as an amount of accumulated dust on the surface – for example, with the filter sampling method. The trigger values for cleaning the air handling systems are $2g/m^2$ and $5g/m^2$ in the cleanliness categories P1 and P2, respectively (FiSIAQ, 2001).

Work is under way at the Lawrence Berkeley National Laboratory in California to define a framework for residential commissioning (see www.epb.lbl.gov, accessed in March 2005). This whole-house commissioning approach looks at building envelope, air distribution systems, cooling equipment and combustion appliances. Concerning air distribution systems, Wray et al (2003) have identified applicable metrics for duct leakage, air handler airflow rate, and distribution system airflow rates; they also recommend a set of field procedures to evaluate those performance metrics.

Future challenges

This review of major contributions regarding health and energy issues related to duct systems points out the significant progress made in research and industry over the past 15 years, as well as the inertia to see this progress actually implemented in the field.

Many studies in the late 1990s have shown that mechanical ventilation and air treatment (humidity and temperature) do not necessarily ensure symptomless occupants. The studies have also shown that ventilation systems and their components may act as a source of odorous substances of supply air. In the worst cases, some niches in ventilation systems have offered a growth medium for microbial colonization, leading to exposure to airborne fungal spores and bacteria. To avoid these additional exposures, systematic work needs to be done. In some countries (for example, Sweden, Finland and Germany), new regulations and guidelines are set to guide tenants and landlords to select proper design values for the installation and maintenance of HVAC systems. The Finnish cleanliness classification of air-handling systems aims to ensure good indoor air quality of supply air. To fulfil the demands of the procedure, manufacturers have developed techniques to produce clean and odourless cleanliness-classified components. Many clients in these markets have, however, decided to have clean ventilation systems. This, in turn, has produced a demand for clean components. Therefore, it is possible to build new clean ventilation systems; however, a bigger challenge remains for improving existing systems. A lot of work needs to be done to avoid even the most common problems. Some countries have already started systematic work towards proper performance of systems through regular inspection of the systems and their maintenance.

As for energy performance, research has very much focused on field characterization. In some cases, this has led to the development or emergence of new technologies, such as certified products for air tightness, aerosol duct-sealing, hybrid ventilation, or DC fans (see Chapter 6).

Although many existing systems have major flaws, the good news is that today's technology can overcome most of the issues raised in field characterization work and control procedures. One key issue at this point is to stimulate the market so that this technology is actually used in the field. The Swedish approach through the AMA guidelines and OVK procedures appears to have great potential in this respect. Another key issue very much linked to the first one lies in the organization of constructing a ductwork system. To this end, the use of checklist tools, such as that developed in the AIRWAYS project (Malmström et al, 2002), seems promising – although this particular tool needs to be validated. Note that quality management tools by CETE de Lyon have confirmed the potential of such approaches (Garin, 2000).

This chapter points out the need for good maintenance, which is all the more difficult since ductwork is difficult to access. Note that ductwork does not have to be hidden, and exposed systems are, of course, much easier to inspect and maintain. This converges with the trend of some architects who consider ductwork as an aesthetic element in their creative work. Why not?

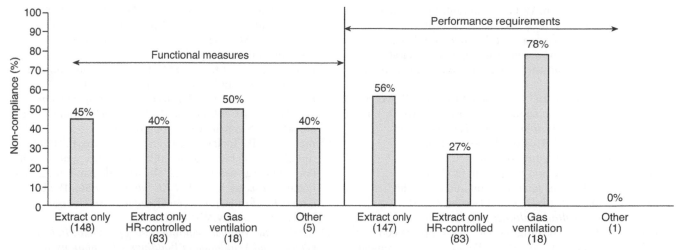

Source: Carrié and Garin (2004), based on 260 inspections; sample size in parentheses

Figure 3.9 *Distribution of non-compliant ventilation systems of new multi-family buildings and grouped individual houses in France*

In the context of sustainable development, comfort (for example, acoustics; see Ling, 2001), health issues and energy conservation, there is a need for progress, as well as innovative research and development. To our knowledge, life-cycle analyses of whole ductwork systems that determine their environmental impact 'from cradle to grave' have not yet been performed.

Acknowledgements

This research for this chapter was funded, in part, by the Centre for the Study of Urban Planning, Transportation and Public Facilities (CERTU) in Lyon, France, and the University of Kuopio in Kuopio, Finland.

References

Ager, B. P. and Tickner, J. A. (1983) 'The control of microbiological hazards associated with air-conditioning and ventilation systems', *Annals of Occupational Hygiene*, vol 27, pp341–358

Ahearn, D. G., Price, D. L., Simmons, R. B. and Crow. S. A. (1992) 'Colonization studies of various HVAC insulation materials', in Coda et al (eds) *Proceedings of IAQ '92 Environments for People*, ASHRAE, Atlanta, US, pp101–105

Ahearn, D. G., Simmons, R. B., Switzer, K. F., Ajello, L. and Pierson, D. L. (1991) 'Colonization by *Cladosporium* spp. of painted metal surfaces associated with heating and air conditioning systems', *Journal of Industrial Microbiology*, vol 8, pp277–280

Andersson, J. V. (1997) 'Swedish duct leakage status', *Supplement to Proceedings of 18th AIVC Conference*, Athens, Greece

Anon (1995) *Initial Evaluation of Methods for Sampling Dust from Air Conveyance System (Duct) Components*, Interim data summary report, EPA Contract No 68–D4–0005, 14 September

ASHRAE (American Society of Heating, Refrigerating, and Air-Conditioning Engineers) (2001) *ASHRAE Handbook: Fundamentals*, ASHRAE, Atlanta, US

Asikainen V., Björkroth M., Holopainen R. and Pasanen P. (2002) 'Oil residues on HVAC components', in Levin H. (ed) *Indoor Air 2002: Proceedings of the Ninth International Conference on Indoor Air Quality and Climate*, vol 1, Monterey, CA, pp356–361

Asikainen V., Holopainen, R., Majanen A., Seppänen, O., Seppälä, A., Jalonen, T. and Pasanen, P. (2003) 'The verifying concept for the cleanliness of HVAC systems', in Tham, K. W. et al (eds) *Proceedings of Healthy Buildings*, vol 2, Stallion Press, Singapore, pp321–326

Axelrad, R. (1995) 'Building operation and maintenance', in Maroni, M., Seifert, B. and Lindvall, T. (eds) *Air Quality Monographs, vol 3: Indoor Air Quality – A Comprehensive Reference Book*, Elsevier, Amsterdam, pp657–696

Babawale, Z. A., Serive-Mattei, L. and Littler, J. (1993) 'Domestic ducted forced air heating: Duct system leakage and heating energy use', *Building Services Engineering, Research & Technology*, vol 14, pp129–135

Barles, P., Lemaire, M.-C. and Larsson R. (2001) 'Testing a method for checking the performance of ventilation systems in commercial buildings in France', *Proceedings of 22nd Air Infiltration and Ventilation Centre Conference*, Bath, UK

Bernstein, R. S., Sorenson, W. G., Garabrant, D., Reaux, C. and Treitman, R. D. (1983) 'Exposure to respirable, airborne *Penicillium* from a contaminated ventilation system: Clinical, environmental and epidemiological aspects', *American Industrial Hygiene Association Journal*, vol 44, pp161–169

Björkroth, M., Torkki, A. and Seppänen, O. (1997a) 'Effect of pollution from ducts on supply air quality', in Woods, J. E., Grimsrud, D. T. and Boschi, N. (eds) *Proceedings of Healthy Buildings '97*, Healthy Buildings/IAQ '97, Washington, DC, vol 1, pp581–586

Björkroth, M., Torkki, A. and Seppänen, O. (1997b) 'Components of the air handling unit and air quality', in Woods, J. E., Grimsrud, D. T. and Boschi, N. (eds) *Proceedings of Healthy Buildings '97*, Healthy Buildings/IAQ '97, Washington, DC, vol 1, pp599–603

Black, M. S. and Bayer, C. W. (1988) 'Pollutants measurement methods used in IAQ evaluations of three office buildings', *Proceedings in ASHRAE IAQ'88: Engineering Solutions to Indoor Air Problem*, ASHRAE, Atlanta, US, pp317–353

Bluyssen, P. M. (1990) *Air Quality Evaluated by a Trained Panel*, PhD Thesis, Technical University of Denmark, Laboratory of Heating and Air Conditioning, Copenhagen, Denmark

Bluyssen, P. M., De Oliveira Fernandes, E., Groes, L., Clausen, G., Fanger, P. O., Valbjørn, O., Bernhard, C. A. and Roulet, C. A. (1996) 'European indoor air quality audit project in 56 office buildings', *Indoor Air*, vol 6, pp221–238

Boverket (1994) *General Guidelines 1992: 3E – Checking the Performance of Ventilation Systems*, The Swedish National Board of Building Housing and Planning, Karlskrona

Burge, H. A., Feeley, J. C., Kreiss, K., Milton, D., Morey, P. R., Otten, J. A., Peterson, K. and Tulis, J. J. (eds) (1989) *Guidelines for the Assessment of Bioaerosols in the Indoor Environment*, American Conference of Governmental Industrial Hygienists, Cincinnati, Ohio, US

Carrié, F. R. and Barhoun, H. (2002) 'Field measurement of the airtightness and delivery effectiveness of HVAC duct systems', *Proceedings of EPIC 2002 AIVC (Third European Conference on Energy Performance and Indoor Climate in Buildings and 23rd Conference of the Air Infiltration and Ventilation Centre)*, Lyon, France, vol 2, pp399–404

Carrié, F. R., Andersson, J. and Wouters, P. (1999) *Improving Ductwork – A Time for Tighter Air Distribution Systems*, Air Infiltration and Ventilation Centre, Coventry, UK

Carrié, F. R., Bossaer, A., Andersson, J. V., Wouters, P. and Liddament, M. (2000) 'Duct leakage in European buildings: Status and perspectives', *Energy and Buildings*, vol 32, pp235–243

Carrié, F. R. and Garin, D. (2004) 'Checking the compliance of residential ventilation systems in France', *Proceedings of 25th AIVC Conference*, Prague, Czech Republic, pp261–266

Chang, J. C. S, Foarde, K. K. and VanOsdell, D. W. (1996) 'Assessment of fungal (*Penicillium chrysogenum*) – growth on three HVAC duct materials', *Environmental International*, vol 22, pp425–431

Christensson, B. and Krants, S. (1994) 'Glass fibre emission from air filters', in Jansson, A. and Olander, L. (eds) *Proceedings of the Fourth International Symposium on Ventilation for Contaminant Control*, Arbetsmiljöinstitutet, Stockholm, Sweden, Part 1, pp437–440

Cummings, J. B., Tooley, J. J., Jr. and Dunsmore, R. (1990) 'Impacts of duct leakage on infiltration rates, space conditioning energy use, and peak electrical demand in Florida homes', *Proceedings of ACEEE Summer Study*, Pacific Grove, California, American Council for an Energy Efficient Economy, Washington, DC

D2 (2003) *The National Building Code of Finland: Indoor Climate and Ventilation in Buildings – Regulations and Guidelines*, Ministry of the Environment, Finland (in Finnish)

Davis, B. E. and Roberson, M. R. (1993) 'Using the "pressure pan" technique to prioritize duct sealing efforts: A study of 18 Arkansas homes', *Energy and Buildings*, vol 20, pp57–63

Delmotte, C., van Orshoven, D. and Wouters, P. (2001) 'Why and how is airtightness stimulated in the proposed new energy performance regulation for the Flemish Region?', *Proceedings of the 22nd Air Infiltration and Ventilation Centre Conference*, Bath, UK

Delp, W. W., Matson, N. E, Dickerhoff, D. J. and Modera, M. P. (1998a) 'Field investigation of duct system performance in California small commercial buildings (round II)', *Proceedings of the ACEEE Summer Study*, pp3.105–3.116

Delp, W. W., Matson, N. E. and Modera, M. P. (1998b) 'Exterior exposed ductwork: Delivery effectiveness and efficiency', *ASHRAE Transactions*, vol 104(II), pp709–721

Delp, W. W., Matson, N. E., Tschudy, E., Modera, M. P. and Diamond, R. C. (1998c) 'Field investigation of duct system performance in California light commercial buildings', *ASHRAE Transactions*, 104(II), pp722–742

Downing, C. C. and Bayer, C. W. (1991) 'Operation and maintenance of indoor air quality', in Geshwiler, M., Montgomery, L., Moran, M. and Johnson, J. (eds) *Proceedings of IAQ '91 Healthy Buildings*, ASHRAE, Atlanta, US, pp372–374

Ducarme, D., Wouters, P. and L'Heureux, D. (1995) 'Evaluation of an IR-controlled ventilation system in an occupied office building', *16th AIVC Conference – Implementing the Results of Ventilation Research*, AIVC, Palm Springs, CA, US, pp517–526

Elixmann, J. H., Jorde, W. and Linskens, H. F. (1987) 'Changes of fungal growth under different climatic conditions in air conditioning filters', *Abstracts of XIV International Botanical Congress*, Berlin, Germany, p149

EN 12097–R5 (2003) *A Draft of Standard Ventilation for Buildings – Ductwork – Requirements for Ductwork Components to Facilitate Maintenance of Ductwork Systems*, European Committee for Standardization

Esmen, N. A., Whittier, D., Kahn, R. A. Lee, T. C., Sheenan, M. and Kotsko, N. (1980) 'Entrainment of fibres from air filters', *Environmental Research*, vol 22, pp450–460

Eurovent 2/2 (1996) *Air Leakage Rate in Sheet Metal Air Distribution Systems*, Eurovent/Cecomaf, Paris, France

Fanger, P. O., Lauridsen, J., Bluyssen, P. and Clausen, G. (1988) 'Air pollution sources in offices and assembly halls, quantified by the olf unit', *Energy and Buildings*, vol 12, pp7–19

Finke, U. and Fitzner, K. (1993) 'Ventilation and air-conditioning systems – investigations to the odour and possibilities of cleaning', in Kalliokoski, P., Jantunen, M. and Seppänen, O. (eds) *Proceedings of Indoor Air '93*, Helsinki, Finland, vol 6, pp279–284

FiSIAQ, (2001) 'Classification of indoor climate 2000: 2001 target values, design guidance and product requirements', *FiSIAQ Publication*, vol 5E, Espoo, Finland

Fisk, W. J., Delp W. W., Diamond, R. C., Dickerhoff, D. J., Levinson, R., Modera, M. P., Nematollahi, M. and Wang, D. (2000) 'Duct systems in large commercial buildings: Physical characterization, air leakage and heat conduction gains', *Energy and Buildings*, vol 32, pp109–119

Fitzner, K., Muller, B., Kuchen, V. and Lussky, J. (1999) 'Definition of cleanliness for two components', *Airless Report Task 2 Maintenance*, Herman-Rietschel Institute for Heating and Air-Conditioning of Technical University of Berlin, Berlin, Germany

Foarde, K. K., VanOsdell, D. W. and Chang, J. C. S. (1996a) 'Amplification of *Penicillium chrysogenum* on three HVAC duct materials', in Yoshizawa, S., Kimura, K.-I., Ikeda, K., Tanabe, S. and Iwata, T. (eds) *Proceedings of Indoor Air '96*, Nagoya, Japan, vol 3, pp197–202

Foarde, K. K., VanOsdell, D. W. and Chang, J. C. S. (1996b) 'Evaluation of fungal growth on fiberglass duct materials for various moisture, soil, use, and temperature conditions', *Indoor Air*, vol 6, pp83–92

Franconi, E. (1998) 'Measuring advances in HVAC distribution system design', *Proceedings ACEEE*, pp3.153–3.157

Fransson, J. I. (1996) 'Particle deposition in ventilation air supply ducts', in Yoshizawa, S., Kimura, K.-I., Ikeda, K., Tanabe, S. and Iwata, T. (eds) *Proceedings of Indoor Air '96*, Nagoya, Japan, pp717–722

Gallo, F. M. (1999) 'Rationale for HVAC maintenance requirements in ASHRAE 62', in Raw, G., Aizlewood, G., and Warren, P. (ed) *Proceedings of the Eighth International Conference on Indoor Air Quality and Climate*, vol 1, pp183–187

Gamboa, R. R, Gallagher, B. P. and Matthews, K. R. (1988) 'Data in glass fiber contribution to the supply air stream from fiberglass duct liner and fiberglass duct board', in

Geshwiler, M., Montgomery, L. and Moran, M. (eds) *Proceedings of IAQ'88, Engineering Solutions to Indoor Air Problems*, ASHRAE, Atlanta, US, pp25–33

Garin, D. (2000) *Réglementation technique des bâtiments d'habitation. Mise en œuvre d'une démarche visant à éviter les non-conformités récurrentes*, CETE de Lyon, Report DVT 00 230 (in French)

Gilmour, I. J., Boyle, M. J., Streifel, A. and McComb, R. C. (1995) 'The effects of circuit and humidifier type on contamination potential during mechanical ventilation: A laboratory study', *American Journal of Infection Control*, vol 23, pp65–72

Girman, J., Truter, R. and McCarthy, J. (1993) 'Maintaining clean HVAC systems', in Levin, H. (ed) *Workshop Summary of Indoor Air '93*, Helsinki, Finland

Hagström, K., Holmberg, R., Lehtimäki, M., Railio, J. and Siitonen, E. (1996) 'Hiukkassuodatuksen perusteet ja optimaalinen valinta' ('Basics of particle filtration and optimal choosing of a filter'), Invent Technology Program, Report 48, TEKES, Federation of Finnish Metal and Electrotechnical Industries, Helsinki, Finland (in Finnish)

Halas, J. J. (1991a) 'Reflections on steam-humidified room air, Part I', *Engineering Systems*, January–February, pp72–75

Halas, J. J. (1991b) 'Reflections on steam-humidified room air, Part II', *Engineering Systems*, March, pp98–100

Holopainen, R., Asikainen, V., Pasanen, P., Majanen, A., Seppälä, A., Jalonen, T. and Seppänen, O. (2003a) 'A cleaning concept for HVAC system', in Tham K. W., Sekhar, C. and Cheong, D. (eds) *Proceedings of Healthy Buildings*, vol 3, pp453–459

Holopainen, R., Asikainen, V., Pasanen, P. and Seppänen, O. (2002a) 'The field comparison of three measuring techniques for evaluation of the surface dust level in ventilation ducts', *Indoor Air*, vol 12, pp47–54

Holopainen, R., Asikainen, V., Tuomainen, M., Björkroth, M., Pasanen, P. and Seppänen, O. (2003b) 'Effectiveness of duct cleaning methods on newly installed duct surfaces', *Indoor Air*, vol 13, pp212–222

Holopainen, R., Narvanne, J., Pasanen, P. and Seppänen, O. (2002b) 'A visual inspection method to evaluate cleanliness of newly installed air ducts', in Levin H. (ed) *Proceedings of the Ninth International Conference on Indoor Air Quality and Climate*, vol 1, pp682–687

Holopainen, R., Palonen, J. and Seppänen, O. (1999) 'The effect of duct cleaning on perceived air quality in two office buildings', in Raw, G., Sekhar, C. and Cheong, D. (eds) *Proceedings of the Eighth International Conference on Indoor Air Quality and Climate*, vol 3, pp37–42

Holopainen, R., Tuomainen, M., Asikainen, V., Pasanen, P., Säteri, J. and Seppänen, O. (2002c) 'The effect of cleanliness control during installation work on the amount of accumulated dust in ducts of new HVAC installations', *Indoor Air*, vol 12, pp191–197

Hugenholtz, P. and Fuerst, J. A. (1992) 'Heterotrophic bacteria in an air handling system', *Applied and Environmental Microbiology*, vol 12, pp3914–3920

Hujanen, M., Seppänen, O. and Pasanen, P. (1991) 'Odor emission from the used filters of air handling units', in Geshwiler, M., Montgomery, L., Moran, M. and Johnson, J. (eds) *Proceedings of IAQ '91 Healthy Buildings*, ASHRAE, Atlanta, pp329–333

Hung, L. L., Yang, C. S., Lewis, F. A. and Zampiello, F. A. (1995) 'Seasonal variation of microorganisms in air handling units – A three-year study', in Marowska, M., Bofinger, N. D. and Maroni, M. (eds) *Proceedings of Indoor Air: An Integrated Approach*, Elsevier Science Publishers, Amsterdam, pp481–484

HVCA (Heating and Ventilating Contractor's Association) DW/143 (1983) *A Practical Guide to Ductwork Leakage Testing*, HVCA, London, UK

HVCA DW/144 (1998a) *Specifications for Sheet Metal Ductwork*, HVCA, London, UK

HVCA TR/17 (1998b) *Cleanliness of Ventilation Systems*, HVAC, London, UK

Iida, Y., Tanaka, T. and Kimura, K. (1996) 'Characterization of fungi in airconditioners in dwellings', in Yoshizawa, S., Kimura, K.-I., Ikeda, K., Tanabe, S. and Iwata, T. (eds) *Proceedings of Indoor Air '96*, Nagoya, Japan, vol 3, pp221–226

Irving, S. (1994) *Air to Air Heat Recovery in Ventilation*, AIVC Technical Note 45, Air Infiltration and Ventilation Centre, Coventry, UK

Ishikawa, K., Iwata, T., Ito, H., Kumagai, K., Kimura, K. and Yoshizawa, S. (1996) 'Field investigation on the effects of duct cleaning on indoor air quality with measured results of TVOC and perceived air quality', in Yoshizawa, S., Kimura, K.-I., Ikeda, K., Tanabe, S. and Iwata, T. (eds) *Proceedings of Indoor Air '96*, Nagoya, Japan, vol 2, pp809–814

Ito, H., Yoshizawa, S., Kumagai, K. and Shizaka, K. (1996) 'Dust deposit evaluation of air conditioning duct' in S. Yoshizawa, K-I. Kimura, K. Ikeda, S. Tanabe, and T. Iwata (eds) *Proceedings of Indoor Air '96*, Nagoya, Japan, vol 3, pp965–970

Jacob, T. R., Hadley, J. G., Bender, J. R. and Eastes, W. (1993) 'Airborne glass fiber concentrations during manufacturing operations involving glass wool insulation', *American Industrial Hygiene Association Journal*, vol 54, pp320–326

Jaffrey, T. S. A. M. (1990) 'Levels of airborne man-made mineral fibres in UK dwellings. I: Fibre levels during and after installation of insulation', *Atmospheric Environment*, vol 24A, pp133–141

Jagemar, L. (1994) 'Energy efficient HVAC systems in office buildings – with emphasis on air distribution systems', *ACEEE Summer Study*, Pacific Grove, CA, pp123–132

Jagemar, L. (1996) *Design of Energy Efficient Buildings – Applied on HVAC Systems in Commercial Buildings*, PhD Thesis, Chalmers University of Technology, Göteborg, Sweden

Jump, D. A. and Modera, M. P. (1994) 'Impacts of attic duct retrofits in Sacramento houses', *Proceedings of the 1994 ACEEE Summer Study on Energy Efficiency in Buildings*, American Council for an Energy Efficient Economy, Washington, DC, vol 9, pp195–203

Kalliokoski, P., Kujanpää, L., Pasanen, A.-L. and Pasanen, P. (1995) 'Cleaning of ventilation systems and its effect on air exchange rates in single family houses', in Maroni, M. (ed) *Proceedings of Healthy Buildings '95*, Milan, Italy, vol 3, pp1525–1529

Kemp, S. J., Kuehn, T. H., Pui, D. Y. H., Vesley, D. and Streifel, A. J. (1995a) 'Filter collection efficiency and growth of microorganisms on filters loaded with outdoor air', *ASHRAE Transactions*, vol 101, pp228–238

Kemp, S. J., Kuehn, T. H., Pui, D. Y. H., Vesley, D. and Streifel, A. J. (1995b) 'Growth of microorganisms on HVAC filters under controlled temperature and humidity conditions', *ASHRAE Transactions*, vol 101, pp305–316

Kjaer, U. D. and Nielsen, P. A. (1993) 'Adsorption studies on dust samples from indoor air environment', in Saarela, K., Kalliokoski, P. and Seppänen, O. (eds) *Proceedings of Indoor Air '93*, Helsinki, Finland, vol 2, pp579–584

Kjaerboe, P. and Strindehag, O. (1993) 'Collection of PAH in air filters', in Seppänen, O., Ilmarinen, R., Jaakkola, J. J. K., Kukkonen, E., Säteri, J. and Vuorelma, H. (eds) *Proceedings of Indoor Air '93*, Helsinki, Finland, vol 6, pp503–508

Kolari, S. (2003) *Ilmanvaihtojärjestelmien Puhdistuksen Vaikutus Toimistorakennusten Sisäilman Laatuun ja Työntekijöiden Työoloihin* [the effect of ventilation system cleaning on indoor air quality and perceived work environment in office buildings], VTT Publications, Espoo, Finland, available at http://www.vtt.fi/inf/pdf/publications/2003/P497.pdf

Kolari S., Luoma M., Ikäheimo M., Pasanen P. (2002) 'The effect of duct cleaning on indoor air quality in office buildings' in H. Levin (ed) *Proceedings of the 9th International Conference on Indoor Air Quality and Climate*, Indoor Air '02, vol 1, pp694–699

Krzyzanowski, M. E. (1992) 'Use of airborne particle counting to evaluate indoor air quality for remediation control', in Geshwiler, M., Montgomery, L. and Moran, M. (eds) *Proceedings of IAQ '92 Environments for People*, ASHRAE, Atlanta, US, pp415–425

Küchen, V. (1998) *Konzentration an Staub und Microorganismen in Lüftungskanälen von Raumlufttechnischen Anlagen. Felduntersuchungen unter Einsatz unterschiedlicher Staubmessverfaren*, Masters thesis, Technical University of Berlin, Germany (in German)

Kuehn, T. H., Yang, C.-H. and Kulp, R. N. (1996) 'Effects of fan cycling on the performance of particulate air filters used for IAQ control', in Yoshizawa, S., Kimura, K.-I., Ikeda, K., Tanabe, S. and Iwata, T. (eds) *Proceedings of Indoor Air '96*, Nagoya, Japan, vol 4, pp211–216

Kumagai, K., Yoshizawa, S., Itoh, H., Shizawa, K. and Ikeda, K. (1997) 'The effect of air duct cleaning on TVOC and fungi' in J. Woods, D. Grimsrud, N. Boschi (eds) *Proceedings of Healthy Buildings '97*, Buildings /IAQ '97, Washington, DC, vol 1, pp611–616

Laatikainen, T. (1993) *Ilmanvaihtokanavaan kertyvät epäpuhtaudet ja kertymiseen vaikuttavat tekijät (Factors Influencing Accumulation of Impurities in Air Ducts)*, Master thesis, University of Kuopio, Kuopio (in Finnish)

Laatikainen, T., Pasanen, P., Korhonen, L., Nevalainen, A. and Ruuskanen, J. (1991) 'Methods for evaluating dust accumulation in ventilation ducts', in Geshwiler, M., Montgomery, L., Moran, M. and Johnson, J. (eds) *Proceedings of IAQ '91 Healthy Buildings*, ASHRAE, Atlanta, US, pp379–382

Lahtivuori, O. (1996) *Tilluftsaggregatets och tilluftsfiltreringens inverkan på föroreningar i tilluftskanalen samt på inneluftens partikelkoncentration (The Effect of the Supply Air Handling Unit and the Supply Air Filtration on the Supply Air Duct Pollutants and the Particle Concentration in the Indoor Air)*, Masters thesis, Helsinki University of Technology, Faculty of Mechanical Engineering, Espoo, Finland (in Swedish)

Laitila, P. and Kosonen, R. (1995) 'Classes for on-site inspection of HVAC-control systems', in Flatheim, G., Berg, K. R. and Edwardsen, K. I. (eds) *Proceedings of Indoor Air Quality in Practice*, Norwegian Society of Chartered Engineers, Norway, pp421–431

Lavoie, J. and Lazure, L. (1994) 'Guide for the prevention of microbial growth in ventilation systems', *Technical Guide, RG–089*, Institut de Recherche en Santé et en Sécurité du Travail du Québec, Montreal, Canada

Leovic, K. W., White, J. B. and Sarsony, C. (1993) 'EPA's indoor/air pollution prevention workshop', *Proceedings of Air and Waste Management Association's 86th Annual Meeting*, Air and Waste Management Association, Pittsburgh, US

Levin, H. and Moschandreas, D. (1990) 'Source assessment', *Proceedings of IAQ '90*, ASHRAE, Ottawa, Canada, pp461–462

Liddament, M. (1996) *A Guide to Energy-Efficient Ventilation*, AIVC, Coventry, UK, available in AIVC CD, no 7

Ling, M. K. (2001) *Acoustics and Ventilation*, AIVC Technical Note 52, AIVC, Coventry, UK

Loyd, S. (1992) *Ventilation System Hygiene – A Review*, BSRIA Technical Note TN/18/19.1, Berkshire, UK

Loyd, S. (1997) *Guidance to the Standard Specification for Ventilation Hygiene*, BSRIA, Berkshire, UK

Luoma, M. (2000) 'Protecting ventilation ducts and accessories against dirt during the construction work', in Seppänen, O. and Säteri, J. (eds) *Proceedings of Healthy Buildings 2000*, Helsinki, vol 2, pp145–150

Luoma M. and Kolari S. (2002) 'Development of a clean installation method for ventilation systems', in Levin, H. (ed) *Indoor Air 2002: Proceedings of the Ninth International Conference on Indoor Air Quality and Climate*, vol 1, pp711–715

Luoma, M., Pasanen, A.-L. and Pasanen, P. (1993) *Ilmastointilaitosten Puhdistustekniikka, (Cleaning of Air Conditioning Systems)*, Technical Research Centre of Finland, Research notes 1942, Espoo, Finland (in Finnish)

Malmström, T.-G. (2002) *A Review of International Literature Related to Ductwork for Ventilation Systems*, AIVC Technical Note 56, Air Infiltration and Ventilation Centre, Coventry, UK, available in AIVC CD no 7

Malmström, T.-G., Andersson J., Carrié, F. R., Wouters, P. and Delmotte, C. (2002) *Source Book for Efficient Air Duct Systems in Europe*, Final report of the AIRWAYS European project, available in AIRWAYS CD, BBRI, Belgium, also available in AIVC CD no 6

Månsson, L.-G. (1999) 'The situation of compulsory performance checking of ventilation systems after 6 years in force', in Raw G, Sekhar, C. and Cheong, D. (ed) *Proceedings of the Eighth International Conference on Indoor Air Quality and Climate*, vol 4, pp549–554

Martikainen, P., Asikainen, A., Nevalainen, A., Jantunen, M., Pasanen, P. and Kalliokoski, P. (1990) 'Microbial growth on ventilation filter materials', *Proceedings of Indoor Air '90*, Toronto, Canada, vol 3, pp203–206

Martiny, H., Möritz, M. and Rüden, H. (1994) 'Deposit of bacteria and fungi in different materials or air-conditioning systems', *Proceedings of the 12th ISCC*, Yokohama, Japan, pp271–274

Maus, R., Goppelsröder, A. and Umhauer H. (1996) 'Viability of microorganisms in fibrous air filters', in Yoshizawa, S., Kimura, K.-I., Ikeda, K., Tanabe, S. and Iwata, T. (eds) *Proceedings of Indoor Air '90*, Nagoya, Japan, vol 3, pp137–142

McGuinness, M. (1993) 'Safety and health considerations for HVAC operations and maintenance activities', in Geshwiler, M., Montgomery, L. and Moran, M. (eds) *Proceedings of IAQ '93, Operating and Maintaining Buildings for Health, Comfort, and Productivity*, ASHRAE, Atlanta, US, pp105–109

McJilton, C. E., Reynolds, S. J., Streifel, A. J. and Pearson, R. L. (1990) 'Bacteria and indoor odour problems – three case studies', *American Industrial Hygiene Association Journal*, vol 51, pp545–549

McWilliams, J. (2003) *Review of Airflow Measurement Techniques*, Annotated Bibliography no 12, AIVC, Coventry, UK, available in AIVC CD no 7

Modera, M. P. (1993) 'Characterizing the performance of residential air distribution systems', *Energy and Buildings*, vol 20, pp65–75

Modera, M. P., Dickerhoff, D. J., Nilssen, O., Duquette, H. and Geyselaers, J. (1996) 'Residential field testing of an aerosol-based technology for sealing ductwork', *Proceedings of ACEEE Summer Study*, Pacific Grove, CA, US

Mølhave, L. and Thorsen, M. (1991) 'A model for investigation of ventilation systems as sources for volatile organic compounds in indoor climates', *Atmospheric Environment*, vol 25A, pp241–249

Morey, P. R. (1992) 'Microbial contamination in buildings: Precautions during remediation activities', in Geshwiler, M., Montgomery, L. and Moran, M. (eds) *Proceedings of IAQ '92 Environments for People*, ASHRAE, Atlanta, US, pp94–100

Morey, P. R. and Feeley, J. C. (1990) 'The landlord, tenant and investigator: Their needs, concerns and viewpoints', in Morey, P. R., Feeley, J. C. and Otten, J. A. (eds) *Biological Contaminants in Indoor Environments*, ASTM, Philadelphia, US, pp1–20

Morey, P. R. and Shattuck, D. E. (1989) 'Role of ventilation in the causation of building-associated illness', *Occupational Medicine: State of Art Reviews*, Hanley and Belfus Inc, Philadeplhia, US, vol 4, pp625–642

Morey, P. R. and Williams, C. M. (1990) 'Porous insulation in buildings: A potential source of microorgamisms', in Walkingshaw, D.S. (ed) *Proceedings Indoor Air '90*, Toronto, Canada, vol 4, pp529–533

Morey, P. R. and Williams, C. M. (1991) 'Is porous insulation inside HVAC system compatible with healthy building?', in Geshwiler, M., Montgomery, L., Moran, M. and Johnson, J. (eds) *Proceedings of IAQ '91 Healthy Buildings*, ASHRAE, Atlanta, US, pp128–135

Möritz, M. (1996) *Verhalten von Mikroorganismen auf Luftfiltern in raumlufttechnischen Anlagen in Abhändigkeit von den klimatischen Bedingungen der Lufttemperatur und der relativen Luftfeuchtikeit*, PhD Thesis. Report D 83, Free University of Berlin, Technical University of Berlin, Berlin, Germany (in German)

Morrison, G. C. and Hodgson, A. T. (1996) 'Evaluation of ventilation system materials as sources of volatile organic compounds in buildings', in Yoshizawa, S., Kimura, K.-I., Ikeda, K., Tanabe, S. and Iwata, T. (eds) *Proceedings of Indoor Air '96*, Nagoya, Japan, vol 3, pp585–590

Müller, B., Fitzner, K. and Küchen, V. (1999) 'Airless, a European project on HVAC systems: Maintenance of HVAC systems task two', in Raw, G., Aizlewood, C. and Warren, P. (eds) *Proceedings of Indoor Air '99*, Construction Research Communications, London, vol 1, pp355–360

NADCA (National Air Duct Cleaning Association) (1995) '*Mechanical Cleaning of Non-Porous Air Conveyance System Components*', NADCA, US

NADCA (2003) *HVAC Inspection Manual: Procedures for Assessing the Cleanliness of Commercial HVAC System*, NADCA, Washington, DC

Nathanson T. (1999) 'The Canadian initiatives to ensure good IAQ', in Raw, G., Aizlewood, G., and Warren, P. (eds) *Indoor Air 1999: Proceedings of the Eighth International Conference on Indoor Air Quality and Climate*, vol 1, pp522–525

National Research Council (1983) *An Assessment of the Health Risks of Morpholine and Dimethylaminoethanol*, National Academy Press, Washington, DC

Neumeister, H. G., Möritz, M., Schleibinger, H. and Martiny, H. (1996) 'Investigation of allergic potential induced by fungi on air filters of HVAC systems', in Yoshizawa, S., Kimura, K.-I., Ikeda, K., Tanabe, S. and Iwata, T. (eds) *Proceedings of Indoor Air '96*, Nagoya, Japan, vol 3, pp125–130

Nielsen, J., Valbjørn, O., Gravesen, S. and Mølhave, L. (1990) *Stöv i ventilationsanlaeg (Dust in Ventilation System)*, SBI rapport 206, Statens byggeforskningsinstitut, Copenhagen, Denmark (in Danish)

Nyman, E. and Sandström, N. Å. (1991) *Microorganismer i ventilationsystem (Microorganisms in Ventilation Systems)*, Byggforskningsrådet, Report R45, Stockholm (in Swedish)

Ottney, T. C. (1993) 'Particle management for HVAC systems', *ASHRAE Journal*, vol 35, pp26–34

Parat, S., Fricker-Hidalgo, H., Perdrix, A., Bemer, A., Pelisser, N. and Grillot, R. (1996) 'Airborne fungal contamination in air-conditioning systems: Effect of filtering and humidifying devices', *American Industrial Hygiene Association Journal*, vol 57, pp996–1001

Parker, D., Fairey, P. and Gu, L. (1993) 'Simulation of the effects of duct leakage and heat transfer on residential space-cooling energy use', *Energy and Buildings*, vol 20, pp97–114

Pasanen, A.-L., Kalliokoski, P., Pasanen, P., Jantunen, M. and Nevalainen, A. (1991a) 'Laboratory studies on the relationship between fungal growth and atmospheric temperature and humidity', *Environment International*, vol 17, pp225–228

Pasanen, P. (1995) 'Impurities in ventilation ducts', in E. L. Besch (ed) *Engineering Indoor Environments*, American Society of Heating, Refrigerating and Air-Conditioning Engineers, Atlanta, US, pp149–153

Pasanen, P.(1996) 'Cleanliness of ventilation systems in Finnish office buildings' in I. Holcatova, M. Belohlavkova and P. Gajdos (eds) *Proceedings of Buildings for Healthy Living*, Charles University, Prague, pp216–224

Pasanen, P. (1998) *Emission from Filters and Hygiene of Air Ducts in the Ventilation Systems of Office Buildings*, PhD thesis, Kuopio University Publications, Kuopio, Finland

Pasanen, P. (1999) 'Verification of cleanliness of HVAC systems', in Loyd, S. (ed) *Proceedings of VHExCo99*, The International Ventilation Hygiene Conference and Exhibition, Birmingham, UK

Pasanen, P., Hujanen, M., Kalliokoski, P., Pasanen, A L., Nevalainen, A. and Ruuskanen, J. (1991b) 'Criteria for changing ventilation filters', in Geshwiler, M., Montgomery, L., Moran, M. and Johnson, J. (eds) *Proceedings of IAQ '91 Healthy Buildings*, ASHRAE, Atlanta, US, pp383–385

Pasanen, P., Kalliokoski, P. and Pasanen, A.-L. (1997) 'The effectiveness of some disinfectants and detergents against microbial growth', *Building and Environment*, vol 32, pp281–287

Pasanen, P., Nevalainen, A., Ruuskanen, J. and Kalliokoski, P. (1992) 'The composition and location of dust settled in supply air ducts', *Proceedings of 13th AIVC Conference: Ventilation for Energy Efficiency and Optimum Indoor Air Quality*, AIVC, Coventry, UK, pp481–488

Pasanen, P. O., Pasanen, A.-L. and Kalliokoski, P. (1995) 'Hygienic aspects of processing oil residues in ventilation ducts', *Indoor Air*, vol 5, pp62–68

Pasanen, P., Nevalainen, A., Kalliokoski, P., Martikainen, P., Asikainen, A., Tarhanen, J. and Jantunen, M. (1990a) *Ilmanvaihtolaitteiden suodattimien aiheuttamat haju – ja mikrobihaitat toimistotiloissa (Odour and Microbial Problems Due to Supply Air Filters)*, Report series in Environmental Sciences, 6/1990, University of Kuopio, Kuopio (in Finnish)

Pasanen, P., Tarhanen, J., Kalliokoski, P. and Nevalainen, A. (1990b) 'Emissions of volatile organic compounds from air conditioning filters of office buildings', in Walkinshaw, D. (ed) *Proceedings of Indoor Air '90*, Toronto, Canada, vol 3, pp183–186

Pejtersen, J. (1996a) 'Sensory air pollution caused by rotary heat exchangers', in Yoshizawa, S., Kimura, K.-I., Ikeda, K., Tanabe, S. and Iwata, T. (eds) *Proceedings of Indoor Air '96*, Nagoya, Japan, vol 3, pp259–264

Pejtersen, J. (1996b) 'Sensory pollution and microbial contamination of ventilation filters', *Indoor Air*, vol 6, pp239–248

Pejtersen, J., Bluyssen, P., Kondo, H., Clausen, G. and Fanger, P. O. (1989) 'Air pollution sources in ventilation systems', in Kulic, E., Todorovic, B. and Novak, P. (eds) *Proceedings of CLIMA 2000, Air Conditioning Components and Systems*, Yugoslav Committee of Heating, Refrigerating and Air Conditioning, and Union of Mechanical and Electrical Engineers and Technicians of Serbia, Beograd, vol 3, pp139–144

Pejtersen, H., Clausen, G. and Fanger, O. (1992) *Olf og energi – Fase 2. Olf vaerdier for og efter rensning as ventilationsanlaeg*, Laboratory of Heating and Air Conditioning, Technical University of Denmark

Pejtersen, J., Oeie, L., Skar, S., Clausen, G. and Fanger, P. O. (1990) 'A simple method to determine the olf load in a building' in D. Walkinshaw (ed) *Proceedings of Indoor Air '90*, Canada Mortgage and Housing Corporation, Ottawa, vol 1, pp537–542

Pejtersen, J. et al (1991) 'Air pollution sources in kindergartens', in Geshwiler, M., Montgomery, L., Moran, M. and Johnson, J. (eds) *Proceedings of IAQ '91 Healthy Buildings*, ASHRAE, Atlanta, US, pp221–224

Pellikka, M., Jantunen, M. J., Kalliokoski, P. and Pitkänen, E. (1986) 'Ventilation and bioaerosols', in Goodfellow, H. D. (ed) *Ventilation '85*, Elsevier Science Publishers, Amsterdam, pp441–450

Pittomvils, J., Hens, H. and Bael, F. V. (1996) 'Evaluation of ventilation in very low energy houses', in *17th AIVC Conference – Optimum Ventilation and Airflow in Buildings*, Gothenburg, Sweden, pp513–520

Porrez J. and Wouters P. (2001) 'Lessons learned from the application of the Swedish Boverket–OVK procedure in Belgium', *Proceedings of the 21st AIVC Conference*, Air Infiltration and Ventilation Centre, The Hague, The Netherlands, Paper 57

Price, B. and Crumb, K. S. (1992) 'Exposure inferences from airborne asbestos measurements in buildings', in Geshwiler, M., Montgomery, L. and Moran, M. (eds) *Proceedings of IAQ '92 Environments for People*, ASHRAE, Atlanta, US, pp63–68

Proctor, J. P. and Pernick, R. K. (1992) 'Getting it right the second time: Measured savings and peak reduction from duct and appliance repairs', *Proceedings of ACEEE Summer Study*, Pacific Grove, California, American Council for an Energy Efficient Economy, Washington, DC

Reddy, T. A., Kissock, J. K., Katipamula, S. and Claridge, D. E. (1994) 'An energy delivery efficiency index to evaluate simultaneous heating and cooling effects in large commercial buildings', *Journal of Solar Engineering*, vol 116, pp79–87

Reponen, T., Nevalainen, A. and Raunemaa, T. (1989) 'Bioaerosol and particle mass levels and ventilation in Finnish homes', *Environmental International*, vol 15, pp203–208

Robertson, J. (1988) 'Source, nature and symptomology of indoor air pollutants: "sick building syndrome"', *Symposium on Air Filtration, Ventilation and Moisture Transfer*, Building Envelope Coordinating Council, Fort Worth, Texas, US, pp243–250

Rothenberger, S. J., Nagy, P. A., Picrell, J. A. and Hobbs, C. H. (1989) 'Surface area, adsorption and desorption studies on indoor dust samples', *American Industrial Hygiene Association Journal*, vol 50, pp15–23

RT 2000 (2000) 'Arrêté du 29 Novembre 2000 relatif aux caractéristiques thermiques des bâtiments nouveaux et des parties nouvelles de bâtiment', *Journal Officiel de la République Française*, no 277, 30 November

Samini, B. (1990) 'Contaminated air in a multi storey research building equipped with 100 per cent fresh air supply ventilation systems', in Walkinshaw, D. (ed) *Proceedings of Indoor Air '90*, Toronto, Canada, vol 4, pp571–581

Schleibinger, H., Böck, R. and Rüden, H. (1995) 'Air filters of HVAC systems a possible source of volatile organic compounds if microbiological origin (mVOCs)?', in Flatheim, G., Berg, K. R. and Edwardsen, K. I. (eds) *Proceedings of Indoor Air Quality in Practice*, Norwegian Society of Chartered Engineers, Norway, pp506–517

Schumate, M. W. and Wilhelm, J. E. (1991) 'Air filtration media evaluations of fiber shedding characteristics under laboratory conditions and in commercial installations' in F. M. Coda et al (eds) *Proceedings of IAQ '91 Healthy Buildings*, ASHRAE, Atlanta, GA, pp337–341

Sherman, M. H. and Walker, I. S. (1998) 'Can duct tape take the heat', *Home Energy*, vol 15, pp14–19

Shukuya, M. and Hammache, A. (2002) *Introduction to the Concept of Exergy – For a Better Understanding of Low-Temperature Heating and High-Temperature Cooling Systems*, VTT Research notes 2158, VTT, Espoo, Finland, also available on AIVC CD no 6, December 2002

Siegel, J (2002) 'Particulate fouling of HVAC heat exchangers', PhD thesis, University of California, Berkeley, CA

Siegel, J. A. and Walker, I. S. (2001) 'Biological fouling of HVAC heat exchangers', *ASHRAE IAQ 2001*, ASHRAE, Atlanta, GA, pp1–9

Siegel, J., Walker, I. and Sherman, M. (2002) 'Dirty air conditioners: Energy implications of coil fouling', *Proceedings of ACEEE Summer Study*, Pacific Grove, California, American Council for an Energy Efficient Economy, Washington, DC, vol 1, pp 287–300

Sugawara F. (1996) 'Microbial contamination in ducts of air conditioning systems', in Yoshizawa, S., Kimura, K.-I., Ikeda, K., Tanabe, S. and Iwata, T. (eds) *Proceedings of Indoor Air '96*, Nagoya, Japan, vol 3, pp161–166

Sundell, J., Andersson, B., Andersson, K. and Lindvall, T. (1993) 'Volatile organic compounds in ventilation air in buildings at different sampling points in the buildings and their relationship with the prevalence of occupant symptoms', *Indoor Air*, vol 3, pp82–93

Sverdrup, C. F. and Nyman, E. (1990) 'A study of microorganisms in the ventilation systems of 12 different buildings in Sweden', in Walkinshaw, D. (ed) *Proceedings of Indoor Air '90*, Toronto, Canada, vol 4, pp 583–588

Teijonsalo, J., Pasanen, P. and Seppänen, O. (1993) 'Filters of air supply units as sources of contaminants', in Seppänen, O., Ilmarinen, R., Jaakkola, J. J. K., Kukkonen, E., Säteri, J. and Vuorelma, H. (eds) *Proceedings of Indoor Air '93*, Helsinki, Finland, vol 6, pp533–538

Thorstensen, E., Hansen, C., Pejtersen, J., Clausen, G. and Fanger, P. O. (1990) 'Air pollution sources and indoor air quality in schools' in D. Walkinshaw (ed) *Proceedings of Indoor Air '90*, Canada Mortgage and Housing Corporation, Ottawa, vol 1, pp531536

Torkki, A. and Seppänen, O. (1996) 'Olfactory and chemical emissions of ventilation ducts', in Yoshizawa, S., Kimura, K.-I., Ikeda, K., Tanabe, S. and Iwata, T. (eds) *Proceedings of Indoor Air '96*, Nagoya, Japan, vol 3, pp995–1000

Tuomaala, P. and Paananen, J. (1991) 'Kokemuksia toimistorakennusten ilmastoinnista' [Survey of air conditioning systems in office buildings] in *LVIS 2000 Research Program: Report 12*, Helsinki University of Technology, Espoo, Finland, pp135142

Valbjørn, O., Nielsen, J., Gravesen, S. and Mølhave, L. (1990) 'Dust in ventilation ducts', in Walkinshaw D. (ed) *Proceedings of Indoor Air 1990*, Toronto, Canada, vol 3, pp361–142

Van der Wal, J. F., Moons, A. M. M. and Steenlage, R. (1989) 'Thermal insulation as a source of air pollution', *Environment International*, vol 15, pp409–412

VDI (Verein Deutscher Ingenieure) 6022 (1998) *Hygienic Standards for Ventilation and Air Conditioning Systems, Part 1*, Offices and Assembly halls, Düsseldorf, Germany

VVS AMA 98 (1998) *Allmän material – och arbetsbeskrivning för VVS – tekniska arbeten*, AB Svensk Byggtjänst, Stockholm (in Swedish)

Walker, I. S., Dickerhoff, D. J. and Sherman, M. H. (2002) 'The DeltaQ method of testing the air leakage of ducts', *Proceedings ACEEE Summer Study 2002*, American Council for an Energy Efficient Economy, Washington, DC, also available as Report LBNL-49749

Walker, I., Sherman, M., Siegel, J., Wang, D., Buchanan, C. and Modera, M. (1998) *Leakage Diagnostics, Sealant Longevity, Sizing and Technology Transfer in Residential Thermal Distribution Systems: Part II. Residential Thermal Distribution Systems Phase VI Final Report*, Lawrence Berkeley National Laboratory, LBNL–42691, Berkeley, CA, US

Walker, I. S., Wray, C. P., Dickerhoff, D. J. and Sherman, M. H. (2001) *Evaluation of Flow Hood Measurements for Residential Register Flows*, Lawrence Berkeley National Laboratory, LBNL-47382, Berkeley, CA, US

Wallin, O. (1994) *Computer Simulation of Particle Deposition in Ventilating Duct Systems*, Royal Institute of Technology (KTH), Sweden

Wouters, P., Delmotte, C., Faysse, J.-C., Barles, P., Bulsing, P., Filleux, C., Hardegger, P., Blomsterberg, A., Pennycook, K., Jackman, P., Maldonado, E., Leal, V. and De Gids, W. (2001) *Towards Improved Performances of Mechanical Ventilation Systems*, Final report of the TIP-Vent European project, available in TIP-Vent CD, BBRI, Belgium; also available in AIVC CD no 6

Wray, C. P., Walker, I. and Sherman, M. (2003) *Guidelines for Residential Commissioning*, Lawrence Berkeley National Laboratory, LBNL-48767, Berkeley, CA, US

Xu, T., Carrié, F. R., Dickerhoff, D., Fisk, W., McWilliams, J., Wang, D. and Modera, M. (2002) 'Performance of thermal distribution systems in large commercial buildings', *Energy and Buildings*, vol 34, pp215–226

Xu, T., Modera, M., Wray, C. and Diamond, R. (2003) 'Metrics and diagnostics for characterizing thermal distribution systems in commercial buildings', in Diamond, R. C., Wray, C. P., Dickerhoff, D. J., Matson, N. E. and Wang, D. (eds) (2003) *Thermal Distribution Systems in Commercial Buildings*, Appendix 1, Lawrence Berkeley National Laboratory LBNL-51860, Berkeley, CA, US

4

Building Air Tightness: Research and Practice

M. H. Sherman and W. R. Chan

Introduction

'Air tightness' is the property of building envelopes that is most important to understanding ventilation. It is quantified in a variety of ways, all of which typically go under the label of 'air leakage'. In this chapter, we will review the state of the art of air tightness research. Before reviewing what is known about air tightness, we will summarize the key roles that air tightness plays in understanding ventilation.

Air tightness is important from a variety of perspectives, but most of them relate to the fact that air tightness is the fundamental building property that affects infiltration. There are numerous definitions of infiltration; but, fundamentally, infiltration is the movement of air through leaks, cracks or other adventitious openings in the building envelope.

The modelling of infiltration (and, thus, ventilation) is a separate topic; but almost all infiltration models require a measure of air tightness as a starting point. While the magnitude of infiltration depends upon the pressures across the building envelope, the air tightness does not, making air tightness a quantity worth knowing in its own right for such reasons as stock characterization, modelling assumptions or construction quality.

Infiltration (and, therefore, air tightness) is important because it affects building energy use, and the transport of contaminants between indoor air and outdoor air (that is, ventilation). From an energy standpoint alone, it is almost always desirable to increase air tightness; but if infiltration provides a useful dilution of indoor contaminants, indoor air quality may suffer. In many countries, infiltration is the dominant source of outdoor air. Providing appropriate indoor air quality (IAQ) at minimal energy costs is a complex optimization process that includes, but may not be dominated by, air tightness concerns. A high degree of air tightness will provide insufficient air through infiltration and thus necessitates a designed ventilation system.

In buildings with designed ventilation systems, especially those with heat recovery, air tightness may be a determining factor in the performance of that system. For example, unbalanced ventilations systems such as exhaust fans require that make-up air comes through building

---

BOX 4.1 NOMENCLATURE FOR CHAPTER 4

A	Area (m^2)	Re	Reynolds number
C_d	Discharge coefficient	S	S number
C	Power-law coefficient ($m^3/s \cdot Pa^n$)	ΔP	Pressure drop (Pa)
d	Diameter of pipe (m)	μ	Viscosity of fluid (kg/m·s)
l	Length (along-flow path) of pipe (m)	υ	Kinematic viscosity of fluid (μ/ρ) (m^2/s)
m	Mass flow correction (2.28)	ϕ	Exponential form factor
n	Power-law exponent	ρ	Density (kg/m^3)
Q	Airflow (m^3/s)		

BOX 4.2 BLOWER DOOR BACKGROUND

'Blower door' is the popular name for a device that is capable of pressurizing or depressurizing a building and measuring the resultant airflow and pressure. The name comes from the fact that in the common utilization of the technology, there is a fan (that is, blower) mounted in a door; the generic term is 'fan pressurization'. Blower door technology was first used in Sweden around 1977 as a window-mounted fan (as reported by Kronvall, 1980) and to test the tightness of building envelopes (Blomsterberg, 1977). That same technology was being pursued by Caffey (1979) in Texas (again as a window unit) and by Harrje et al (1979) at Princeton University (in the form of a blower *door*) to help find and fix the leaks.

   During this period the diagnostic potentials of blower doors became apparent. Blower doors helped Harrje et al (1979) to uncover hidden *bypasses* that accounted for a much greater percentage of building leakage than did the presumed culprits of window, door and electrical outlet leakage. The use of blower doors as part of retrofitting and weatherization became known as *house doctoring* both by Harrje and Dutt (1981) on the US East Coast and by Diamond et al (1982) on the West Coast. This, in turn, led Harrje (1981) to creating instrumented audits to computerized optimizations.

   While it was well understood that blower doors could be used to measure air tightness, the use of blower door data could not be generally used to estimate real-time airflows under natural conditions or to estimate the behaviour of complex ventilation systems. When compared with tracer-gas measurements, early modelling work by Caffey (1979) was found wanting. There was a rule of thumb, which Sherman (1987) attributes to Kronvall and Persily, that seemed to relate blower door data to seasonal air change data in spite of its simplicity. Modelling of infiltration, however, is discussed elsewhere.

---

leaks. Overly leaky or overly tight buildings could reduce the effectiveness of such systems.

   When poor air tightness allows air to be drawn in from contaminated areas, indoor air quality can be reduced even though total ventilation may increase. These contaminated areas could be attics, crawl spaces or even outdoor air. Sometimes the building envelope itself may be a source of contamination because of mould or toxic materials.

   Moisture is a special class of contaminant because it commonly exists in both liquid and vapour form and is a limiting factor in the growth of moulds and fungus. Poor air tightness that allows damp air to come in contact with cool surfaces is quite likely to lead to the growth of microbiologicals. In cold climates, poor air tightness can lead to the formation of ice in and on exterior envelope components.

   Often the most noticeable impact of poor air tightness is draught and noise. Tight buildings provide increased comfort levels to occupants, which, in turn, can have impacts upon energy use and upon the acceptability of the indoor environment.

Measurement fundamentals

From a measurement standpoint, air tightness means measuring the flow through the building envelope as a function of the pressure across the building envelope.

This relationship often fits a power law, which is the most common way of expressing the data. The power-law relationship has the form:

$$Q = C\Delta P^n \tag{1}$$

where C (m^3/s–Pa^n) is the flow coefficient and n is the pressure exponent. The pressure exponent is normally found to be in the vicinity of 0.65, but has the limiting values of 0.5 and 1 from simple physical considerations. Because of the non-linear nature of this expression there are some interesting challenges in understanding any measured data; these issues will be addressed in subsequent sections.

   In her general study of airflow measurement, McWilliams (2002) reviews the techniques for measuring air tightness. The vast majority of techniques fall into the category of 'fan pressurization' in which a fan (or blower) is used to create a steady state pressure difference across the envelope. The flow through the fan is measured at a variety of pressures. The most common incarnation of the fan pressurization technique for dwellings and small buildings is known as a blower door. Although other methods for measuring air tightness have been examined, we will concern ourselves principally with fan pressurization techniques.

The hydrodynamics of leaks

Before discussing measurement techniques in any more detail, it is important to understand the physical properties of what we are measuring – namely, the leaks themselves.

Although the power law has been found to be a reasonably good empirical description of the flow versus pressure relationship, it does not correspond to any physical paradigm. There are physical paradigms that could be (and have been) applied to the problem of air tightness:

- If the leak is very short, frictional forces in the leak itself can be ignored and the leak may be treated as an orifice in which the flow is proportional to the square root of the pressure drop. The higher the flow rate (that is, the Reynolds number), the longer the leak can be; it may also still be treated as an orifice.
- If the flow rate (the Reynolds number) is low enough, the flow will be dominated by laminar frictional losses and the flow will be linearly proportional to the pressure drop.

Comparing with the power law, the first case corresponds to an exponent of 0.5, while the second case corresponds to an exponent of 1. The fact that measured data typically results in an intermediate value indicates that neither of these two limits is a good explanation.

The Reynolds number of a typical leak is below that at which fully developed turbulent flow is an issue; but the length of many such leaks is such that laminar friction is neither negligible nor dominant. The problem becomes one of developing laminar flow in short pipes.

Sherman (1992) used the standard techniques for developing laminar flow to characterize the problem of short circular pipes. In such a development, the pressure drop is the sum of that associated with the acceleration of the fluid and friction losses of the form:

$$\Delta P = \frac{128\mu l}{\pi d^4}Q + \frac{8\rho m}{\pi^2 d^4}Q \tag{2}$$

This expression can be used to derive a quadratic relationship for flow as a function of pressure; but the more interesting result is that it can be manipulated into a power-law formulation:

$$Q \propto S^n \tag{3}$$

Where S is a dimensionless pressure:

$$S = \frac{m\rho d^4}{4096\,\mu^2 l^2}\Delta P \tag{4}$$

Where the exponent can be determined from S (or vice versa):

$$n = \frac{1}{2}\left(1 + (1 + 8S)^{-1/2}\right) \tag{5}$$

If the leak were a single circular pipe, this derivation could, in principle, be used to determine the diameter and length of the leak; but real envelopes are much more complicated. Walker et al (1997) expanded this derivation to look at more general crack geometries and the issue of series and parallel leaks.

The above analysis assumes a smooth pipe. As shown by Kula and Sharples (1994), among others, roughness can have a substantial impact and must be considered if the parameters of this model are to be interpreted physically. The form of the model would only need to be changed if the roughness induced a transition to fully developed turbulence in the leaks that dominate the flow; but that has not been reported for real buildings.

The benefit of this analysis is not so much in providing an ability to infer the geometry of leaks, but to confirm that a power-law formulation is a robust description upon which to base data analyses. It also tells us that the exponent is pressure dependent. This dependency is low, so that over a narrow range of pressures the exponent can be assumed to be fixed. If the pressure ranges over an order of magnitude, however, one cannot assume that it is a constant.

Fan pressurization measurement techniques

The fan pressurization technique has been around a long time and there are many standard test methods that describe its use, such as ASTM (1998, 2002), CAN/CGSB (1986) and ISO (1996). The basic technique involves measuring the steady-state flow through the fan necessary to maintain a steady pressure across the building envelope.

The first-level reporting of this data is generally the same. One reports the pressure and volumetric flow at whatever measurement stations were chosen. If necessary, the raw readings from the equipment are corrected for zero offsets, temperature, altitude, etc. Such corrections are standard experimental practice, but will depend upon the details of the apparatus and experimental layout.

What separates the different test methods and protocols derived from them is the analysis of pressure–flow data. The simplest and most often used protocol is simply to measure at a single pressure. The pressure chosen is conventionally 50Pa, so much of the published data quotes airflow at 50Pa.

A metric airflow at 50Pa has much to recommend. 50Pa is high enough to overpower pressure noise and zero drifts caused by wind or stack effects. Thus, it is reasonably precise and therefore reproducible. The simplicity of a single-point measurement and its reproducibility are why it is the most popular measurement.

Unfortunately, the flow at 50Pa is not the quantity of interest if one is trying to understand what envelope airflows are under natural driving pressures. The average pressure across a leak in a building envelope is closer to 1Pa than to 50Pa. To have an accurate estimate of air tightness, it is necessary to determine it at normal pressures. Furthermore, higher pressures can induce non-linear effects such as valving that would not be relevant for normal pressures.

Depending upon the metric chosen, such reference pressures would be in the 1–4Pa range; but because these pressures are the size of the natural pressure variations, it is very difficult to get a precise measurement of airflow. One must sacrifice precision to get accuracy or accuracy to get precision.

In order to mitigate these errors, many test methods require that the flow be measured over a range of pressures and then extrapolated to the reference pressure of interest using the power law. Because of the non-linearities of the power law and the biases that can be associated with pressure measurements, care must be taken not to introduce unnecessary errors into the data analysis. Modera and Wilson (1990) looked at the impact that wind pressure variations have on the analysis of pressurization data and examined methods to mitigate them by using pressure averaging.

Sherman and Palmiter (1995) have examined the errors associated with analysing fan pressurization data, including precision, bias and modelling errors. They studied the overall uncertainty for a variety of analysis strategies and recommended optimal strategies for selecting instrumentation and pressure stations.

Multi-zone pressurization techniques

The discussion above has focused on single-zone pressurization techniques. Although such tests comprise the vast majority of tests, in many circumstances the actual configuration is not single zone. Some of this is due to a true multi-zone nature; but some of this can also be due to the fact that there is no true air barrier between the 'inside' and the 'outside'.

Attached housing has leakage paths both to outside and to other dwelling units. Even detached housing can have multi-zone properties when buffer spaces partially connect to the living area and partially connect to outside.

For detached housing, the experimental problem can often be solved by making a determination of what constitutes the air barrier and then opening up doors and windows that are not part of the air barrier; thus reducing the configuration to a single zone.

For apartments and other attached dwelling units, it is sometimes desirable to separately know the leakage to the outside and the leakage to other adjacent units. Although not used widely, there are measurement approaches for determining these. Most methods, such as that used by Levin (1988) in Sweden, require access to adjacent units and often multiple blower doors. Some researchers – for example, Shaw (1980) – have used a single blower door and auxiliary pressure measurements to infer component leakage.

Duct leakage measurements

Duct leakage measurement techniques are a spin-off from envelope air-tightness techniques. There are significant differences because of the fact that ducts operate under externally applied pressure differences. When the air handling system is not operational, duct leakage looks quite similar to envelope leakage and may represent a quarter of the total envelope leakage.

The topic of air distribution leakage is too broad to be reviewed herein. Francisco (2001) reviewed five measurement techniques that were under evaluation; but the field is active and there have been developments since then. Carrié et al (1997) have looked at some duct leakage issues in a European context. A new standard test method in the US (ASTM, 2004) makes use of the novel DeltaQ method for determining leakage.

Air tightness metrics

Etheridge (1977) has been a proponent of the quadratic representation of flow, but most researchers use the power law. In both cases, however, the representation is a two-parameter model, with a recognition that these parameters may vary when the range of applied pressure becomes large. Since Sherman (1992) showed that these representations can be interchanged, we will only discuss the common power-law representation.

Although there is general agreement that the power law is a good descriptor of air tightness data, there is no real agreement on the best metrics to use in quoting air tightness data. The best way to quote air tightness data will depend upon what it is used for. Issues such as how many parameters to be used and whether or not air tightness data should be normalized by the size of the building are important when deciding upon the optimal metric.

The exponent: The second metric

Whenever a two-parameter description of air tightness is used, the second parameter is always the power-law exponent, n. The exponent is critical for extrapolating measurements from one pressure regime to another. When the actual measurements are made in the pressure regime for which the data is desired – as often happens for 50Pa metrics – extrapolation is not necessary and high accuracy determination of the exponent is unnecessary. For such cases it is often sufficient to use the average exponent found from large datasets, which has been found by Orme et al (1994) to be approximately 0.65.

The exponent is also interesting from a research and/or diagnostic perspective because it provides an indication of the relative size of the dominant leaks. If the leakage paths are dominated by large, short leaks (for example, orifices) one would expect the exponent to be closer to 0.5; if the leakage is dominated by long path leaks, one would expect the exponent to be closer to 1.

When making measurements before and after some retrofit or other sealing operation, it is especially important to consider changes in the exponent. The exponent can be different before and after such an operation. If an extrapolation is done without taking this into account, the change in air tightness can be significantly mis-estimated. Usually it is easier to seal the large leaks, which tends to imply that a post-sealing measurement will tend to have a higher exponent.

The main air leakage metric

Whether found by extrapolation, interpolation or direct measurement, the principle metric used to quantify air tightness is the airflow through the envelope at a specific reference pressure. The most common reference pressures are 50 and 4, but 1, 10, 25 and 75Pa are used as well. The airflow is often denoted with the reference pressure as a sub-script (for example, Q_{50} or Q_{25}).

75Pa was once suggested as a reference pressure because other envelope components are sometimes tested at this pressure (for example, windows; see Henry and Patenaude, 1998). In practice, this pressure is too high to use, both because some components may change under that much pressure and because the pressurization equipment is often too small to achieve that pressure directly. The airflow required to reach this pressure may itself be a problem because of the flow required or in severe climates.

50Pa, by contrast, is the most common pressure to measure airflow. This has been the traditional value since blower door techniques became popular. It is low enough for standard blower doors to achieve in most houses and high enough to be reasonably independent of weather influences. When single-point measurements are made, it is almost always at 50Pa.

25Pa is a standard reference pressure for measuring duct leakage (Cummings et al, 1996). It is sometimes used as an envelope reference pressure for that reason. It is also occasionally used as an alternative single-point pressure station when the equipment cannot reach 50Pa.

10Pa is used as the reference pressure in the Canadian definition of equivalent leakage area, but not normally directly as a flow rate.

4Pa is similarly used as the reference pressure in the American Society for Testing and Materials (ASTM E779-99) definition of effective leakage area (ELA) and in the American Society of Heating, Refrigerating and Air-Conditioning Engineers (ASHRAE) standards that reference it. ELA can be defined as the area (of unity discharge coefficient) that would have the same flow rate at the specified reference pressure:

$$Q = \text{ELA} \cdot \sqrt{\frac{2Pr}{\rho}} \qquad (6)$$

where 4Pa is chosen as the reference pressure (P_r) for weather-induced pressure.

1Pa is the lowest of the reference pressures used in the literature. At a pressure of unity, the power-law coefficient is equal to the flow rate. This form appears to make this metric independent of the power-law exponent; but because of the non-linearities and cross-correlations associated with the measurement process, this is an illusion based on the system of units used. Furthermore, extrapolation of the measured data, which is normally collected at much higher pressures, is more uncertain than for any other reference pressure.

Flow rate at a specified pressure and leakage area at a specified pressure contains the same information, just in different forms. Flow rate formulations are easier for those performing the measurements because they relate more directly to their equipment. Leakage area formulations are sometimes more intuitive for the occupant or owner because they can imagine a number of holes in their structure of a certain size.

Norms and normalization

The metrics above all refer to the total amount of leakage of the tested envelope. For setting norms or standards, or for comparing one structure to another, it is often desirable to normalize this total by something that scales with the size of building. In that way, buildings of different sizes can be evaluated to the same norm.

There are three quantities commonly used to normalize air leakage: building volume, envelope area and floor area. Each has advantages and disadvantages and each is useful for evaluating different issues.

Building volume is particularly useful when normalizing airflows. When building volume is used to normalize such data, the result is usually expressed in air changes per hour (ACH) at the reference pressure; ACH_{50} is probably the most common air tightness metric reported. Many people find this metric convenient since infiltration and ventilation rates are often quoted in air changes per hour.

Envelope area is particularly useful if one is attempting to define the quality of the envelope as a uniform 'fabric'. Dividing (especially a leakage area) by the envelope area makes the normalized quantity a kind of porosity. Although this normalization can sometimes be the hardest to use, it can be particularly useful in attached buildings were some walls are exposed to the outdoors and some are not.

Floor area can often be the easiest to determine from a practical standpoint. Because usable living space scales most closely to floor area, this normalization is sometimes viewed as being more equitable. This normalization is used most often with ELA measurements and can be converted to a different kind of dimensionless leakage, such as the normalized leakage used by ASHRAE (2001).

Air tightness data

Air tightness data can be expensive to collect. The larger and more complex the building, the more difficult and time consuming it is to collect the data. Furthermore, air tightness in large buildings was not, until recently, thought to be as important a consideration as for dwellings. Thus, the majority of existing data are for dwellings and, more specifically, for single-family homes. We will review these first and then move on to the other kinds of data.

Single-family houses

Air Infiltration and Ventilation Centre (AIVC) numerical database

A report by Orme et al (1994) describes the Air Infiltration and Ventilation Centre (AIVC) air tightness numerical database. Over 2000 measurements on single- and multi-family dwellings are summarized. These data were collected from ten countries as listed in Table 4.1. Mean airflow rates at 50Pa are shown by country in the report; but it should be emphasized that they only act as guidelines because air tightness can vary a lot from building to building.

Expected values for air tightness have been developed for a number of generic forms of construction – namely, timber frame and block-and-brick for low rises; concrete/curtain wall for high rises; and concrete panel and metal panel for industrial buildings. For each of these construction types, the effects of a number of building characteristics on air tightness are tabulated. For example, the 'basic leakage' for a low-rise building with a timber frame is suggested to be 3 ACH_{50}. If no vapour barrier is present, the dwelling is expected to be leakier and the air leakage value should be increased by 3 ACH_{50}. On the other hand, if the dwelling has gasket window/door frames, then 1 ACH_{50} should be subtracted from the default value.

Apart from these generic air leakage guidelines, Orme et al (1994) also summarized 1758 power-law exponent measurements from Canada, The Netherlands, New Zealand, the UK and the US. The distribution of power-law exponent is roughly normal with a mean value of approximately 0.66. The authors did not observe a meaningful relationship between ACH_{50} and the corresponding power-law exponent.

Factors that affect air tightness include age of construction, building type (single-family versus multi-family dwellings), climate and construction materials. Many of the findings are confirmed by recent studies, which are discussed in more detail below.

Whole-building measurements

Air tightness measurements of single-family dwellings are by far the most abundant among the different building types. Many studies measured air tightness as a starting point and then made use of the findings to address problems such as ventilation, energy cost and indoor air quality. Some research also focuses on the air tightness of energy-efficient dwellings and techniques to achieve a higher level of air tightness. Air tightness is known to vary substantially among dwellings. This is not only true in countries where the climate is relatively mild, such as that in the US (Sherman and Dickerhoff, 1998) and the UK (Stephen, 1998); a wide variation has also been observed in more severe zones, such as in Canada (Parent et al, 1996) and Sweden (Kronvall and Boman, 1993). A tenfold difference between the leakiest and tightest dwellings has been observed in those studies where the size of sample is relatively large. The same level of variation in air tightness is evident even among new dwellings according to studies in Canada (Hamlin and Gusdorf, 1997), Belgium (Wouters et al, 1997), and the US (Sherman and Matson, 2001).

BOX 4.3 ERROR ANALYSIS OF PRESSURIZATION DATA

It is almost impossible to do a good job of analysing measurement data without an understanding of the uncertainties that go along with the measurements. Standard texts describe considerations of precision and accuracy, as well as error propagation and robustness; such information will not be repeated here. Sherman and Palmiter (1995) have used these techniques to develop specific expressions for fan pressurization and to optimize the measurement process.

Few of the references in this section, however, report rigorous uncertainty analyses. In fact, some of the relatively early publications have included incorrect error analyses because they failed to properly account for the fact that the non-linear nature of the power law makes parameter errors highly correlated. When this error happens during an extrapolation, it greatly increases the apparent error – for example, in the effective leakage area (ELA).

Most of the reported data are based on single-point measurements and assumed exponents. Using extant exponent data as a prior in a Bayesian analysis, it is possible, in principle, to estimate the extrapolation bias caused by the assumed exponents; but this kind of analysis is very rare.

In looking at large datasets, one hopes that the central limit theorem will apply and that all of the biases and other uncertainties will be reflected in the standard deviations of the data themselves.

Trends by building characteristics

Among the largest database to date on the air tightness of single-family dwellings is the Lawrence Berkeley National Laboratory (LBNL) Residential Diagnostics Database, which has over 70,000 measurements from across the US. Data collection is an ongoing effort by the Energy Performance of Buildings Group at LBNL. A recent analysis of the database by Chan et al (2005) summarizes the measurements in terms of year of construction, size of dwelling, presence of heating ducts, and floor/basement construction type. The database also contains measurements from two special groups of houses – namely, energy-efficiency programmes and weatherization programmes for low-income families.

Among the building characteristics mentioned, the year of construction and the size of dwellings are found to be the most influential factors related to air leakage. The distribution of normalized leakage is roughly lognormal. Regression analyses show that year of construction and size of dwelling can explain some of the observed variation in the floor area-normalized leakage.

Using regression analysis, additional variables were tested to see if their inclusions improve prediction. The location of the dwelling, the presence of heating ducts and the floor/basement construction type were all found to be statistically insignificant. The result is a simple model that can predict the air leakage distribution for a housing stock in the US using only distributions of year of construction and size of dwellings as inputs.

Many studies have observed similar trends by comparing the air leakage of dwellings built at different periods of time. An analysis based on over 2000 houses showed a consistent increase in air tightness across all regions of Canada (Hamlin and Gusdorf, 1997). Kronvall and Boman (1993) found similar results from an analysis of 50 single-family houses in Sweden. The authors observed an over twofold reduction in the mean ACH_{50} of houses built before 1940 and those that were built during 1976–1988.

In countries where the maximum allowable air leakage for new dwellings is written into building codes (for example, in Sweden), air tightness improvement over time is more obvious. However, in milder-climate countries where there is no air tightness standard or code on new dwellings, newer dwellings are not necessarily more airtight than older ones. Stephen (1998) analysed the air tightness measurements of 471 UK dwellings carried out by the Building Research Establishment (BRE) and found no apparent systematic differences in terms of year of construction. On the other hand, voluntary changes in construction practices in the US have resulted in tighter buildings. Analysis on an earlier version of the LBNL Residential Diagnostics Database by Sherman and Dickerhoff (1998) showed a clear decrease in air leakage from the oldest constructions compared to those that were built around 1980. After that, air leakage is fairly constant with year built.

Age of dwelling is a measure of deterioration from wear and tear, which can induce air leakage. This is different from using year of construction as the measure that captures the possible influence from change of building practices on air tightness. Recent constructions, however, appear to be fairly resistant to age-induced leakages. A study by Bossaer et al (1998) showed that among 51 Belgian dwellings built between 1990 and 1995, there was no meaningful relationship between

Table 4.1 *List of data sources and sample sizes in the Air Infiltration and Ventilation Centre (AIVC) database and more recent studies in some countries*

Country	AIVC database			Recent studies	
	Source	Size		Source	Buildings tested
Belgium	Belgium Building Research Institute (BBRI)	57		Bossaer et al (1998)	200 single-family dwellings and apartment units
				Pittomvils et al (1996)	6 low-energy houses
Canada	Canada Mortgage Housing Corporation (CMHC)	475		Gusdorf (2003)	37,490 dwellings, mostly single-family
				Hamlin and Gusdorf (1997)	2263 single-family (incl. 63 energy-efficient) dwellings
				Buchan et al (1996)	11 log houses
				Parent et al (1996); Proskiw (1998)	47 single-family dwellings
				Buchan (1992); Fugler and Moffatt (1994)	Basements and crawl spaces
				Scanada (2001)	Attached garages
				Elmahdy (2003); Proskiw (1995a, 1995b)	Windows
				Fugler (1999)	Attics
				Petrone Architects (2000)	Air barriers
				Air-Ins Inc. (1998a, 1998b)	Building materials and joints
France	Centre Scientifique et Technique du Batiment (CSTB)	66		Litvak et al (2000)	37 single-family dwellings
Germany	Not contained in database	0		Zeller and Werner (1993)	48 single-family dwellings and apartment units
Netherlands	Netherlands Organisation for Applied Scientific Research (TNO)	303		No new data published	
New Zealand	Building Research Association of New Zealand (BRANZ)	83		No new data published	
Norway	Norges byggforskningsinstitutt (NBI)	40		No new data published	
Sweden	National Swedish Institute for Building Research (SIB)	144		Sikander and Olsson-Jonsson (1998)	3 single-family dwellings
Switzerland	National Energy Research Fund (NEFF), Swiss Federal Laboratories for Materials Testing and Research (EMPA)	37		No new data published	
UK	Building Research Establishment (BRE)	385		Stephen (1998)	96 single-family dwellings
				Lowe et al (1997)	15 two-storey dwellings
				McGrath and McManus (1996)	Basements
US	Lawrence Berkeley National Laboratory (LBNL)	435		Sherman and Matson (2001)	70,000 single-family dwellings
				Desjarlais et al (1998); Yuill and Yuill (1998)	Exterior envelopes
				Kosny et al (1998); Petrie et al (2003)	Insulated concrete form systems
				Brennan et al (1990)	Crawl spaces
				Louis and Nelson (1995)	Windows
				Wilcox and Weston (2001)	Air barriers

Source for AIVC database: Orme et al (1994)

duration of occupancy and air tightness. Similarly, Proskiw (1995b) measured the air tightness of 24 houses over periods of up to three years and observed no significant degradation.

The influence of building geometry on air tightness has been studied by Bassett (1985) from measurements on 80 single-family houses in New Zealand. The author showed that envelope area-normalized airflow rate increases as the geometry of the envelope becomes more complex. Envelope complexity is defined as the joint length between wall, floor and ceiling, divided by the envelope area. Chan et al (2005) also observed that floor area-normalized leakage is a function of dwelling size. While it is speculated that larger dwellings tend to have better constructions and, therefore, tighter building envelopes, the explanation can also be that larger dwellings have more favourable surface area to volume ratios and/or less envelope complexity.

Dwellings in severe climates such as Sweden, Norway and Canada are known to be more airtight than those that are located in milder climates such as the US and the UK. The main reasons for tighter construction are to conserve energy costs and to maintain thermal comfort. Within Canada, Hamlin and Gusdorf (1997) observed consistent regional difference in the air leakage of houses built at different periods of time. For a qualitative sense of how the air tightness of dwellings from different countries compares, Orme et al (1994) showed up to twofold to threefold differences in mean ACH_{50} among the ten countries listed in the AIVC numeric database. The data used to compute those mean values included both single-family and multi-family dwellings and are not adjusted for other influential factors, such as year of construction. The findings, nonetheless, support the general notion that dwellings in more severe climates are more airtight.

Energy-efficiency dwellings

Few energy-efficiency programmes in the US have a specific air leakage performance requirement. As a result, the air tightness of energy-efficiency programme houses is not always guaranteed, even though common practices of these programmes, such as caulking and weather stripping, are known to help reduce air leakage. Persily (1986) measured the air tightness of 74 passive solar homes located throughout the US and found little difference in air tightness when compared to other dwellings in the country. At that time, availability of data on conventional houses was quite limited and cannot be considered as representative of the US. It is, nonetheless, a surprising finding, as noted by the author, because the passive solar homes were designed to consume relatively low levels of energy for space conditioning and were therefore expected to be more airtight.

More recently, Sherman and Matson (2001) compared the air leakage of new energy-efficient houses against other new conventional homes. They found that energy-efficient houses are built tighter, in general; but the key benefit is that these programmes promote consistency in construction practice. This is demonstrated by less variation in the air tightness of houses built under energy-efficiency programmes compared to the others. In Canada, Hamlin and Gusdorf (1997) found houses that met the energy-efficiency R-2000 Standard are at least twice as airtight as new conventional houses in most regions of the country. However, the gap between the two is narrowing as builders and house buyers are now generally more aware of the problems associated with excessive air leakage.

There are also examples where consistency in construction practice is not ensured by the energy-efficiency programme. In another air tightness comparison between 47 energy-efficient residential buildings in New York State and 50 nearby conventional houses as controls, the two groups had similar standard deviations (Matson et al, 1994).

The air tightness of low-energy houses is particularly important when the dwellings are equipped with a heat recovery ventilation system in order to achieve energy efficiency. Pittomvils et al (1996) studied the air tightness of six low-energy houses in Belgium for this reason and found ACH_{50} that ranged from 3.8 to 4.9. Despite the fact that these values are half of those from conventional Belgian dwellings (Bossaer et al, 1998), air leakage at these levels still compromises the fractional reduction in ventilation-related building load. In Germany, Zeller and Werner (1993) measured the air tightness of 48 dwellings where some of them are designed to be low energy. About 40 per cent of the dwellings tested have ACH_{50} greater than 3, at which the ventilation system cannot be run energy efficiently.

Key leakage pathways

The types of leakage problems have much to do with how the dwellings were constructed. In a project that studied the effectiveness of various retrofitting strategies, Lowe et al (1997) found that one of the most important factors is the method used to construct the walls. Load-bearing masonry walls with timber-framed roofs and intermediate floors are common forms of construction in the UK. If plasterboard-on-dabs is used, all of the leakage paths in the house will become interconnected, which makes air sealing difficult. Conversely, houses built with wet plastered masonry wall can potentially be quite airtight (Lowe et al, 1994).

A UK study by Stephen (1998) on the BRE database found that timber-framed structures are, on average, tighter than masonry ones. However, the timber-framed houses tested also tend to be newer. So after adjusting for the age of dwellings, the difference in their air leakage appears to be less significant.

In a research project for which the goal was to give guidance in choosing appropriate materials for air barrier systems, Air-Ins Inc (1998a) tested 36 common building materials for air leakage using a laboratory test chamber.

Only half of the samples were found to be in compliance with the Canada National Building Code limit of 0.02 l/s/m² at 75Pa. The testing found little homogeneity within different samples of the same material.

The use of polyethylene air barriers is a common practice to reduce air leakage from walls. A recent study by Wilcox and Weston (2001) measured the air tightness of four pairs of new California homes built with and without spun-bonded polyolefin housewrap. The authors found that houses with housewrap are, on average, 13 per cent tighter than their counterparts. It is expected that the impact of a housewrap air barrier would be significantly greater if the air barrier were installed as part of a continuous pressure envelope instead of as an external finish, as done in the study. Yuill and Yuill (1998) also found that the technique of using housewrap over untapped extruded polystyrene foam sheathing has the highest flow resistance among the different materials studied. However, a longevity study by Air-Ins Inc (1998b) showed that spun-bonded olefin paper can fail to stretch around joints under high temperature and can break away.

There are other alternatives to improve the air tightness of timber-frame buildings. Sikander and Olsson-Jonsson (1998) tested diffusion-permitting polymer-based fibre sheets (sometimes known as 'windproof' sheets) and gypsum board panels on three detached houses and a test structure in the laboratory. Measurements showed that it is possible to meet the Sweden Building Regulations, provided that the technical designs and quality of contractor work are of a high standard. Likewise, Proskiw (1998) concluded that both polyethylene air barriers and airtight drywall approaches can meet the requirement of the Canadian R-2000 Standard based on measurements of 17 dwellings taken over a period of 11 years. However, a study by Air-Ins Inc (1998a) found that some types of perforated polyethylene are permeable to air.

A longevity study on the behaviour of various air barrier connection techniques submitted to pressure and temperature differentials showed that silicone-base sealant and adhesive tape are the most durable (Air-Ins Inc, 1998b). On the other hand, open cell gaskets, mineral wool and perforated polyethylene should not be used due to their high permeability. Spun-bonded olefin and acrylic sealant can exhibit problems at high temperatures. There are now recommendations on specific assembly instructions for rigid air barriers by the Canada Mortgage and Housing Corporation (CMHC) (Petrone Architects, 2000).

Recent laboratory studies by Kosny et al (1998) on insulated concrete form (ICF) systems suggested that dwellings of this sort can be more airtight than wood-frame constructions. Petrie et al (2003) tested two identical houses located side by side with the only difference that one had ICF as the exterior walls and the other had conventional wood-framed exterior walls. Air leakage measurements showed that the ICF house was 6–23 per cent less leaky than the wood-framed one, depending upon the components sealed and the climate condition during the test.

A few studies in Canada and the US have shown that log houses can also be reasonably airtight (Buchan et al, 1996). Lateral joints were often found not to be the major leakage source. Instead, smoke pencil tests suggested that significant leakage occurred at the corners, the transitions between log walls and other building components, around doors and windows, and around other wall penetrations.

Apart from leakage through walls, other important components contributing to air leakage include windows and doors, flues and fireplaces, heating ducts, and the connections to the attic, basement, crawl space and garage. Effective leakage areas of many of these building components, including walls, are tabulated in Chapter 26 of the 2001 *ASHRAE Handbook of Fundamentals*. About half of the data have been updated by Colliver et al (1994) from the previous version in 1989. The authors found that the central tendency of most estimates remained unchanged, but the variability observed was much wider as more data were included.

Window air leakage appears to be most studied and some suggested that reductions have been successful. In Canada, a study by Henry and Patenaude (1998) tested 35 windows for their air leakage at cold temperatures. They found that the majority of windows met or exceeded the highest levels of air leakage performance of Canadian window standards at normal temperatures, and many did very well even at the lowest temperatures tested. There have also been many studies on the impact of window air leakage on other problems, such as heat transfer (Haile et al, 1998) and condensation (Elmahdy, 2003). Desjarlais et al (1998) found that the air leakage of windows can be further reduced by 60–80 per cent when an additional storm window is added.

Despite this, current window-testing standards do not include air leakage from the joint between window and wall assemblies or from the sides of the windows. Louis and Nelson (1995) presented a test methodology for quantifying this portion of air leakage. Measurements from a few case studies show that the extraneous air leakage from window perimeters is often higher than the air leakage through the window unit. Proskiw (1995a) showed that conventional rough-opening sealing methods (that is, packed fibre glass) can contribute up to 14 per cent of the total leakage of a single-family detached dwelling. This source of air leakage can be greatly reduced by using alternative sealing methods, such as casing tape, poly-return, polywrap and foamed in-place urethane.

Dumont (1993) reports detailed measurements of air tightness revealing significant leakage at many of the component interfaces in buildings. Through visualization with smoke and through reductive sealing methods, Pittomvils et al (1996) found that the connections between wall and roof and at the top of the roof are common sources of leaks among the six low-energy houses studied in Belgium. A solution to this problem has been addressed in a summary report by Adalberth (1997), which provides some guidelines to practitioners on how to achieve good air tightness. The document not only includes drawings and specifications, but also suggests suitable materials and a quality assurance system for meeting the goal.

Research on attic-related heat and moisture flows has been under way for over a decade in Canada. Among the first effort was quantifying the attic interface leakage areas by method of subtraction (that is, house ELA, including the attic interface, minus house ELA with the attic equally depressurized). The attic interface leakage areas were found to be fairly uniform, with an average ELA_{10} of 330cm^2 among the 20 houses tested. Only tightly built R-2000 houses had an interface leakage area of 20cm^2. Wouters et al (1997) also found insulated attics to be a significant source of air leakage (one third of the total) in new Belgian dwellings.

Significant interface leakage at crawl space has also been observed. Brennan et al (1990) compared the ELA of the crawl space of nine dwellings against the rest of the building envelopes and found that even with passive vents closed, crawl spaces are much leakier. Among the ten houses measured in British Columbia, Fugler and Moffatt (1994) found that the interface leakage between crawl space and the rest of the house is more pronounced with the presence of forced-air systems, instead of radiant heating. Air leakage from basements can also bring moisture and soil contaminants into the living space. McGrath and McManus (1996) used tracer gas techniques to measure the airflow through the basement ceiling to the room above in two homes in the UK. Through visual inspection, the reason for leakiness was the cracks between the floor boards and between the floor and wall.

Houses built slab-on-grade or that have fully conditioned basements are known to have much less floor leakage. Sherman and Dickerhoff (1998) and Stephen (1998) observed that this group of houses are 6 per cent and 27 per cent more airtight, respectively, than those that were built with crawl spaces or had unconditioned basements. In the interest of reducing radon exposure, sub-slab polyethylene air barriers have been shown to be very effective in making concrete basement floors airtight (Yuill et al, 2000a). After proper installation, the effective leakage area of the slab dropped to undetectable levels. Buchan et al (1992) measured the air leakage of 13 heated basements and 1 crawl space, where preserved wood foundations were used. Test results showed that the foundations were, in general, tightly constructed and that most of the air leakage occurred around the windows and headers in the basement.

Air leakage between garages and houses has been found to be significant among the 25 Canadian dwellings tested (Scanada Consultants Limited, 2001). The technique used to measure the interface leakage area is similar to that described above for attic measurements – the difference between depressurization of the house with the garage door opened and with the garage simultaneously depressurized. The average ELA_{10} is found to be 140cm^2, which is about 13 per cent of the total air leakage. This is roughly proportional to the ratio of interface area to house envelope area, meaning that the house–garage interface is built with the same tightness as the rest of the house envelope.

Studies by Bossaer et al (1998) and Pittomvils et al (1996) on Belgian dwellings also revealed similar observations. Bossaer et al (1998) determined the room-by-room airflow rates at 50Pa by means of compensating flow meter. The average garage interface air leakage among 26 dwellings tested accounts for about one third of the total leakage. Pittomvils et al (1996) also found the interface between garages and the houses to be quite leaky even among the six low-energy houses tested.

Implications of air tightness measurements

Studies on the relationship between air tightness, ventilation and energy use have revealed the interdependency of these factors. For example, Yoshino and Zhao (1996) made recommendations on the optimum air tightness for dwellings by using various ventilation systems in different climatic regions of Japan. Sherman and Matson (1997) estimated the energy liability associated with providing the current levels of ventilation in US dwellings, and found substantial energy savings by tightening building envelopes while maintaining adequate ventilation. Zmeureanu (2000), on the other hand, found that by considering the life-cycle energy consumption, the initial cost of renovation and the carbon dioxide tax credits, increase in air tightness of existing houses is not always cost effective in Montreal, Canada.

Whole-building air tightness measurements provide useful information about the energy demand of dwellings. However, the correlation between the measured air tightness of houses and indoor air quality is less clear. Parent et al (1996) found that the carbon dioxide levels measured in 30 single-family dwellings in Canada during the heating season have little to do with their respective air tightness. Bossaer et al (1998) found that the air tightness of rooms can vary greatly in a given house, which can partly explain why whole-building air tightness is a poor predictor for indoor air quality.

Multi-family dwellings

The problem of air leakage in multi-family dwellings is more complex due to the partition wall between units and the sheer size of the building envelope. Furthermore, there are additional leakage pathways to be considered – for example, adjacent units, stairwell doors, garbage chutes and elevator shafts. If the fan pressurization method is used, multiple blower doors and/or very large-scale equipment will be needed. Not only is the test procedure more time and labour intensive, it also requires more cooperation from residents in order to access multiple units simultaneously. Some of the studies discussed below used tracer gas methods to measure inter-zonal airflow. Even though the measurements themselves are not a direct measure of air tightness of the units tested, some of the findings provide insights about the relative importance of various leakage pathways in the building.

Relative to the amount of data on single-family dwellings, there are fewer measurements on the air leakage of multi-family dwellings. Table 4.2 shows some of the major studies available from various countries. While the list does not include all past measurements, it captures most of the recent studies on the air leakage of various types of multi-family dwellings.

Lower-rise building measurements

Measurements of air leakage of multi-family dwellings can be divided into whole-building envelope measurements, zonal measurements (floor by floor or unit by unit) and component leakage measurements. Most data are available on unit-by-unit bases. Levin (1991) summarized the air leakage of 53 units measured under the Stockholm Project, many of which are quite airtight (0.45–0.9 ACH_{50}). The air tightness of a number of apartment units in this study was measured under the condition that the adjacent units were also pressurized. Using this method, the internal air leakage between apartment units was found to account for 12–33 per cent of the total air leakage at 50Pa. Similar relative leakage to internal walls has been reported by Lagus and King (1986), Reardon et al (1987) and Love (1990) in Canada, and Cornish et al (1989) in the UK, of which the dwellings tested were all row house (terraced house) types.

Boman and Lyberg (1986) analysed 150 units from some three-storey buildings and found that they were similar to single-family dwellings in air tightness. But such is not always the case. Later studies by Blomsterberg et al (1995) and Kronvall and Boman (1993) also carried out in Sweden suggested that multi-family dwellings have lower ACH_{50} than single-family ones. The authors attributed this to the fact that multi-family dwellings have higher volume-to-surface area ratios and, therefore, lower ACH_{50} values. Litvak et al (2000) and Murakami and Yoshino (1983) also observed multi-family building units to be more airtight than single-family ones in France and Japan, respectively. Despite this, the air tightness of many multi-family dwellings still does not meet building codes and standards in many countries. In Canada, for example, the air tightness of ten typical mid-size buildings tested did not meet the requirements of the 1995 Canadian National Building Code (Nichols and Gerbasi, 1997).

By using a multi-tracer measurement system, Palmiter et al (1995) found significant flow from ground-floor units directly into the top-floor units in some three-storey buildings due to the stack effect. The average flow measured in common walls with plumbing and electrical utilities running from the ground floor to the top was larger than most of the horizontal inter-zonal flows. The building tested was of a standard wood-frame construction, with a slab-on-grade foundation. An earlier study by Cornish et al (1989) in the UK and Dietz et al (1985) in the US also found similar stack-induced leakage between units.

Reardon et al (1987) found that the units on the upper level were much leakier than those below. The reason for this is that the structure was built with a concrete lower level and a wood-frame upper level. Furthermore, the lower units had one less air leakage pathway – the roof top. Vertical distribution of leakage is a concern because, according to a modelling parametric study by Sateri et al (1995), this is the most important factor affecting infiltration.

Recent studies in countries where measurements on multi-family dwellings were not previously available, such as in France (Barles and Boulanger, 2000) and Lithuania (Juodis, 2000), found that there is large variation in the air tightness of units in the same building. At the most extreme, a tenfold difference was observed.

Table 4.2 *List of recent studies on the air leakage of multi-family dwellings*

Country	Source	Buildings tested	# Units tested
Canada	Nichols and Gerbasi (1997)	10 mid-size buildings	–
	Gulay et al (1993)	10 high-rises	12
	Shaw et al (1991)	1 five-storey buildings	10
	Love (1990)	9 row houses	42
	Shaw et al (1990)	2 high-rises	2
	Shaw (1980)	5 high-rises	-
	Reardon et al (1987)	2 row houses	3
Finland	Kovanen and Sateri (1997)	3 buildings	8
France	Barles and Boulanger (2000)	3 buildings	35
	Litvak et al (2000)	Multi-family dwellings	26
Japan	Murakami and Yoshino (1983)	7 buildings	16
Lithuania	Juodis (2000)	High-rises	33
Russia	Armstrong et al (1996)	12 buildings	50
Sweden	Blomsterberg et al (1995)	3 buildings	6
	Levin (1991)	7 buildings	53
	Boman and Lyberg (1986)	Three-storey buildings	150
	Lundin (1981)	2 terraced houses	2
UK	Cornish et al (1989)	Large panel system dwellings	9
US	Palmiter et al (1995)*	Three-storey houses	6
	Flanders (1995)	3 quadraplexes	7
	Lagus and King (1986)	4 row houses	24
	Dietz et al (1985)*	2 quadraplexes	8
	Zuercher and Feustel (1983)	1 high rise	–

Note: * The study used a tracer gas method to measure infiltration, and not air tightness directly

Flanders (1995) compared the air leakage of multi-family units measured using various fan pressurization protocols based on standards by the International Organization for Standardization (ISO), the American Society for Testing and Materials (ASTM) and the Canadian General Standard Board (CGSB). Flanders concluded that most of the protocols used gave similar flow coefficient and exponent values when the weather condition was calm, but uncertainty increased when windier. Flanders (1995) recommended that the door of the adjacent units should be left opened, instead of closed, when carrying out a blower test if the units could not be pressurized simultaneously.

High-rise building measurements

Most of the studies mentioned above are low-rise multi-family dwellings. Air leakage of high-rise buildings has been measured in a relatively large-scale study in Canada (Gulay et al, 1993) and in Russia (Armstrong et al, 1996). Recent measurements by Barles and Boulanger (2000) in France and Juodis (2000) in Lithuania also included some high-rise residential buildings. The Canadian study included measurements on whole-building leakage, floor-by-floor leakage, unit leakage and component leakage. Findings confirmed that the air leakage rates for the high-rise residential buildings far exceeded National Research Council Canada (NRCC) proposed guidelines.

Whole-building air leakage tests require access to every unit and room located within the perimeter of the building. This method requires the most cooperation from tenants and owners. It also requires access to large-scale fan pressurization equipment. Parekh (1992) measured the whole-building air leakage of two buildings before and after the air sealing of the building envelope and observed 32 per cent and 38 per cent reduction in air leakage. Parekh (1992) also suggested some guidelines for the qualitative assessment of the air leakage characteristics of the building envelope with regard to the following components: windows; external doors; the building envelope; elevator shafts and services shafts; and miscellaneous components, including exhaust fan dampers and ducts. In their summary report, Gulay et al (1993) tabulated the percentage distribution of the whole-building leakage by component, estimated based on NRCC guidelines: windows, 42 per cent; doors, 26 per cent; vertical shafts, 14 per cent; and building envelopes, 6 per cent.

Shaw (1980) and Shaw et al (1990, 1991) used similar methods to measure the whole-building air tightness of four high-rise apartment buildings. They found the pressure difference across the envelope to decrease with building height due to large flow resistance in the stairwell. The airflow corresponding to a height-averaged pressure difference of 50Pa ranged from 1.8 l/s/m^2 to 3.6 l/s/m^2. The value reported by Gulay et al (1993), which was measured before air sealing work, lay at approximately 2.15 l/s/m^2.

Armstrong et al (1996) measured the air leakage of 50 apartments located in 12 buildings and found a correlation between ELA$_4$ and the apartment volume. This correlation was particularly profound when the blower door tests were carried out with the major leakage pathways sealed, such as the windows, the balcony door, and the kitchen and bathroom exhaust grilles. Windows and patio doors were found to contribute less than one third of the total ELA under a 'vents-sealed' condition. These results were, unfortunately, compromised by variation in the incremental sealing techniques and by non-uniform outside pressure on the envelope of the tested apartment.

Leakage characteristics of stairwells have been studied by Zuercher and Feustel (1983) on a nine-storey student dormitory. Power-law coefficients and exponents were reported from the pressurization and depressurization tests carried out under various door (including emergency door) operation conditions. Tracer gas measurements were also carried out to study the influence of wind and stack effect upon air infiltration.

Smoke control is another common concern in high-rise buildings. Tamura and Shaw (1981) measured the pressure differences and flow velocities in various parts of two high-rise buildings. Results demonstrated that the performance of the smoke shaft in venting the fire floor can be seriously impaired by the extraneous leakage flow into the smoke shaft through the shaft wall construction from other floors. Related studies regarding the ventilation and infiltration characteristics of lift shafts and stairwells have recently been summarized by Limb (1998).

Key leakage pathways

Shaw (1980) used an airtight test chamber to measure the leakage through windows, walls, balcony doors and various joints. Most of the air leakage values vary widely from building to building, and even within the same unit. Of all the windows tested, only one third of them passed the ASHRAE 90-75 standard. A larger fraction (two-thirds) of balcony doors meets the standard. The major air leakage sources in exterior walls are found to be floor–wall joints, windows and window sills.

Kovanen and Sateri (1997) measured the component leakage of two multi-family dwellings using direct (pressure chamber) and indirect (reductive sealing) methods. The main leakage route was found, again, to be the balcony door. Three out of eight apartment units became less airtight after renovation that was carried out without special attention to envelope sealing, even though the air tightness of the windows and apartment doors improved in every apartment. The most problematic component appeared to be the balcony door.

Measurements of the equivalent leakage areas of ten suite-access doors in some mid- to high-rise apartment buildings in Canada were taken to understand their ventilation characteristics (Wray, 2000). The leakages were found to be highly variable and did not meet smoke-control requirements. Often the primary ventilation air supply to apartment units is designed to be from the corridor through the door to the unit, which might explain the high leakage observed.

Murakami and Yoshino (1983) tested the component leakage of a few apartment units and rooms and found that there are many background leakages other than windows, doors, ventilation inlets and pipe openings. For example, in a bedroom tested, the leakage through ceiling, ceiling–wall, and floor/wall joints together accounted for three-fifths of the total leakage. Installed windows were often found to have excess air tightness not meeting their expected performance.

Shaw et al (1991) found that the exterior wall of a five-storey apartment is nine times more leaky than the floor–ceiling separations. Leakage to left and right partitions was somewhere in between the two extremes. Good agreement between the summations of individual leakage component and the measured overall leakage for a unit was observed.

Trends by building characteristics

Juodis (2000) and Hill (2001) did not find year built to be a determining factor for the air tightness of multi-family residential buildings. The study by Kronvall and Boman (1993), however, found the opposite. This difference can perhaps be explained by the fact that the later study focused on Swedish dwellings where building codes have more stringent specifications regarding air tightness over the years. Boman and Lyberg (1986) found that if older buildings have been retrofitted or weather-stripped, the age effect may become less significant.

Boman and Lyberg (1986) also found that the presence of a fireplace tends to correlate with higher air leakage in both single-family and multi-family dwellings. For dwellings that were built between 1940 and 1960,

Table 4.3 *List of Studies on the Air Tightness of Non-Residential Buildings*

Country	Source	Buildings tested
Belgium	Wouters et al (1988)	45 schools
Canada	Proskiw and Parekh (2001)	1 shopping mall
		1 indoor swimming pool
	Zhang and Barber (1995)	1 swine building
	Shaw (1981)	9 supermarkets
	Shaw and Jones (1979)	11 schools
	Tamura and Shaw (1976)	8 office buildings
France	Litvak et al (2001)	2 office buildings
		4 schools
		4 hotels
		2 multi-use halls
	Fleury et al (1998)	4 industrial buildings
Japan	Hayakawa and Togari (1990)	3 office buildings
Sweden	Lundin (1986)	9 industrial buildings
UK	Potter et al (1995)	12 office buildings
		12 industrial buildings
	Jones and Powell (1994)	3 industrial buildings
	Perera and Parkins (1992)	6 industrial buildings
	Perera et al (1990)	10 office buildings
US	Bahnfleth et al (1999)	1 office building
		1 library wing
	Cummings et al (1996)	69 small Commercial buildings
	Persily and Grot (1986)	8 office buildings

those with a fireplace have an averaged normalized leakage area nearly twice that of those without a fireplace. Blomsterberg et al (1995) found that apartments with passive stack ventilation are much tighter than the ones with exhaust ventilation.

Shaw et al (1991) observed that the overall air tightness values of four buildings with different wall constructions are not very different from each other. This is because the air tightness value of a wall assembly is largely dependent upon how well the vapour barrier/interior component is installed, instead of the wall construction type. Lundin (1981) found significant air leakage induced by the air/vapour barrier that breaks at the walls which separate apartment units. As a result, apartment-separating walls should be connected to the inside of the exterior wall in order to ensure a continuous air/vapour barrier enclosing the entire wooden frame.

To summarize, there are more leakage pathways in multi-family dwellings than in houses. There are also more structural differences between the different types of multi-family dwellings, from row houses to high-rise apartments. All these factors lead to more complex sets of measurement techniques and air leakage data. Many studies have identified potential important sources of air leakage and suggested ways to avoid them. However, the diverse sets of data also mean that a comprehensive and quantitative air leakage evaluation of multi-family dwellings remains a challenge.

Non-residential buildings

A recent analysis on existing air tightness data of 139 commercial and institutional buildings by Persily (1999) found that non-residential buildings are often not airtight enough. About half of the data analysed were part of a study conducted by Cummings et al (1996) on small, predominately one-storey commercial buildings. The rest include office, industrial and retail buildings, as well as schools, from Canada, Sweden, the UK and the US. No correlation between air tightness and building age or wall construction was observed. Part of the reason was that there was simply not enough data for trends to be identified. There were some indications, however, that taller buildings tend to have more airtight envelopes. This might be a result of more careful design and construction necessary to deal with more demanding structural requirements, such as increased wind loads and the control of rain penetration.

Few other measurements have been made available and they are listed in Table 4.3, together with those included in Persily's (1999) analysis. Even with the new additions, air tightness measurements of non-residential buildings remain scarce and they do not adequately represent the existing building stock. A recent literature review by Proskiw and Phillips (2001) summarized most of the same data as Persily's, but with the addition of measurements made in Canada. The bulk of their report focused on test methods and specifications for large buildings.

One of the earliest efforts was by Tamura and Shaw (1976), who tested eight new office buildings in Ottawa, Canada. More recently, Shaw and Reardon (1995) went back to six of these buildings that are still in use to determine the changes in their air tightness. Comparisons indicated that as a result of various retrofit measures applied, all but two building envelopes became more airtight than 20 years ago. The improvement in the overall air tightness value at 50Pa ranges from 25–43 per cent of its original value. There were two exceptions. One received no retrofit measure and deteriorated by 23 per cent with time. The other exception had all joints in the curtain wall re-caulked in 1990, and the building showed no change in air tightness, which suggested that the retrofit measure was just sufficient to offset the effect of aging. This study demonstrated that significant improvements can be realized in the overall air tightness by retrofit measures.

The experimental set-ups used by Tamura and Shaw (1976) and Shaw and Reardon (1995) were identical, and involve pressurizing the test building using the building's supply air system and measuring the corresponding pressure differences across the building envelope. In the US, Persily and Grot (1986) tested the air tightness of seven federal buildings in a similar manner. The difference between the two test methods lay in the way that the airflow through the air-handler system was measured. While the former used a pair of total pressure-averaging tubes together with a static pressure probe to measure airflow, the latter used a constant-injection tracer gas technique.

Persily and Grot (1986) also found that the federal buildings tested were comparable in air tightness to the Canadian buildings. They also commented that it was probably more appropriate to normalize the airflow by wall area only, instead of including roof area, because the roofs were constructed to be impervious to air. Normalizing the leakage rate with the wall area alone would lead to higher apparent air leakage values than if the roof area were included.

In countries such as the UK where many buildings are naturally ventilated, an alternative approach is needed. Measurements by BRE (Perera et al, 1990; Perera and Parkins, 1992) and Building Services Research and Information Association (BSRIA) (Potter et al, 1995) were obtained by attaching an external large-scale fan to the building. While the low-energy office building tested by BRE had the air tightness average of those tested in North America, most conventional office buildings were found to be leakier by a few fold. Litvak et al (2001) found that only 2 out of the 12 buildings sampled are in compliance with the French RT2000 regulation, which is a new thermal regulation that takes into account energy for ventilation. Most of the large commercial buildings tested had air tightness in the range of those tested in North America.

Hayakawa and Togari (1990) developed a simple test method that utilizes buoyancy caused by the stack effect instead of using fans to pressurize the test building. While the stack effect is active, the test building can be pressurized or depressurized by opening doors and windows on the bottom floor or top floors. Under calm weather conditions, Hayakawa and Togari measured the equivalent effective leakage area for three high-rise office buildings. This method was found to be effective if no large unknown cracks were present and the friction resistance of the airflow in the building was small.

The study by the Florida Solar Energy Center tested 69 small commercial buildings and found that a large fraction of them were leakier than the residential homes in the area (Cummings et al, 1996). Strip mall units were found to be 2.5 times leakier than detached buildings. The reason for this is that the attached units were often well connected to each other above the ceiling level.

A study by Shaw and Jones (1979) measured the air tightness of 11 Canadian schools and found lower values than those of office buildings (Tamura and Shaw, 1976). The results indicated that there was no meaningful relation between total energy consumption and the measured air leakage rate. Instead, poor workmanship and sealing were observed to be the cause of high air leakage. The air tightness of 45 Belgian schools tested by Wouters et al (1988) revealed a much wider range of values, even among the newly constructed schools.

The air tightness of industrial buildings has been tested by a few researchers using similar large-scale fan pressurization methods. The buildings tested by Lundin (1986) in Sweden were found to be a few fold tighter than those in the UK (Perera and Parkins, 1992; Potter and Jones, 1992; Jones and Powell, 1994) and in France (Fleury et al, 1998). A wider range in air tightness values was also observed in the UK and France compared to those in Sweden.

Bahnfleth et al (1999) attempted to measure the envelope air leakage of one floor of a university library by the floor-by-floor blower door method. However, they found that it was impossible to adequately seal a single floor in order to isolate it from its neighbours. Proskiw and Parekh (2001) proposed a new air tightness procedure to separate the exterior envelope air leakage from the interior partition air leakage in a multi-zone building. The preliminary test result at an indoor swimming pool, which was attached to a recreational complex, showed that this procedure seems to offer advantages over those of the pressure-masking technique.

To answer the need of assessing the installation of air barriers during construction, Knight et al (1995) developed test equipment that is capable of handling all materials and design configurations involved. The end product was called a Pressure Activated Chamber Test System, which used a soap solution to visualize the leaks present. The authors tested the equipment at three swimming pools, two healthcare facilities and a seven-storey building, and found the test procedure to be effective in identifying leaks.

Since many air leakage problems are caused by poor design and workmanship, practical guidelines for designers, contractors and developers have been made available by various agencies. For example, CMHC recommended certain jointing materials, primary air barriers and prefabricated assemblies that are effective in controlling air leakage in high-rise commercial buildings (Canam

Building Envelope Specialists Inc, 1999). National Institute of Standards and Technology (NIST) published a document on envelope design guidelines for federal office buildings to ensure thermal integrity and air tightness (Persily, 1993). Aside from its guidelines (Perera et al, 1994), BRE also developed a tool for predicting the air tightness of office building envelopes either at the design stage or before a major refurbishment (Perera et al, 1997). Comparison with ten office buildings in the BRE database showed reasonable agreement between measurements and predictions.

Trends by building characteristics

Unlike residential buildings where multiple studies have suggested that new dwellings are built tighter, Potter et al (1995) concluded otherwise from a comparison of office buildings built before and after 1990. Similarly, Cummings et al (1996) found that the small commercial buildings tested did not demonstrate a clear age trend. Shaw (1981) noticed that newly constructed supermarkets were found to be generally much leakier than older ones, which could be explained by the opening around the receiving doors with hydraulic ramp.

No significant trend has been observed between air leakage and construction materials of commercial buildings. However, building type can be an important factor because of the differences in typical architecture according to their functions. For example, when compared against hotels and schools, office buildings and multi-use halls appear to be leakier because of the presence of suspended ceilings (Litvak et al, 2001). The air leakages of the supermarkets and malls tested by Shaw (1981) were also found to be higher and more variable than schools and high-rise office buildings measured in Canada.

Ideally, air-conditioned buildings should have minimal air infiltration and naturally ventilated buildings should have air infiltration under occupant control. By comparing among the 12 buildings tested for air tightness, Potter et al (1995) found that the four naturally ventilated buildings tend to be tighter than the remaining eight, which have air conditioning. This shows that construction practices and defects often have a greater impact upon air tightness than building design.

Key leakage pathways

Air leakage at suspended ceilings where electrical, lighting and ventilation equipment are housed has been found to be significant among many of the 12 non-residential buildings tested in France (Litvak et al, 2001). A study by the Florida Solar Energy Center (Cummings et al, 1996) also found similar results among smaller commercial buildings. Some studies from the UK have shown that even the roof tops of large buildings are not guaranteed to be impervious to air infiltration (Perera and Parkins, 1992; Potter et al, 1995). This is somewhat counter-intuitive because most would assume that rain penetration problems would have prevented any buildings from having a leaky roof top.

Cracks along the top edge of most operable windows were also found to be an important source at a building tested, which was known to have air leakage-induced problems (Perera and Parkins, 1992). When compared to the ASHRAE window leakage standard of 0.77 l/s/m, Persily and Grot (1986) also found that many of the windows tested in the federal buildings exceeded that standard. However, it should be noted that the window leakage standard excluded leakage through the window frame and sash, which the test procedure included.

In relative terms, Potter et al (1995) found exposed cavities to be more problematic than windows. This means that electrical and service penetrations through the structure into the cavity are in need of careful sealing. Cummings et al (1996) found that this problem is particularly disastrous among small commercial buildings where cavities are commonly used as ducts or plenums.

Duct leakage among commercial buildings is profound even after accounting for the fact that they have a greater surface area than those in residential buildings (Cummings et al, 1996). The duct systems tested were about 70 times leakier than the Sheet Metal and Air Conditioning Contractors National Association (SMACNA) standard. Depending upon where the ducts are located, the impact on energy consumption can vary. However, excessive air leakage associated with the ventilation system among non-residential buildings is quite common. Among the 11 schools tested, Shaw and Jones (1979) found that 15–43 per cent of the overall air leakage can be attributed to the air intake and exhaust openings. The leakage through roof ventilators among leaky UK industrial buildings was found to be a bit less significant at 9 per cent (Jones and Powell, 1994).

Despite the fact that the test results on loading doors among UK industrial buildings were satisfactory, Potter and Jones (1992) noticed a wide variation in the quality of the roller shutter doors among the 12 industrial buildings tested. As improvements in the air tightness of other parts of the building progress, this leakage component should not be neglected. Recent work by Yuill et al (2000b) estimated the power-law coefficients of automatic doors as a function of door type and rate of use.

Another common air leakage pathway is elevator shafts since they are normally vented to atmosphere (Potter et al, 1995). It is therefore essential for elevator doors to be fitted with adequate seals. In an effort to isolate one floor from the others, Bahnfleth et al (1999) found numerous holes and cracks that could not be reached and sealed in elevator shafts. Among other leakage components such as the stairway, a literature search by Edwards (1999) concluded that the data on air leakage associated with elevator shafts is very limited. Data on many other important leakage pathways, such as underground parking, garage access doors and garbage chutes, is also lacking. The current state of knowledge on non-residential buildings is limited, but the new air-tightness regulation in UK that requires all new buildings to be tested is likely to bring about more air leakage data on wider range of buildings.

Dynamic airflow

Before concluding this state-of-the-art review, it would be remiss not to mention some of the more innovative techniques for measuring air tightness, even if they have not generated a lot of data. The discussion so far, and the vast majority of published air-tightness work, is on steady-state flow. The closest that most cracks and leaks actually get to a steady state is during fan pressurization tests. In this section we will review the issues associated with non-steady flow in relation to air tightness.

When considering time-varying airflows, there are two regimes, which we will call pseudo steady state and unsteady. The difference comes about because the change in airflow (or driving pressure) is either long or short compared to the characteristic time of the problem at hand. The characteristic time can be estimated by the time it takes sound to cross the leak, or to cross the building. It can also be estimated by the time for a boundary layer (or jet) to form, or for the flow to accelerate to steady state.

In a pseudo steady-state flow, the driving pressures are changing slowly enough so that the individual leaks are presumed to be instantaneously in equilibrium. Because the air leakage is inherently non-linear, pseudo steady state can generate complex phenomena despite the assumption of equilibrium.

Siren (1997) has shown that turbulence can cause a 5 per cent bias in the power-law flows using pseudo steady-state assumptions due to non-linearities. Whether a 5 per cent bias from turbulence is acceptable will depend upon the intended use of the data. Measurements by Sharples and Thompson (1996) confirm that there is no large difference due to these non-linearities, but the study does not contain an error analysis sufficient to separate out a 5 per cent bias from a null result.

AC pressurization

Siren (1997) and Sharples and Thompson (1996) refer to the well-known phenomena that occur when the flow actually begins to reverse (that is, fluctuate). The issue of how to treat fluctuating airflows from the perspective of ventilation is beyond the scope of this section; but the physical principles of fluctuating pressures led to the development of a dynamic air tightness measurement technique known as AC pressurization.

Sherman and Modera (1986) describe the physics of AC pressurization. The system operates by putting a sinusoidal volume change (of order 1Hz) on the inside of the building and measuring the pressure response. At this frequency, the flow is pseudo steady with respect to flow through cracks, but is fast enough to allow compression in the building and, thus, phase shifts from which information can be extracted. The approach breaks down when any individual opening becomes sufficiently large that it can be considered unsteady. This typically can happen for open windows or un-dampered chimneys, but not for more normal building leaks.

In a pair of papers, Dewsbury (1996a, 1996b) examined additional analysis approaches involving low frequencies: Fourier analysis and non-linear optimization strategies. The lower the frequency, the less susceptible the analysis is to the effects of inertia and flexing of the envelope. Since low frequencies imply low pressure, signal-to-noise ratios can become an issue.

Because of its relative complexity compared to fan pressurization, AC pressurization has not seen widespread

---

BOX 4.4 PULSE PRESSURIZATION

AC pressurization has no DC component and uses repeated sinusoidal variations; but Sherman and Modera (1988) have also devised a dynamic air tightness measurement approach that has a single perturbation. Pulse Pressurization works by providing a pressure pulse to the inside of a building (for example, from a compressed air tank) and then watching the pressure decay. The power-law equation predicts a finite recovery time for such a decay and can be used to analyse the data to determine leakage and volume.

Like AC Pressurization, the limitation of this procedure is when unsteady flow develops. The problem for pulse pressurization comes not from large external openings, but from the need for the pressure to be the same throughout the volume of the space. This tends to limit the application to small homes or apartments unless multiple injectors are used.

use. It has, however, been used in some special circumstances when fan pressurization was undesirable.

Summary

The physics of air leakage through building components is non-linear. The non-linearity of the process can lead to some challenging measurement and interpretation problems. The fundamental forms of the air leakage equations are not *a priori* clear; but there is general agreement that a power-law formulation is theoretically justifiable and empirically valid.

There is less consensus on how to report air leakage data, and several metrics are commonly in use. The difference of opinion comes, in part, from the fact that different quantities are useful for different purposes. Assuming a power-law description, all two-parameter (un-normalized) formalizations are interchangeable. Single parameter forms provide less accuracy, but can be useful for specific purposes.

Regardless of the parameterization chosen, air leakage data show large variability even within ostensibly homogeneous populations. It is not atypical to see log-normal distributions where the leakiest 10 per cent of the buildings are an order of magnitude leakier than the tightest 10 per cent. The large variation can be attributable to variations in workmanship, variations in structure use and maintenance, and variations in renovation and repair activities.

Despite the variance, there are some very general and not overly surprising trends that can be gleaned from the data. The air leakage characteristics of single-family dwellings are better understood than multi-family dwellings or non-residential buildings because more measurements are available. Dwellings in more severe climates, such as in Sweden and Canada, have shown to be more airtight than in the US and the UK, where the climate is milder. In countries where there is a demand for tighter envelopes driven by building codes or energy savings, new constructions have been shown to be more airtight than older ones. Dwellings of different construction types have different envelope air-tightness properties, but some air leakage pathways are common among many dwellings, such as the connections between building materials and components. Leakage to attics, basements, crawl spaces and garages is significant and raises additional energy and health concerns. Many studies have addressed the effectiveness of air barriers and building materials to minimize leakage; but it is often the quality of workmanship and careful design that are the determining factors in achieving desirable air tightness.

When compared to single-family dwellings, individual units in multi-family dwellings tend to be more airtight. However, this does not mean that multi-family buildings are sufficiently airtight, particularly among high-rise buildings. Despite the fact that air leakage to the exterior still dominates, studies have also revealed significant air leakage between units in multi-family dwellings. Stack-induced vertical airflow between units and in elevator shafts and stairwells are among some of the concerns. Partly limited by the number of measurements available, few trends have been observed between building characteristics and air tightness. The task of identifying air leakage trends is further complicated by large variations in air tightness found between units in a same building. Many of the findings observed among single-family dwellings also apply to multi-family dwellings. For example, dwellings with fireplaces tend to be leakier, and the integrity of the air barrier system is crucial to ensure air tightness of the unit.

Office buildings, industrial buildings, schools and retail stores are among the few non-residential building types for which air tightness measurements are available. Since measurements in these buildings often required large-scale equipment, a few alternative methods have been proposed so that measurements can be made more easily and can be less costly. However, these methods remains in research status and are not yet widely used in practice. In fact, the most recent measurements were collected almost exclusively by using large-scale fan pressurization. It is evident that commercial buildings are rarely airtight enough. There is a slight geographical difference in the air tightness of buildings in Sweden (most tight), the UK (most leaky), and North America (somewhere in between). On the other hand, air tightness is unrelated to age or construction materials. Suspended ceilings, exposed cavities and ventilation ducts are among the key leakage pathways. Due to the architectural differences of different building types, some tend to be leakier than others. But until more data has been collected, these trends are only scattered observations that cannot be generalized to the various commercial building types. Nonetheless, learning from air-leakage studies is useful in yielding practical guidelines to help designers and builders control air leakage in commercial buildings more effectively, and in informing energy-use and indoor air quality analysis.

Acknowledgements

The research for this chapter was supported by the Assistant Secretary for Energy Efficiency and Renewable Energy, Building Technology Program, of the US Department of Energy, under Contract No DE-AC03-76SF00098.

References

Adalberth, K. (1997) 'Airtight building – A practical guide', in *Proceedings: Ventilation and Cooling, 18th Air Infiltration and Ventilation Centre Conference*, Athens, Greece, 23–26 September

Air-Ins Inc (1998a) *Air Permanence of Building Materials*, Research Highlights, Technical Series 98–108, Canada Mortgage and Housing Corporation, Canada

Air-Ins Inc (1998b) *Airtightness Tests on Components Used to Join Different or Similar Materials of the Building Envelope*, Research Highlights, Technical Series 98–121, Canada Mortgage and Housing Corporation, Canada

Armstrong, P., Dirks, J., Klevgard, L., Matrosov, Y., Olkinuora, J. and Saum, D. (1996) 'Infiltration and ventilation in Russian multi-family buildings', in *Proceedings: Profiting from Energy Efficiency, Summer Study on Energy Efficiency in Buildings*, American Council for an Energy Efficient Economy, Washington, DC

ASHRAE (American Society of Heating, Refrigerating and Air-Conditioning Engineers) (1988) *ASHRAE Standard 119: Air Leakage Performance for Detached Single-Family Residential Buildings*, ASHRAE, Atlanta, GA

ASHRAE (2001) *ASHRAE Handbook of Fundamentals*, ASHRAE, Atlanta, GA, Chapter 26

ASTM (American Society of Testing and Materials) (1992a) *ASTM Standard E1186–87: Practices for Air Leakage Site Detection in Building Envelopes*, ASTM Book of Standards, vol 4, no 11, ASTM, Philadelphia, PA

ASTM (1992b) *ASTM, E1258–88: Standard Test Method for Airflow Calibration of Fan Pressurization Devices*, ASTM Book of Standards, vol 4, no 11, ASTM, Philadelphia, PA

ASTM (1998) *ASTM Standard E1186-98: Standard Practices for Air Leakage Site Detection in Building Envelopes and Air Retarder Systems*, American Society for Testing and Materials, Philadelphia, PA

ASTM (2000) *ASTM Standard E779–99: Test Method for Determining Air Leakage by Fan Pressurization*, ASTM Book of Standards, vol 4, no 11, ASTM, Philadelphia, PA

ASTM (2002) *ASTM Standard E1827–96: Standard Test Methods for Determining Airtightness of Buildings Using an Orifice Blower Door*, ASTM Book of Standards, vol 4, no 11, ASTM, Philadelphia, PA

ASTM (2004) *ASTM, E1554: Standard Test Method for Determining External Air Leakage of Air Distribution Systems*, ASTM Book of Standards, vol 4, no 11, ASTM, Philadelphia, PA

Bahnfleth, W. P., Yuill, G. K. and Lee, B. W. (1999) 'Protocol for field testing of tall buildings to determine envelope air leakage rate', *ASHRAE Transactions*, vol 105, no 2, pp27–38

Barles, P. and Boulanger, X. (2000) 'Airtightness and under-pressures measurements in French apartments', in *Proceedings: Innovations in Ventilation Technology, 21st Air Infiltration and Ventilation Centre Conference*, Hague, The Netherlands, 26–29 September

Bassett, M. (1985) 'The infiltration component of ventilation in New Zealand houses', in *Proceedings: Ventilation Strategies and Measurement Techniques, 6th Air Infiltration and Ventilation Centre Conference*, The Netherlands, 16–19 September

Blomsterberg, A. (1977) *Air Leakage in Dwellings*, Department of Buildings Construction Report No 15, Swedish Royal Institute of Technology, Sweden

Blomsterberg, A., Carlsson, T. and Kronvall, J. (1995) 'Short term and long term measurements of ventilation in dwellings', in *Proceedings: Implementing the Results of Ventilation Research, 16th Air Infiltration and Ventilation Centre Conference*, Palm Springs, US, 19–22 September

Boman, C. A. and Lyberg, M. D. (1986) 'Analysis of air change rates in Swedish residential buildings', in Trechsel, H. R. and Lagus, P. L (eds) *Measured Air Leakage of Buildings*, ASTM STP 904, American Society for Testing and Materials, Philadelphia, US, pp399–406

Bossaer, A., Demeester, J., Wouters, P, Vandermarke, B. and Vangroenweghe, W. (1998) 'Airtightness performance in new Belgian dwellings', in *Proceedings: Ventilation Technologies in Urban Areas, 19th Air Infiltration and Ventilation Centre Conference*, Oslo, Norway, 28–30 September

Brennan, T., Pyle, B., Williamson, A., Balzer, F. and Osborne, M. (1990) 'Fan door testing on crawl space buildings', in Sherman, M. H. (ed) *Air Change Rate and Airtightness in Buildings*, ASTM STP 1067, American Society of Testing and Materials, Philadelphia, US, pp146–150

Buchan, Lawton, Parent Limited (1992) *Airtightness and Air Quality in Preserved Wood Foundations*, Final report, Canada Mortgage and Housing Corporation, Canada

Buchan, Lawton, Parent Limited, Jools Development and Drerup Armstrong Limited (1996) *Air Leakage Performance of 11 Log Houses in Eastern Ontario and Western Quebec*, Research report, Canada Mortgage and Housing Corporation, Canada

Caffey, G. E. (1979) 'Residential air infiltration', *ASHRAE Transactions*, vol 85, no 9, pp41–57

CAN/CGSB (Canadian General Standards Board) (1986) *CAN/CGSB Standard 149: Determination of the Airtightness of Building Envelopes by Fan Depressurization Method*, Canadian General Standards Board, Canada

CAN/CGSB (1996) *CAN/CGSB-149.15-96: Determination of the Overall Envelope Airtightness of Buildings by the Fan Pressurization Method Using the Building's Air Handling Systems*, Canadian General Standards Board, Canada

Canam Building Envelope Specialists Inc (1999) *Practical Guidelines for Designers, Contractors and Developers on the Installation of Air Leakage Control Measures in New and Existing High-Rise Commercial Buildings*, Research Highlights Technical Series 99–119, Canada Mortgage and Housing Corporation, Canada

Carrié, F. R., Wouters, P., Ducarme, D., Andersson, J., Faysse, J. C., Chaffois, P., Kilberger, M. and Patriarca, V. (1997) 'Impacts of air distribution system leakage in Europe: The SAVE duct European programme' in *Proceedings: Ventilation and Cooling, 18th Air Infiltration and Ventilation Centre Conference*, Athens, Greece, 23–26 September, pp651–660

Chan, W. R., Nazaroff, W. W., Price, P. N., Sohn, M. D. and Gadgil, A. J. (2005) 'Analyzing a database of residential air leakage in the United States', *Atmospheric Environment*, vol 39, pp3445–3455.

Colliver, D. G., Murphy, W. E. and Sun, W. (1994) 'Development of a building component air leakage data base', *ASHRAE Transactions*, vol 100, no 1, pp292–305

Cornish, J. P., Henderson, G., Uglow, C. E., Stephen, R. K., Southern, J. R. and Sanders, C. H. (1989) *Improving the Habitability of Large Panel System Dwellings*, Report 154, Building Research Establishment, UK

Cummings, J. B., Withers, C. R., Moyer, N. A., Fairey, P. W. and McKendry, B. B. (1996) *Uncontrolled Airflow in Non-Residential Buildings*, Final report, FSEC-CR-878-96, Florida Energy Solar Center, US

Desjarlais, A. O., Childs, K. W. and Christian, J. E. (1998) 'To storm or not to storm: Measurement method to quantity impact of exterior envelope airtightness on energy usage prior to construction', in *Proceedings: Thermal Performance of Exterior Envelopes of Buildings VII*, Clearwater Bay, Florida, 6–10 December, pp607–618

Dewsbury, J. (1996a) 'AC pressurization: Analysis of non-linear optimization', *Building Services Engineering Research and Technology*, vol 17, no 2, pp65–71

Dewsbury, J. (1996b) 'AC pressurization: Fourier analysis and the effect of compressibility', *Building Services Engineering Research and Technology*, vol 17, no 2, pp73–77

Diamond, R. C., Dickinson, J. B., Lipschutz, R. D., O'Regan, B. and Schole, B. (1982) *The House Doctor's Manual*, Report 3017, Lawrence Berkeley National Laboratory, California, US

Dietz, R. N., D'Ottavio, T. W. and Goodrich, R. W. (1985) 'Multizone infiltration measurements in homes and buildings using a passive perfluorocarbon tracer method', *ASHRAE Transactions*, vol 91, no 2, pp1761–1776

Dumont, R. S. (1993) 'An overview of insulation and airtightness in wood frame construction', in *Proceedings: Innovative Housing*, Canada National Research Council, Vancouver, Canada, 21–25 June 1993

Dumont, R. S. and Snodgrass, L. J. (1990) 'Investigation of chimney backflow conditions: A case study in a well-sealed house', *ASHRAE Transactions*, vol 96, no 1, pp53–59

Edwards, C. (1999) *Modeling of Ventilation and Infiltration Energy Impacts in Mid and High-Rise Apartment Buildings*, Report, Sheltair Scientific Limited, Vancouver, Canada

Elmahdy, A. H. (2003) 'Quantification of air leakage effects on the condensation resistance of windows', *ASHRAE Transactions*, vol 109, no 1, pp600–606

Etheridge, D.W. (1977) 'Crack flow equations and scale effect', *Building and Environment*, vol 12, pp181–189

Flanders, S. N. (1995) 'Fan pressurization measurements by four protocols', in *Proceedings: Implementing the Results of Ventilation Research, 16th Air Infiltration and Ventilation Centre Conference*, Palm Springs, 19–22 September

Fleury, E., Millet, J. R., Villenave, J. G., Veyrat, O. and Morisseau, C. (1998) 'Theoretical and field study of air change in industrial buildings', in *Proceedings: Ventilation Technologies in Urban Areas, 19th Air Infiltration and Ventilation Centre Conference*, Oslo, Norway, 28–30 September, pp57–65

Francisco, P. (2001) 'The current state of duct leakage measurements: Field evaluation of five techniques,' *Home Energy*, March/April, pp33–35

Fugler, D. W. (1999) 'Conclusions from ten years of Canadian attic research', *ASHRAE Transactions*, vol 105, no 1, pp819–825

Fugler, D. W. and Moffatt, S. D. (1994) 'Investigation of crawl space performance in British Columbia', *ASHRAE Transactions*, vol 100, no 1, pp1411–1419

Gettings, M. B. (1989) 'Blower-door directed infiltration reduction procedure description and field test', *ASHRAE Transactions*, vol 95, no 1, pp58–63

Gulay, B. W., Stewart, C. D. and Foley, G. J. (1993) *Field Investigation Survey of Airtightness, Air Movement and Indoor Air Quality in High Rise Apartment Buildings*, Summary report, Canada Mortgage and Housing Corporation, Canada

Gusdorf, J. (2003) Personal communication, October 2003; see also EnerGuide for Houses Program, Natural Resources Canada, www.oee.nrcan.gc.ca/energuide/home.cfm

Haile, S., Bernier, M. A., Patenaude, A. and Jutras, R. (1998) 'The combined effect of air leakage and conductive heat transfer in window frames and its impacts on the Canadian energy rating procedure', *ASHRAE Transactions*, vol 104, no 1a, pp176–184

Hamlin, T. and Gusdorf, J. (1997) *Airtightness and Energy Efficiency of New Conventional and R-2000 Housing in Canada, 1997*, CANMET Energy Technology Centre, Department of Natural Resources, Canada

Harrje, D. T. (1981) 'Building envelope performance testing,' *ASHRAE Journal*, March, pp39–41

Harrje, D. T., Blomsterberg, A. and Persily, A. K (1979) *Reduction of Air Infiltration Due to Window and Door Retrofits*, CU/CEES Report 85, Princeton University, US

Harrje, D. T. and Dutt, G. S. (1981) 'House doctors program: Retrofits in existing buildings,' in *Proceedings: Second Air Infiltration and Ventilation Centre Conference*, pp61–72

Harrje, D. T., Dutt, G. S. and Beya, J. E. (1979) 'Locating and eliminating obscure, but major energy losses in residential housing,' *ASHRAE Transactions*, vol 85, no 2, pp521–534

Hayakawa, S. and Togari, S. (1990) 'Simple test method for evaluating exterior wall airtightness of tall office buildings', in Sherman, M. H. (ed) *Air Change Rate and Airtightness in Buildings*, ASTM STP 1067, American Society for Testing and Materials, Philadelphia, US, pp231–245

Henry, R. and Patenaude, A. (1998) 'Measurements of window air leakage at cold temperatures and impact on annual energy performance of a house', *ASHRAE Transactions*, vol 104, no 1b, pp1254–1260

Hill, D. (2001) 'Valuing air barrier', *Home Energy*, September/October, pp29–32

ISO (International Organization for Standardization) (1996) *ISO Standard 9972: Thermal Insulation – Determination of Building Air Tightness – Fan Pressurization Method*, ISO, Geneva, Switzerland

Jones, P. J. and Powell, G. (1994) 'Reducing air infiltration losses in naturally ventilated industrial buildings', in *Proceedings: The Role of Ventilation, 15th Air Infiltration and Ventilation Centre Conference*, Buxton, UK, 27–30 September, pp397–409

Juodis, E. (2000) 'Energy saving and airtightness of blocks of flats in Lithuania', *Indoor Built Environment*, vol 9, pp143–147

Knight, K., Knight, G., Sharp, J., Guerriero, J. and Phillips, B. (1995) *Evaluating Test Equipment for Air Tightness of Construction Details*, Report, Canada Mortgage and Housing Corporation, Canada

Kosny, J., Christian, J. E. and Desjariais, A. O. (1998) 'Performance check between whole building thermal performance criteria and exterior wall measured clear wall R-value, thermal bridging, thermal mass and airtightness', *ASHRAE Transactions*, vol 104, no 2, pp1379–1389

Kovanen, K. and Sateri, J. (1997) 'Airtightness of apartments before and after renovation', in *Proceedings: Ventilation and Cooling, 18th Air Infiltration and Ventilation Centre Conference*, Athens, Greece, 23–26 September, pp117–125

Kronvall, J. (1980) *Air Tightness Measurements and Measurement Methods*, Report D8, Swedish Council for Building Research, Stockholm, Sweden

Kronvall, J. and Boman, C. A. (1993) 'Ventilation rates and air tightness levels in the Swedish housing stock', in *Proceedings: Energy Impact of Ventilation and Air Infiltration, 14th Air Infiltration and Ventilation Centre Conference*, Copenhagen, Denmark, 21–23 September

Kula, H. G. and Sharples, S. (1994) 'Airflow through smooth and rough cracks', in *Proceedings: The Role of Ventilation, 15th Air Infiltration and Ventilation Centre Conference*, Buxton, UK, 27–30 September, pp710–717

Lagus, P. L. and King, J. C. (1986) 'Air leakage and fan pressurization measurements in selected naval housing', in Trechsel, H. R. and Lagus, P. L. (eds) *Measured Air Leakage of Buildings*, ASTM STP 904, American Society for Testing and Materials, Philadelphia, pp5–16

Levin, P. (1988) 'Air leakage between apartments', in *Proceedings: Effective Ventilation, Ninth Air Infiltration and Ventilation Centre Conference*, Gent, Belgium, 12–15 September, pp251–263

Levin, P. (1991) *Building Technology and Airflow Control in Housing*, Report D16, Swedish Council for Building Research, Stockholm, Sweden

Limb, M. J. (1998) *Ventilation and Infiltration Characteristics of Life Shafts and Stair Wells*, Annotated Bibliography 4, Air Infiltration and Ventilation Centre, Coventry, UK

Litvak, A., Boze, D. and Kilberger, M. (2001) 'Airtightness of 12 non-residential large buildings results from field measurement studies', in *Proceedings: Market Opportunities for Advanced Ventilation Technology, 22nd Air Infiltration and Ventilation Centre Conference*, Bath, UK, 11–14 September

Litvak, A., Kilberger, M. and Guillot, K. (2000) 'Field measurement results of the airtightness of 64 French dwellings', in *RoomVent 2000: Seventh International Conference on Air Distribution in Rooms*, University of Reading, UK, and Elsevier Science Publishers, pp1093–1098

Louis, M. J. and Nelson, P. E. (1995) 'Extraneous air leakage from window perimeters', in Modera, M. P. and Persily, A. K. (eds) *Airflow Performance of Building Envelopes, Components and Systems*, ASTM STP 1255, American Society for Testing and Materials, Philadelphia, pp108–122

Love, J. A. (1990) 'Airtightness survey of row houses in Calgary, Alberta', in Sherman, M. H. (ed) *Air Change Rate and Airtightness in Buildings*, ASTM STP 1067, American Society for Testing and Materials, Philadelphia, pp194–210

Lowe, R. J., Curwell, S. R., Bell, M. and Ahmad, A. (1994) 'Airtightness in masonry dwellings: Laboratory and field experience', *Building Services Engineers Research Technology*, vol 15, no 3, pp149–155

Lowe, R. J., Johnston, D. and Bell, M. (1997) 'Airtightness in UK dwellings: A review of some recent measurements', in *Proceedings: The Second International Conference on Buildings and the Environment*, Paris, 9–12 June

Lundin, L. I. (1981) 'Airtightness in terraced houses', in *Proceedings: Building Design for Minimum Air Infiltration, 2nd Air Infiltration and Ventilation Centre Conference*, Sweden, 21–23 September, pp185–195

Lundin, L. I. (1986) 'Air leakage in industrial buildings – description of equipment', in Trechsel, H. R. and Lagus, P. L. (eds) *Measured Air Leakage of Buildings*, ASTM STP

904, American Society for Testing and Materials, Philadelphia, pp101–105

Matson, N. E., Feustel, H. E., Warner, J. L. and Talbott, J. (1994) 'Climate-based analysis of residential ventilation options: New York analysis', in *Proceedings: The Role of Ventilation, 15th Air Infiltration and Ventilation Centre Conference*, Buxton, UK, 27–30 September

McGrath, P. T. and McManus, J. (1996) 'Air infiltration from basements and sub-floors to the living space', *Building Services Engineers Research Technology*, vol 17, no 2, pp85–87

McWilliams, J. A. (2002) *Review of Airflow Measurement Techniques*, Annotated Bibliography 12, Air Infiltration and Ventilation Centre, Coventry, UK

Modera, M. P. and Wilson, D. J. (1990) 'The effects of wind on residential building leakage measurements', in Sherman, M. H. (ed) *Air Change Rate and Airtightness in Buildings*, ASTM STP 1067, American Society for Testing and Materials, Philadelphia, US, pp132–145

Murakami, S. and Yoshino, H. (1983) 'Air-tightness of residential buildings in Japan', in *Proceedings: Air Infiltration Reduction in Existing Buildings, Fourth Air Infiltration and Ventilation Centre Conference*, Elm, Switzerland, 26–28 September

Nichols, L. and Gerbasi, D. (1997) *Field Investigation into Airtightness, Air Movement and Quality of Inside Air in Medium-Sized Residential Buildings and an Energy Audit of These Buildings*, Research Highlights, Technical Series 97–106, Canada Mortgage and Housing Corporation, Canada

Orme, M., Liddament, M. and Wilson, A. (1994) *An Analysis and Data Summary of the AIVC's Numerical Database*, Technical Note 44, Air Infiltration and Ventilation Centre, Coventry, UK

Palmiter, L., Heller, J. and Sherman, M. (1995) 'Measured airflow in a multifamily building', in Modera, M. P. and Persily, A. K. (eds) *Airflow Performance of Building Envelopes, Components, and Systems*, ASTM STP 1255, American Society for Testing and Materials, Philadelphia, pp7–22

Parekh, A. (1992) 'Power demand and energy savings through air leakage control in high-rise residential buildings in cold climates', in *Proceedings: Thermal Performance of the Exterior Envelopes of Buildings V*, Clearwater Beach, Florida, 7–10 December, pp632–642

Parent, D., Stricker, S. and Fugler, D. (1996) 'Ventilation in houses with distributed heating systems', in *Proceedings: Optimum Ventilation and Airflow Control in Buildings, 17th Air Infiltration and Ventilation Centre Conference*, Gothenburg, Sweden, 17–20 September, pp186–195

Perera, M. D. A. E. S., Henderson, J. and Webb, B. C. (1997) 'Predicting envelope air leakage in large commercial buildings before construction', in *Proceedings: Ventilation and Cooling, 18th Air Infiltration and Ventilation Centre Conference*, Athens, Greece, 23–26 September, pp205–211

Perera, M. D. A. E. S. and Parkins, L. (1992) 'Airtightness of US buildings: Status and future possibilities', *Environmental Policy and Practice*, vol 2, no 2, pp143–160

Perera, M. D. A. E. S., Stephan, R. K. and Tull, R. G. (1990) 'Airtightness measurements in two UK office buildings', in Sherman, M. H. (ed) *Air Change Rate and Airtightness in Buildings*, ASTM STP 1067, American Society for Testing and Materials, Philadelphia, pp211–221

Perera, M. D. A. E. S., Turner, C. H. C. and Scivyer, C. R. (1994) *Minimizing Air Infiltration in Office Buildings*, Report 265, Building Research Establishment, UK

Persily, A. K. (1986) 'Measurements of air infiltration and airtightness in passive solar homes', in Trechsel, H. R. and Lagus, P. L. (eds) *Measured Air Leakage of Buildings*, ASTM STP 904, American Society for Testing and Materials, Philadelphia, pp46–60

Persily, A. K. (1993) *Envelope Design Guidelines for Federal Office Buildings: Thermal Integrity and Airtightness*, National Institute of Standards and Technology, US

Persily, A. K. (1999) 'Myths about building envelopes', *ASHRAE Journal*, vol 41, no 3, pp39–47

Persily, A. K. and Grot, R. A. (1986) 'Pressurization testing of federal buildings', in Trechsel, H. R. and Lagus, P. L. (eds) *Measured Air Leakage of Buildings*, ASTM STP 904, American Society for Testing and Materials, Philadelphia, pp184–200

Petrie, T. W., Kosny, J., Desjarlais, A. O., Atchley, J. A., Childs, P. W., Ternes, M. P. and Christian, J. E. (2003) *How Insulating Concrete Form versus Conventional Construction of Exterior Walls Affects Whole Building Energy Consumption: Results from a Field Study and Simulation of Side-by-Side Houses*, Report, Oak Ridge National Laboratory, US

Petrone Architects (2000) *Rigid Air Barrier Assemblies*, Research Highlights, Technical Series 00–111, Canada Mortgage and Housing Corporation, Canada

Pittomvils, J., Hens, H. and Van Bael, F. (1996) 'Evaluation of ventilation system in very low energy houses', in *Proceedings: Optimum Ventilation and Airflow Control in Buildings, 17th Air Infiltration and Ventilation Centre Conference*, Gothenburg, Sweden, 17–20 September

Potter, I. N. and Jones, T. J. (1992) *Ventilation Heat Loss in Factories and Warehouses*, Technical Note TN 7/92, Building Services Research and Information Association, Bracknell, UK

Potter, I. N., Jones, T. J. and Booth, W. B. (1995) *Air Leakage of Office Buildings*, Technical Note TN 8/95, Building Services Research and Information Association, Bracknell, UK

Proskiw, G. (1995a) 'Air leakage characteristics of various rough-opening sealing methods for windows and doors', in Modera, M. P. and Persily, A. K. (eds) *Airflow Performance of Building Envelopes, Components and Systems*, ASTM STP 1255, American Society for Testing and Materials, Philadelphia, pp123–134

Proskiw, G. (1995b) 'Measured airtightness of 24 detached houses over periods of up to three years', in Modera, M. P. and Persily, A. K. (eds) *Airflow Performance of Building Envelopes, Components and Systems*, ASTM STP 1255, American Society for Testing and Materials, Philadelphia, pp248–265

Proskiw, G. (1998) 'The variation of airtightness of wood-frame houses over an 11-year period', in *Proceedings: Thermal Performance of the Exterior Envelopes of Buildings VII*, Clearwater Beach, Florida, 6–10 December, pp745–751

Proskiw, G. and Parekh, A. (2001) 'A proposed test procedure for separating exterior envelope air leakage from interior partition air leakage', in *Proceedings: Performance of Exterior Envelopes of Whole Buildings VIII*, Integration of Building Envelopes, Clearwater Beach, Florida, 2–7 December

Proskiw, G. and Phillips, B. (2001) *Air Leakage Characteristics, Test Methods and Specifications for Large Buildings*, Report, Canada Mortgage and Housing Corporation, Canada

Reardon, J. T., Kim, A. K. and Shaw, C. Y. (1987) 'Balanced fan depressurization method for measuring component and overall air leakage in single- and multifamily dwellings', *ASHRAE Transactions*, vol 93, no 2, pp137–152

Sateri, J., Heikkinen, J. and Pallari, M. (1995) 'Feasibility of ventilation heat recovery in retrofitting multi-family buildings', in *Proceedings: Implementing the Results of Ventilation Research, 16th Air Infiltration and Ventilation Centre Conference*, Palm Springs, US, 19–22 September

Scanada Consultants Limited (2001) *Air Infiltration from Attached Garages in Canadian Houses*, Research Highlights, Technical Series 01–122, Canada Mortgage and Housing Corporation, Canada

Sharples, S. and Thompson, D. (1996) 'Experimental study of crack flow with varying pressure differentials', in *Proceedings: Optimum Ventilation and Airflow Control in Buildings, 17th Air Infiltration and Ventilation Centre Conference*, Gothenburg, Sweden, 17–20 September, pp243–253

Shaw, C. Y. (1980) 'Methods for conducting small-scale pressurization tests and air leakage data of multi-storey apartment buildings', *ASHRAE Transactions*, vol 86, no 1, pp241–250

Shaw, C. Y. (1981) 'Air tightness: Supermarkets and shopping malls', *ASHRAE Journal*, March, pp44–46

Shaw, C. Y. and Jones, L. (1979) 'Air tightness and air infiltration of school buildings', *ASHRAE Transactions*, vol 85, no 1, pp85–95

Shaw, C. Y. and Reardon, J. T. (1995) 'Changes in airtightness levels of six office buildings', in Modera, M. P. and Persily, A. K. (eds) *Airflow Performance of Building Envelopes, Components, and Systems*, ASTM STP 1255, American Society for Testing and Materials, Philadelphia, pp47–57

Shaw, C. Y., Gasparetto, S. and Reardon, J. T. (1990) 'Methods for measuring air leakage in high-rise apartments', in Sherman, M. H. (ed) *Air Change Rate and Airtightness in Buildings*, ASTM STP 1067, American Society for Testing and Materials, Philadelphia, pp222–230

Shaw, C. Y., Magee, R. J. and Rousseau, J. (1991) 'Overall and component airtightness values of a five-story apartment building', *ASHRAE Transactions*, vol 97, no 2, pp347–353

Sherman, M. H. (1987) 'Estimation of infiltration from leakage and climate indicators', *Energy and Buildings*, vol 10, pp81–86

Sherman, M. H. (1988) 'Estimation of infiltration from leakage and climate indicators', *Energy and Buildings*, vol 10, pp81–86

Sherman, M. H. (1992) 'A power-law formulation of laminar flow in short pipes', *Fluids Engineering*, vol 114, pp601–605

Sherman, M. H. (1993) 'A power law formulation of laminar flow in short pipes', *Journal of Fluids Engineering*, vol 114, pp601–605

Sherman, M. H. and Dickerhoff, D. (1998) 'Air-tightness of US dwellings', *ASHRAE Transactions*, vol 104, no 2, pp1359–1367

Sherman, M. H. and Matson, N. E. (1997) 'Residential ventilation and energy characteristics', *ASHRAE Transactions*, vol 103, no 1, pp717–730

Sherman, M. H. and Matson, N. E. (2001) 'Air tightness of new houses in the US', in *Proceedings: Market Opportunities for Advanced Ventilation Technology, 22nd Air Infiltration and Ventilation Centre Conference*, Bath, UK, 11–14 September

Sherman, M. H. and Modera, M. P. (1986) 'Low frequency measurement of the leakage in enclosures', *Review of Scientific Instruments*, vol 57, no 7, pp1427–1430

Sherman, M. H. and Modera, M. P. (1988) 'Signal attenuation due to cavity leakage', *Journal of the Acoustic Society of America*, vol 84, no 6, pp2163–2169

Sherman, M. H. and Palmiter, L. (1995) 'Uncertainty in fan pressurization measurements', in Modera, M. P. and Persily, A. K. (eds) *Airflow Performance of Building Envelopes, Components, and Systems*, ASTM STP 1255, American Society for Testing and Materials, Philadelphia, pp266–283

Sikander, E. and Olsson-Jonsson, A. (1998) 'Airtightness of timber frame buildings not having a plastic film vapor barrier', in *Proceedings: Ventilation Technologies in Urban Areas, 19th Air Infiltration and Ventilation Centre Conference*, Oslo, Norway, 28–30 September

Siren, K. (1997) 'A modification of the power law equation to account for large-scale wind turbulence', in *Proceedings: Ventilation and Cooling, 18th Air Infiltration and Ventilation Centre Conference*, Athens, Greece, 23–26 September, pp557–561

Stephen, R. K. (1998) *Airtightness in UK Dwellings: BRE's Test Results and Their Significance*, Report 359, Building Research Establishment, UK

Tamura, G. T. and Shaw, C. Y. (1976) 'Studies on extorter wall air tightness and air infiltration of tall buildings', *ASHRAE Transactions*, vol 82, no 2, pp122–134

Tamura, G. T. and Shaw, C. Y. (1981) 'Field checks on building pressurization for smoke control in high-rise buildings', *ASHRAE Journal*, February, pp21–25

Walker, I. S., Wilson, D. J. and Sherman, M. H. (1997) 'A comparison of the power law to quadratic formulations for air infiltration calculations', *Energy and Buildings*, June, pp293–299

Wilcox, B. A. and Weston, T. A. (2001) 'Measured infiltration reduction in California production houses using house-wrap', in *Proceedings: Performance of Exterior Envelopes of Whole Buildings VIII, Integration of Building Envelopes*, Clearwater Beach, Florida, 2–7 December 2001

Wouters, P., Bossaer, A., Demeester, J., Ducarme, D., Vandermarke, B. and Vangroenweghe, W. (1997) 'Airtightness of new Belgian dwellings: An overview picture', in *Proceedings: Ventilation and Cooling, 18th Air Infiltration and Ventilation Centre Conference*, Athens, Greece, 23–26 September

Wouters, P., L'Hcurcux, D., Voordecker, P. and Bossicard, R. (1988) 'Ventilation and air quality in Belgian buildings: A state of the art', in *Proceedings: Effective Ventilation, 9th Air Infiltration and Ventilation Centre Conference*, Gent, Belgium, 12–15 September, Paper 23

Wray, C. P. (2000) *Suite Ventilation Characteristics of Current Canadian Mid- and High-Rise Residential Buildings*, Report 43254, Lawrence Berkeley National Laboratory, California, US

Yoshino, H. and Zhao, Y. (1996) 'Numerical study of building airtightness for optimum indoor air quality and energy efficiency', in *Proceedings: Fifth International Conference on Air Distribution in Rooms*, Yokohama, Japan, 17–19 July 1996, vol 2, pp23–30

Yuill, G. K. and Yuill, D. P. (1998) 'Development of a field procedure to measure the airtightness of wall construction elements of houses', in *Proceedings: Thermal Performance of Exterior Envelopes of Buildings VII*, Clearwater Bay, Florida, 6–10 December, pp753–765

Yuill, G. K. and Associates (Man.) Limited (2000a) *Airtightness of Concrete Basement Slabs*, Research Highlights, Technical Series 00–130, Canada Mortgage and Housing Corporation, Canada

Yuill, G. K., Upham, R. and Hui, C. (2000b) 'Air leakage through automatic doors', *ASHRAE Transactions*, vol 106, no 2, pp145–160

Zeller, J. and Werner, J. (1993) 'Airtightness of buildings, measurements on houses with a very low heating energy consumption', in *Proceedings: International Symposium Energy Efficient Buildings: Design, Performance and Operation – CIB Working Commission W67 Energy Conservation in the Built Environment and IEA–SHC Working Group Task XIII on Low Energy Buildings*, Leinfelden-Echterdingen, Germany, 9–11 March

Zhang, Y. and Barber, E. M. (1995) 'Infiltration rates for a new swine building', *ASHRAE Transactions*, vol 101, no 1, pp413–422

Zmeureanu, R. (2000) 'Cost-effectiveness of increasing airtightness of houses', *Journal of Architectural Engineering*, vol 6, no 3, pp87–90

Zuercher, C. H. and Feustel, H. (1983) 'Air infiltration in high-rise buildings', in *Proceedings: Air Infiltration Reduction in Existing Buildings, Fourth Air Infiltration and Ventilation Centre Conference*, 26–28 September, Paper 9

5
Ventilation Performance Indicators and Targets

Peter G. Schild

Introduction

Performance indicators are essential, not only for the assessment of ventilation in individual rooms or buildings, but also at a societal level in monitoring and directing governmental policy. At the latter macroscopic scale, ventilation performance indicators are useful for monitoring the evolution of the ventilation industry, and to define agreed targets or objectives for future research and technical development activities. Such data, when properly defined and validated, will be a useful tool for steering research programmes and economic development policy. The lack of reliable data on trends in energy technology parameters can be a major constraint in a decision-making process, and can thus hinder societal progress.

Although each country or organization can define its own indicators, there is, in principle, an obvious added value when it is harmonized internationally. The objective is to develop appropriate indicators, and then to discuss and, ideally, to come to a first agreement on the definition of relevant or critical indicators that characterize a particular technology and issues related to that technology and its market evolution.

A general discussion on technology indicators for ventilation is made in the following section. Two practical proposals of sets of technology indicators are given in the section on 'Available data and suggested targets'.

The hierarchy of ventilation principles

Performance indicators can be grouped into two main types: indoor environment constraints (for example, air quality and noise) and energy-performance indicators. The indoor environmental indicators are the primary constraints on the ventilation system, while energy is of a secondary consequence.

The system of indoor environmental indicators

The aspects of the indoor environment that can be influenced by ventilation have been systematized in Figure 5.1. This pyramid can be likened to Maslow's hierarchy of human needs, where the basic physiological needs must first be met before addressing psychosocial needs (such as aesthetics). All of the five human senses (that constitute the 'physiological environment') are influenced by ventilation. Furthermore, many aspects of the psychosocial sphere are heavily influenced by ventilation, such as the need to control one's environment and pleasure experienced by dynamic stimuli (for example, changing draught in hot weather.

Basic requirements for the physiological indoor environment are well documented in standards and building regulations. For example, requirements for thermal comfort are given in the 1998 European Committee for Standardization (CEN) CR 1752, the 1990 International Organization for Standardization (ISO) 7730 and the 1992 American Society of Heating, Refrigerating and Air-Conditioning Engineers (ASHRAE) Standard 55, among others. These indoor environmental targets have a large bearing on the air-conditioning industry, which has a turnover of US\$28 billion per year in equipment alone (Nicol, 2003). Similarly, minimum requirements for ventilation flow rates are suggested in the 1998 CEN CR 1752 and the 2001 ASHRAE 62.

However, these basic requirements for the indoor environment are still open to debate. For example, strict application of the criteria for 'optimal' thermal comfort

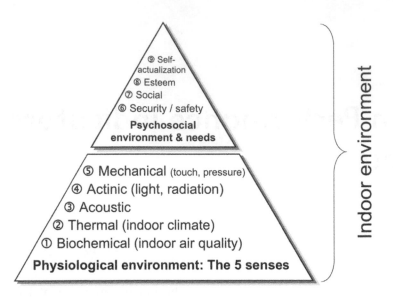

Source: chapter author

Figure 5.1 *The indoor environment: A hierarchy of human constraints for a ventilation system, based on the different forms and levels of human needs and experience*

can lead to buildings with a static thermal environment and highly expensive air conditioning. Moreover, it can be argued that a very closely controlled thermal environment is unnecessary since humans derive pleasure from the opportunity of responding to a stimulating, dynamic environment (Baker and Standeven, 1995, 1996). Therefore, having strict requirements for thermal comfort is, to a degree, self-fulfilling since they constrain the adaptive opportunities for occupants (for example, open windows, draw blinds, switch on window fan and control lighting). An example of this is high-rise air-conditioned office buildings with no openable windows, in which occupants more commonly exhibit symptoms of sick building syndrome (SBS).

The hierarchy of energy-efficient principles

There is a natural order in which energy-efficiency measures are conducted. The main principle is the *trias energetica*, whereby the energy demand is first reduced, then renewables are utilized to fill in as much as possible of the remaining demand. Finally, low-exergy principles are applied to the whole system. This is illustrated in Figure 5.2. This hierarchy shows the relative importance of different energy performance indicators, and is the main foundation of this chapter. One of the most important energy-efficiency measures is to minimize air change rates; this is clearly shown in the hierarchy.

Application of the hierarchy ventilation performance indicators

A discussion about technology indicators in relation to ventilation is less straightforward than other technologies such as wind turbines, thermal insulation or window performance. There are many reasons for this:

- The optimization challenges for ventilation technology involve various issues:
  - correct measurement of ventilation needs – for example, by air quality sensors, presence detection and humidity control;
  - adjusting airflow rates in accordance with changes in ventilation needs – for example, by demand-controlled ventilation;
  - reducing uncontrolled ventilation – for example, better building air tightness and better ductwork air tightness;
  - minimizing the required energy for air handling in terms of heating, cooling, humidification, dehumidification and air cleaning;
  - minimizing the required energy for transporting the air through the building – for example, low pressure distribution networks, energy-efficient fans and hybrid ventilation;
  - avoiding and/or minimizing negative effects due to ventilation – for example, noise, visual aesthetics, space use and costs increase;

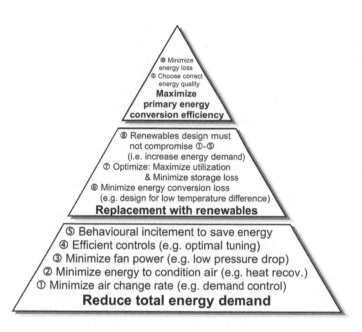

Figure 5.2 *Hierarchy of energy-conscious design principles for ventilation*

- developing building standards and regulations towards performance-oriented requirements.
- There are major climate differences throughout Europe. Due to the significant variations in climate, certain technologies that are appropriate in one region will not necessarily be appropriate at all in another climate. The reason for this is not only the difference in energy demand for climatizsation, but also major cultural differences in comfort concerns (such as draughtiness and overheating).
- There are major differences in building and ventilation traditions in Europe.
- There are major differences in expectations for indoor climate and ventilation. Differences, for example, in building style, living patterns and building materials, in combination with differences in culture, result in major variations in the sensation of ventilation and indoor climate, including expressed needs.
- There are major differences in the philosophy of national building regulations.

As a result, it is clear that technology indicators of ventilation include much more than energy-related indicators. Even more importantly, using primarily quantitative indicators may be neither evident, nor appropriate. Nevertheless, an attempt to suggest such quantitative targets is presented in the section on 'Available data and suggested targets'.

General recommendations

Considerations/recommendations in relation to technology indicators can be subdivided into a number of sub-groups:

- source control in buildings;
- ventilation and related indoor air quality (IAQ) needs;
- the building envelope and ventilation;
- energy for conditioning of supply air;
- energy for transporting air;
- ventilation and summer comfort;
- envelope components integrating ventilation aspects; and
- related aspects.

Source control in buildings

In most buildings, ventilation should be primarily done for evacuating/diluting the pollutants associated with occupants. However, research during the last two decades (European Community (EC) projects IAQ Office; IAQ database project) has clearly shown that many buildings have other major sources of pollutants – for example, certain building materials, cleaning- and maintenance-related activities, air distribution systems and smoking.

Substantial efforts (by industry and in the framework of academic research) have been made in the area of source control; but further developments are required in the following areas:

- low emission materials;
- air distribution systems designed and operated with concern for pollution control;
- design for maintenance;
- isolation of materials with harmful emissions;
- local control over the environment; and
- regulations to control the emissions from furniture and building materials.

Recommendations concerning indoor air quality requirements can be found in CEN and ISO standards, as well as in various national documents and regulations. Issues or concern include:

- the attention to control the local environment;
- smoke-free environment;
- special rooms for smokers with adequate ventilation for the prevention of smoke dispersion to other rooms; and
- attention in standards and regulations to maintenance and commissioning.

In CEN CR 1752, the lowest recommended airflow rate that fulfils existing requirements, using carbon dioxide (CO_2) as an indoor air quality indicator, is 4l/s per person.

Improvements in terms of control of volatile organic compounds (VOCs), radon, formaldehyde and dust particles are expected in the near future.

Ventilation and related indoor air quality (IAQ) needs

As far as required airflow rates are concerned, there are several issues of concern:

- *Optimal matching between ventilation needs and ventilation provisions.* It is not an easy matter to define an optimal airflow rate. Of course, more ventilation with clean air means more dilution of indoor pollutants and a lower indoor pollution level due to indoor pollution sources. However, more ventilation in most cases requires more energy. Therefore, an optimization is required. A major problem is the fact that optimization is far from evident. This is illustrated by the fact that large differences exist in the design values found for ventilation. As an example, CEN CR 1752 gives airflow rates per person varying from 4–10l/s.
- *Adjusting airflow rates in accordance with changes in ventilation needs.* The required airflow rate at any given time is a function of the pollution level in the room at that time. Therefore, detecting the IAQ level

and, as a function of this level, steering the airflow rate is very important. Research has been carried out in several projects (for example, IEA ECBCS, Annex 18; EC CONVENT). At present, various sensors for obtaining information on the indoor climate exist and are used in a number of ventilation systems. In certain countries (particularly France), demand-controlled sensors are widely used, in particular humidity-controlled ventilation. However, progress on development and market penetration of new indoor climate sensors (in terms of relevant indicators and cost considerations) is clearly needed.

- *Ventilation performances in terms of air cleaning.* The assumption of clean outdoor air becomes less and less evident, especially given increased urbanization:
  - more intensive traffic and, hence, higher outdoor air pollution;
  - more buildings being built close to highways because of better accessibility and optimization of space; and
  - increased sensitivity regarding environmental agents – for example, microbiological agents.

Therefore, the development and optimization of air-cleaning devices is a priority. In particular, the development of low pressure-drop air-cleaning systems is important in order to allow for, and/or stimulate, the use of natural or hybrid ventilation systems. This has been studied in the EC NATVENT project.

The building envelope and ventilation

Building air tightness is an important aspect; it has a substantial impact on energy use, but also on the correct distribution of air within buildings and on the indoor climate, in general.

At present, certain countries already have a long history of concern for achieving air tightness (for example, Sweden) whereas others are imposing requirements for certain types of buildings (for example, the UK for larger office buildings). Thus, the air tightness of the building envelope, expressed with the n_{50} value, differs among all European countries. The current values vary in the range 1.5–15.0 air changes per hour (ACH). Each country/climatic region has to set its own targets.

Energy performance requirements will gradually lead to more airtight buildings in which purpose-provided ventilation is a necessity.

Given the large diversity in climate and building tradition, it is not evident how to define clear target values for the air tightness. An example of the approach used in Norway can be found below.

Optimizing air tightness, and relevance of this to building regulations

The target values or the *optimum tightness* will depend upon the following parameters:

*   type of building;
*   ventilation system;
*   heating and/or cooling season for the region;
*   cost of better tightness; and
*   cost of energy.

This calculation of the optimum tightness should be done for different kinds of buildings in different climate regions.

As an example, we have performed a calculation for detached houses in the Nordic region. The following parameters were chosen:

*   type of building: detached houses;
*   ventilation system: balanced ventilation with heat recovery;
*   heating season: 91,200 degree hours;
*   cost of better tightness: results from Swedish and Norwegian experiment buildings;
*   cost of energy: 0.08 Euros per kWh;
*   lifetime: 60 years.

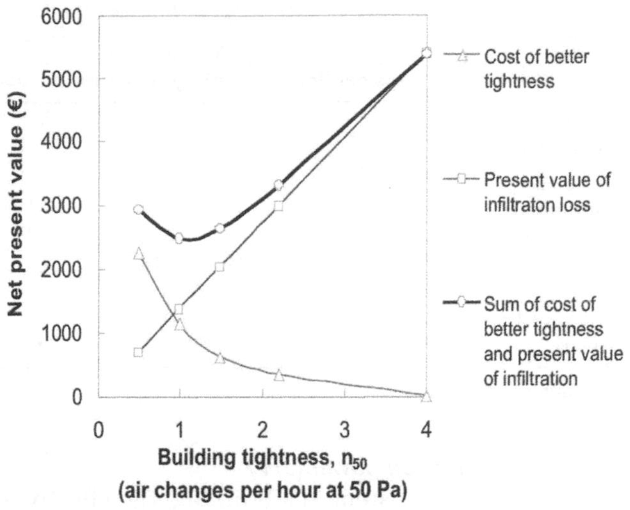

Energy for conditioning supply air

The energy use for ventilation conditioning includes energy for heating and energy for cooling, humidification and dehumidification.

Heating

The amount of fresh air can be optimized by varying the flow rate in accordance with the ventilation requirement at any given time (see the earlier section on 'Ventilation and related indoor air quality (IAQ) needs'). The lower the fresh airflow rate, the lower the required energy.

For a given airflow rate, heating energy consumption can be reduced by applying heat recovery. Heat exchangers have traditionally had a thermal efficiency of around 50–60 per cent. Due to the use of counter-flow heat exchangers, efficiencies above 80 per cent have become possible and are already applied on a large scale in temperate regions of Europe. In sub-arctic regions, highly efficient heat exchangers need frost protection, which reduces their heat recovery efficiency.

Other possibilities for heat recovery are combinations with a heat pump, whereby the ventilation exhaust air is used as a cold source for a heat pump for (domestic) hot water.

Source: chapter author

Figure 5.3 *Optimum air tightness for detached houses with balanced ventilation in a Nordic climate*

Cooling, humidification and dehumidification

The most important savings can be obtained by designing a building that does not have overheating problems and, as a result, has no need for active cooling.

The cost of sensor technology is a critical aspect. For example, it is expected that if sensors cost less then 10 Euros, the estimated number of applications in ventilation systems in The Netherlands will, in ten years, grow to 250,000 sensors a year.

Energy for transporting air

There is increasing interest in limiting energy use in terms of primary energy. As a result, electricity consumption becomes, in most cases, an issue of concern. Mechanical ventilation systems use fans for transporting energy. In poorly designed systems with classical fans, the primary energy use for transporting the air can be substantial.

Optimization of the primary energy use can be achieved through various means:

*   use of low pressure-drop ductwork, thus reducing fan energy to transport the air;
*   use of energy-efficient fans (for example, direct current fans for dwelling applications);
*   demand-controlled ventilation;
*   combined use of natural and mechanical ventilation, also called hybrid ventilation – studied in projects such as IEA ECBCS, Annex 35, HYBVENT (Heiselberg, 2002) and EC RESHYVENT.

Table 5.1 *Future targets for specific fan power (SFP) and ventilation system efficiency*

Fan energy	Target 2005			Target 2020		
	Residential	Tertiary, 11 hours	Tertiary, 24 hours	Residential	Tertiary, 11 hours	Tertiary, 24 hours
Main target:						
SFP (specific fan power), kW/(m³/s)	1.5	2.0	1.5	0.75	1.0	0.75
Secondary targets:						
η_{tot} = total fan system efficiency	0.2	0.45	0.5	0.3	0.55	0.6
Δp_{tot} (duct system + air handling unit (AHU)) (Pa)	300	900	750	225	550	450

Source: chapter author

Definition of specific fan power (SFP)

For each ventilation system in a building, the effectiveness of the system can be evaluated with the SFP value:

$$SFP = \frac{\Sigma P}{q_{max}} \quad (kW/(m^3/s)) \; or \; (W/(m^3/s)) \qquad (1)$$

where:

ΣP sum of electrical power to the fans in the system (kW active power)

q_{max} moved airflow in the system = highest value of supply or extract airflow (m³/s).

SFP depends upon the total pressure drop in the system and the total efficiency of the fan system:

$$SFP = \frac{\Delta P_{tot}}{\eta_{tot}} \quad (kPa) \qquad (2)$$

where:

Δp_{tot} pressure drop in the ventilation system plus pressure drop in the air handling unit (kPa), dimensionally equivalent to (kW/(m³/s))

η_{tot} total efficiency of the fan system (fan, motor, drive).

In Norway, for example, the consumption of electrical power to fans in commercial buildings (not industry) was estimated in 1998 to be 15–20 per cent of the total energy consumption for these buildings (depending upon running time per 24 hours).

The average SFP for modern Norwegian commercial buildings has been measured and reported in two projects:

1   Oslo area in Norway, measured by NBI in 1998: four buildings and ten systems built in 1985–1996, with balanced ventilation:
  → Flow rate: 1–5m³/s; average SFP: 3.4 kW/m³/s;

2   Bergen area in Norway, measured by Thunes Partners A/S in 2001: 12 buildings, 21 systems built in 1995–2000, with balanced ventilation:
  → Average SFP: 3.0 kW/m³/s.

Targets for SFP

Best practice recommended values for new buildings in Scandinavia are currently:

- constant air volume flow rate (CAV) systems: running time less than 11 hours per 24 hours: SFP_{max} = 2.0 kW/m³/s;
- CAV systems: running time 24 hours per 24 hours: SFP_{max} = 1.5 kW/m³/s;
- variable air volume flow rate (VAV) systems: SFP_{VAV} = SFP_{CAV} + 1, calculated with 100 per cent flow.

Recommended target values for new installations in existing buildings can be 1 kW/m³/s higher than for new buildings.

These target values correspond to 2002 Norwegian recommendations. The values are roughly the same as 1991 Swedish recommended values and 1995 Danish recommended values.

Future targets for best practice ventilation system efficiency are suggested in Table 5.1. These targets are very feasible and not particularly ambitious. Best practice low-energy buildings today are on a level of about 1.0–2.0kW/(m³/s); future targets will supposedly be as low as 0.5kW/(m³/s) in 2020. For domestic buildings, available technology today can achieve 0.1kW/(m³/s), and it is technically be possible to set an ambitious target of 0.02kW/(m³/s) by 2020.

Ventilation and summer comfort

Since the end of the 1980s, interest in ventilation as an instrument for reducing summer cooling and/or improving summer comfort has become very high. This has been studied in various European and International Energy Agency (IEA) projects (for example, EU PASCOOL, EU AIOLOS, EU NATVENT, IEA HYBVENT and EU RESHYVENT).

It is not possible to set energy technology indicators for ventilation technology in relation to ventilation for summer comfort since this is essentially a challenge of integrating good building design, good HVAC technology and good operation.

Envelope components integrating ventilation aspects

The integration of ventilation systems in the building envelope can, if well designed and operated, result in a number of advantages. Examples of such envelope components are:

- dynamic insulation systems; and
- active façades and dynamic façades.

Related aspects

Several other issues are receiving increased attention, such as the following:

- *Ventilation performance in relation to acoustic performance.* A large majority of ventilation systems contribute to a higher acoustical load in buildings. Improvement of acoustical performance is therefore important. Again, the target values heavily depend upon the country, building type and function of the room.
- *Draught-free ventilation systems.* For many ventilation systems (particularly natural ventilation systems), draught problems are an issue of concern. Optimization of air supply devices is therefore needed.
- *Commissioning of ventilation systems.* There is increased evidence that many ventilation systems are not functioning as expected (experience gained from projects such as SAVE DUCT, EC TIP-Vent, EC NATVENT and IEA Annex 40 'Commissioning'). The development of intelligent systems and/or procedures for commissioning is therefore very important.
- *Operation and maintenance.* Experience shows us that many systems are poorly maintained or not at all. In such cases, the expected performance is not achieved during the lifetime of the system.

Health and productivity versus energy consumption

More than 90 per cent of the total operating costs of commercial office buildings in industrialized countries comprises the salary cost of workers. It can thus be argued that energy conservation measures (for example, limiting ventilation rates) that worsen worker heath or performance are not profitable at a socio-economic level. On the other hand, the need for energy conservation and its environmental benefits cannot be entirely measured in fiscal terms.

A large number of field and laboratory studies have been conducted to evaluate the magnitude of productivity loss due to substandard ventilation. A literature survey has shown that the majority of the 500 studies analysed indicate an average productivity loss of 10 per cent due to poor IAQ. They estimated that if every non-industrial building was properly ventilated, total productivity and health benefits for the US would be US$62.7 billion per year, or US$910 per worker and per year. Based on the available literature and analyses of statistical and economic data, Fisk (2000) estimated that, for the US, the potential annual savings and productivity gains due to good ventilation are US$6 billion to $14 billion due to reduced respiratory disease and US$2 billion to $4 billion from reduced allergies.

In conclusion, improving IAQ can provide significant and very profitable benefits in the form of improved workforce health and productivity; but such measures can come into conflict with the need for energy conservation.

Available data and suggested targets

This section gives two alternative practical sets of performance indicators.

- one set that has been developed in the framework of the European RESHYVENT project (see the following section on 'Basic indicators and recommendations for dwellings');
- another set that has been prepared by the Norwegian Building Research Institute (see the later section on 'Detailed system of indicators for ventilation in general').

Basic indicators and recommendations for dwellings

In the framework of the European RESHYVENT project, the information as given in Table 5.2 has been reported with respect to the present ventilation market and the targets for ventilation as set by the RESHYVENT project.

Table 5.2 *Summary of quantified objectives for dwellings (research project RESHYVENT)*

	Market average 2001	Targets for 2012	
Energy use for domestic ventilation in gigajoules (GJ)/dwelling:			
• severe climate	43	7.5	
• cold climate	33	5.8	
• moderate climate	22	4.9	
• mild climate	9	1.2	
• warm climate	5	0	
Indoor air quality:			
• CO_2 hours > 700 ppm	2000 (kppm.h)	500 (kppm.h)	
• CO_2 hours > 1400 ppm	500 (kppm.h)	100 (kppm.h)	
Thermal comfort (percentage of room < 15%; T_{out} = 0° C)	50–75%	85–95%	
Noise	>35dBA	< 35dBA	
Airflow stability: percentage of time according design	6–12 % (poor)	25–50% (good)	
Maximal additional costs and payback time:		Costs/dwelling (Euros)	Pay-back (years)
• severe climate		2000	5.6
• cold climate		1800	6.6
• moderate climate		1500	8.8
• mild climate		1500	19.2
• warm climate	–	1000	20.0
Total energy use for domestic ventilation in European Union (EU)	EU: 2797 PJ/year EU plus candidates: 3850 PJ/year	EU: 2733 PJ/year EU plus candidates: 3784 PJ/year	
Total CO_2 emission ventilation-related energy	EU: 153 Mton/year EU plus candidates: 215 Mton/year	EU: 149.4 Mton/year EU plus candidates: 211.9 Mton/year	
Total energy saving (total market penetration: EU 2%; candidate EU 0.25%)	–	EU: 64.1 PJ/year EU plus candidate EU: 65.6 PJ/year	
Total CO_2 saving (total market penetration: EU 2%; candidates 0.25%)	–	EU: 3.6 Mton/year EU plus candidates: 3.7 Mton/year	

Source: chapter author

Table 5.3 *Classification of important system characteristics as a function of climate (research project RESHYVENT)*

	Mild and warm	Moderate	Cold	Severe
Building type	Dwellings	Dwellings – apartments	Apartments	Single-family houses
Use of renewable energies	Photovoltaics (PV)	PV; wind	PV with/without heat recovery; wind	Wind; heat recovery
Summer comfort	Crucial	Limited	No	No
Winter comfort	Important	Important	Important	Crucial
Optimal supply	Important	Important	Crucial	Crucial
Optimal exhaust	Crucial	Crucial	Crucial	Crucial

Source: chapter author

BOX 5.1 SYSTEM OF INDICATORS FOR VENTILATION TECHNOLOGY

A: Global indicators for ventilation technology

Worldwide/regional impact (environmental/social assessment):
- greenhouse gas (GHG) emissions and climate change;
- social indicators: production employment.

B: Energy-efficient ventilation technology

1 Minimize ventilation air change rate:
- source control:
    - indoor air quality (IAQ) (that is, chemical);
    - climatization (that is, thermal);
- ventilation efficiency:
    - spatial efficiency;
    - displacement ventilation;
    - minimize short circuiting;
    - duct air tightness;
    - building air tightness;
- temporal efficiency:
    - demand-controlled ventilation.
2 Minimize energy use for conditioning-required supply air (that is, to offset ventilation loss/gain, not space heating/cooling):

- preheating;
- precooling;
- infiltration heat loss/gain.
3 Minimize energy for transporting air (that is, fan energy):
- fan efficiency;
- low pressure-drop components.
4 Efficient controls:
- sensor/actuator/communications technology;
- control strategies technology.
5 Behavioural incitement to save energy:
- dwelling area.
6 Constraints for energy-efficient ventilation technology:
- minimum ventilation airflow rates for IAQ;
- noise.

C: Replacement with renewables

- Integration of solar energy.
- Integration of wind energy.
- Integration of other renewables.

D: Maximize primary energy conversion

- Minimize exergy loss.
- Choose correct energy quality.

Detailed system of indicators for ventilation in general

Box 5.1 presents a suggested structure for organizing a hierarchy of energy indicator targets, reflecting the principles illustrated in Figures 5.1 and 5.2. The same structure is used in Table 5.4.

Summary

This chapter has given an overview of the hierarchy of factors that, in combination, influence ventilation performance.

Given the wide variety of building traditions on Earth, as well as significant differences in climatic conditions and expectations in relation to ventilation and indoor climate, we cannot expect to develop harmonized quantitative targets for future policy-making related to ventilation.

Given the significant uncertainty in our knowledge of actual performance (such as air change losses and infiltration) in individual buildings, it is doubly difficult to accurately evaluate the performance indicators for the whole building mass. This lack of accurate data can hamper strategic governmental planning for improving the energy efficiency of ventilation. This chapter has therefore focused on identifying major issues of concern. Nevertheless, existing research gives us enough quantitative insight to be able to suggest two incomplete practical sets of technology indicators: a basic one for dwellings and a detailed one for buildings in general.

Table 5.4 *Energy indicator targets*

Indicator	Present (2004) Residential	Present (2004) Tertiary	Target (2020) Residential	Target (2020) Tertiary
A. GLOBAL INDICATORS FOR VENTILATION TECHNOLOGY Worldwide/regional impact (environmental/social assessment)				
GHG emissions & climate change Social indicators: Production employment in ventilation industry				
CO_2 emission due to ventilation (Mton/y in EU)	153 (yr 2001)	–	149.4 (yr 2017)	–
B. ENERGY-EFFICIENT VENTILATION TECHNOLOGY				
Delivered energy for ventilation (various indicators can be used, ideally GJ/person for dwellings, and GJ/m² for tertiary)	GJ/dwelling: 43 severe 33 cold 22 moderate 9 mild 5 warm	Typical range for EU: 70–150 MJ/m²·y	GJ/dwelling 7.5 severe 5.8 cold 4.9 moderate 1.2 mild 0 warm	< 100 MJ/m²·y
1. Minimize ventilation air change rate				
Indoor Air Quality, IAQ (i.e. chemical)				
Environmental tobacco smoke	smoke exposure > 1000 hours	smoke exposure > 1000 hours	smoke exposure < 400 hours	smoke exposure < 10 hours
Indoor respiratory particulates PM2.5 e.g. Average internal heat gains from electrical equipment & lighting	–	–	–	–
Ventilation efficiency Average total-system net contaminant removal efficiency, accounting for all recirculation paths in system (ventilation efficiency. 100% is equivalent to fully mixed)	<100%	<100%	>110%	>110%
Climatization (i.e. thermal)				
Spacial efficiency				
• Ventilated area per person	?	?	< 20 m²	depends on bldg type. Office < 10m²
• Displacement ventilation Market penetration of displacement vent (including cascade ventilation)	< 5 %	< 5%	> 25%	> 25%
• Minimise short circuiting – Duct air tightness Average duct air tightness (Eurovent class. Class A is worst)	existing bldgs < A but new bldgs ~A n50=1 15	existing bldgs < A but new bldgs ≤B	B	B
– Building air tightness				
Temporal efficiency • Demand-controlled ventilation Utilisation of potential annual reduction air exchange relative to continuous ventilation (CAV); window opening excluded	< 5%	< 5%	new bldgs: > 20%	new bldgs: > 30% exist bldgs: > 15%
2. Minimize energy use for conditioning-required supply air (i.e. to offset ventilation loss/gain, not space heating/cooling)				
Preheating Energy to air preconditioning (refrigeration) kWh/degree day over 20°C per person or m²	–	–	–	–
Ventilation heat recovery Percentage retrofit existing buildings	< 5%	not known	> 15%	> 15%
Percentage of new buildings	< 10%		> 90%	> 90%
Typical net heat recovery efficiency for ventilation unit	40–75%		> 70%	> 70%
Precooling Energy to air preconditioning (refrigeration) kWh/degree day over 28°C per person or m²	–	–	–	–
Percentage cooling demand provided by "passive means" or "ambient energy"				

Infiltration heat loss/gain. See page below

Indicator / Target metric				
3. Minimize energy for transporting air, i.e. fan energy				
Specific Fan Power, SFP (kW/(m³/s)) VAV = variable air operation, 11h/day CAV = continous operation 24h/day	balanced vent: 2.7	balanced vent: 3.4	Target 2005: 1.5 Target 2020: 0.75	VAV 2005: 2.0 CAV 2005: 1.5 VAV 2020: 1.0 CAV 2020: 0.75
Fan efficiency Secondary target: Total fan system efficiency,	?	35%	Target 2005: 0.2 Target 2020: 0.3	VAV 2005: 0.45 CAV 2005: 0.50 VAV 2020: 0.55 CAV 2020: 0.60
Low pressure-drop components Secondary target: Total pressure drop for air path through bldg (incl. internally in air handling unit, AHU) (Pa)	balanced vent: > 300	balanced vent: 1000	Target 2005: 300 Target 2020: 225	VAV 2005: 900 CAV 2005: 750 VAV 2020: 550 CAV 2020: 450
4. Efficient controls				
Sensor / actuator / communications technology			–	–
Control strategies technology Air flow stability, % of time according to design or demand	6–12 % (poor)	– –	25–50% (good)	– –
5. Behavioural incitement to save energy				
Design features that reduce behavioural energy wastage Percentage of apartment blocks that have individual heating energy metering in each apartment	< 5%	n/a	> 10%	n/a
6. Constraints for energy-efficient ventilation technology				
Minimum ventilation airflow rates for IAQ CO_2 concentrations (kppm·h) CO_2 hours over 700 ppm CO_2 hours over 1400 ppm	– 2000 500		– 50 100	–
? Noise Indoor noise level from mechanical ventilation equipment	> 35 dBA	> 35 dBA	bedroom: < 25 dBA living areas: < 30 dBA 'wet' rooms: < 35 dBA	office: < 35 dBA Different requirements for bldgs
C. REPLACEMENT WITH RENEWABLES Percentage of total delivered energy for ventilation from renewables	< 5%	< 5%	?	?
1. Integrate solar energy *Solar energy for air transport* *Solar energy for conditioning supplied air* Delivered solar energy (e.g. PV)	–	–	–	–
2. Integrate wind energy *Exploitation of wind* Average percentage of transporting energy provided by wind (wind-augmented)	< 5%	< 5%	new bldgs: > 10% (2015)	new bldgs: > 10% (2015)
3. Other delivered renewables				
D. MAXIMIZE PRIMARY ENERGY CONVERSION Preference should be given to ventilation technologies that minimize primary energy conversion, e.g. use of waste heat with district heating. However, no indicators are suggested here.	n/a	n/a	n/a	n/a

Source: chapter author

Acknowledgements

This chapter is based on invaluable contributions from P. Wouters, International Network for Information on Ventilation and Energy Performance (INIVE EEIG), www.inive.org; W. De Gids, Netherlands Organisation for Applied Scientific Research, Delft (TNO), www.tno.n;, F. Durier, Centre Technique des Industries Aérauliques et Thermiques, France (CETIAT), www.cetiat.fr; T. Hestad and J. Brunsell, Norwegian Building Research Institute (NBI), (Oslo & Trondheim, Norway, www.byggforsk.no; J. R. Millet, Centre Scientifique et Technique du Bâtiment, France (CSTB), www.cstb.fr; and P. Op't Veld, Cauber-Huygen.

This work has been supported by ENOVA through Contract SID02/1755, Air Infiltration and Ventilation Centre (AIVC), and the Research Council of Norway (SIP contract).

References

ASHRAE (American Society of Heating, Refrigerating and Air-Conditioning Engineers) (1992) *ASHRAE Standard 55-1992: Thermal Environmental Conditions for Human Occupancy*, ASHRAE

ASHRAE (2001) *ASHRAE Standard 62-2001: Ventilation for acceptable IAQ.*

Baker, N. and Standeven, M. (1995) *PASCOOL – Comfort Group – Final report*, European Commission, Brussels

Baker, N. and Standeven, M. (1996) 'Thermal comfort for free running buildings', *Energy and Buildings*, vol 23, pp175–182

Bluyssen P. M., de Oliveira Fernandes, E., Fanger, P. O., Groes, L., Clausen, G., Roulet, C. A., Bernhard, C. A. and Valbjørn, O. (1995), *European Audit Project to Optimize Indoor Air Quality and Energy Consumption in Office Buildings*, Final report, TNO Building and Construction Research, Delft

Bluyssen, P. M., de Oliveira Fernandes, E. and Molina, J. L. (1999) 'SOPHIE: A European database on indoor air pollution sources: Marketing and organisational matters', *Proceedings of Indoor Air 1999*, BRE, Edinburgh

BRECSU (Building Research Energy Conservation Support Unit) (1999) *Natural Ventilation in Offices – NATVENT; A Better Way to Work*, BRECSU, BRE, Garston

CEN (European Committee for Standardization) CR 1752 (1998) *Ventilation for Buildings: Design Criteria for the Indoor Environment*, CEN, Brussels

Delmotte, C. (1999) www.tip–vent.com, BBRI

Dubrul, C. (1998) *Inhabitants' Behaviour with Regard to Ventilation*, IEA ECBCS, Annex 8

ECA–IAQ (European Collaborative Action on Urban Air, Indoor Environment and Human Exposure) (1992) *Guidelines for Ventilation Requirements*, Report no 11, EUR 14449 EN, Office for Official Publications of the European Commission, Luxembourg

ECA–IAQ (1996) *Indoor Air Quality and the Use of Energy in Buildings*, Report no 17, EUR 16367 EN, Office for Official Publications of the European Commission, Luxembourg

ECA–IAQ (1997) *Evaluation of VOC Emissions from Building Products – Solid Flooring Materials*, Report no 18, Office for Official Publications of the European Communities, Luxembourg

Fanger, P. O. (1988) 'Introduction of the Olf and the Decipol units to quantify air pollution perceived by humans indoors and outdoors', *Energy and Buildings*

Fisk, W. J. and Rosenfeld, A. H. (1997) 'Estimates of improved productivity and health from better indoor environment', *Indoor Air*, vol 7, pp158–177

Fisk, W. J. (2000) 'Review of health and productivity gains from Better IEQ', *Proceedings of Healthy Buildings*, vol 4

Heiselberg, P. (ed) (2002) *Principles of Hybvent Ventilation*, Final report from IEA ECBCS Annex 35, Copenhagen, www.hybvent.civil.auc.dk

Humphreys, M. A. (1992) 'Thermal comfort in the context of energy conservation', in Roaf and Handcock (eds) *Energy Efficient Buildings*, Blackwells, Oxford, UK

Humphreys, M. A. and Nicol, J. F. (1995) 'An adaptive guide-line for UK office temperatures', in Nicol, Humphreys, Sykes and Roaf (eds) *Standards for Thermal Comfort – Indoor Air Temperature Standards for the 21st Century*, E and FN Spon, London, UK

IEA (International Energy Agency) (1998) *Inhabitants' Behaviour with Regard to Ventilation*, Summary document, Energy Conservation in Buildings and Community Systems Programme, Annex 8, AIVC, Coventry, UK

International Journal of Indoor Air Quality and Climate (1997) no 3, September, Copenhagen

ISO (International Organization for Standardization) (1994) *ISO 7730: Moderate Thermal Environments – Determination of the PMV and PPD Indices and Specification of the Conditions for Thermal Comfort*, 5 December

Liddament, M. (1996) *A Guide to Energy Efficient Ventilation*, Air Infiltration and Ventilation Centre, Coventry, UK

Mansson, L. G. (1992) *Demand Controlled Ventilating Systems – Source Book*, IEA Annex 18, Swedish Council for Building Research, Stockholm, Sweden

Nicol, F. (2003) 'The dialectics of thermal comfort', Inaugural lecture, 19 February

Norlen, U. and Andersson, K. (eds) (1993) *The Indoor Climate in the Swedish Housing Stock*, Document D10:1993, Swedish Council for Building Research, Stockholm, Sweden

Øie, L. (1998) *The Role of Indoor Building Characteristics as Exposure Indicators and Risk Factors for Development of Bronchial Obstruction in Early Childhood*, PhD thesis, Trondheim, Norway

RESHYVENT (2001–2004) *EU Cluster Project on Demand-Controlled Hybrid Ventilation in Residential Buildings*, www.reshyvent.com/

Santamouris, M. (1999) *Energy and Climate in the Urban Environment*, James & James, London

Seppänen, O. (1998) *Ventilation Strategies for Good Indoor Air Quality and Energy Efficiency, IAQ and Energy '98 ASHRAE Conference*, New Orleans, US, October 1998

Seppänen, O. and Säteri, J. (1998) 'Experiences of the Finnish classification of indoor climate, construction and finishing materials', *Proceedings of the Second International Conference on Indoor Climate and Energy Efficiency*, EPIC, Lyon

Wheeler, A. (1999) 'A view of IAQ as the century closes', *ASHRAE Journal*, November

6

Heat Recovery

Peter G. Schild

Introduction

A heat recovery unit transfers heat (some units also transfer moisture) from the exhaust air stream to the supply air stream, thus reducing the heat loss due to ventilation, as well as reducing the need to condition the cold supply air. Conversely, in hot and humid outdoor conditions, a heat recovery unit can keep heat (and also moisture in some) outside, thus reducing air-conditioning costs.

General definitions

General definitions found in this chapter include the following:

- *air handling unit (AHU)*: packaged unit with fans for balanced ventilation, filters, perhaps heating/cooling batteries, dehumidifier/humidifier, etc.;
- *heat recovery unit*: air handling unit that also incorporates a heat exchanger and its auxiliary equipment (controls, etc.);

- *heat exchanger*: a component within the heat recovery unit that transfers heat between two fluid flow streams (that is, air);
- *bypass*: alternative passage for airflow avoiding the heat exchanger, thus avoiding heat recovery (can be useful in summer);
- *total enthalpy*: specific heat content per kilogram of dry air; this is the sum of *sensible enthalpy* of the moist air (temperature dependent) and the latent heat of its water vapour content.

Conditions of use

Heat recovery may be used in balanced ventilation systems (that is, fan-powered supply and exhaust airflows). The building should be satisfactorily airtight – air leakages constitute an extra heat loss since they do not pass through the heat recovery unit. For dwellings, the infiltration rate should not exceed 10–20 per cent of the flow rate through the heat recovery unit.

Use

Heat recovery is equally appropriate for buildings with any space heating system. Correctly dimensioned and maintained heat recovery units with high efficiency will pay for themselves in a few years in terms of reduced ventilation and space heating costs. This profitability is higher if the exhaust fan is located before the heat exchanger. It should be possible to reduce the heat recovery efficiency outside the heating season to prevent overheating indoors. Some heat exchangers can also recover moisture. It may be desirable to recover moisture this way in buildings with central humidification in winter in order to reduce humidification costs. For AHUs with

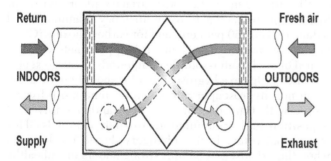

Return · Fresh air

INDOORS · OUTDOORS

Supply · Exhaust

Source: Norwegian Building Research Institute

Figure 6.1 *Heat recovery unit*

cooling (air conditioning), moisture recovery may be desirable in summer (when the outdoor air is hot and humid) to reduce the cooling energy needed for dehumidification. If the exhaust air has water-soluble odours/pollutants, one should, nevertheless, use a heat exchanger that does not recover moisture – that is, totally separate air streams. Heat recovery units require regular inspection and maintenance, though anyone with normal technical aptitude can do this on the condition that a proper operation and maintenance manual is available.

Types and areas of use

Main types

There are two main types of heat recovery unit: *regenerative* (cyclic) and *recuperative* (static).

Regenerative heat recovery

Regenerative heat exchangers transfer heat via heat-accumulating surfaces that are repeatedly exposed to either the exhaust air or the supply air stream. The heat-accumulating surfaces are normally metal. These heat exchangers can also recover moisture. Undesirable leakage between the two air streams can occur, though this problem can be reduced by judiciously locating the fans to counteract the leakage.

Rotary heat exchangers

These consist of a rotor wheel (thermal wheel) with many small parallel channels through which air flows. While one half of the wheel is being heated up by the exhaust air, the other half is releasing stored heat to the supply air. There are two types:

- *hygroscopic rotors* (enthalpy wheels) – that is, the rotor of a material that adsorbs moisture (vapour and liquid), usually a surface of porous aluminium oxide; 'sorption wheels' are the most effective type at moisture recovery;
- *non-hygroscopic rotors* (condensation wheels); these can only transfer moisture if the moisture in the exhaust air condenses.

The recovery efficiency of rotary heat exchangers is controlled by regulating the rotor speed or intermittent operation. Bypass control may also be used. The alternating flow direction through the rotor tends to keep it clean of dust. However, to prevent recirculation of pollutants, a purge sector can be used (see Figures 6.2 and 6.4). The disadvantage of a *purge sector* is that it slightly reduces

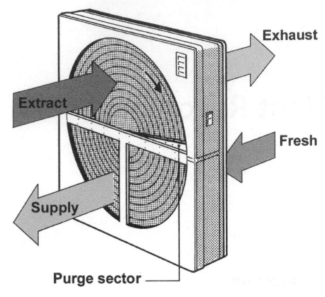

Source: Norwegian Building Research Institute

Figure 6.2 *Rotary heat exchanger*

heat recovery efficiency. To prevent fouling of the rotor when it is static (for example, during summer), it should be operated periodically. Normally freezing occurs only below approximately –20° C for hygroscopic types (this depends upon the humidity of extract air and the heat recovery efficiency).

The fans should be located so that the leakage is driven from supply to exhaust air stream (see Figure 6.3). Without a purge sector, at least 2–4 per cent of the extract air is recirculated. This is partly caused by leakage and partly caused by carry-over, the residual volume of air in the cells of the rotor, which increases with rotor speed and thickness. In small AHUs, this carry-over can be as large as 10 per cent. A purge sector can reduce this below 1 per cent. The degree of transfer of odours and pollutants depends upon each individual compound's diffusion coefficient and upon the rotor surface's adsorptivity. For hygroscopic rotors, the transfer efficiency of ammonia and butane exceeds 50 per cent, and for carbon dioxide (CO_2), petroleum and cooking odours, it is approximately 10 per cent. For such water-soluble compounds, a purge sector is of little help, so a hygroscopic rotor should not be used. Even non-hygroscopic rotors can become more hygroscopic with time due to corrosion (and possibly dust deposition and accumulation) in their channels. This means that rotary heat exchangers should not be used when the exhaust air contains strong odours or unhealthy pollutants.

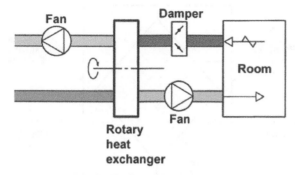

Source: Norwegian Building Research Institute

Figure 6.3 *Location of fans and damper to reduce recirculation*

Reciprocating heat exchangers

Reciprocating heat exchangers consist of two separate heat-accumulating chambers and a motorized damper to alternate flow direction at regular intervals, normally each minute. Each chamber has many parallel plates or a material similar to rotary heat exchangers. Unwanted recirculation from the exhaust to supply air stream is similar to a rotary heat exchanger (1–6 per cent, including external short circuiting from exhaust to fresh air intake). Risk of frosting is very similar to that of rotary heat exchangers.

Recuperative heat recovery

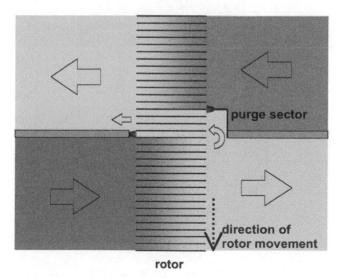

Source: Norwegian Building Research Institute

Figure 6.4 *Principle of operation of the purge sector in a rotary heat exchanger (cross-section)*

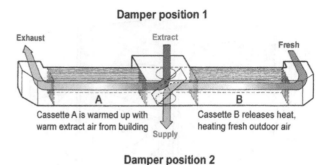

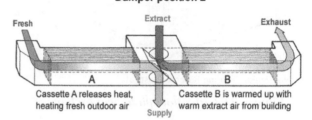

Source: Norwegian Building Research Institute

Figure 6.5 *Example of a reciprocating heat exchanger*

Recuperative heat exchangers transfer heat across a dividing plate by means of thermal conduction (plate or tube heat exchanger) or with an intermediate fluid (run-around heat exchanger, heat pipe or heat pump). Since the two air streams are kept separate, these exchangers can theoretically have zero transfer of odours, though in practice, plate heat exchangers, which are the most common type, typically have 1–3 per cent recirculation due to internal leakage.

Plate heat exchangers

Plate heat exchangers consist of parallel plates (flat or corrugated) that separate the supply and exhaust air streams (see Figure 6.6). Heat is transferred through the plates by conduction. If the plates' temperature drops below 0° C, ice can grow in the exhaust air paths in the exchanger, and it will eventually become blocked. When heat recovery is not desirable, in summer, a bypass damper may be used or for small residential units, the exchanger may be replaced by one with only one plate ('summer cassette'). Due to condensation in the exchanger during winter, a condensate drain must be provided. The condensate must be protected from frost and have a gradient along its entire length, and must have a drain trap with sufficient height in relation to the maximal operating pressure within the AHU (see Figure 6.8).

Traditionally, cross-flow heat exchangers are the most common (see Figure 6.6). Counter-flow heat exchangers

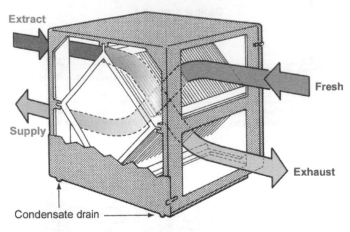

Source: Norwegian Building Research Institute

Figure 6.6 *Cross-flow plate heat exchanger*

(see Figure 6.7) are a more recent development – they have higher recovery efficiency, but are more susceptible to frosting.

Tube heat exchangers

Tube heat exchangers function in much the same way as plate heat exchangers, where tubes replace plates. These are easier to clean than plate heat exchangers and can be equipped with an automatic washing mechanism. The

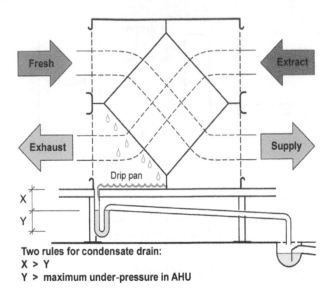

Source: Norwegian Building Research Institute

Figure 6.8 *Correct condensation collection and drain trap*

tubes can be made of glass, giving good corrosion resistance. The risk of becoming blocked due to frosting is less than plate heat exchangers, and the internal leakage is generally less.

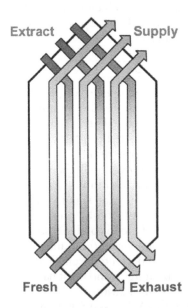

Source: Norwegian Building Research Institute

Figure 6.7 *Principle of a counter-flow plate heat exchanger*

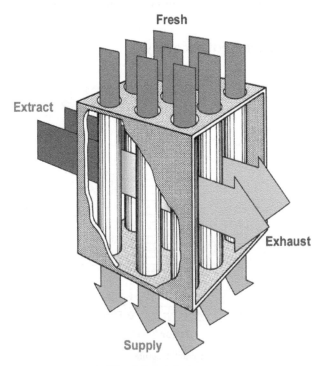

Source: Norwegian Building Research Institute

Figure 6.9 *Tube heat exchanger*

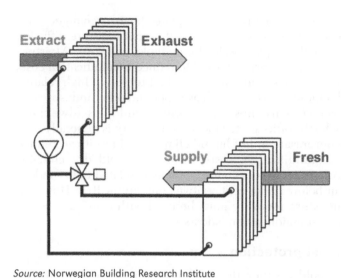

Source: Norwegian Building Research Institute

Figure 6.10 *A run-around heat exchanger system*

Run-around heat exchangers

These consist of two batteries (coils), one in each air stream, connected by a fluid circuit of water/glycol or water/alcohol (see Figure 6.10). The concentration of glycol that is required (for frost protection) depends upon the temperature operating range, but is usually 30–40 per cent. The heat recovery efficiency is reduced with increasing glycol concentration. In large systems, brine is used instead of water/glycol. The advantage with this type of heat exchanger is that there can be a large distance between the supply and exhaust ducts, and heat can be reclaimed from multiple exhaust ducts by means of individual batteries. This system is appropriate in cases with heavily polluted extract air since there is no risk of air leakage from the exhaust-to-supply air streams (stainless steel, copper or plastic batteries give corrosion resistance). A three-way valve is used to control the heat recovery efficiency – this is used for frost protection.

Heat pipe

Vertical heat pipes function in much the same way as run-around heat exchangers. The working medium is a refrigerant that evaporates under heat and condenses when cooled. No pump is needed. The heat recovery efficiency is higher in colder weather. Heat recovery is controlled, if necessary, by a bypass. There are two types:

1   vertical heat pipes (see Figure 6.11); and
2   horizontal (or slightly sloping) heat pipes, which are less common; natural circulation is achieved by a wick

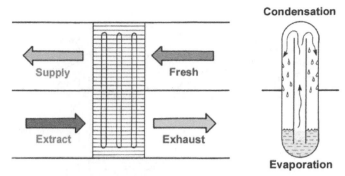

Source: Norwegian Building Research Institute

Figure 6.11 *Vertical heat pipe*

inside the tube, along which the condensate is conducted.

Heat pump-based heat recovery

AHUs can have an inbuilt heat pump. The heat pump consists of two batteries (one in each air stream, just as a run-around) connected by a refrigerant circuit with a motorized compressor and a pressure reduction valve. The compressor's power consumption represents 20–30 per cent of the heat extracted from the exhaust air stream, and is also released as heat in the supply air stream. Heat pumps do not transfer moisture. The heat recovery efficiency is normally controlled by regulating the compressor speed or by diverting gas from the compressor's pressure side to its suction side. Defrosting is done by periodical inverting or simply by changing the exhaust air set-point to just over 0° C. Due to high installation cost, heat pumps are primarily used in buildings that need cooling since the same system can be used for cooling in summer and for heat reclaim in winter.

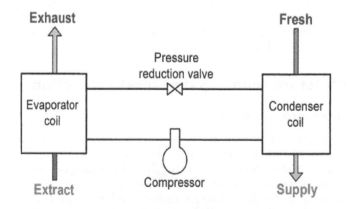

Source: Norwegian Building Research Institute

Figure 6.12 *Heat pump-based heat recovery*

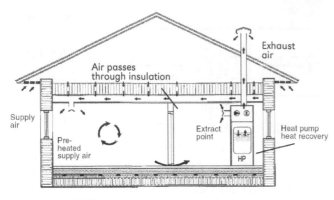

Source: Brunsell (1994)

Figure 6.13 *Counter-flow dynamic insulation system combined with exhaust air heat pump*

Dynamic insulation

Dynamic insulation effectively reduces fabric heat loss by passing ventilation air through the building's thermal envelope. The only viable application of dynamic insulation (counter-flow; see Figure 6.13) involves a conventional extract ventilation system that creates an under-pressure in the building, drawing air inwards through the building fabric. The effective U-value of the fabric reduces with increasing flow rate, hence the term 'dynamic insulation'. This reduces the building's combined ventilation and fabric heat loss by up to 23 per cent (Jensen, 1993), which is not better than a conventional balanced ventilation system with up to 50 per cent heat recovery efficiency. Brunsell (1994) has evaluated the system's performance in residential buildings.

It also possible to recover heat by means of outward airflow (co-flow), but this is not recommended due to risk of moisture damage caused by condensation in the insulation. Table 6.1 lists the advantages and disadvantages of dynamic insulation. It is unlikely that dynamic insulation will offer a practical alternative to conventional heat recovery methods in the near future.

Moisture recovery and frost protection

Moisture recovery

Some heat recovery units recover water vapour (also known as 'enthalpy recovery'). These are primarily regenerative heat recovery units. Non-hygroscopic units recover moisture whenever condensation occurs, while hygroscopic units recover moisture under all conditions (with about the same efficiency for moisture and sensible heat). Though normal recuperative heat exchangers do not

recover moisture, some plate heat exchangers are constructed with materials that permit moisture diffusion through the plate walls (for example, cellulose). Moisture recovery is normally used to reduce humidification costs in buildings where it is important to keep a high relative humidity in winter, such as paper and other industries, as well as museums. For 'normal' buildings (dwellings, schools, offices, etc.) moisture recovery is generally only appropriate in hot humid climates (where it can reduce dehumidification costs) or in very cold dry climates (where it reduces comfort problems related to dry indoor air below 25 per cent relative humidity, or RH). However, moisture recovery must be used with caution to avoid over-humidification indoors.

Frost protection

In cold weather, the water vapour in the extract air can fall below its dewpoint temperature and condense within the heat exchanger. Very dry exhaust air can have a dewpoint temperature below 0° C, when the condensate freezes. The degree to which this ice causes operational problems depends upon the type of heat exchanger and operating conditions.

Frost protection is particularly important for (static) recuperative heat exchangers of high efficiency, especially when the extract air is moist (for example, swimming pools) and in cold climates. Ice build-up is generally not a problem for regenerative heat exchangers due to the alternating air streams through the exchanger. As a simple rule of thumb, ice accumulation can become noticeable in regenerative heat exchangers when the average temperature of the outside air and indoor air streams into the exchanger drops below 0° C (for example, +20° C extract air and –20° C intake air).

Common frost protection methods include bypass, preheat and periodic stopping of the supply fan. Recuperative units need a condensate drain, while regenerative units generally do not. The heat recovery efficiency is reduced whenever frost protection is active (see Figure 6.16), so it is used only when necessary. If a preheat battery is employed, it should have sufficient capacity for expected coldest outdoor temperatures. Frost protection should be controlled automatically – for example, with humidity or temperature sensors to detect the conditions that cause ice growth.

Heat recovery performance

Definitions of heat recovery efficiency

The efficiency of a heat recovery system must be known in order to calculate energy savings and profitability.

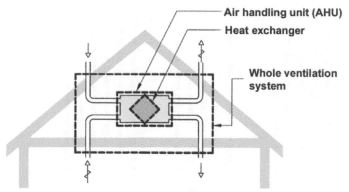

Source: chapter author

Figure 6.14 *Illustration of nested system boundaries*

There are different definitions of efficiency for heat exchangers, depending upon where you define the system boundary – that is, either the heat exchanger itself, or the AHU, or the entire ventilation system (see Figure 6.14).

Furthermore, efficiency can be measured for temperature (that is, sensible enthalpy), moisture or total enthalpy. It can also be measured under steady-state conditions or under varying conditions over a whole heating season (that is, annual mean value). Some common definitions of efficiency are given below:

- *Heat exchanger sensible heat recovery efficiency* = the temperature ratio defined in equation (1) (valid for balanced mass flows).
- *AHU net sensible heat recovery efficiency* = sensible heat recovery for the AHU as a whole, corrected for system losses (air leakages, recirculation and the component of energy used by fans, defrost batteries, etc., that is lost as heat in the exhaust). It is equal to the unit's exhaust temperature ratio – equation (2) – if the system is balanced and there is no condensation in the heat exchanger, and there is negligible recirculation from fresh to exhaust. For a more rigorous definition and equation, see Nordtest (2004).
- *AHU supply temperature ratio* – equation (3). This only gives an indication of the AHU's thermal comfort properties (that is, supply temperature) and is not an accurate measure of heat recovery efficiency.
- *AHU net moisture recovery efficiency* = same as equation (2), but where humidity ratio ($kg_{water\ vapour}$/$kg_{dry\ air}$) is measured instead of temperature. It accounts for recirculation. See also Nordtest (2004).
- *AHU seasonal mean net heat recovery efficiency* = degree day weighted mean value of AHU's net sensible heat recovery efficiency during the heating

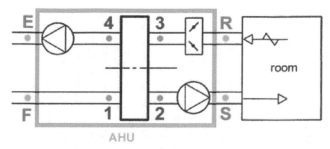

Source: Norwegian Building Research Institute

Figure 6.15 *Measurement points for determining heat recovery efficiency*

season. It is based on measurements of heat recovery efficiency at different outdoor temperatures. It is valid for a specific combination of:

- flow rate;
- local climate; and
- building's balance point temperature, which describes when the building's heating season starts/stops.

Measuring a heat exchanger's efficiency

The parameters in equations (1), (2) and (3) can be temperature (T °C), humidity ratio (w kg/kg), sensible enthalpy ($T(1006+1805w)$ J/kg) or total enthalpy ($T(1006+1805w)+2501000w$ J/kg):

$$\eta_{heatx} \approx \frac{T_2 - T_1}{T_3 - T_1} \tag{1}$$

$$\eta_{AHU,\,exhaust} \approx \frac{T_R - T_E}{T_R - T_F} \tag{2}$$

$$\eta_{AHU,\,supply} \approx \frac{T_S - T_F}{T_R - T_F} \tag{3}$$

Seasonal mean net heat recovery efficiency

Figure 6.18 illustrates the influence of outdoor temperature on the heat recovery efficiency of four different heat recovery units, all with ideal frost protection strategies. In cold weather, condensation in the heat exchanger increases the sensible heat recovery efficiency. This effect is more pronounced for more efficient heat exchangers. However, at lower outdoor temperatures and higher efficiencies, the condensation will freeze and block the unit. This is avoided by various frost protection strategies that reduce the heat recovery efficiency in cold weather.

Table 6.1 Comparison of different types of heat recovery systems for balanced ventilation

Type of heat exchanger	Relative cost*	Temperature ratio	Seasonal net efficiency	Controls (winter/summer) and comments	Advantages	Disadvantages
Regenerative						
Rotary	1.00	70–80%	50–70%	• Rotational speed control • Bypass damper • Moisture recovery efficiency with hygroscopic rotor: 70–85%	• High heat recovery efficiency • Can have equally high moisture recovery efficiency • Efficiency easily controlled • Frost protection not needed above approximately −15°C (especially for dry indoor air)	• Risk of recirculation of odour and pollutants • Risk of smoke spreading during fire • Supply and exhaust ducts must be gathered to the same location • Has moving components requiring maintenance • Risk of fouling of rotor when static
Reciprocating	1.15	75–90%	70–85%	• Change alternating time interval; high efficiency at one-minute intervals; longer intervals less efficiency	• Highest heat recovery efficiency • Can have reasonable moisture recovery efficiency • Efficiency easily controlled • Frost protection not needed above approximately −15°C (especially for dry indoor air) • Damper is the only moving component	• Risk of recirculation of odour and pollutants • Risk of smoke spreading during fire • Supply and exhaust ducts must be gathered to the same location • Large space requirement • Must be located adjacent to external wall
Recuperative						
Plate or tube heat exchanger Plate: cross-flow or counter-flow	1.05	Cross: 50–70% Counter: 80–90%	45–60% 60–75%	• Bypass damper • Some small units have replacement summer cassette	• Minimal recirculation of odour and pollutants • Bypass damper is only moving component • Mechanically reliable • Counter-flow plate heat exchanger has high heat-recovery efficiency	• Supply and exhaust ducts must be gathered to the same location • Bypass requires extra space • Requires frost protection • Must have condensate drain • Risk of fire spreading
Run-around	1.45	50–65%	45–55%	• Change fluid flow rate • Shunt bypass control of fluid flow • Periodic stop of pump	• Zero recirculation of odour and pollutants • Supply and exhaust ducts can have different locations; relatively easy to install in existing buildings • Heat recovery efficiency is easily controlled	• Has moving components requiring maintenance • Requires frost protection and fluid must have antifreeze • Requires piping between batteries
Heat pipe	1.25	55–65%	50–55%	• Partly self-regulating • Bypass damper • Changing slope of heat pipes (applies to 'horizontal' type)	• Zero recirculation of odour and pollutants • No moving components • No pipe connections • Less risk of freezing than plate heat exchanger due to more uniform temperature distribution	• Supply and exhaust ducts must be gathered to the same location • Requires frost protection • Control of heat recovery efficiency is complicated • Bypass requires extra space

Type				Advantages		Disadvantages
Heat pump based	2.0~3.0	60–80%	35–60%	• Zero recirculation of odour and pollutants • Supply and exhaust ducts can have different locations if heat-pump system is split; relatively easy to install in existing buildings • Can provide cooling in summer	• Control compressor power	• Requires a lot of space • Has moving components; higher risk of failure • Control of heat recovery efficiency is complicated • Requires frost protection (or limit cooling of exhaust to +5°C) • Risk of noise problems • High investment and running costs
Dynamic insulation	Up to 50 %	Up to 50 %		• Heat recovery possible • Counter-flow system is well suited combined with exhaust air heat pump		• Requires careful design and installation, adding to capital and construction costs • Co-flow systems risk moisture condensation damage, thus not recommended • Counter-flow systems recover less heat than ordinary air-to-air heat recovery • Inevitable clogging of the insulation, emission of insulation fibres or gases

Note: * The stated relative installation costs are for an 8000 cm^3/h air handling unit (AHU) in an office building. The AHU normally constitutes approximately 25 per cent of the total cost of the whole ventilation and mechanical cooling system in such buildings, typically 7 to 8 Euros per m^3/h.

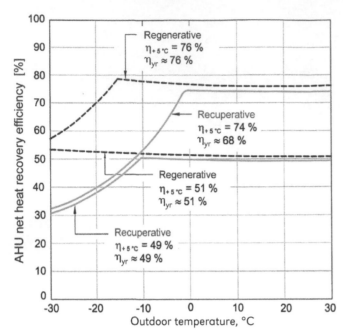

Note: The recuperative units have the same frost protection battery capacity limit; the regenerative units reduce their effective heat recovery efficiency without a heating battery. The annual net sensible heat recovery efficiency (η_{yr}) applies to a modern house in Norway.

Source: Norwegian Building Research Institute

Figure 6.16 *The effect of outdoor temperature on net sensible heat recovery efficiency for four different heat recovery units with good frost protection*

The AHU's seasonal mean net heat recovery efficiency is used to calculate annual savings in heating costs. It can be calculated with the help of an annual distribution curve for outdoor temperature. The mean efficiency during the heating season is practically equal to the degree day weighted mean value of the AHU's net sensible heat recovery efficiency during the season (with extract air temperature as a base). The calculation method is described in Nordtest (2004). Figures 6.17 and 6.18 are examples of such distribution curves for a house and office building, respectively. The lighter curve (*supply air balance-point temperature*) is the supply air temperature above which space heating is not required – in other words, solar and internal gains (people and equipment), together with eventual heat recovery, are enough to keep the room temperature at the required level. The *supply air balance-point temperature* depends upon the building's U-value, air tightness, and internal and solar gains. The four areas ①, ②, ❶ and ❷ represent four energy quantities:

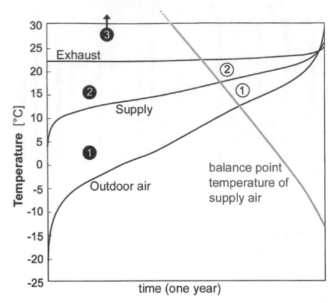

Source: Norwegian Building Research Institute

Figure 6.17 *Annual temperature distribution curve: House*

- Area ① + ② + ❶ + ❷ represents the theoretical maximum energy loss due to ventilation without heat recovery.
- Area ① + ② is the part of the aforementioned ventilation heat loss that consists of internal and solar gains.
- Area ❶ + ❷ is the part of the aforementioned ventilation heat loss that is heat from bought energy if there is no heat recovery.
- Area ① + ❶ is heat that is recovered by the heat exchanger, of which ❶ represents savings in space heating costs; however, ① does not lead to cost savings, but merely overheating in summer.
- Area ❷ is bought energy to heat the supply air up to room temperature.
- Area ❸ applies to buildings that have low internal gains relative to the heat losses, such that the supply air can be heated above room temperature in winter without causing overheating.

Calculating energy savings and profitability

Savings in energy consumption

Accurate method

The energy savings provided by a heat recovery unit can be accurately estimated by using building thermal simula-

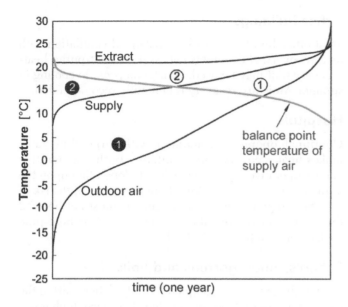

Source: Norwegian Building Research Institute

Figure 6.18 *Annual temperature distribution curve: Office*

tion software (such as EnergyPlus, ESPr, TRNSYS and BSim). Ideally, the software should be able to account for the change in heat recovery efficiency as a function of outdoor temperature (see Figure 6.16). It is also important that the heat recovery efficiency is correctly specified: if it models the individual components of the AHU separately (for example, heat exchanger, fans, etc.), then the *heat exchanger's sensible efficiency* should be used. If the model does not calculate fan power separately (this includes simpler methods such as EN 832), then the *AHU's net sensible heat recovery efficiency* should be used.

Simplified method

As an alternative to using software, the following simple equation may be used to estimate energy savings compared to natural ventilation:

$$\text{kWh}_{saved} = 3 \cdot n \cdot V \cdot \bar{\eta} \cdot k \cdot (22 - \bar{T}_u) \qquad (4)$$

where:

- 3 = constant $(1.2\text{kg/m}^3 \times 1050\text{J/kgK} \times 0.001\text{kW/W/} 3600\text{s/h} \times 8760\text{h/year})$ $(\text{J/m}^3\text{K})$;
- n = mean air change rate during the heating season due to mechanical ventilation (infiltration not included); the mean value takes into account periods of reduced flow rate or no ventilation (for example, nights and weekends) (h^{-1});
- V = ventilated building volume (m^3);

- $\bar{\eta}$ = AHU's *seasonal mean net sensible heat recovery efficiency* at relevant airflow rate and local climate (fraction between 0.0 and 1.0); for mechanical exhaust ventilation of dwellings (with no heat recovery) $\bar{\eta} \approx -0.05$;
- $\bar{T}_u$ = annual mean outdoor temperature (°C);
- k = fraction of recovered heat that constitutes a reduction in heating costs – that is, not surplus heat (fraction 0 to 1):

$$k \approx a(1 - b \cdot \bar{\eta}^3) \qquad (5)$$

- Constants a and b depend upon the building construction, use and internal/solar gains (see Table 6.2).

Table 6.2 *Examples of values of constants a and b used in a simplified estimation of energy savings for a Nordic climate with U-values: walls = 0.22W/m²K; windows = 1.6W/m²K; roof and ground = 0.15W/m²K*

Building type	a	b
Detached house	0.87	0.03
Apartment	0.83	0.06
Kindergarten	0.87	0.14
School	0.79	0.30
Office building	0.77	0.27

Source: Schild (2003)

Enthalpy or sensible heat recovery?

In buildings with central humidification, energy savings can be estimated by calculations based on total enthalpy recovery efficiency (see the earlier section on 'Measuring a heat exchanger's efficiency'). However, simplified calculations using total enthalpy recovery efficiency can lead to overestimation of cost savings. It is preferable to conduct separate calculations of sensible heat savings (using *net sensible recovery efficiency*) and latent heat savings (using *net moisture recovery efficiency*) and to add the two. Latent heat calculations need only be done for buildings with a central humidification (or equivalent) that has a lower running cost due to moisture recovery in the heat exchanger.

Choice of heat recovery for different building types

Choosing between different types of heat recovery systems is a question of profitability and of functional qualities, such as leakage (recirculation), moisture recovery and frost protection (see Table 6.1). Fire safety can be a decisive factor in large buildings to prevent risk of smoke and fire spreading via the ventilation system. Profitability increases with increasing flow rate and operational hours. Savings are greatest for swimming pools, industry and hospitals, less for offices, and least for dwellings.

Location and ductwork requirements

The best location for the AHU is in a dedicated plant room or other technical room (washroom or store) or a warm loft with good access. The chosen location should also give a short duct system (especially the fresh air duct) and wide ducts. In cold climates, locating the AHU outdoors or in a cold loft (outside the building's insulation envelope) is not recommended. If the supply and exhaust ducts do not gather to the same location, then a run-around or split heat-pump system must be chosen. The fresh air and exhaust air ducts must always be insulated along their entire length, and the supply and extract ducts must be insulated where they pass through unheated or especially hot zones (for example, attic). Duct insulation should be ≥ 50mm or 100–150mm in especially cold climates.

Houses

Balanced ventilation with heat recovery provides the best combination of good air quality and energy efficiency in dwellings, especially multi-storey houses in cold climates. The house should be airtight (≤ 2 air changes at 50Pa pressure blower door test). Special attention must be given to prevent noise (sound attenuator in the supply duct after the AHU; wide ducts ideally never less than 125mm in diameter; quiet supply diffusers) and draught problems (in cold climates, an after-heat battery should be considered; but the set-point must not exceed 19° C).

Apartment buildings

In new apartment buildings, and existing buildings with balanced ventilation, heat recovery is almost standard today. For existing apartment buildings that were constructed with only an exhaust system, it is generally not profitable to install balanced ventilation with heat recovery due to lack of space for new supply ducts. In this case, it is more economic to install an exhaust air heat pump that can heat the domestic hot water or preheat water in the radiator heating system. Nevertheless, balanced ventilation ensures the best air quality and least draught problems.

Office buildings

Most office buildings now have balanced ventilation with heat recovery. If cooling is necessary, a heat-pump heat-recovery unit may be used, which can provide cooling in summer. Moisture recovery is generally recommended.

Hospitals

Continual operation combined with large flow rates makes heat recovery very profitable. Both regenerative and recuperative types are pertinent, depending upon the application. For ventilation of zones with a high risk of infection, regenerative heat exchangers must not be used due to risk of recirculation – run-around or heat-pipe exchangers can be used instead.

Schools, kindergartens and halls

For buildings that are occupied for only a short part of the day and have a high occupancy density (hence, high internal gains), the profitability of heat recovery is less distinct. Moisture recovery could, at times, lead to too high humidity indoors, so it is not necessarily beneficial.

Industry

For industrial applications, perhaps the greatest concern is the influence of pollutants, both on the reliability of the heat recovery system and due to the risk of cross-contamination (recirculation). For rotary heat exchangers, the risk of cross-contamination can be reduced with a purge sector and correct location of the fans (see Figure 6.3). Dry particles pose less of a fouling risk than sticky particles. The rotary wheel is also protected by a filter on the extract air side. Measures against specific pollutants are described as follows:

- *Oil or fat vapours* can lead to significant dust accumulation at the front face of the heat exchanger, as well as in ducts. A filter must be placed in front of the heat exchanger. If the extract air contains oil aerosols with particles, the system must be designed for cleaning with solvents and a pressure hose. Tube heat exchangers are often the best choice since they are easier to keep clean.
- *Fibrous materials* – for example, from textiles, mineral wool or glass fibre – can be removed with a fine gauze in the extract duct. In addition, a normal filter should be located after the gauze.
- *Solvents*: regenerative heat exchangers with hygroscopic or sorptive material must not be used if the extract air contains ammonia, formaldehyde or solvents. Even non-hygroscopic aluminium heat

exchangers become increasingly hygroscopic over time due to fouling and oxidation of the aluminium.

- *Paint and lacquer*: the degree of fouling and blockage depends upon the type of paint and how dry the particles are. Specialized filtration is required in the extract duct.
- *Salts* present a significant risk of corrosion. The AHU, together with the heat exchanger and condensate drain, must be of a corrosion-resistant material.
- *Welding smoke*: the most economic strategy is to clean the heat exchanger whenever necessary. Filtration of the extract air is of little practicability due to rapid blocking. Cyclonic separators are an alternative. Rotary heat exchangers have good self-cleaning properties if they run continually.

Commissioning, operation and maintenance

Hand-over test

Heat exchangers should be checked as part of the hand-over procedure for the ventilation system. The choice of measurement method and the extent of the tests depend upon the size of the system and should be agreed upon before the tests. For larger AHUs, the casing leakage should be documented and comply with limits set in EN 1886. Other tests for larger AHUs are described in EN 13053. Installation checks for residential units are described in EN 14134. The hand-over tests should also check that the controls are functioning properly. See, for example, CIBSE (2003).

Function testing

The heat exchanger's performance is normally checked by a simple measurement of the temperature ratio. This involves measuring the supply and exhaust flow rates, and measuring the air temperature in both air streams before and after the heat exchanger. Due to uneven temperature distribution in the heat exchanger, many temperature measurements must be made across the flow area. It is also possible to measure the temperature after a fan or other mixing device (subtracting the temperature rise through the fan). Measurements should be conducted under conditions that do not cause condensation in the heat exchanger.

For some types of heat exchanger, it is also appropriate to measure the moisture recovery efficiency. Other function tests include measuring after-heat power, pressure drop and ventilation noise level in some selected rooms and outside.

Operation, inspection and maintenance (O&M) instruction documentation

Reliable and economic functioning of the heat recovery system can only be achieved if there are good routines for operation, inspection and maintenance (O&M). This requires proper O&M documentation tailored for the individual ventilation system.

Acknowledgements

Illustrations in this chapter are reproduced with permission from the Norwegian Building Research Institute. François Durier, Miroslav Jitca and Peter Op't Veld have provided useful comments.

The work for this chapter has been supported by Enova through contract SID02/1755, Air Infiltration and Ventilation Centre (AIVC), and the Research Council of Norway.

References

Brunsell, J. T. (1994) 'The performance of dynamic insulation in two residential buildings', *Proceedings of 15th AIVC Conference*, vol 1, Buxton, UK, pp285–288

CIBSE (Chartered Institution of Building Services Engineers) (2003) *CIBSE Commissioning Code M: Commissioning Management*, www.cibse.org

EN 308 (undated) *EN 308: Heat Exchangers – Test Procedures for Establishing Performance of Air to Air and Flue Gases Heat Recovery Devices*, CEN

EN 1886 (undated) *EN 1886: Ventilation for Buildings – Air Handling Units – Mechanical Performance*, CEN

EN 13053 (undated) *EN 13053: Ventilation for Buildings – Air Handling Units – Ratings and Performance for Units, Components and Sections*, CEN

EN 14134 (undated) *EN 14134: Ventilation for Buildings – Performance Testing and Installation Checks for Residential Ventilation Systems*, CEN

Irving, S. (ed) (1994) *AIVC Technical Note 45: Air-to-Air Heat Recovery in Ventilation*, December, www.aivc.org

Jensen, L. (1993) 'The energy impact of ventilation and dynamic insulation', *Proceedings of 14th AIVC Conference*, Copenhagen, Denmark, pp251–260

Nordtest (2004) *Nordtest Method NT VVS 130: Air/Air Heat Recovery Units – Aerodynamic and Thermal Performance*, www.nordtest.org

Orme, M. (1998) *AIVC Technical Note 49: Energy Impact of Ventilation – Estimates for Service and Residential Sectors*, AIVC

Schild, P. G. (2003) Project report 341-2003, Norwegian Building Research Institute, Oslo

7

Hybrid Ventilation in Non-Residential Buildings

Per Heiselberg

Introduction

Hybrid ventilation is a new ventilation concept that utilizes and combines the best features of natural and mechanical ventilation systems. Hybrid ventilation provides opportunities for innovative solutions to the problems of mechanically or naturally ventilated buildings: solutions that simultaneously improve the indoor environment and reduce energy demand. Natural and mechanical ventilation have developed separately over many years and the potential for further improvements is limited. But the combination of natural and mechanical ventilation opens a new world of opportunities.

This chapter focuses on the application of hybrid ventilation in office and educational buildings and is, to a large extent, based on the work of the international task-shared project Annex 35 Hybrid Ventilation in New and Retrofitted Office Building (Heiselberg, 2002) initiated by the International Energy Agency (IEA) under the implementing agreement Energy Conservation in Buildings and Community Systems (ECBCS).

Hybrid ventilation systems can be described as systems that provide a comfortable internal environment using both natural ventilation and mechanical systems, but using different features of these systems at different times of the day or season of the year. In hybrid ventilation, mechanical and natural forces are combined in a two-mode system where the operating mode varies according to the season and within individual days. Thus, the active mode reflects the external environment and takes maximum advantage of ambient conditions at any point in time. The main difference between a conventional ventilation system and a hybrid system is the fact that the latter has an intelligent control system that can switch automatically between natural and mechanical modes in order to minimize energy consumption.

There are multiple motivations for the interest in hybrid ventilation. The most obvious are:

- Hybrid ventilation has access to both ventilation modes in one system, exploits the benefits of each mode and creates new opportunities for further optimization and improvement of the overall quality of ventilation.
- Advanced hybrid ventilation technology fulfils the high requirements on indoor environmental performance and the increasing need for energy savings and sustainable development by optimizing the balance between indoor air quality, thermal comfort, energy use and environmental impact.
- Hybrid ventilation results in high user satisfaction because of the use of natural ventilation, the high degree of individual control of the indoor climate (including the possibility of varying the indoor climate – adaptive comfort), as well as a direct and visible response to user interventions.
- Hybrid ventilation technology offers an intelligent and advanced ventilation solution for the complex building developments of today – that is, user transparent and sustainable.

Naturally, expectations of hybrid ventilation performance will vary between different countries because of climate variations, energy prices and other factors. In countries with cold climates, hybrid ventilation can avoid the trend to use mechanical air conditioning in new buildings, which has occurred in response to higher occupant expectations, the requirements of codes and standards, and, in some cases, higher internal gains and changes in building design. In countries with warm climates, it can reduce the reliance on air conditioning and reduce the cost, energy penalty and consequential environmental effects of full year-round air conditioning.

Both natural and mechanical ventilation have advantages and disadvantages. For natural ventilation systems, one of the major disadvantages is the uncertainty in performance, which results in an increased risk of draught problems and/or low indoor air quality in cold climates and a risk of unacceptable thermal comfort conditions during summer periods. On the other hand, air-conditioning systems often lead to complaints from the occupants, especially in cases where individual control is not possible. Hybrid ventilation systems have access to both ventilation modes and therefore allow the best ventilation mode to be chosen depending upon the circumstances.

The focus on the environmental impacts of energy production and consumption has provided an increased awareness of the energy used by fans, heating/cooling coils and other equipment in ventilation and air-conditioning systems. An expectation of a reduction in annual energy costs has also been an important driving force for the development of hybrid ventilation strategies. Available data from case studies provided in the international project IEA ECBCS-Annex 35 (Heiselberg, 2002) show that a substantial energy saving has been achieved in a number of buildings, mainly because of a very substantial reduction in energy use for fans and a reduced energy use for cooling as a result of careful attention to minimizing pressure differences in the system, to minimizing internal and external heat gains and to demand control.

Buildings with natural ventilation are associated with less sick building syndrome (SBS) symptoms than buildings with traditional ventilation systems, (Seppänen and Fisk, 2002). Natural ventilation is well accepted by occupants and the natural ventilation mode should therefore be used when the climatic conditions allow it. In addition, the high degree of user control in hybrid ventilation systems has an influence on the perceived indoor environmental quality. Another aspect is that hybrid ventilation implies less noise (provided that there are no outdoor sources of heavy noise), which may also improve the perceived quality. In IEA-ECBCS Annex 35 the high degree of user control in the investigated buildings was greatly appreciated by the occupants. In one of the cases (Rowe, 2002), investigations via occupant questionnaires showed that the perceived performance increased as a function of perceived indoor air quality and thermal comfort.

Estimating the initial cost of hybrid ventilation systems in buildings can be quite difficult as the installation often consists of both mechanical installations and building elements. Part of the investment in mechanical equipment is often shifted towards a larger investment in the building itself: increased room air volume per person, a shape favourable to air movement, a more intelligent façade/window system, etc. On the other hand, the building might provide more usable (rentable) space, as space for plant rooms, stacks for ventilation channels, etc., is not needed. Recently a method for the calculation of life cycle costs (LCCs) of natural ventilation systems has been developed, (Vik, 2003), which takes all of these issues into consideration. This method can also be applied to buildings with hybrid ventilation systems.

In IEA–ECBCS Annex 35, the reference cost range provided by the participants was used to compare the initial costs of hybrid ventilation systems, and buildings with hybrid ventilation, with the initial cost of traditional systems and buildings, see (van der Aa, 2002a). The life cycle costs for hybrid ventilated buildings were often lower than for reference buildings; but the relationship between initial, operating and maintenance costs was different.

Hybrid ventilation strategies

There is a wide range of hybrid ventilation strategies and the concepts vary widely in the level of building integration and industrialization (de Gids, 2001; Wouters et al, 1999 and 2000). In order to characterize a hybrid ventilation strategy it is necessary to describe the hybrid ventilation principle, the control strategy for indoor air quality (IAQ) and summer comfort, and the specific boundary conditions and components, as well as the level of building integration. Typical hybrid ventilation strategies are described in Chapter 3.

Hybrid ventilation principles

The main hybrid ventilation principles are:

*   *Natural and mechanical ventilation.* This principle is based on two fully autonomous systems where the control strategy either switches between the two systems, or uses one system for some tasks and the other system for other tasks. It covers, for example, systems with natural ventilation in intermediate seasons and mechanical ventilation during midsummer and/or midwinter; systems with mechanical ventilation during occupied hours and natural ventilation for night cooling; or systems with a mechanical

system for task ventilation and/or cooling and a natural system for building ventilation.

• *Fan-assisted natural ventilation.* This principle is based on a natural ventilation system combined with an extract or supply fan. It covers natural ventilation systems which, during periods of weak natural driving forces or periods of increased demands, can enhance pressure differences by mechanical (low-pressure) fan assistance.

• *Stack- and wind-supported mechanical ventilation.* This principle is based on a mechanical ventilation system that makes optimal use of natural driving forces. It covers mechanical ventilation systems with very small pressure losses where natural driving forces can account for a considerable part of the necessary pressure.

Indoor air quality (IAQ) control

Acceptable indoor air quality can be achieved by either of the above ventilation principles. Demand-controlled systems are important during periods of heating and cooling demand since the control strategy needs to focus on achieving an optimal equilibrium between IAQ and energy use. The level of demand control can vary from manual by occupants, simple timer control and motion detection, to direct measurement of IAQ.

Control of summer comfort

Acceptable thermal comfort conditions can be achieved by passive means (free cooling with outdoor air, free cooling in embedded ducts, night cooling, etc.) or by a combination of passive means and active cooling (cooled ceilings, air conditioning) during extreme weather conditions. The control of room temperature during occupied hours can be either manual or automatic.

Specific boundary conditions and ventilation components

There are only a few real hybrid ventilation components. In most cases, hybrid ventilation systems consist of a combination of components, which can be used in purely natural systems or in purely mechanical systems. However, the availability of appropriate components is essential for the successful design and operation of a hybrid ventilation system to handle the specific boundary conditions of the site and the building, such as draught control, security, air preheating, outdoor air pollution, and noise and fire regulations. To facilitate combining natural and mechanical forces in the air distribution system, appropriate components can include:

• low-pressure ductwork;
• low-pressure fans with advanced control mechanisms, such as frequency control, airflow control, etc.;
• low-pressure static heat exchangers and air filters; and
• wind towers, solar chimneys or atria for exhaust, as well as underground ducts, culverts or plenums to precondition supply air.

To facilitate the control of thermal comfort, indoor air quality and airflow in the building, appropriate components can include:

• manually operated and/or motorized windows, vents or special ventilation openings in the façade and in internal walls;
• room temperature, carbon dioxide (CO_2) and/or airflow sensors; and
• a control system with a weather station.

Level of building integration

Hybrid ventilation strategies can vary widely in the level of integration with the building. An integrated approach is a necessity for all hybrid ventilation systems; but the integration of the building and the ventilation system is more important when natural ventilation plays a dominant role. In extreme cases, the whole architectural concept and installation design is fully linked and integrated. In this case, a very close collaboration between architects and mechanical engineers is essential.

Hybrid ventilation design

Today, the construction industry is in the early stages of reinventing the design process that was used before the advent of mechanical systems. Design teams including both architects and engineers are formed and the building design is developed in an iterative process from the conceptual design ideas to the final detailed design. Building energy use and the sizes of mechanical equipment are reduced without the use of sophisticated technologies, but only through an effective integration of the architectural design and the design of mechanical systems. This design approach is necessary for a successful application of hybrid ventilation systems both with regard to optimization of life cycle cost, environmental performance and indoor environmental quality, (Heiselberg, 2000; Tjelflaat, 2001).

Design challenges

The purpose of the ventilation system in many projects is not only to control indoor air quality, but also during the

Table 7.1 *Issues of concern in optimizing hybrid ventilation for indoor air quality control and natural cooling*

Indoor air quality control	Natural cooling
Limitation of pollution sources (building materials, equipment, local exhaust, etc.)	Limitation of heat load (low-energy equipment, solar shading, daylight)
Choice of appropriate indoor air quality targets and related airflow rates	Choice of appropriate thermal comfort targets (minimum and maximum values)
Optimum air supply to occupants and removal of pollutants (ventilation efficiency)	Optimum air supply to occupants (temperature efficiency)
Minimize heating and cooling energy (heat recovery, passive heating, passive cooling, etc.)	Minimize cooling load (thermal mass, night ventilation)
Minimize fan energy (low-pressure ductwork and components, natural driving forces, etc.)	Minimize fan energy (low-pressure duct-work and components, natural driving forces, etc.)
Adapt airflow rates to indoor air quality needs (control strategy, demand-controlled ventilation)	

Source: Heiselberg (2002)

summer in an energy-efficient way to achieve thermal comfort through natural cooling. In the design of hybrid ventilation systems, it is often necessary to separate the design of ventilation for indoor air quality control and the design of ventilation as a natural cooling strategy during the summer. The major reason for this is the fact that devices for indoor air quality control and thermal comfort control are, in general, quite different, and that the potential barriers and problems to be solved, including the optimization challenge, also are fundamentally different (see Table 7.1).

Ventilation for indoor air quality control

When optimizing ventilation for indoor air quality control, the challenge is to achieve an optimal equilibrium between indoor air quality, thermal comfort, energy use and environmental impact during periods of heating and cooling demands.

First of all, this includes minimizing the necessary fresh airflow rate by reducing pollution sources and by optimal demand control of airflow rates for the occupants. Second, it includes reducing heating and cooling demands by heat recovery, passive cooling and/or passive heating of ventilation air. Finally, it includes reducing the need for fan energy by using low-pressure ductwork and other components, as well as optimizing natural driving forces from stack effect and wind. During periods without heating and cooling demands, there is no need to reduce airflow rates as more fresh air will only improve the indoor air quality. The optimization challenge then becomes primarily a question of minimizing the use of fan energy. In addition to the challenges mentioned above, ventilation should, of course, be provided without creating comfort problems, such as draught, high temperature gradients or noise.

With regard to indoor air quality control, a hybrid ventilation system based on stack- and wind-assisted mechanical ventilation, or on natural and mechanical ventilation with a seasonal changeover operation strategy, does not differ from a traditional demand-controlled mechanical ventilation system. However, for a hybrid ventilation system based on fan-assisted natural ventilation, or on natural and mechanical ventilation with a continuous changeover operation strategy, it is necessary to consider the dynamic behaviour of the airflow rate in the natural ventilation mode and the buffer effect of the building, which causes the indoor air quality to vary around an average value. The acceptable magnitude and duration of variations must be defined in order to determine when assisting fans are started or when the system switches to the mechanical mode. A strategy for switching back to the natural ventilation mode must also be defined.

Ventilation for temperature control

When optimizing ventilation as a natural cooling strategy, the challenge is to achieve an optimal equilibrium between cooling capacity, cooling load, thermal mass and thermal comfort.

First of all, this includes reducing internal and external heat loads by using low-energy equipment, by utilizing daylight and by effective solar shading. Second, it includes exploiting the thermal mass of the building, which absorbs and stores heat during occupied hours and is cooled during unoccupied hours by night ventilation. Finally, it includes reducing the need for fan energy by using low-pressure ductwork and other components, as well as optimizing natural driving forces derived from the stack effect and wind. The major issues of concern with regard to thermal comfort are avoiding excessively low temperatures at the start of the working hours (appropriate night cooling strategy) and achieving an acceptable temperature increase during working hours (solar shading and thermal mass). The ventilation airflow rate needed for natural cooling is, in general, much higher than the ventilation airflow rate needed for indoor air quality control.

With regard to temperature control, it is important that a hybrid ventilation system based on stack- and wind-assisted mechanical ventilation is capable of adapting to window opening to improve occupant comfort tolerance and to reduce fan energy. As the stack effect is limited, it is also important that the ventilation system is designed for optimum use of wind effect. For a hybrid ventilation system based on natural and mechanical ventilation, the natural mode dominates for temperature control in cold climates. The size of the mechanical system (designed for indoor air quality control) is usually not enough to achieve acceptable temperature control; but in periods that lack adequate natural driving forces, it can provide a valuable supplement. In warm climates the natural mode is mainly useful for temperature control in the intermediate season, while it can give a valuable supplement to night ventilation during the summer period. For a hybrid ventilation system based on fan-assisted natural ventilation, an optimum use of the wind effect is important.

Design procedure

Effective ventilation of indoor spaces has the best chance of success when the design process is carried out in a logical, consecutive manner with increasing detail richness towards the final design and in the framework of a design procedure. In the case of hybrid ventilation, the need for a design procedure is even more evident due to the comprehensive design team, where users, building owner, architect, civil engineer and indoor climate and energy counsellor must all be involved – simultaneously.

The hybrid ventilation process is very dependent upon the outdoor climate and the microclimate around the building, as well as the thermal behaviour of the building; therefore, it is essential that these factors are taken into consideration in the first design step. The output from the first step is a building orientation, design and plan that minimize the thermal loads on the building during overheated periods, which together with the selected ventilation strategy make it possible to exploit the dominating driving forces (wind and/or buoyancy) at the specific location, and which ensure a proper air distribution through the building. It is also important that issues such as night-cooling potential, noise and air pollution in the surroundings, as well as fire safety and security, are taken into consideration.

In the second design step the natural ventilation mode of the hybrid system is designed. The location and size of openings in the building, as well as features to enhance the driving forces such as solar chimneys and thermal stacks, are designed according to the selected strategy for both day- and night-time ventilation. Passive methods to heat and/or cool the outdoor air are considered, as well as heat recovery and filtration. Appropriate control strategies for the natural ventilation mode are determined and decisions are made regarding the level of automatic and/or manual control and user interaction.

In the third step the necessary mechanical systems to fulfil comfort and energy requirements are designed. These can range from simple mechanical exhaust fans to enhance the driving forces to balanced mechanical ventilation or full air-conditioning systems. The hybrid ventilation and the corresponding whole-system control strategy are determined to optimize the energy consumption, while maintaining acceptable comfort conditions.

The hybrid ventilation design procedure differs from the design procedure for conventional heating, ventilation and air-conditioning (HVAC) systems in the way that the design in all design phases needs to focus on all three steps in the integrated design approach. The design procedure can consist of the following phases: conceptual design phase, basic design phase, detailed design phase and design evaluation.

The conceptual design phase includes decisions on building form, size, function and location. Targets are set for indoor air quality, thermal comfort and energy use, as well as cost limits. The conceptual design of the hybrid ventilation system is based on these considerations, as well as guidelines and experiences from previous buildings. The natural ventilation principle (stack and/or wind driven, single sided and/or cross-ventilation) to be used is decided together with the principle of the necessary additional mechanical systems.

In the basic design phase the building heat, sun and contaminant loads are estimated and the hybrid ventilation system layout is designed. The necessary airflow rates, as well as expected indoor air quality and temperature levels, are calculated. A rough yearly energy consumption is calculated, together with the necessary peak power demands. If the results do not meet the targets, the building and its systems will have to be redesigned before entering the next phase.

In the detailed design phase, contaminants and thermal loads are re-evaluated and source control options are considered and/or optimized. The type and location of hybrid ventilation system components are selected, as well as the control strategy and sensor location. Based on hour-by-hour calculations through a design year, the

BOX 7.1 LIBERTY TOWER OF MEIJI UNIVERSITY

Location	Tokyo, Japan
Architect	Nikken Sekkei Ltd
HVAC Engineers	Nikken Sekkei Ltd
More information	Chikamoto and Kato (2002)

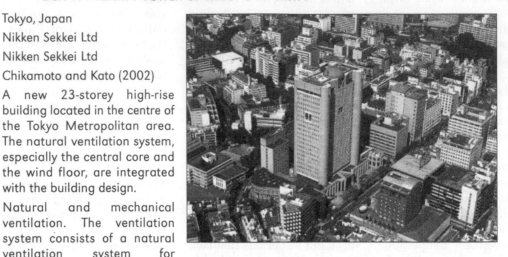

Building description and system integration A new 23-storey high-rise building located in the centre of the Tokyo Metropolitan area. The natural ventilation system, especially the central core and the wind floor, are integrated with the building design.

Ventilation principle and components Natural and mechanical ventilation. The ventilation system consists of a natural ventilation system for controlling indoor air quality and temperature in intermediate seasons, and a mechanical air-conditioning system for the rest of the year when the outdoor climate is not comfortable.

In the natural ventilation system, air enters via perimeter counter units on every floor and is exhausted through openings at the top of the centre core. The central core is designed to utilize the stack effect at each floor, and above the centre core a wind floor is designed to enhance driving forces from the wind. As the wind floor is open to four directions, the driving force is expected to be stable throughout the year regardless of wind direction

Control of indoor air quality and summer comfort Outdoor air intake control is based on CO_2 and temperature sensors and is controlled via a building energy management system (BEMS). The system includes automatically controlled natural ventilation windows at night with an automatic intake of outdoor air and wind floor outlets. In the mechanical air-conditioning system, the supplied airflow rate is controlled by a variable air volume flow rate (VAV) system, where the fresh airflow rate is automatically controlled based on indoor CO_2 concentration and the airflow rate and inlet temperature are controlled by room temperature and humidity sensors.

Building performance The use of the natural ventilation system reduces the annual average energy use for cooling the building by 17 per cent, ranging from 90 per cent in April (spring) to a minimum of 6 per cent in July (summer), and continues to reduce cooling to about 62 per cent in November (autumn). The wind floor design on the 18th floor increases the ventilation rate by an average of 30 per cent. Acceptable thermal comfort and indoor air quality (IAQ) conditions were achieved. The most significant problem was encountered close to the low-positioned openings, where occupants experienced draughts. Another problem was high pressure loss in overflow ducts with smoke and fire dampers between rooms and the centre core.

Source: Chikamoto and Kato (2002)

whole system (building and technical systems) is optimized with regard to indoor climate, energy consumption and costs. Finally, in the design evaluation phase, detailed predictions of indoor air quality and thermal comfort are performed to establish if the design fulfils the targets of the project.

Practical solutions

Quite a number of hybrid ventilated buildings have been built around the world. In this section, the characteristics of typical hybrid ventilation systems are illustrated by a few built examples. Specific components for hybrid ventilation are not yet available; but typical solutions are described.

BOX 7.2 B&O HEADQUARTERS

Location	Struer, Denmark
Architect	KHR Architects A/S
HVAC Engineers	Birch and Krogboe A/S
More information	Hendriksen et al (2002)

Building description and system integration   A new three-storied and narrow (width 8.3m) open-plan office building located in the outskirts of a small city. The natural ventilation system is integrated with the building design.

Ventilation principle and components   Fan-assisted natural ventilation. Inlets are low-positioned narrow hatches (windows) located in front of the floor slab.

Inlet air is preheated with a ribbed pipe to improve thermal comfort and is distributed by the displacement principle to improve ventilation efficiency. Air flows through the occupied space to the staircase and is extracted from the top of the staircase. If needed, the system is assisted with a fan located in a cowl on the roof. The glass façade with ventilation hatches is facing north and has no solar shading. The south façade has a moderate window area with user-controlled windows, which are automatically controlled for night cooling.

Control of indoor air quality and summer comfort   The ventilation system is demand controlled by temperature and CO_2 sensors in the offices and run with a constant airflow rate of 1.5 air changes per hour (ACH) in daytime and 3.0 ACH in summer during night cooling. The constant flow is achieved by measurement of the air speed in the extract hood. When ventilation is needed, the hatches and the dampers in the extract cowl open. If the necessary ventilation airflow rate is not achieved by natural means, the fan speed is controlled. The hatches on each floor are controlled by the temperature and CO_2 level of the floor. The ventilation is controlled by a building management system with a weather station on the roof. If the external temperature is below 5 °C the ventilation system is shut down to protect the ribbed pipes and the window hatches from freezing. The ventilation system is also shut down in case of rain or high wind speeds.

Building performance   The initial cost of the system is only about 60 per cent of a conventional system. The displacement air distribution principle works quite well, resulting in a high ventilation efficiency. Energy use of the assisting fans is very low (1.7 kWh/m^{2}year) and accounts for only about 3 per cent of the electrical energy use. The measured indoor air quality (IAQ) (CO_2 used as an indicator) was very high. The energy demand for heating was remarkably higher than expected. This can be related to the large areas for transmission heat loss, a very large glazed area towards the north, and an infiltration rate that is larger than expected.

Source: Hendriksen et al (2002)

Existing office and educational building examples

The examples in Boxes 7.1 to 7.5 show the state of the art with regard to ventilation systems, control strategies, ventilation components, thermal comfort, IAQ and energy use. The examples include both office and educational buildings.

Box 7.3 MEDIÅ SCHOOL

Location	Grong, Norway
Architect	Letnes Arkitekter A/S
HVAC Engineers	SINTEF and NTNU
More information	Tjelflaat (2002)

Building description and system integration	A new single-storey school building located in a small town. The hybrid ventilation system is fully integrated with the building design.
Ventilation principle and components	Stack- and wind-supported mechanical ventilation. This is a balanced low-pressure mechanical system with both air supply and extract in the classrooms. Air is taken from an inlet tower at some distance from the building that utilizes wind forces where the inlet fan is located. Air flows through an underground culvert with a large thermal mass to reduce daily temperature swings and is distributed via a purpose-made basement corridor to low-positioned supply air terminal devices in classrooms to increase ventilation efficiency. Air is extracted from classrooms through a high-positioned hatch into a purpose-made lightwell corridor and exhausted through a roof tower with outlet valves that ensure suction by wind from any direction. A heat recovery unit and a low-pressure exhaust fan are located in the tower. The system also includes filtering, preheating of the ventilation air and heat recovery with bypass, which is located in the basement between the underground culvert and the basement corridor. The flow is driven by low-pressure fans in the supply and extract, supported by wind and stack effects. Window opening is possible and the ventilation system will normally adapt to it.
Control of indoor air quality and summer comfort	Ventilation is demand controlled by a CO_2-sensor in each classroom. If the CO_2 level exceeds the set-point, the extract hatch is opened and adjusted by a motor. The supply fan is controlled by the pressure in the basement supply corridor to 2Pa over-pressure compared to the external. The extract fan is controlled to maintain a 5Pa pressure drop between the basement supply corridor and the lightwell extract corridor to avoid over-pressure in the building. Both fans are frequency controlled. The ventilation is controlled by the building management system.
Building performance	The performance shows acceptable indoor air quality and thermal comfort. The cooling effect of the underground culvert and the basement corridor is higher than expected. The energy use corresponds to the reference consumption in Norway, but is higher than predicted. The reasons for this are initial failures in the control system (now solved) and under-prediction of the energy loss by cold bridges and energy use by BEMS, pumps, etc.

Source: Tjelflaat (2002)

Ventilation components

The availability of appropriate components is essential for the successful design and operation of a hybrid ventilation system. Some components are fully integrated with the building design, while others are off-the-shelf components or traditional ventilation components adjusted for use in hybrid ventilation.

In hybrid ventilation, building-integrated components for the natural ventilation mode play a very important role. Apart from strongly characterizing the building architecture, these components, if well designed, allow supply, distribution and extraction of air from the building at very low pressure drops. Almost all new-built designs apply building-integrated components. Besides building-integrated components, there exist only a few

BOX 7.4 WILKINSON BUILDING

Location	Sydney, Australia
Architect	Not known
HVAC Engineers	McConnel, Smith and Johnson, Sydney
More information	Rowe (2002, 2003)
Building description and system integration	A five-storey cellular office building located in a suburban area of Sydney. Built in 1978 and renovated in 1997, there is no integration of hybrid ventilation systems and building design.
Ventilation principle and components	Natural and mechanical ventilation. The natural

ventilation is mainly wind driven by window airing and with the possibility of cross-flow through corridors. Each cellular office is equipped with an occupant-controlled supplementary variable refrigerant flow heating/cooling fan-coil system. Air distribution is by the mixing principle. There is no night ventilation and solar shading is manually controlled.

Control of indoor air quality and summer comfort	The control strategy is laissez faire. Window airing and door opening is controlled manually by occupants. The fan-coil units are operated by occupants, but are disabled at 9.00 pm and at midnight to save operation costs. The system is, however, immediately available for restart if required. Fan-coil units are controlled by room temperature with the set-point controlled by the user. The occupant can also control fan speed and airflow direction.
Building performance	Occupants perceive thermal comfort and air quality to be better than reported by occupants in 36 other (mainly air-conditioned) settings. Air quality depends upon window adjustment; but an average CO_2-concentration of 930 ppm was measured in winter. Fan-coil units tend to default to 'off' – that is, if conditions in a room are acceptable then the system is not turned on. Occupants prefer to open windows and doors in pleasant weather, but close them when hot dry or warm humid winds occur in summer or on colder days in winter. Energy use over four years has averaged about one quarter of what is typical of spaces with conventional mechanical heating, cooling and ventilation.

Source: Rowe (2002)

real hybrid ventilation components. Most hybrid ventilation systems apply a combination of components, which can be used in purely natural systems or in purely mechanical systems. However, to handle the specific boundary conditions of the site and the building, such as draught control, security, air preheating, outdoor air pollution and noise or fire regulations, some adaptation is essential for the successful performance of a hybrid ventilation system. The following subsections briefly summarize a number of typical components used in hybrid ventilation systems.

Enhancing natural driving forces

The effect of wind forces can be enhanced by the use of wind towers (Tjelflaat, 2002). Openings in the towers can be controlled by a weather station; for intake towers, openings open on the windward side and for exhaust towers, openings open on the leeward side. A special solution for the Liberty Tower of Meiji University (Chikamoto and Kato, 2002) is the wind floor, where large openings in four directions allow wind to cross the floor and thereby enhance the driving forces from the wind on the exhaust stack. It is expected that the ventilation flow rate is increased by 30 per cent. Wind turrets and roof

BOX 7.5 TÅNGA SCHOOL

Location	Falkenberg, Sweden
Architect	Christer Nord-ström Arkitekt-kontor AB
HVAC Engineers	J.& W. Consulting Engineers
More information	Blomsterberg et al (2002a, 2002b)
Building description and system integration	A two-storey school building located in a residential area in the city of Falkenberg. Built in 1968 and renovated in 1997 there is some integration of hybrid ventilation systems and building design.
Ventilation principle and components	Fan-assisted natural ventilation. Three air intakes are positioned

below windows in each room and air is preheated by convectors. Air is distributed by the mixing principle. Air is exhausted below the ceiling and evacuated through a vertical ventilation duct. Local dampers are mounted both in the air intakes and in the exhaust duct from each room. To increase stack effect, a 6m high passive stack and solar chimney have been installed on the roof with assisting fan and a central damper mounted in parallel. The pressure drop in the air distribution system is very low. Low-pressure vents in the façade and low-pressure exhaust air terminal devices are used. There are no filters, large ventilation ducts and no heat recovery. Window airing is possible at any time. Night cooling can be used.

Control of indoor air quality and summer comfort

The control system is a combination of individual and central control. A CO_2 sensor in each room controls the local inlet and outlet dampers. At a CO_2 level of 1000 ppm or less the local dampers are set to a minimum open position, which can be varied as a function of the outdoor temperature. At low outdoor temperatures the airflow rate is therefore automatically limited to prevent excessive energy use and problems with dry indoor air. If the CO_2 level exceeds 1000 ppm, this is indicated by a signal lamp in the classroom. At CO_2 levels above 1500 ppm, the local dampers open 100 per cent. The teacher can always override the local control system and manually change the position of the local dampers between 50–100 per cent. At low temperature differences the fan is started and the central damper is closed. The fan is frequency controlled. The fan speed is controlled by the pressure difference across the fan, which increases as the temperature difference decreases.

Building performance

Occupants perceived the indoor climate as rather good. The CO_2 concentration is mostly around 1000 ppm or lower and only for short periods (10 to 20 minutes) is higher, but very seldom above 1500 ppm. The personnel appreciate that the system can be operated manually and do so fairly often. 30 per cent reduction in energy use for space heating, 55 per cent reduction in use of electricity for ventilation and 45 per cent reduction in use of electricity for lighting were realized.

Source: Blomsterberg et al (2002b)

cowls must be designed to ensure negative pressure at the opening for all wind directions, both to enhance the natural driving forces but also to avoid backflow in the system (Hendriksen et al, 2002). The stack effect can be enhanced by stacks and/or solar chimneys (Blomsterberg et al, 2002a, 2002b).

Ensuring low pressure drops

In hybrid ventilation, low pressure drops in the air paths are ensured by avoiding the use of ducts, by using other components to transport air – for example, corridors, stairwells and atria (Chikamoto and Kato, 2002; Hendriksen et al, 2002; Meinhold and Rösler, 2002; Principi et al, 2002; Tjelflaat, 2002), or by using large ducts. Windows or similar openings result in very small pressure losses. In cases where these openings are not suitable – that is, because of the need for preheating, air filtration, sound attenuation or security – it is important that low-pressure intake vents and exhaust terminal devices with low pressure-drop dampers are used (Blomsterberg et al, 2002b, Elmualim et al, 2003; Heijmans et al, 2002a; van der Aa, 2002b). This is often ensured by choosing large component sizes, as is also the case with static heat exchangers and air filters.

Specially designed low-pressure components are being introduced on the market; but the development is not as far as the development within residential hybrid ventilation initiated by the European research projects TIP-Vent (TIP-Vent, 2001) and RESHYVENT (van der Aa and Op´t Veld, 2004).

Ventilation openings

Ventilation openings for air intake must, besides having a low pressure drop, also provide air according to the need of the occupants and without creating unnecessary draught risk in the occupied zone. Zeidler et al (1998) investigated the cooling capacity for an office room, with two windows close to the occupied zone. It was concluded that the draught risk was too high for outdoor air temperatures below about 10–12° C, depending upon the window construction, and that the cooling capacity was in the range of 25–40W/m^2 for a room depth of 4.5m. It was emphasized that windows of different sizes and positions with adjustable window panes were needed in future designs to provide occupants with opportunities for adjusting volume flow and thermal comfort conditions.

Heiselberg et al (2001), Heiselberg and Bjorn (2002) and Wildeboer and Fitzner (2002) investigated comfort conditions for airflow through different types of window openings. It was concluded for a single-sided ventilation strategy that the high-positioned bottom-hung window was the best choice in winter because air was supplied outside the occupied zone and was easily controlled by changing the opening angle. A side-hung window close to the occupied zone was not a good choice as air was supplied directly to the occupied zone and was difficult to control because the amount of air and the velocity levels

increase very rapidly with increasing opening angles. In summer with small temperature differences, the bottom-hung window will not be able to supply enough air to the room, but will have to be combined with a side-hung window. For a cross- or stack-ventilation strategy, the bottom-hung window was the best choice in winter because the air travels the largest distance and mixes with room air before it reaches the occupied zone. For the side-hung window, the problems are even worse compared to the single-sided situation because the pressure difference is increased.

Trickle ventilators are also used in hybrid ventilation systems, and with the development of both passive and electronic self-regulating trickle ventilators with direction-sensitive flow sensors they offer good opportunities for controlling both airflow and comfort. Low-positioned adjustable intake grilles are often used in cases where preheating and air filtration are needed, either as stand-alone units (Elmualim et al, 2003) or in combination with perimeter radiators.

Energy conservation

The possibility of heat recovery depends upon the ventilation principle. For systems based on the natural and mechanical ventilation principle, mechanical ventilation systems are often equipped with heat recovery since this is frequently the main reason for choosing this principle, especially in cold climates. For other hybrid ventilation principles, heat recovery can only be used in systems with central intake and exhaust of air, and the typical solution involves heat exchangers with water circulation (Tjelflaat, 2002).

Passive cooling, including automatically controlled solar-shading devices, is adopted in many buildings, as well as the use of underground ducts, culverts or plenums to precondition the supply air (Narita and Kato, 2002; Schild, 2002; Tjelflaat, 2002). Underground culverts can be quite effective for preheating intake air in winter, especially for culverts in the basement of buildings, and precooling in summer, and the performance is often better than expected from predictions (Wachenfeldt, 2003).

Air filtering

Filtering of intake air is not very common in existing buildings with hybrid ventilation. Traditional back filters are often used and the pressure loss across the filter is reduced by a large face area (Tjelflaat, 2002), which is not always possible to implement.

Underground culverts and air intake ducts have a measurable filtration effect due to settlement of particles, especially large particles (Schild, 2001). For a 60m culvert, the number of particles with sizes above 0.3μm was reduced by 85 per cent, and for particles above 10μm by 95 per cent.

In order to increase the application of hybrid ventilation in the urban environment, solutions with a very low pressure loss for filtering of incoming air are needed. In this field there are developments – for example, with electrostatic filtering technology in the European Union (EU) NATVENT project (NATVENT, 1999). However these technologies are not yet main stream technologies.

Draught

To avoid draught risk, buildings located in the coldest areas with distributed air intakes preheat supply air by means of specific components (diffusers or fan coils), or by means of radiators located below windows (Blomsterberg et al, 2002b; Hendriksen et al, 2002; van der Aa, 2002b). In buildings with central intake, air is preheated in the intake duct and supplied to rooms by low-velocity low-level diffusers (Schild, 2002; Tjelflaat, 2002).

In well-insulated and tight buildings with high internal and/or external heat loads, the number of hours during the period of occupation when heating is required can be limited. Therefore, to ensure preheating of intake air it is necessary to separate this function from general room heating, or at least to have a separate control of the inlet air temperature, (Hendriksen et al, 2002; van der Aa, 2002b).

Urban aspects

For urban buildings, the application of hybrid ventilation, especially the natural mode of hybrid ventilation, is affected by the special characteristics of the urban environment. These include, among others, smaller natural driving forces, excess temperatures due to absorption of solar radiation, as well as pollution and noise from traffic and other activities, (Santamouris et al, 2001). Apart from this, the possibility of optimal design of building shape, plan and façades for hybrid ventilation application is often constrained by the dense urban environment.

Natural driving forces are weaker in the urban environment, especially because the wind speed in urban areas is generally much lower than in the countryside (Santamouris et al, 2001). Besides this, shielding from other buildings in the vicinity has a large impact upon the local pressure distribution on the building surfaces, which makes it very difficult to estimate the correct natural driving force and thereby the potential benefit of natural ventilation.

Solar radiation absorbed by urban surfaces results in a temperature increase that is important to take into consideration as it will limit the natural cooling potential of hybrid ventilation. In less dense urban areas, this might only be a local phenomenon for intakes located on south-facing walls, while in dense urban areas it might result in a general increase of outdoor temperatures compared to rural areas (the heat island effect).

In the urban environment, particle and gaseous pollutions from industry, traffic and the burning of fossil fuels decrease the outdoor air quality level. The impact depends upon the distance to the sources, the source characteristics and the pollution transport and mixing. Urban pollution will limit the use of natural ventilation. For example, application of natural ventilation should be avoided during heavy traffic periods or when pollution levels occur above critical levels; the location of intake openings should be avoided close to street level or limited to façades not facing streets; and intake air should be filtered.

Noise from traffic in the urban environment poses similar limitations to the use of natural ventilation. For instance, application of natural ventilation should also be avoided during periods of heavy traffic or when noise levels are above a critical level; the location of intake openings should be turned away from the noise source; and intake openings with good sound insulation should be used.

Hybrid ventilation control

In hybrid ventilation, the control system is as important as the ventilation system itself. Although there is a strong interaction between the ventilation and the control system, the control system is crucial. It is therefore important that the ventilation system and the control system are designed together in one process. Many of the hybrid ventilation components are also integral parts of the building. This demands strong cooperation between the architect, the HVAC engineer and the control engineer. Control algorithms for hybrid ventilation are still under development. Much research is still needed to solve all forthcoming problems in the application of hybrid ventilation in real-life buildings.

Control strategies

Hybrid ventilation systems can be made quite complex. However, it is very important to develop a control strategy and to design a control system that are easy for users to understand and can be operated by the maintenance staff. Therefore, simplicity and transparency of the user/system interface are of the utmost importance. Control system designers need to recognize that most users are not technically literate and are not interested in learning complex operations to suit varying outdoor conditions. They want a system that responds to their needs unobtrusively and allows them to change a condition if it is perceived as unsatisfactory, with rapid feedback.

The control strategy for a building should at least include a winter control strategy, where IAQ is normally the main parameter of concern, and a summer control strategy, where the maximum room temperature is the main concern. It should also include a control strategy, to be used in the interval between winter and summer, where there might occasionally be a heating demand, as well as excess heat in the building.

Demand control is very important in hybrid ventilation systems, and in many cases demand control also proves to be very energy efficient. However, one of the main problems encountered in automatic control of indoor air quality is the cost and reliability of CO_2 sensors used to control the ventilation demand (Willems and Van der Aa, 2002). If a sensor is needed in each building zone, it can become expensive both in initial cost and in regular calibration. In some cases, the ventilation demand was controlled by infrared detection. The major advantage of this system is its relatively low cost (compared to CO_2 sensors) and its autonomy (it can work on a long-life battery; no wiring is required). The major disadvantage is that the airflow is only indirectly correlated to the demand. Sometimes the airflow can be too low or too high. Presence detection can be a good way of controlling the ventilation demand in rooms with low occupancy variation, such as cellular offices. In some cases, it has also been successfully applied in school classrooms. For rooms such as conference rooms, a CO_2 strategy is more suitable because it usually estimates the real needs more accurately. There is a strong need for reliable and cheap CO_2 sensors to be developed.

Both the hybrid ventilation strategy and the control strategy are significantly influenced by the general climate in the region where the building is located. In cold climates, the control strategy should focus on minimizing the ventilation energy needed to achieve good IAQ, and on achieving a good indoor climate in summer and spring without mechanical cooling. In warm climates, the control strategy should focus mainly on reducing the energy consumption for mechanical cooling during summer.

User interaction

One of the advantages of natural ventilation systems is higher user satisfaction due to individual control of windows and indoor environmental conditions (Rowe,

2003). If possible, this feature should be maintained in the control of a hybrid ventilation system even if it could conflict with the possibility of guaranteeing a specific level of indoor thermal comfort or air quality in the rooms. Unfortunately, the relationship between the indoor climate and user acceptance in user-controlled rooms is not well known. Recent research indicates that users are more tolerant of deviations in the indoor thermal climate if the system is controlled by themselves (see de Dear, 1999).

Occupants want to be able to alter conditions quickly in response to unpredictable events (such as glare, draughts or outside noises). If conflicting or unsatisfactory conditions occur, occupants want to decide for themselves how to resolve the conflicts by overriding default settings rather than having conditions chosen for them. Occupants demonstrate a tendency to use supplementary mechanical cooling/heating equipment sparingly and in an energy-efficient way (Rowe, 2002), and prefer to use operable windows and other adaptive behaviours to modify conditions. Most occupants do, however, appear to have an upper 'tolerance' limit, when active intervention will be applied if the opportunity is available. This upper tolerance limit will be very individual and can be different from day to day (Rowe, 2002).

Even though users should have the maximum possibility of controlling their own environment, automatic control is needed to support the users in achieving a comfortable indoor climate and to take over during non-occupied hours. In rooms for several people (for example, open-plan offices) and in rooms occupied by different people (for example, meeting rooms), a higher degree of automation is needed. In some cases (see Meinhold and Rösler, 2002), users strongly appreciated the manual control and refused a fully automatic system; but measurements showed that the mechanical system was seldom applied and that the air quality during some periods was very low. Automatic control is also needed during non-occupied hours to reduce energy use and to precondition rooms for occupation – that is, to provide and control night cooling.

Hybrid ventilation systems can be quite complex systems, which makes it even more important to develop control strategies and design control systems that occupants find easy to use and that the building management team can understand. Most users are not experts and do not want to spend much time investigating how complex controls work.

It is also very important to carefully consider how user interaction is integrated within the control system, both with regard to the type of functions that can be overruled and how and when the automatic control regains control after being overruled by the occupant. For systems with presence detection, the automatic control system usually takes over when the occupants leave the room. For other systems, it can take over after the normal occupation period has ended or after a certain time period, which can be adjusted as a part of the commissioning of the hybrid ventilation system.

Typical strategies

Recommendation for typical control strategies for office and educational buildings are given in Aggerholm (2002a).

Users greatly appreciate the possibility of manually controlling the indoor environment, and the full responsibility of occupants for controlling their own indoor climate during occupied hours can work very well in cellular offices (Rowe, 2002). Individual control can be either manual or motorized. During non-occupied hours, automatic control is needed for cooling the building structure through night ventilation.

In landscape offices, automatic control is needed; but it can be difficult to find an acceptable strategy for window control that satisfies all occupants. If windows are operated automatically during occupied hours, and the external temperature is more than a few degrees lower than the room temperature, there is a great risk of user dissatisfaction due to the sensation of draught. Therefore, it is important that occupants have the opportunity to override the control for openings close to their work station.

If the inlet air is preheated, the best solution is to have separate control of the inlet temperature because preheating the inlet air can be needed even when there is excess heat in the room (van der Aa, 2002b). This is especially important to consider in cases where openings are below windows and perimeter radiators are used for preheating intake air. There is a risk that the inlet temperature setpoint will be raised to compensate for insufficient room heating. This is very critical for systems with low-positioned low-velocity inlet diffusers because the displacement air distribution principle can be destroyed, with very low ventilation efficiency as a result.

The control of outlets and night ventilation seems less problematic.

In classrooms, a simpler control strategy and control system is often installed, with manual control and supplementary window airing in breaks. With the high density of occupancy, the CO_2 level quickly exceeds the limit in winter if assisting fans or mechanical ventilation systems are not operating. In schools where the occupants were responsible for indoor air quality control, there is a great risk of high CO_2 concentrations for some of the time (Meinhold and Rösler, 2002). This risk also applies in classrooms where the inlet air is preheated. Automatic control or a combined manual and automatic control strategy is therefore advisable in school buildings, with the possibility of manual override (Blomsterberg et al, 2002b).

In buildings that also have active mechanical cooling, where the operation mode is automatically switched between hybrid ventilation and mechanical cooling depending upon the temperature or enthalpy difference between external and internal air, there is a risk that once activated, the system will stay in active cooling mode.

Control tasks

The control strategy should determine both time and rate control. It should also determine different control modes in relation to different weather conditions. The actual control strategy should reflect the demands of the building owner, the needs of the users and the requirements in standards and regulations. Recommendations for typical control tasks for office and educational buildings are given in Aggerholm (2002a) and Heiselberg (2002).

Indoor air quality

The control of ventilation for indoor air quality can either be manual, by the occupants, simple timer control, motion detection (occupants present), based on direct measurement of indoor air quality, or a combination of these. For direct measurement of indoor air quality, CO_2 concentration is a useful indicator if occupants are the only or dominating pollutant source. If other significant sources influencing indoor air quality are present – for example, pollutants from materials and cleaning – then the CO_2 concentration in the room may be a less satisfactory indicator.

In small rooms with work desks for one or a few people – for example, cellular offices – it can normally be expected that the occupants will be able to control indoor air quality to their own satisfaction if the ventilation system provides them with the necessary facilities (such as user-controlled windows and vents of different sizes and positions).

In large rooms for many people (for instance, landscape offices) and in rooms occasionally occupied by different people (such as meeting rooms), automatic control of ventilation for indoor air quality is normally needed. The purpose of the control in this case is to reduce the ventilation energy consumption by limiting the operating hours and ventilation rate according to the occupancy pattern. The optimum strategy should have both a good user control to allow occupants to adjust conditions locally at their work station and an automatic back up.

Even during non-occupied hours there might be a need for indoor air quality-controlled ventilation, especially in tight buildings. This includes ventilation after the end of the occupancy period to remove built-up pollution, ventilation during non-occupied periods to remove pollution from materials, and cleaning and ventilation before occupancy to start the occupancy period with fresh air in the building.

Thermal comfort and draught

Room temperature control during occupied hours in summer can be either manual or automatic. Occupants do have a very clear sense of their own thermal comfort, but typically they react too late, when the temperature already is above the acceptable temperature limit. Automatic control of openings, as well as solar shading, can be beneficial since it ensures action as soon as the indoor temperature begins to increase. The need for direct automatic control of room temperature during occupancy is mainly related to large rooms catering for many people and to rooms occasionally occupied by different people. Direct automatic control of room temperature is also necessary if comfort is achieved by mechanical means – for example, mechanical cooling or additional mechanical fan-forced airflow.

During the summer, the normally small difference between indoor and external air temperature on warm summer days has limited potential to reduce room temperature, even if the flow rate is high. In many cases, the body cooling potential of air movement due to open windows might be the most important in relation to thermal comfort. If the external air temperature is higher than the indoor air temperature, external airflow will increase room temperature. This will often be the situation for buildings with efficient night cooling, mechanical cooling and/or efficient solar shading. In such cases, this can be handled through an automatic control system by changing the control mode from temperature control to indoor air quality control.

To avoid sensations of draught, it might be necessary to preheat incoming external air; this might also be necessary even if cooling is needed in the room. Coils or radiators for preheating the supply air should normally be controlled based on the temperature of the inlet air.

It is important that occupants are carefully instructed on how to operate windows when outdoor air temperature is high or when cooling is on.

Night ventilation during summer

The control of night ventilation is of great importance to achieving acceptable thermal comfort during hot summer days in buildings without mechanical cooling, and to reducing energy consumption for mechanical cooling. Building structures should be as cold as possible without creating thermal discomfort in the morning.

The control of night ventilation should normally be automatic; but it is possible to have night ventilation with manual user-controlled windows or hatches in the individual rooms. Manual control by occupants requires clear and easy-to-understand instructions. Automatic control can be local per room or central for the building or a section of the building. Local control is normally only relevant in larger rooms and especially if local fan assistance is used. Central control must normally be based on measured temperatures in representative rooms. The selection of the representative rooms is of great importance.

The actual night ventilation strategy depends upon the system. If fans are included, it is preferable to have a few degrees of cooling potential available from the external air before fans are started because of fan power consumption. Night ventilation must continue until the building is sufficiently cooled or occupied again. If the building structures are cooled to low temperatures, it might be necessary to interrupt night ventilation before the end of the non-occupied period in order to regain acceptable surface temperatures before the start of building occupation.

Natural and mechanical mode switch

One of the main characteristics of a hybrid ventilation system is the ability to switch automatically between natural and mechanical modes in order to optimize the balance between indoor environmental quality and energy use. This challenge differs between the different hybrid ventilation principles.

For a fan-assisted system, the fan can be controlled by the temperature or by the indoor air quality in the rooms, by the pressure in the supply or exhaust ducts, or by the airflow rate through the fan. If the fan is in the natural ventilation flow path, the control can be either on/off, stepped or continuous, depending upon the natural driving forces. If the fan is in parallel to the natural ventilation flow path and uses part of the same flow path, it is difficult to have continuous control and to determine when the conditions allow the fan to be switched off again.

Alternating natural and mechanical ventilation must normally be controlled based on the external temperature and humidity. Alternatively, it can be controlled by a time schedule. Good information for the occupants is needed about the actual mode of the ventilation system.

Table 7.2 *Conclusions and recommendations with regard to sensors in buildings and weather stations*

Sensors in the building

Temperature	Ordinary room and duct temperature sensors are reliable and not expensive. Surface temperature sensors exist; but there is not as much experience with their use in control systems.
CO_2	CO_2 is an indoor air quality (IAQ) indicator of body odour, but is not harmful to people in the concentrations normally found in buildings. CO_2 sensors are quite expensive and need regular calibration (see Willem et al, 2002).
PIR	Infrared presence sensors are reliable and not expensive. They are easy to test and can also be used for other purposes (for example, control of artificial light).
Air speed	Air speed sensors can be used to measure the airflow rate in ducts. Air speed sensors are quite expensive and need regular cleaning and calibration.

Weather station

External temperature	External temperature sensors are reliable and not expensive. Often, the problem is finding a position to install them where the temperature is not influenced by the building or solar radiation.
Wind	Traditionally, wind speed is measured with a cup anemometer and wind direction is measured with a wind vane. A new type without moving parts is available where both speed and direction is measured by using the Doppler effect in two directions.
Solar radiation	Solar radiation sensors do not need to be very accurate for control purposes. It is preferable to have a sensor on the upper part of each main façade.
Precipitation	Precipitation sensors are reliable and not expensive. They normally only need to produce an on/off signal for overrule purposes.

Source: Heiselberg (2002)

Sensors

To fulfil the determined control strategy, sensors are needed in the building to measure temperature, indoor air quality and occupancy. Sensors are also needed to measure actual weather conditions.

In hybrid ventilation, demand control of indoor air quality is very important for increasing the energy efficiency of the system. At present, CO_2 is the most promising indicator of air quality in buildings, and to ensure satisfactory conditions an air quality sensor in every building zone is the optimum solution. CO_2 sensors have been applied in a number of buildings, (Blomsterberg et al, 2002b; Hendriksen et al, 2002; Schild, 2002; Tjelflaat, 2002; van der Aa, 2002b) and with regular calibration they have performed well. In Willems and Van der Aa (2002), a market survey of more than 30 CO_2 sensors is documented. The survey shows that the main disadvantage of existing sensors on the market, with acceptable quality for indoor air quality measurements, is the price, which is about 300–500 Euros. The report also shows that there is a potential for devising low-cost and reliable sensors, if the market is developed.

Table 7.2 shows conclusions and recommendations with regard to sensors in buildings and weather station from IEA–ECBCS Annex 35 (Heiselberg, 2002).

Hybrid ventilation airflow process

In hybrid ventilation, natural and mechanical airflow processes are combined and knowledge of the natural ventilation airflow process is necessary to estimate the need for mechanical support and the most beneficial strategy for combining driving forces. The key difference between natural and mechanical ventilation airflow processes lies in the fact that neither volume flow rate nor flow direction at the ventilation openings is predetermined in the former system. Natural ventilation driving forces are highly unsteady, and both flow rates and airflow directions can vary considerably during the running period and not necessarily in phase with the occupants' needs.

In designing the building, the natural ventilation system and the control system, this can be handled to some extent. In hybrid ventilation, mechanical support systems (fans) are installed to ensure that the occupants' needs are fulfilled – for example, so that too low volume flow rates and/or unwanted flow directions in occupied hours are avoided. However, estimating the need and optimizing the use of mechanical support require a detailed knowledge of the hybrid ventilation airflow process.

The airflow process can be divided into different elements from airflow around buildings, airflow through openings and airflow in rooms, to airflow between rooms in a building.

Pressure distribution on the building envelope

Calculation of the pressure distribution created by wind on the building envelope is crucial in hybrid ventilation analysis, in particular for prediction of airflow rates in natural ventilation mode, but also for estimation of the need and determination of the strategy for mechanical support.

Usually climate data is only available from meteorological stations. The local wind conditions are strongly affected by the microclimate around the buildings, which is again affected by landforms, vegetation and other surrounding buildings. The data from meteorological stations are therefore not always representative for the building location and a transformation of meteorological data into local input data is necessary. Second, the pressures on the building surfaces, which depend upon local wind speed, wind direction, roughness of the environment and local obstacles, must be determined.

There are several methods to transform meteorological wind velocities to local data, (Sherman, 1980; ASHRAE, 2001; BS5925, 1991). Heijmans and Wouters (2002b) compared measurements of local wind conditions in three different positions close to a building with wind directions from meteorological data and predicted local wind velocities from meteorological data by the above-mentioned methods. Differences in wind direction were seen both between the individual measurement positions and between measured local data and the meteorological data. The differences in velocity levels were found to be up to ±25 per cent depending on the method used.

Transformation of wind speed and direction to wind pressures acting over building surfaces, where ventilation openings are intended, appear in natural ventilation calculations in the form of dimensionless wind pressure coefficients. The values change according to wind direction, building surface orientation, and topography and roughness of the terrain in the wind direction. Typical data for simple solid models is given in tabular form in the literature (Orme and Leksmono, 2002), and the use of these pressure coefficients is largely a matter of convenience since this data measured for wind-loading purposes is readily available. Heijmans and Wouters (2002b), compared predicted pressure coefficients (Knoll et al, 1995; Orme and Leksmono, 2002) with measured values for two different buildings. Large variations were seen for certain wind directions as the prediction methods were not able to take local obstacles into consideration. More detailed methods through wind tunnel experiments and computational fluid dynamics (CFD) predictions are needed to obtain more accurate results.

However, in a design situation, it is the uncertainty in predicting thermal comfort and indoor air quality that is problematic, rather than the uncertainty of driving forces and ventilation capacity. Therefore, Heijmans and Wouters (2002b) investigated the impact of the uncertainties of wind velocities, wind directions and pressure coefficients on the prediction of thermal comfort conditions. They showed that the uncertainties had an important impact on the thermal comfort prediction, but noted that the impact was at the same level as a variation of 10 per cent on the assumption of the internal thermal load.

Airflow characteristics of openings

Computation of natural ventilation airflow through small openings is most commonly done using the orifice flow equation and a discharge coefficient, which is related to both the contraction of the streamlines near the opening and to the turbulent pressure losses, and quantifies the airflow efficiency of an opening or, alternatively, the airflow resistance of openings. Correct estimation of the discharge coefficient, C_d, is important. Many of the discharge coefficient values used are derived from data traditionally used for fluid flow in pipes; it can also be determined experimentally. In the case of discharge coefficients for window or door openings, a value of 0.65 for a sharp-edged rectangular opening is often used. Entry conditions, such as incidence of openings to the approaching wind, can significantly influence discharge due to momentum effects at windward openings. In addition, downstream conditions significantly influence the discharge, but are rarely accounted for (Aynsley, 1988).

Results from laboratory experiments show that the discharge coefficient cannot be regarded as constant for window openings (Heiselberg et al, 2001; Heiselberg, 2002), but varies between 0.6 and 1.0 according to pressure difference across the opening, temperature difference (which modifies the flow through the opening and thereby affects the discharge coefficient and opening type), area and local geometrical conditions. The current approach is therefore not very useful for predicting flow rates through external openings in hybrid ventilation systems.

There is no established theoretical justification for assuming that the wind pressure coefficient approach and the modified Bernoulli's equation, which are applicable for air filtration through cracks, should be applicable for calculating flows through large external openings in buildings. Unfortunately, this question has significant implications for the methods of calculating wind-induced airflow rates in multi-zone methods and the performance of natural and hybrid ventilation systems. Treatment of various openings in analysis methods needs to be based on the physics of the flows through the openings. Several authors have performed investigations by wind tunnel experiments, computational fluid dynamics and theoretical analyses to develop new approaches to large opening modelling. So far, no major breakthrough has been achieved; but more knowledge of the governing airflow phenomenon has been gathered.

In a series of wind tunnel experiments, the airflow through openings in a circular disk and a cylinder were investigated (Sandberg, 2002, 2004; True et al, 2003). The main results and conclusions from this work were:

- The approaching flow has a choice – to pass through the opening or not; therefore, the velocity field is split into two parts: the streamlines passing through the opening (the catchment area) and those outside the opening. In between, there is a stagnation point. In the standard approach, the catchment area is infinite. The catchment area depends upon the resistance to the wind offered by the building – the larger the resistance, the higher the airflow rate through the opening.
- Flow contact exists between large openings in opposite sides of a room, and airflow through the second opening is partly momentum driven. Therefore, for the same opening area, the pressure loss is smaller across the leeward than across the windward opening. For different opening areas, the larger pressure difference occurs across the smallest opening, as expected. However, the flow rate depends upon the order of the openings.

Cross-ventilation through a full-scale building model was investigated in a wind tunnel by Sawachi et al (2004) and focused on investigation of the influence of the airflow field around the building, including the wind angle and the inclination of incoming airflow into the opening. The main results and conclusions from this work were:

- As the wind direction changes from perpendicular to parallel to the opening, the discharge coefficient decreases. At first, this is caused by a change in the flow direction through the opening, and at a certain wind direction, bidirectional flow occurs over a portion of the opening. Finally, when the wind direction is close to parallel with the opening, there is an exchange of air through the openings driven by pressure fluctuations (pumping). Then, bidirectional flow occurs over the whole opening.
- In an opening located close to the boundary of a façade, bidirectional flow over a portion of the opening may occur even for wind perpendicular to

the opening, which reduces the net flow rate.
- The largest pressure loss (smallest discharge coefficient) occurs across the windward opening.
- A tentative relationship between the discharge coefficient and the difference of the wind pressure coefficient across the opening was found.

Kurabuchi et al (2004) and Ohba et al (2004) have proposed a new model – the local dynamic similarity model – to predict ventilation flow rates for cross-ventilation. The model relates the value of the discharge coefficient to the dimensionless indoor pressure P_R^*, which is expressed by the ratio of the ventilation driving force to the difference between total pressure and wind pressure at the opening. Before the ventilation airflow reaches the opening, the total pressure of the approach flow is preserved almost completely regardless of the approaching flow angle if flow separation does not occur before it reaches the opening. The pressure loss at the opening depends upon the contraction of ventilation airflow and upon static pressure loss consumed for the production of turbulent kinetic energy. When P_R^* is high, total pressure at the opening inlet will be consumed during the change of flow direction and deceleration of the flow. This will cause a decrease of the discharge coefficient. For low values of P_R^*, the discharge coefficient will approach a constant value. In a series of laboratory experiments, various types of openings were evaluated and approximate expressions of discharge coefficient as a function of P_R^* were derived.

Airflow through buildings

The airflow process in and between rooms in a building is very important for the performance of hybrid ventilation systems but is not very well understood. However, a better understanding is needed before major improvements in design and analytical methods can be achieved.

Airflow in hybrid ventilated buildings can exhibit non-linear dynamical phenomena, such as the existence of multiple steady-state solutions, and/or periodic and a-periodic flows. This means that, for a given set of boundary conditions, the airflow pattern in the space can adopt one of several possibilities, or that the pattern can oscillate between one solution and another. These multiple-solution characteristics have recently been identified for simple building geometries by both small-scale experiments and CFD predictions by several researchers (Hunt and Linden, 2000; Heiselberg et al, 2003b). The multiple steady-state solutions are associated with the sensitivity of the initial conditions, including the 'history' or preceding flow pattern, and it is necessary to understand and incorporate this in the design to cover both normal operating conditions and to deal with smoke and fire.

Hybrid ventilation modelling

Different phases in the design process call for different types of design and analysis methods. Guidelines, decision tools, experience from colleagues and catalogues on products are useful in the conceptual design phase. In this phase, input data are not well known and/or can vary within large ranges, and output only needs to be accurate enough to make principle decisions on which systems and/or combination of systems are appropriate to use in the given situation.

In the basic design phase, analytical calculations and simulation programmes are used to develop the design. Input data are known with much better accuracy and output data should be detailed enough to convince the designer that the system can fulfil the energy targets and the comfort requirements for the building. In the detailed design phase, the individual components are designed and the system and control strategies are optimized with regard to energy consumption and comfort conditions. The design methods are the same as for the basic design phase, but input data on the building and individual components are well known in this phase and output therefore becomes accurate enough to perform a system optimization. Finally, detailed simulation methods or physical models are used to evaluate the final design. These analysis methods are expensive and time consuming to use. They require very detailed input data and are able to give precise predictions on the performance (energy, IAQ and thermal comfort) of the building and the ventilation system.

In this chapter, analysis methods are defined as the physical descriptions and computational algorithms for ventilation, heat and pollutant transport, while analysis tools are defined as the computer software packages or design tools that engineers and architects can use. Focus is only on airflow aspects – that is, methods to predict the airflow rate through openings in the building envelope (ventilation rate), to predict the airflow between zones in a building, and to predict the air distribution (airflow patterns) in the building or in individual rooms. These types of analyses are fundamental for predicting and optimizing thermal comfort, indoor air quality and energy consumption.

Requirements for analysis methods

Since the hybrid ventilation process and the thermal behaviour of the building are linked, the development of design methods for hybrid ventilation must take both aspects into consideration and include efficient iteration schemes. This is the case for all types of methods, from simple decision tools, analytical methods and multi-zone methods, to detailed CFD analysis methods.

The ideal analysis method for hybrid ventilation systems should include modelling of the natural ventilation mode and modelling of the mechanical ventilation mode, as well as modelling of the control strategy. It should be able to answer such questions as: when do the natural driving forces fail to fulfil ventilation demands; when is mechanical ventilation more energy efficient than natural ventilation; and how does occupant interaction affect performance?

A model that combines a thermal simulation model with a multi-zone airflow model will allow the thermal dynamics of the building to be taken into account and will considerably improve the prediction of the performance of hybrid ventilation. Such a model will be capable of predicting the yearly energy consumption for hybrid ventilation and will therefore be the most important design tool for hybrid ventilation systems.

Simple analytical and empirical methods

Simple approaches vary, from back-of-the-envelope computations and simple design graphs to basic spreadsheet programmes. Most analytical solutions have been based on the conventional macroscopic approach for analysing natural ventilation and are derived from the Bernoulli equation. The Bernoulli equation (which is based on the conservation of energy) is used to calculate air velocities in openings, while the law of mass conservation applied to an enclosure allows calculation of the mass flow.

Analytical solutions exist for simple natural and hybrid ventilated buildings (Li, 2002a). For a temperature-given problem, the solution for the ventilation flow rate can be expressed as a function of indoor/outdoor air temperature differences, wind pressure, flow rate from mechanical fans, the size of ventilation openings and so on. For a heat source problem, an analytical solution can be expressed for ventilation flow rate as a function of heat source strength, wind pressure, flow rate from mechanical fans, the size of ventilation openings and so on.

Examples for which simple analytical models have been derived for hybrid ventilation include a single-zone building with two or three openings and a fan incorporating:

* combined stack- and supply fan-driven flows with fully mixed conditions; and
* combined stack- and exhaust fan-driven flows with fully mixed conditions (Li, 2002a, 2002b; Leung, 2003).

Simple analytical solutions can be very useful in providing an in-depth understanding of the governing parameters, of the possibility of multiple steady-state solutions and of dynamic phenomena. They also frequently provide a test case for more complicated numerical methods, such as the multi-zone method.

Li et al (2002b) studied a simple building with either an exhaust fan or a supply fan, together with natural ventilation. It was shown that the pressures induced by fans or natural forces alone cannot be added linearly. Linear addition is only valid when the mechanical fans are dominant. The linear assumption for estimating the combined pressures, which is used to derive the simple quadratic formula for calculating combined flow rate, is not correct and can introduce errors of up to more than 40 per cent. This explains why the quadratic formula does not give a good prediction of the combined flow rate when the natural force and the mechanical force are similar in magnitude. It is noted that in smoke-control design of buildings, a linear addition model is often used to calculate the combined pressure (forces). There is still a lack of information on negative pressures in atria created by atrium exhausts.

Analytical solutions can also be used for basic vent sizing. Simple non-dimensional methods developed by Etheridge (2002) are good examples. Such sizing techniques were compared with experimental studies by Fracastoro et al (2002a), who showed a reasonable agreement. Generally, the majority of methods cannot be used explicitly in sizing ventilation openings and ducts, etc. In other words, they can only be used to evaluate the result of specific configurations. To overcome this, an 'inverted' approach or 'loop' method has been developed by Axley (2000, 2001). This is a more generalized version of an explicit method described by the Chartered Institute of Building Services Engineers (CIBSE, 1997). The development of computer programmes for sizing openings and ducts for both natural and hybrid ventilation is still needed.

A simplified method has been developed that enables the designer to quickly determine, at a first design stage, the permeability of the building envelope in order to satisfy the required air change rate in the building (Fracastoro et al, 2002b, Fracastoro and Perino, 2002). The method requires only information about building typology and surrounding terrain, along with the test reference year (TRY) of the location. The first two pieces of information yield two non-dimensional coefficients, which are introduced in a simplified model to determine the effective pressure difference across the envelope as a function of outdoor–indoor temperature difference and wind velocity. A graphical procedure is then used, given the overall permeability of the building envelope, to obtain the time fraction during the heating season for which natural ventilation exceeds the required air change rate. If this fraction is less than one, this information allows the designer to determine the required increase in building envelope permeability using natural airing devices, or to determine the time fraction when mechanical ventilation will be necessary.

Short-time window airing is a typical strategy to boost ventilation during periods of high loads. A simple two-zone model has been developed (Fracastoro et al, 2002a; Perino and Heiselberg, 2003) to predict the time-dependent buoyancy-driven airflow rate by window airing. The model has been verified by laboratory experiments and compared with CFD prediction, (Heiselberg et al, 2003a; Perino and Heiselberg, 2003). The work shows short-term airing is an effective ventilation strategy – during the first ten minutes after opening the window, high ventilation efficiency and high air change rates result.

Multi-zone or network methods

Buildings are represented by a number of zones. Network methods can predict overall ventilation flow rates for the entire building, the ventilation rate of each zone and individual airflow rates and directions through each opening. The methods are able to take into account the effect of outdoor climate, the location and size of each opening and stack, and wind-driven and mechanically driven ventilation. However, they cannot predict detailed flow patterns in each zone of the building. They are compatible with most multi-zone thermal-modelling programmes.

Network methods are based on the application of the Bernoulli equation to determine the pressure difference and, hence, flow rate across each opening in the flow network. Indoor air velocities are assumed to be negligible; however, Axley et al (2002b) and Carrilho da Graca and Linden (2002) show that this assumption is not valid when ventilation openings are relatively large and there is a strong wind effect. As an alternative method in such cases, Kato (2004) has proposed a flow network model based on the power balance – that is, the power loss along a stream tube in and around the building. The flow rate through each opening is generally expressed as a simple function of the pressure difference, such as the power-law relationship. This pressure difference can be a result of wind pressure, stack pressure, fan-induced pressure or a combination of all of these.

There are two main types of multi-zone methods:

1 the zonal pressure-based approach; and
2 the loop pressure equation-based approach.

In a zonal pressure-based approach, a mass balance is applied to each zone. This leads to a set of simultaneous non-linear equations, the solution of which gives the internal zonal pressures (Walton, 1995; Feustel, 1999). In the loop pressure equation approach, balance equations are written for the changes in pressure that occur along each ventilation loop of a building ventilation system, following a ventilation flow path from the inlet to exhaust and back to the inlet again. The sum of these pressure changes around any loop must be equal to zero and the solution will be the flow rates in the loops (Nitta, 1994; Axley, 2001; Axley et al, 2002a; Li, 2002d).

Traditional mixing ventilation systems utilize supply jets to create global circulation in a room, which results in a rather uniform air temperature distribution. In hybrid ventilation, the natural ventilation part is driven by thermal buoyancy, where air heated by heat sources in the room rises to the upper part of the space and a vertical temperature gradient is established. In natural ventilation, the thermal driving force is a function of the temperature difference between indoor and outdoor air, and the thermal stratification directly affects the ventilation flow rate, as well as the airflow direction at some ventilation openings. Most existing network models assume that the air temperature in each zone is uniformly constant. However, for hybrid ventilation systems, this assumption is not valid. Li et al (1999) and Li (2002b) have shown that the effect of thermal stratification on airflow can be very significant and lead to substantial underestimation of the neutral level in a building. The authors have also identified the conditions under which thermal stratification is important and how it should be considered in hybrid ventilation analysis.

Combined thermal and airflow modelling methods

Prediction of the yearly performance of hybrid ventilation systems requires combined thermal and airflow modelling methods. A number of different tools have been developed or are under development (for example, ESP-r; CHEMIX; TRNSYS+COMIS; IDA; and HYBCELL 1.0).

However, the ability of these tools to predict hybrid ventilation performance is still limited and several questions need to be answered:

- How well is the hybrid ventilation system modelled?
- How well are the hybrid ventilation control strategies modelled?
- Can general guidelines on simulation of control strategies be developed?
- How robust are the tools' predictions – that is, how dependent upon the tool used are the conclusions?

In order to evaluate how well the simulation tools model hybrid ventilation systems and control strategies, a test simulation of ventilation in a typical single-zone classroom using four different tools was performed in IEA–ECBCS Annex 35 (Delsante et al, 2002; Delsante and Aggerholm, 2002). The purpose was not to perform an inter-programme comparison, but to test the robustness of any conclusions drawn about the relative merits of different ventilation systems or control strategies. Clearly, if different tools yield significantly different conclusions, then any conclusion drawn from a single tool must be viewed with caution. The results show that the different tools were quite consistent with respect to IAQ performance and fan energy. There were some deviations regarding heating energy, possibly because of quite different implementations of the heating controller, which underlines the importance of modelling control strategies appropriately in simulations of buildings with hybrid ventilation systems.

A combined model (TRNSYS+COMIS) was used to simulate the performance of the hybrid ventilation system in a school building in Germany (Meinhold and Rösler, 2002; Seifert et al, 2002). Comparing simulation results with measurements for the building, modelled with 129 zones, showed very good results. Deviations in temperatures were in the range of 0.2–1.0K. For a more detailed investigation of the night-cooling situation, CFD simulation for a large central space (an atrium) in the building was included in the coupling (Albrecht et al, 2002). It was concluded that the fully coupled (macroscopic and microscopic models) calculation from the scientific point of view was the best way of representing all of the phenomena and processes in a hybrid ventilated building. However, from the practical point of view, the expenses are very high and only a team of experts are able to handle the coupling of simulations tools. Therefore, it was concluded that fully coupled building simulation is not, at the moment, useful for design purposes.

A similar simulation exercise has been performed on a hybrid ventilated school building in Norway using ESP-r (Jeong and Haghighat, 2002; Wachenfeldt, 2003). The simulation results were validated extensively against monitored and measured data in the real building. All results indicate that the performance can be accurately predicted in the simulation model. It was possible to identify and quantify all major energy transport mechanisms in the building. Similar accuracy cannot be expected in a design situation with very limited knowledge about the 'future' building as the model was calibrated with respect to both thermal bridges and the earth layer surrounding the embedded inlet duct. In addition, the existing fan frequency records enabled very good predictions of the airflow rate in the system at all times. In normal cases, such inputs can carry very high uncertainties.

The simulation exercises showed that the combined thermal and airflow modelling method is a powerful tool for analysing hybrid ventilated buildings and that its accuracy is sufficient for designing hybrid ventilation system. However, there is still a need for improvement, especially with regard to the simulation of control strategies, the simulation of wind impact upon the ventilation (opening) performance, and the heat transfer in integrated building elements, such as culverts and stacks, where the airflow is far from fully turbulent.

Probabilistic methods

A deterministic approach implies that all input parameters and model coefficients are 100 per cent certain with zero spread. In practice, this is not the case – for example, inhabitant behaviour and internal loads may vary significantly, and external loads such as wind, external temperature and solar radiation are obviously stochastic in nature. One reason for ignoring randomness is the fact that mechanically ventilated heavy buildings are often highly 'damped' and shielded from external loads. These kinds of buildings will also control the influence of the internal load effectively by means of the building energy management system and the HVAC system. However, lighter constructions that are naturally or hybrid ventilated can be very sensitive to stochastic load variations.

When using a probabilistic method, some or all of the input parameters are modelled either as random variables or as stochastic processes, described by statistics – that is, mean values, standard deviations, auto-correlation functions, etc. (Brohus et al, 2002a), and the results are the corresponding statistics of the output expressed as probabilities of occurrences. A stochastic method is, thus, a formulation of a physical problem, where the randomness of the parameters is taken into account. In principle, any of the above-mentioned analytical methods can be applied stochastically.

The advantage of probabilistic methods is the possibility of not only designing for peak load and estimating annual energy consumption based on a reference year, but also to examine the range of variation and to quantify the uncertainty (Brohus et al, 2002b). Probabilistic methods can be used as a tool to evaluate the trade-off between economy (cost, energy and environment) and risk (expectations not met, violation of regulations, etc.) on a firm foundation. It allows a more courageous design, which may, in turn, result in increased user satisfaction, energy savings and 'greener' buildings due to the fact that the uncertainty can be calculated and not just assumed roughly. For instance, a building owner can prescribe the probability allowed for a certain parameter to exceed the bounds of a design interval. In the project IEA–ECBCS Annex 35, the Monte Carle simulation approach was demonstrated in single and multi-zone models of a hybrid ventilated building (Brohus et al, 2002c), as well as in an analytical method based on stochastic differential equations (Brohus et al, 1999).

Obviously, consideration of randomness increases the level of complexity of the analysis and the expense in terms of central processing unit (CPU) time. Therefore, it should primarily be applied to more complex or unusual design cases. Another application is to use probabilistic modelling to gain further knowledge and to develop simple, easy-to-use deterministic models. Alternatively, the software applied in the calculations can be supplemented by a user-friendly interface hiding the underlying mathematics, especially if appropriate default values are available. Probabilistic modelling can be very useful in considering the effects of occupant behaviour.

Future challenges and trends in research and development

Hybrid ventilation is a very new technology and several challenges still need to be solved with the assistance of research and development. These challenges lie in all areas, from design methods, ventilation components, control systems and sensors, to analysis and evaluation methods.

The application of hybrid ventilation requires a careful design in the early design phases, and the scarcity of simple and fast design tools is one of the most important issues.

It is often impossible, in practice, to distinguish between building and hybrid ventilation system performance. A poor ventilation performance is, in many cases, caused by a poor building performance or a poor integration of the ventilation system and the building design. Strategies to integrate hybrid ventilation within whole-building concepts, as well as technologies to exploit the benefits, are needed.

Until now, many hybrid ventilation components have been designed for a specific building project, and the use of advanced technologies to develop hybrid ventilation-specific components and systems could significantly improve the performance and, consequently, the uptake of hybrid ventilation. The needed components include intake components with the possibility of preheating and air filtration that are controllable and have a low pressure drop; new low pressure-loss methods for air filtration; low pressure fans with high efficiency; overflow diffusers with sound attenuation that are fire safe; and exhaust towers with efficient use of wind pressure. For domestic ventilation, this development has occurred in a more systematic fashion, such as in the EU RESHYVENT (Residential Hybrid Ventilation) project (van der Aa and Op´t Veld, 2004).

Further development of robust control strategies that are able to handle user interaction is important, as successful user interaction is one of the major selling points of hybrid ventilation. In hybrid ventilation, demand control is essential for the performance, and more reliable and cheap CO_2 sensors (or alternative sensors for demand control) are very important for reducing both initial and running costs. In many systems, airflow and fan speed are controlled based on pressure differences. However, the needed accuracy is much higher than the capabilities of the current sensors. There is also a need for developing more accurate, reliable and cheaper pressure sensors.

With regard to modelling and analysis of hybrid ventilation, a number of aspects of different models are identified for further development or improvement. Since integrated multi-zone airflow and thermal models are the most promising engineering methods for predicting yearly performance of natural and hybrid ventilation systems, it is crucial to ensure that the fundamental governing equations are correct for airflows through buildings with large openings (Axley et al, 2002). It is necessary to further understand wind-driven flows through large openings in enclosures. It is expected that large eddy simulation and wind tunnel testing will continue to play an important role, in this respect, due to the very unsteady nature of the flows. It is not expected that large eddy simulation will become a daily engineering tool for ventilation engineers in the near future; but it can be a very useful tool for improving the fundamental understanding of airflows and for assisting in the development of improved engineering methods. Integration of generalized control strategies within integrated airflow/thermal models is important to assist designers in choosing the most effective control strategies for a particular design.

Reliable performance assessment of hybrid ventilation and other innovative ventilation systems is important for a wide application and for development of the technology and new products. In relation to implementing the EU Energy Performance of Buildings Directive in European countries, it will be important that methods are developed which allow and account for the advantages of applying hybrid ventilation systems, as well as other energy-efficient and renewable technologies.

It is also important that the predicted performance of buildings with hybrid ventilation systems corresponds to the actual performance of the real building. A number of parameters that influence the performance are not very well defined, such as human behaviour (especially in relation to occupant control), internal heat loads and wind effects; therefore, the actual building performance can very considerably. When combining existing physical models, such as multi-zone methods with stochastic models, to develop probabilistic methods, it is important to include the effects of stochastic parameters, such as human behaviour, in order to obtain an expected range of building performance and a quantification of the associated uncertainty.

Acknowledgements

All participants in IEA ECBCS Annex 35 Hybrid Ventilation in New and Retrofitted Office Buildings are greatfully acknowledged for their contributions. Much of the knowledge included in this chapter is created by the task-shared work of the participants and through mutual discussions in many meetings.

References

Aggerholm, S. (2002a) 'Control of hybrid ventilation systems', *International Journal of Ventilation*, vol, 1, February, Hybvent – Hybrid Ventilation Special Edition, pp65–75

Aggerholm, S. (2002b) *Hybrid Ventilation and Control Strategies in the Annex 35 Case Studies: Technical Report in Principles of Hybrid Ventilation*, Hybrid Ventilation Centre, Aalborg University, Aalborg

Albrecht, T., Gritzki, R., Grundman, R., Perschk, A., Richter, W., Rösler, M. and Seifert, J. (2002) 'Evaluation of a coupled calculation for a hybrid ventilation building from a practical and scientific point of view', in *Proceedings of the 4th International Forum on Hybrid Ventilation*, HybVent Forum 02, Montreal, Canada, May

ASHRAE (American Society of Heating, Refrigerating and Air-Conditioning Engineers) (2001) *ASHRAE Handbook Fundamentals*, ASHRAE, Atlanta

Axley, J. W. (2000) 'Design and simulation of natural ventilation systems using loop equations', in *Proceedings of Healthy Buildings 2000*, vol 2, pp475–480

Axley, J. W. (2001) *Residential Passive Ventilation Systems: Evaluation and Design – A Critical Evaluation of the Potential for Adapting European Systems for Use in North America and Development of a General Design Method*, AIVC Technical Note 54, AIVC, www.aivc.org

Axley, J. W., Emmerich, S., Dols, S. and Walton, G. (2002a) 'An approach to the design of natural and hybrid ventilation systems for cooling buildings', in *Proceedings of Indoor Air 2002, Ninth International Conference on*

Indoor Air Quality and Climate, Monterey, California, 30 June–5 July

Axley, J. W., Wurtz, E. and Mora, L. (2002b) 'Macroscopic airflow analysis and the conservation of kinetic energy', in *Proceedings of Roomvent 2002, Seventh International Conference on Air Distribution in Rooms*, Copenhagen, 8–11 September

Aynsley, R. M. (1988) 'A resistance approach to estimating airflow through buildings with large openings due to wind', *ASHRAE Transactions*, vol 94, part 2, pp1661–1669

Blomsterberg, Å., Wahlström, Å. and Sandberg, M. (2002a) 'Hybrid ventilation in a retrofitted school – monitoring and evaluation of indoor climate and energy use', in *Proceedings of the Fourth International Forum on Hybrid Ventilation, HybVent Forum 02*, Montreal, Canada, May

Blomsterberg, Å., Wahlström, Å. and Sandberg, M. (2002b) *'Tånga School': Case Study Report in Principles of Hybrid Ventilation*, Hybrid Ventilation Centre, Aalborg University, Aalborg

Brohus, H., Frier, C. and Heiselberg, P. (1999) 'Probabilistic analysis methods for hybrid ventilation – preliminary application of stochastic differential equations', in *Proccedings of the First International Forum on Hybrid Ventilation, HybVent Forum 99*, Sydney, Australia, September

Brohus, H., Frier, C. and Heiselberg, P. (2002a) *Stochastic Load Models Based on Weather Data: Technical Report in Principles of Hybrid Ventilation*, Hybrid Ventilation Centre, Aalborg University, Aalborg

Brohus, H., Frier, C. and Heiselberg, P. (2002b) *Quantification of Uncertainty in Thermal Building Simulation by Means of Stochastic Differential Equations: Technical Report in Principles of Hybrid Ventilation*, Hybrid Ventilation Centre, Aalborg University, Aalborg

Brohus, H., Frier, C. and Heiselberg, P. (2002c) *Stochastic Single and Multizone Models of a Hybrid Ventilated Building – A Monte Carlo Simulation Approach: Technical Report in Principles of Hybrid Ventilation*, Hybrid Ventilation Centre, Aalborg University, Aalborg

BS5925: 1991 (1991) *Code of Practice for Ventilation Principles and Designing for Natural Ventilation*, British Standards Institution, UK

Carrilho da Graca, G. and Linden, P. F. (2002) 'Contribution to simplified modelling of building airborne pollutant removal', in *Proceedings of Indoor Air, Ninth International Conference on Indoor Air Quality and Climate*, Monterey, California, July

Chikamoto, T. and Kato, S. (2002) *The Liberty Tower of Meiji University: Case Study Report in Principles of Hybrid Ventilation*, Hybrid Ventilation Centre, Aalborg University, Aalborg

CIBSE (Chartered Institute of Building Services Engineers) (1997) 'Natural ventilation in non-domestic buildings', in *CIBSE Applications Manual, AM10: 1997*, CIBSE, London, UK

de Dear, R. (1999) 'Adaptive thermal comfort in natural and hybrid ventilation', in *Proceedings of First International Forum on Hybrid Ventilation*, Sydney, Australia, September

de Gids, W. (2001) 'Hybrid ventilation concepts – classification and challenges', in *Proceedings of the Fourth International Conference on Indoor Air Quality, Ventilation and Energy Conservation in Buildings*, vol 1, Hunan, China, pp133–139, 2–5 October

Delsante, A. and Aggerholm, S. (2002) *The Use of Simulation Tools to Evaluate Hybrid Ventilation Control Strategies: Technical Report in Principles of Hybrid Ventilation*, Hybrid Ventilation Centre, Aalborg University, Aalborg

Delsante, A., Aggerholm. S., Citterio, M., Cron, F. and Mankibi, M. E. (2002) 'The use of simulation tools to evaluate ventilation systems and control strategies', in *Proceedings of the Fourth International forum on Hybrid Ventilation, HybVent Forum 02*, Montreal, Canada, May

Elmualim, A. A., Awbi, H. B., Fullford, D. and Wetterstad, L. (2003) 'Performance evaluation of a wall-mounted convector for pre-heating naturally ventilated spaces', *International Journal of Ventilation*, vol 2, no 3, December, pp213–222

Etheridge, D. W. (2002) 'Nondimensional methods for natural ventilation design', *Building and Environment*, vol 37, pp1057–1072

Feustel, H. E. (1999) 'COMIS – An international multizone airflow and contaminant transport model', *Energy and Buildings*, vol 30, pp87–95

Fracastoro, G. V. and Perino, M. (2002) 'Natural versus mechanical ventilation – a tool to help making a choice', *International Journal of Ventilation*, vol 1, no 2, pp101–108

Fracastoro, G. V., Mutani, G. and Perino, M. (2002a) 'Experimental and theoretical analysis of natural ventilation by windows opening', *Energy and Buildings*, vol 34, pp817–827

Fracastoro, G. V., Perino, M. and Mutani, G. (2002b) *A Simple Tool to Assess the Feasibility of Hybrid Ventilation Systems, Technical Report in Principles of Hybrid Ventilation*, Hybrid Ventilation Centre, Aalborg University, Aalborg

Heijmans, N. and Wouters, P. (2002a) *IVEG Office Building: Case Study Report in Principles of Hybrid Ventilation*, Hybrid Ventilation Centre, Aalborg University, Aalborg

Heijmans, N. and Wouters, P. (2002b) *Impact of the Uncertainties on Wind Pressures on the Prediction of Thermal Comfort Performances, Technical Report in Principles of Hybrid Ventilation*, Hybrid Ventilation Centre, Aalborg University, Aalborg

Heiselberg, P. (2000) 'Design principles for natural and hybrid ventilation', in *Proceedings of Sixth International Conference on Healthy Buildings*, Helsinki, Finland, August, vol 2, pp35–46

Heiselberg, P (ed) (2002) 'Principles of hybrid ventilation', in *IEA–ECBCS Annex 35 Final Report*, Hybrid Ventilation Centre, Aalborg University, Aalborg (the booklet, technical and case study reports, as well as papers from the first, second and fourth Hybvent Forum, are available for download at www.hybvent.civil.auc.dk)

Heiselberg, P. and Bjørn, E. (2002) 'Impact of open windows on room airflow and thermal comfort', *International Journal of Ventilation*, vol 1, no 2, October, pp91–100

Heiselberg, P., Jepsen, L. B., Hyldgård, A., Nielsen, P. V. and Perino, M. (2003a) 'Short-time airing by single sided natural ventilation – Part 1: Measurement of transient airflow rates', in *Proceedings of ISHVAC 2003, The Fourth International Symposium on Heating, Ventilation and Air-Conditioning*, Beijing, China, 9–11 October, vol 1, pp117–124

Heiselberg, P., Li, Y., Andersen, A., Bjerre, M. and Chen, Z. D. (2003b) 'Experimental and CFD evidence of multiple solutions in a naturally ventilated building', *Indoor Air*, vol 13, pp1–12

Heiselberg, P., Svidt, K. and Nielsen, P. V. (2001) 'Characteristics of airflow from open windows'. *Building and Environment*, vol 36, pp859–869

Hendriksen, O., Brohus, H., Frier, C. and Heiselberg, P. (2002) *Bang and Olufsen Headquarters: Case Study Report in Principles of Hybrid Ventilation*, Hybrid Ventilation Centre, Aalborg University, Aalborg

Hunt, G. R. and Linden, P. F. (2000) 'Multiple steady airflows and hysteresis when wind opposes buoyancy', *Air Infiltration Review*, vol 21, no 2, pp1–3

Jeong, Y. and Haghighat, F. (2002) 'Modelling of a hybrid ventilated building – using ESP-r', *International Journal of Ventilation*, vol 1, no 2, pp127–139

Kato, S. (2004) 'Flow network model based on power balance as applied to cross-ventilation', *International Journal of Ventilation*, vol 2, March, pp395–408

Knoll, B., Phaff, J. C. and de Gids, W. F. (1995) 'Pressure simulation program', *Proceedings of 16th AIVC Conference on Implementing the Results of Ventilation Research*, AIVC, www.aivc.org

Kurabuchi, T., Ohba, M., Endo, T., Akamine, Y. and Nakayama, F. (2004) 'Local dynamic similarity model of cross-ventilation – Part 1: Theoretical framework', *International Journal of Ventilation*, volume 2, March, pp371–382

Leung, H. (2003) *Analysis of Natural and Hybrid Ventilation in Simple Buildings*, MSc thesis, University of Hong Kong, Hong Kong, June

Li, Y. (2002a) *Analysis of Natural Ventilation – A Summary of Existing Analytical Solutions, Technical Report in Principles of Hybrid Ventilation*, Hybrid Ventilation Centre, Aalborg University, Aalborg

Li, Y. (2002b) 'Analysis, prediction and design of natural and hybrid ventilation for simple buildings', in *Proceedings of the Fourth International Forum on Hybrid Ventilation, HybVent Forum 02*, Montreal, Canada, May

Li, Y. (2002c) *Integrating Thermal Stratification in Natural and Hybrid Ventilation Analysis: Technical Report in*

Principles of Hybrid Ventilation, Hybrid Ventilation Centre, Aalborg University, Aalborg

Li, Y. (2002d) 'Spurious numerical solutions in coupled natural ventilation and thermal analyses', *International Journal of Ventilation*, vol 2, no 1, pp1–12

Li, Y., Delsante, A. and Chen, L. (1999) 'Consideration of thermal stratification in multi-zone models of natural ventilation', in *Proceedings of the First International Forum on Hybrid Ventilation, HybVent Forum 99*, Sydney, Australia, September

Li, Y. and Heiselberg, P. (2003) 'Analysis methods for natural and hybrid ventilation – a critical literature review and recent developments', *International Journal of Ventilation*, vol 1: Hybvent – Hybrid Ventilation Special Edition, February, pp3–20

Li, Y., Leung, H. and Law, A. Y. M. (2002) 'Combined ventilation flow in buildings driven by combined natural and mechanical forces', in *Proceedings of Roomvent 2002, Eighth International Conference on Air Distribution in Rooms*, Copenhagen, September

Meinhold, U. and Rösler, M. (2002) *Bertolt Brecht Gymnasium: Case Study Report in Principles of Hybrid Ventilation*, Hybrid Ventilation Centre, Aalborg University, Aalborg

Narita, S. and Kato, S. (2002) *Fujita Technology Center: Case Study Report in Principles of Hybrid Ventilation*, Hybrid Ventilation Centre, Aalborg University, Aalborg

NATVENT (1999) *Overcoming Technical Barriers to Low-Energy Natural Ventilation in Office-Type Building in Moderate and Cold Climate*, EU JOULE IV Research Project, BRE, UK

Nitta, K. (1994) 'Calculation method of multi-room ventilation', *Memoirs of the Faculty of Engineering and Design*, Kyoto Institute of Technology, vol 42, pp59–94

Ohba, M., Kurabuchi, T., Endoh, T., Akamine, Y. and Kurahashi, A. (2004) 'Local dynamic similarity model of cross-ventilation – Part 2: Application of local dynamic similarity model, *International Journal of Ventilation*, vol 2, March, pp383–394

Orme, M. and Leksmono, N. (2002) *The AIVC Ventilation Modelling Data Guide (Version 1.0)*, April, AIVC, www.aivc.org

Perino, M. and Heiselberg, P. (2003) 'Short term airing by single sided natural ventilation – Part 2: Comparison of experimental results and model predictions', in *Proceedings of ISHVAC 2003, The Fourth International Symposium on Heating, Ventilation and Air-conditioning*, Beijing, China, 9–11 October, vol 1, pp109–116

Principi, P., Di Perna, C. and Ruffini, E. (2002) *I Guzzini Illuminazione Office Building: Case Study Report in Principles of Hybrid Ventilation*, Hybrid Ventilation Centre, Aalborg University, Aalborg

Rowe, D. (2002) *Wilkinson Building: The University of Sidney – Case Study Report in Principles of Hybrid Ventilation*, Hybrid Ventilation Centre, Aalborg University, Aalborg

Rowe, D. (2003) 'A study of a mixed mode environment in 25 cellular offices at the University of Sydney', *International Journal of Ventilation*, vol 1: Hybvent – Hybrid Ventilation Special Edition, February, pp53–64

Sandberg, M. (2002) 'Airflow through large openings – a catchment problem?', in *Proceedings of Roomvent 2002, Eighth International Conference on Air Distribution in Rooms*, Copenhagen, September

Sandberg, M. (2004) 'An alternative view on the theory of cross ventilation', *International Journal of Ventilation*, vol 2, March, pp409–418

Santamouris (editor) (2000) *Energy and Climate in the Urban Built Environment*, James & James, London, UK

Santamouris, M., Papanikolaou, N., Livada, I., Koromakis, I., Georgakis, C., Argiriou, A. and Assimakopoulos, D. N. (2001) 'On the impact of urban climate on the energy consumption of buildings', *Solar Energy*, volume 70, issue 3, pp201–216

Sawachi, T., Narita, K.-I., Kiyota, N., Seto, H., Nishizawa, S. and Ishikawa, Y. (2004) 'Wind pressure and airflow in a full-scale building model under cross ventilation', *International Journal of Ventilation*, vol 2, March, pp343–358

Schild, P. G. (2001) 'An overview of Norwegian buildings with hybrid ventilation', in *Proceedings of the Second International Forum on Hybrid Ventilation, HybVent Forum 01*, Delft, The Netherlands, May

Schild, P. G. (2002) *Jaer School: Case Study Report in Principles of Hybrid Ventilation*, Hybrid Ventilation Centre, Aalborg University, Aalborg

Seifert, J., Perschk, A., Rösler, M. and Richter, W. (2002) *Coupled Airflow and Building Simulation for a Hybrid Ventilated Educational Building: Case Study Report in Principles of Hybrid Ventilation*, Hybrid Ventilation Centre, Aalborg University, Aalborg

Seppänen, O. and Fisk, W. J. (2002) 'Association of ventilation system type with SBS symptoms in office workers', *Indoor Air*, vol 12, no 12, pp98–112

Sherman, M. H. (1980) *Air Infiltration in Buildings*, Applied Science Division, Lawrence Berkeley Laboratory, University of California, California, US

TIP-Vent (2001) *Towards Improved Performances of Mechanical Ventilation Systems*, Source Book from the EU Joule IV Research Project, TIP-Vent, Belgian Building Research Institute, Brussels

Tjelflaat, P. O. (2001) 'Design procedures for hybrid ventilation', *Second International Forum on Hybrid Ventilation*, Delft, The Netherlands, May

Tjelflaat, P. O. (2002) *Mediå School: Case Study Report in Principles of Hybrid Ventilation*, Hybrid Ventilation Centre, Aalborg University, Aalborg

True, J. P. J., Sandberg, M., Heiselberg, P. and Nielsen, P.V. (2003) 'Wind driven cross-flow analysed as a catchment problem and as a pressure driven flow', *International Journal of Ventilation*, vol 1: Hybvent – Hybrid Ventilation Special Edition, February, pp89–101

van der Aa, A. (2002a) *Cost of Hybrid Ventilation Systems: Case Study Report in Principles of Hybrid Ventilation*, Hybrid Ventilation Centre, Aalborg University, Aalborg

van der Aa, A. (2002b) *Hybrid Ventilation Waterland School Building, The Netherlands – First Results of the Monitoring Phase: Case Study Report in Principles of Hybrid Ventilation*, Hybrid Ventilation Centre, Aalborg University, Aalborg

van der Aa, A. and Op´t Veld, P. (2004) *RESHYVENT – A EU Cluster Project on Demand Controlled Hybrid Ventilation for Residential Buildings*, European Green Cities Network, Budapest, 21–22 April 2004

Vik, T. A. (2003) *Life Cycle Cost Assessment of Natural Ventilation Systems*, PhD thesis, Norwegian University of Science and Technology, NTNU, Trondheim

Wachenfeldt, B. J. (2003) *Natural Ventilation in Buildings – Detailed Prediction of Energy Performance*, PhD Thesis, Norwegian University of Science and Technology, NTNU, Trondheim

Walton, G.N. (1995) 'Airflow network for element based building airflow modelling', *ASHRAE Transactions*, vol 101, part 2, pp611–620

Wildeboer, J. and Fitzner, K. (2002) 'Influence of the height of the supply air opening for natural ventilation on thermal comfort', in *Proceedings of Roomvent 2002, Eighth International Conference on Air Distribution in Rooms*, Copenhagen, September

Willems, E. M. M. and van der Aa, A. (2002) *CO_2-Sensors for Indoor Air Quality – A High-Tech Instrument is Becoming a Mass Product: Market Survey and Near Future Developments: Case Study Report in Principles of Hybrid Ventilation*, Hybrid Ventilation Centre, Aalborg University, Aalborg

Wouters, P., Heijmans, N., Delmotte, C. and Vandaele, L. (1999) 'Classification of hybrid ventilation concepts', in *Proceedings of First International Forum on Hybrid Ventilation*, Sydney, Australia, September

Wouters, P., Heijmans, N., Delmotte, C. and Vandaele, L. (2000) 'Typology of hybrid ventilation systems and practical examples', in *Proceedings of 21st Annual AIVC Conference 'Innovations in Ventilation Technology*, The Hague, The Netherlands, 26–29 September

Zeidler, O. and Fitzner, K. (1998) 'Investigation of the impact of natural ventilation through windows on thermal comfort', in *Proceedings of Roomvent'98*, Stockholm, Sweden, 14–17 June, vol 2, pp323–326

8

Ventilation for Comfort and Cooling: The State of the Art

Matheos Santamouris

Introduction

Buildings are one of the most important economic sectors. The global world's annual output of construction is close to US$3000 billion and represents almost one tenth of the global economy (CICA, 2002). In parallel, buildings represent more than 50 per cent of the national capital investment, while the sector employs more than 111 million employees and accounts for almost 7 per cent of total employment, as well as 28 per cent of global industrial employment. As mentioned by CICA (2002), every job in the construction sector generates two new jobs in the global economy; thus, it can be said that the buildings sector is in a direct or indirect way is linked to almost 20 per cent of global employment.

Concerned with the consumption attributed to the building sector, Baris Der Petrossian (2001) has reported that almost one sixth of the world's resources are consumed by the construction sector, which is responsible for almost 70 per cent of sulphur oxide (SO_x) and 50 per cent of carbon dioxide (CO_2) emissions. According to the United Nations Centre for Human Settlements (Habitat) (UNCHS, 1993), buildings use almost 40 per cent of the world's energy, 16 per cent of the world's fresh water and 25 per cent of the world's forest timber.

As a result of intensive energy conservation measures, the specific energy consumption of buildings spent for heating purposes has almost stabilized or decreased, at least in the developed world. On the contrary, the specific energy needs for cooling has increased in a dramatic way, mainly because the increase of family income in developed countries has made the use of these systems highly popular. Recent statistics show that there are more than 240 million air-conditioning units installed worldwide (IIR, 2002), while the refrigeration and air-conditioning sectors consume about 15 per cent of all electricity used worldwide.

The impact of air conditioning on electricity demand is a significant problem since peak electricity loads are increasing continuously; thus, utilities have to build additional plants. In parallel, important environmental problems are associated with the use of air conditioning.

Passive and hybrid cooling techniques involving heat modulation and dissipation methods and systems, particularly convective cooling techniques, can contribute highly to reducing the cooling load of buildings and to improving thermal comfort during the summer season.

Results of recent research projects (Santamouris and Argiriou, 1997; Santamouris, 2004) have improved knowledge on this specific topic, and developed design tools and advanced techniques to better implement ventilative cooling systems.

In fact, ventilation is very important for the building's energy load. According to Liddament and Orme (1998), air change accounts for approximately 36 per cent of the total space conditioning energy and contributes to almost half of heating equipment losses. Techniques using ventilation for cooling have gained increased interest since the early 1980s (Chandra et al, 1982a, 1982b). Extended monitoring has shown that naturally ventilated buildings typically use less than 50 per cent of the corresponding energy consumption of air-conditioned buildings (Kolokotroni et al, 1996a, b). Research and assessments of passive ventilation cooling techniques in Europe (Kolokotroni et al, 2002) have shown that ventilative cooling techniques may contribute highly to reducing the cooling needs of buildings in Europe.

Sizing of ventilation systems for cooling, as well as selection of the more appropriate strategies to follow, depends upon many climatic, technical, operational,

Table 8.1 *Actual and forecast total air-conditioning sales in the world*

(In thousands of units)	1998 Actual	1999 Actual	2000 Actual	2001 Actual	2002 Projected	2003 Forecast	2004 Forecast	2005 Forecast	2006 Forecast
World total	35,188	38,500	41,874	44,834	44,614	46,243	47,975	50,111	52,287
Japan	7270	7121	7791	8367	7546	7479	7344	7459	7450
Asia (excluding Japan)	11,392	11,873	13,897	16,637	16,313	17,705	19,227	20,890	22,705
Middle East	1720	1804	1870	1915	1960	2010	2060	2112	2166
Europe	1731	2472	2709	2734	3002	3157	3318	3489	3670
North America	10,437	12,408	12,322	11,894	12,521	12,522	12,524	12,525	12,525
Central and South America	1588	1665	2109	1939	1866	1906	1973	2043	2114
Africa	511	670	664	758	781	806	833	861	887
Oceania	539	487	512	593	625	659	693	731	770

Source: JARN and JRAIA (2002)

economic and cultural parameters. Existing tools permit accurate evaluation of the expected performance of the various techniques, while combination methods such a multi-criteria analysis may help to optimize the ventilation system (Blondeau et al, 2002).

This chapter aims to present the more recent progress on the field of convective or ventilative cooling. The main scientific knowledge in the field of natural and mechanical ventilative cooling, as well as in the field of thermal comfort, is also discussed.

Cooling buildings: Recent trends

The continuous improvement of living standards, in association with the increased income of major human groups and non-climatic responsive architecture, has contributed highly to increasing the total sales of air conditioners. Based on recent data, there are more than 240 million air-conditioning units installed worldwide (IIR, 2002), while according to a recent study of the International Institution of Refrigeration (IIR, 2002), the refrigeration and air-conditioning sectors consume about 15 per cent of all electricity utilized worldwide.

The total annual sales of air-conditioning equipment is close to US$60 billion, of which US$20.9 billion are spent for room air units, US$15.7 billion for packaged systems, US$6.5 billion for roof-top units and US$12.3 billion for residential heat pumps (IIR, 2002). Such a number represents almost 10 per cent of the car industry's business.

The air-conditioning market is under continuous expansion. In 1998, the total annual sales of the air-conditioning industry was close to 35,188,000 units; in 2000, it increased to 41,874,000 units, increasing further to 44,614,000 units in 2002 (JARN and JRAIA, 2002), with a predicted level of 52,287,000 units in 2006 (see Table 8.1).

The use of air conditioning in the US and Japan is much higher than in Europe. The European Energy Room Air Conditioners (EERAC) study (Adnot, 1999), carried out by the European Economic Community (EEC), has shown that in Europe, the 'penetration rate' of room air conditioners (using 1997 data) is less than 5 per cent in the residential sector and less than 27 per cent in the tertiary sector (see Table 8.2). The penetration rate in the tertiary sector is almost 100 per cent in Japan and 80 per cent in the US, while almost 85 per cent and 65 per cent of the residential buildings in Japan and the US, respectively, have at least one air conditioner correspondingly.

The number of households in the US with central air conditioning has increased from 17.6 million in 1978 to 47.8 in 1997. In parallel, the number of households with room air conditioners has increased during the same period from 25.1 to 25.8 million (see Table 8.3; EIA, 1997). The energy consumption due to air conditioning has increased, during the same period, from 310,000 billion to 420,000 billion British Thermal Units (Btu). In 1997, American households with air conditioners spent almost US$140 per year for air conditioning, while almost 40 per cent used their air conditioners all summer. Because of the increased efficiency of air conditioners, increase in electricity consumption has not followed the rate of penetration of air conditioners.

Table 8.2 *Penetration of room air conditioners in the tertiary and residential sector in the US, Japan and Europe, 1997*

Country	Tertiary	Residential
Japan	100%	85%
US	80%	65%
Europe	< 27%	< 5%

Source: Adnot (1999)

Table 8.3 *Consumption of electricity for air conditioning and associated factors by survey year*

Survey year	Household electricity consumption for air conditioning (billion Btu)	Number of households with central air conditioning (millions)	Number of households with room air conditioning (millions)	Average seasonal energy efficiency ratio (SEER) of central air conditioning units sold during the year
1978	310,000	17.6	25.1	7.34
1980	320,000	22.2	24.5	7.55
1981	330,000	22.4	26.0	7.78
1982	300,000	23.4	25.3	8.31
1984	320,000	25.7	25.8	8.66
1987	440,000	30.7	26.9	8.97
1990	480,000	36.6	27.1	9.31
1993	460,000	42.1	24.1	10.56
1997	420,000	47.8	25.8	10.66

Source: Energy Information Administration (EIA) (1978–1982, 1984, 1987, 1990, 1993, 1997) *Residential Energy Consumption Surveys*

In parallel, the energy consumption for the cooling purposes of the US commercial sector is close to 250TWh (Terawatt hours) per year, while the corresponding peak power demand for summer cooling is close to 109GW.

There are several problems associated with the use of air conditioners. The most important problem deals with the serious increase of peak electricity loads that oblige utilities to build additional plants in order to satisfy demand. As the use of these plants is for a short period, the average cost of electricity increases considerably. In California (Besant-Jones and Tenenbaum, 2001), the demand for electricity during the summer months of 2002 increased due to air-conditioning loads because of the highest temperatures recorded for 106 years. As a consequence, the supply started to fall below demand and electricity prices increased tremendously. It is characteristic that during 1998–1999 and the first months of 2000, the market clearing price in the day-ahead Cal PX (an energy price index in California) was between US$25 and US$50/MWh; it increased to US$150/MWh during the summer months of 2000 (Besant-Jones and Tenenbaum, 2001).

Southern European countries face a very important increase in their peak electricity load, mainly because of

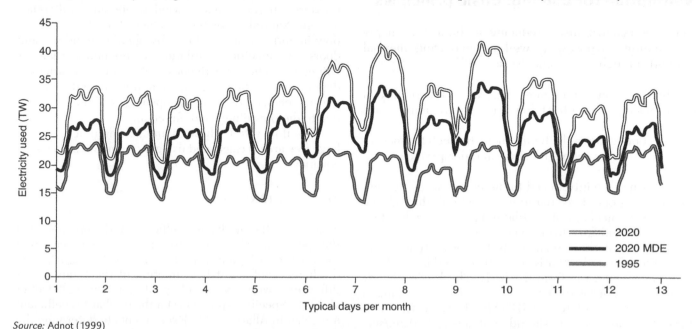

Source: Adnot (1999)

Figure 8.1 *Electricity load curves for 1995 and 2020 in Spain*

the very rapid penetration of air conditioning. Because of high demand for air conditioners, Italy faced substantial electricity problems during the summer of 2003. Figure 8.1 depicts the actual load curves, as well as the foreseen evolution of peak electricity load in Spain (Adnot, 1999). It is evident that an extremely high increase of the peak load is expected, which may require doubling of installed power.

In parallel, important environmental problems are associated with the use of air conditioning. Emissions of refrigerant gases used in air-conditioning installations significantly affect ozone depletion and global warming. Refrigeration and air conditioning-related emissions represent almost 64 per cent of all chlorofluorocarbons (CFCs) and hydrochlorofluorocarbons (HCFCs) produced (AFEAS, 2001). New air conditioners use more efficient refrigerants that have a lower impact on atmospheric ozone depletion.

Problems related to indoor air contamination should not be neglected as well. Cooling coils and condensate trays can become contaminated with organic dust that may lead to microbial growth. The organic dust may also cause mould and fungal growth in fans and fan housing. Inefficient and dirty filters may also lead to unfiltered air in buildings. Contaminated emissions from cooling towers that have not been properly maintained may cause spread of *Legionella* from poorly maintained systems.

Ventilation for cooling: Basic principles

Ventilation contributes to reducing or eliminating energy for cooling purposes, as well as increasing thermal comfort through two mechanisms:

- by removing the higher-temperature indoor air and replacing it with fresh low-temperature ambient air; and
- by cooling down the human body through the mechanisms of convection, radiation and perspiration.

Fresh air may be introduced to the indoor space through the building openings, (natural ventilation), through the use of fans, (mechanical ventilation) or by a combination of openings and fans (hybrid ventilation).

Cooling of the human body by convection occurs when the surrounding air is cooler than the skin and, thus, heat is carried away from the body. The higher the air speed, the higher is the body cooling effect. According to the existing standards (ASHRAE, 1992; ISO, 1994) permitted air speed should not exceed 0.2m/sec. However, recent research trying to better understand

thermal comfort mechanisms in naturally ventilated buildings (Nicol, 2003) has shown that occupants of these buildings may prefer much higher indoor air speeds. Quite recently, the American Society of Heating, Refrigerating and Air-Conditioning Engineers (ASHRAE) has quantified the difference between the thermal response of people living in air-conditioned and naturally ventilated buildings. A new adaptive comfort standard has been proposed.

In parallel, when ceiling fans are used, air blows downward on the human body; thus, higher speeds are allowed and are supported by the human body. This permits an increase in the indoor air temperature of up to 2° C, which results in very important energy conservation.

When the temperature of the surrounding opaque surfaces in a building is lower than skin temperature, the body is losing heat by radiation. Use of ventilation, especially during the night time, may contribute significantly to reducing the interior temperature of the opaque surfaces of a building. Thus, the radiative balance is negative for the human body, while the interior air reduces its temperature because of the convection between the opaque surfaces and the indoor air.

During hot weather or physical exercise, the human body dissipates excess heat through perspiration mechanisms. When air flows around the skin, it contributes to evaporation of human moisture and, thus, to benefits from the associated latent heat.

Natural ventilation is caused by naturally produced pressure differences due to wind, temperature difference or both. Natural ventilation is achieved by allowing air to flow in and out of a building by opening windows and doors or specific ventilation components such as chimneys. The effectiveness of natural ventilation depends upon the wind speed, temperature difference, size and characteristics of the openings and their orientation to the prevailing wind direction.

The flow rate Q through an opening of a relatively large free area is calculated using the common orifice flow equation:

$$Q = C_d A \sqrt{(2\Delta P / \bar{\rho})} \qquad (1)$$

where C_d is the discharge coefficient of the opening, A is the opening area [m²], ΔP is the pressure difference across the opening [Pa] and ρ is the air density [kg/m³]. The discharge coefficient is a function of the temperature difference, wind speed and opening height (Pelletret et al, 1991). Specific expressions for the discharge coefficient are given in Allard (1998). Experiments to determine the discharge coefficients are reported by Flourentzou et al (1998).

Calculation of airflow through large openings is a complicated task. Simplified, network, zonal and computational fluid dynamic (CFD) models may be used to calculate the airflow rate in naturally ventilated buildings. A review of the commonly used models is given by Vollebregt et al (1998) and Allard (1998). Simplified models are based on experimental or simulated data and generally propose simple formulas or graphs for designing the envelopes of naturally ventilated buildings (Chandra et al, 1986; Ernest, 1991; CSTB, 1992; Etheridge, 2002; Fracastoro et al, 2002). These tools must always be used in the limits of their validity. Zonal models are based on the equations of mass and energy conservation. A zone is divided into several macroscopic homogeneous cells in which mass and heat conservation must be obeyed (Lebrun, 1970; Howarth, 1985; Inard and Buty, 1991, 1996; Togari et al, 1993; Rodriguez et al, 1994; Wurtz et al, 1996, Haghighat et al, 2001) Zonal models can quite accurately predict the temperature patterns in a room; but their main limitation is that a pre-knowledge of the flow pattern is necessary.

Network calculation models are the more commonly used tools. Network models are based on the equation of mass conservation, combined with some empirical knowledge. Well-known network models for natural ventilation systems are AIOLOS (Dascalaki and Santamouris, 1998b), COMIS (Allard et al, 1990), CONTAM (Walton 1994) and BREEZE, (BREEZE, 1993). Most of these models can be used for mechanical ventilation calculations as well.

Ventilation and thermal comfort

Energy savings associated with the use of ventilative cooling techniques are fully linked to applied thermal comfort standards. In parallel, thermal comfort is linked, as well, with the perception of indoor air quality in a building and productivity (Humphreys et al, 2002; McCartney and Humphreys, 2002). Existing standards and methods primarily cover thermal comfort conditions under steady-state conditions. The most well-known and widely accepted methods are the 'comfort equation' proposed by Fanger (1972) and the J. B. Pierce two-node model of human thermoregulation (Gagge, 1973; Gagge et al, 1986). Based on these models, several steady-state thermal comfort standards have been established (Jokl, 1987; ASHRAE, 1992; ISO, 1994).

Because of the thermal interaction between a building's envelope, its occupants and the auxiliary system, steady-state conditions, in practice, are rarely encountered in buildings. In particular, indoor temperature in free-floating buildings is far from steady. Monitoring of

passive solar buildings with a constant set-point has shown that there are important indoor fluctuations of between 0.5 and 3.9 °C as a result of the control system (Madsen, 1987). Thus, knowledge of thermal comfort under transient conditions is necessary.

Field studies and basic thermal comfort research (Humphreys, 1975) have shown that there is an important discrepancy in the steady-state models, especially for the zones where no mechanical conditioning is applied. This is mainly due to the temporal and spatial variation of the physical parameters in the building (Baker, 1993). In fact, occupants living on a permanent basis in air-conditioned spaces develop expectations for low temperatures and homogeneity and are critical when indoor conditions deviate from the comfort zone that they are used to. On the contrary, people who live in naturally ventilated buildings are able to control their environment and become used to climate variability and thermal diversity. Thus, their thermal preferences extend to a wider range of temperatures or air speeds. Such an adaptation to the thermal environment has been extensively studied and documented, (Nicol et al, 1995; Brager and De Dear, 1998, 2000; De Dear, 1998; De Dear and Brager, 1998; Rijal et al, 2002).

Field surveys have verified that comfort temperature is very closely related to mean indoor temperature (Nicol et al, 1999; McCartney and Nicol, 2002). Nicol and Humphreys (1973) suggested that such an effect could be the result of the feedback between the thermal sensation of subjects and their behaviour.

The adaptive principle has also been verified through the PASCOOL research project. Based on previous work, the comfort group of the European research project PASCOOL (Baker, 1993; Baker and Standeven, 1994; Standeven and Baker, 1994) has carried out field measurements to understand the mechanisms by which people make themselves comfortable at higher temperatures. It was found that people are comfortable at much higher temperatures than expected, while it was observed that people take a number of actions to make themselves comfortable, including moving to cooler parts of the room. It is characteristic that there were 273 adjustments to building controls and 62 alterations to clothing out of 864 monitored hours.

Various other research studies have verified the adaptive comfort approach. Klitsikas et al (1995) have performed comfort studies in office buildings in Athens, Greece, during the summer period. It was found that the theoretical predicted mean vote (PMV) value is almost always higher or equal to the measured thermal sensation vote, and the subjects felt more comfortable than predicted by the PMV theory. Lin Borong et al (2004) have performed comfort studies in Chinese naturally ventilated buildings. They concluded that the thermal sensation of people has a larger range than that in a stable environment. Comparisons have been performed against the PMV scale and it has been found that the PMV model, when applied to unstable or natural thermal environments to evaluate people's thermal sensation, needs correction. Similar results have been found during a comfort survey under hot and arid conditions in Israel (Becker et al, 2003), in Singapore (Hien and Tanamas, 2002), in Indonesia (Feriadi, 2002a), in Algeria (Belayat et al, 2002) and in Bangladesh,Mallick, 1994).

Humphreys and Nicol (2002) and Parsons et al (1997) have provided some explanations for the errors in the PMV theory. According to the authors, since PMV is a steady-state model there is a theoretical contradiction between the basic assumptions of the model and the imbalance assumed if the body is not comfortable. Another reason is related to the uncertainty and the fuzziness to exactly calculate the metabolic heat and clothing insulation.

Important research has been carried out in order to develop an adaptive comfort standard. Analysis of the data included in the ASHRAE RP-884 database involving data of comfort surveys around the world (De Dear and Brager, 2002) has shown that while PMV predictions fit very well with the preference of occupants in heating, ventilating and air-conditioning (HVAC) buildings, occupants of naturally ventilated buildings prefer a wider range of conditions that more closely reflect outdoor climate patterns (see Figure 8.2).

The same conclusions have been reported from various comfort field studies (Webb, 1959; Nicol, 1973; Humphreys, 1975; Busch, 1992; Nicol and Roaf, 1994; Matthews and Nicol, 1995; Taki et al, 1999; Nicol et al, 1999; Bouden and Ghrab, 2001) As a result of the field studies, it was proposed that optimum comfort temperature is a function of the outdoor temperature, and may be predicted by equations of the following form (Humphreys, 1978; Auliciems and De Dear 1986; Nicol and Raja, 1995):

$$T_{comf} = a\,T_{a,out} + b \tag{2}$$

where $T_{a,out}$ is the mean outdoor air temperature. Thus, De Dear and Brager (2002) have proposed the following expression:

$$T_{comf} = 0.31T_{a,out} + 17.8 \tag{3}$$

while Humphreys (1978), Humphreys and Nicol (2000) and Nicol (2002) have proposed an almost similar expression:

$$T_{comf} = 0.534T_{a,out} + 11.9 \tag{4}$$

Based on these results, a new adaptive thermal comfort for naturally ventilated buildings has been proposed to be integrated within ASHRAE Standard 55 (see Figure 8.3).

Designing for ventilative cooling requires knowledge of the appropriate air speed inside buildings. However,

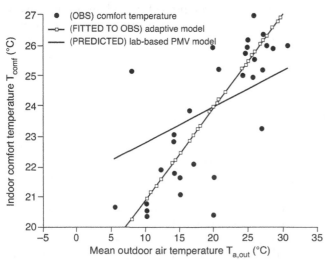

Source: De Dear and Brager (2002)

Figure 8.2 *Observed (OBS) and predicted indoor comfort temperatures from ASHRAE RP-884 database for naturally ventilated buildings*

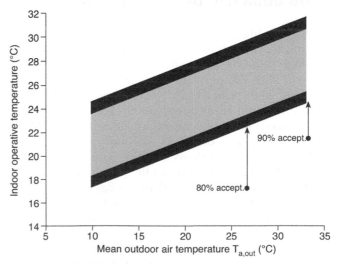

Source: De Dear and Brager (2002)

Figure 8.3 *Proposed adaptive comfort standard (ACS) for ASHRAE Standard 55, applicable for naturally ventilated buildings*

the impact of air movement on thermal comfort is an open research area (Arens et al, 1984; Arens and Watanabe, 1986; Tanabe and Kimura 1994). Air velocity affects both convective and evaporative losses. Recently, studies performed in tropical climates (De Dear, 1991; Mallick 1996; Hien and Tanamas, 2002), confirm that the increase in air velocities, especially at higher temperatures, enhances thermal comfort conditions. According to Kukreja (1978), indoor air speed in warm climates should be set at 1.00–1.50m/s. Hardiman (1992) proposes an air speed of between 0.2–1.5m/s for light activity. Hien and Tanamas (2002) report that undesirable effects of high air movements of above 3m/sec have been observed.

Similar results are also reported from a recent Danish climatic chamber research project (Toffum et al, 2000), where the subjects preferred 28° C when permitted to select their own preferred airspeed than 26° C, with a fixed air speed of 0.2m/sec.

Adaptive and variable indoor temperature comfort standards for air-conditioned buildings may result in remarkable energy savings for cooling (Auliciems, 1989; Milne, 1995; Wilkins, 1995; Hensen and Centrenova, 2001). Estimated energy savings of more than 18 per cent over that from using a constant indoor temperature are reported by Stoops et al (2000), while the corresponding energy savings for UK conditions have been estimated at close to 10 per cent.

Designing for natural ventilation and cooling

Direct ventilative cooling

Natural ventilation may be used directly for cooling purposes when the ambient temperature and humidity are within comfort limits. This technique is a common practice in mild climates; but in hot areas, indirect natural ventilative cooling techniques, such as night ventilation, may be used. The exact boundaries of outdoor temperature and humidity within which indoor comfort can be provided by daytime natural ventilation have been proposed by Givoni (1994). For an indoor wind speed of 2m/sec, the upper suggested outdoor temperature for hot developed countries is close to 32° C.

The potential of direct ventilative cooling techniques has been assessed worldwide through detailed experimental and theoretical studies. The expected reduction of the cooling needs varies as a function of local climatic conditions; but a mean maximum contribution of close to 50 per cent of the needs has frequently been reported. In particular, Chandra et al (1986) have calculated the potential of direct natural ventilation techniques to reducing

the cooling needs of buildings in the US. They concluded that the possible reduction of cooling needs varies between 10–50 per cent as a function of climatic characteristics. Carrol et al (1982) have simulated the impact of natural ventilation in office buildings in the US. Energy savings varied from 25 per cent in humid climates up to 50 per cent in warm climates. Vieira and Parker (1991) have found that in Florida, US, the longer the natural ventilation season can be extended, the lower overall air-conditioning consumption will be. Data from 384 single-family homes, apartments and condominiums have shown that each month, from May to September, a household claimed to use natural ventilation rather than air conditioning, resulted in an average savings of 777 kilowatt hours (kWh).

In Europe, Emmerich et al (2001) have reported that natural ventilation in the UK may provide cooling energy savings of the order of 10 per cent and fan power savings of the order of 15 per cent of annual energy consumption. Cardinale et al (2003) have studied the cooling potential of daytime natural ventilation in major Italian cities. They report energy savings of up to 53 per cent compared to an air-conditioned building. Santamouris and Fleury (1989) have studied the cooling potential of daytime ventilation in Greece and found that it is possible to cover almost 30 per cent of the cooling load of an air-conditioned building. Aynsley (1999) has also studied the possibility of providing comfort using natural ventilation for a building located in the Australian tropics. He concluded that the proper design of building permits thermal comfort to be achieved for most of the hot period.

Important tools to assess the potential of direct ventilative cooling techniques have been developed under the framework of the URBVENT research project by the European Commission. Germano et al (2002) have developed a tool to estimate the natural ventilation potential, as well as the passive cooling potential, of urban buildings. The method uses geographic information system (GIS) techniques, as well as multi-criteria evaluation, and may assist designers integrating natural ventilative cooling techniques in urban buildings. In the framework of the same research project, Ghiaus and Allard (2002) have developed a method to assess the potential of direct natural ventilative cooling using degree hours data.

Architectural integration of openings

The overall architectural design of a building determines its ventilation and passive cooling potential. Vernacular architecture in warm climates is full of ideas and examples on how to better integrate natural ventilation and passive cooling in buildings (Fathy, 1986).

Techniques to enhance natural ventilation in buildings have been well researched, and appropriate strategies involving reduction of the plan depth, maximization of the skin permeability through openings, minimization of internal obstructions, increased openness, orientation to prevailing winds, and use of the stack effect are among the main proposed strategies (Fleury, 1990; Hyde, 2000). The CIBSE *Applications Manual* (CIBSE, 1997) and Martin (1995) describe several natural ventilation configurations incorporating advanced windows, window and vent actuators, thermal chimneys, wind chimneys, atria, etc.

The overall architecture of a building and, in particular, the positioning and shape of the openings, balconies and internal partitions, as well as the shape of the building, play a very important role and determine the air speed and comfort conditions in naturally ventilated buildings. Important research has been carried out in trying to optimize the main architectural parameters (Olgay, 1973; Sobin, 1981; Kindangen et al, 1996, 1997; Chand et al, 1998; Chiang et al, 2000; Prianto et al, 2000; Prianto and Depecker, 2002). Givoni's (1976) pioneering work on the position of openings in naturally ventilated buildings has permitted a better understanding of the specific contribution of windows to different boundary conditions.

Rosenbaum (1999) has summarized some of the main conclusions of research on the position of walls and proposed techniques to enhance airflow through the building's envelope. The main suggestions are:

- In order to enhance cross-ventilation, irregularly shaped or spread-out buildings have to be designed.
- It is better to face the building at an oblique angle to the prevailing wind than to face it directly perpendicular to the wind direction.
- The inlet area should be equal to the outlet area.
- Horizontally shaped windows perform better than vertical windows.

Other building elements may be used to enhance airflow in naturally ventilated buildings. In particular, verandas, balconies and decks may contribute to significantly increasing the air speed inside a building because pressure differences may also increase (Chand, 1973a). As reported, when the angle of incidence ranges from 0–60 degrees, the windward or leeward location of a veranda open on three sides produces an important air motion in a room. In parallel, the use of pelmet-type wind deflectors enhance the air movement at a working plane in a room by about 30 per cent (Chand et al, 1975). In addition, sashes projecting outward result in enhanced indoor air motion compared with those projected inward (Chand

and Bhargava, 1975).

Single-sided naturally ventilated buildings may not present a high airflow rate because of the specific pressure difference. Givoni (1976) has proposed the use of wing walls in windward openings that permit the creation of distinct positive and negative pressures on the openings and, thus, enhance the airflow in the building. Givoni (1976) reported that for oblique winds, the use of wing walls created an average air velocity in the room of about 40 per cent of the outside wind, while when no wing walls were used, the air speed was just 15 per cent of the ambient wind.

Chandra et al (1983), have carried out full-scale experiments to measure the performance of wing walls. Airflow was measured in a room with and without wing walls. They found that the presence of the wing walls considerably increases the inlet air speed to the room.

Atria and courtyards

Ventilated atria and courtyards attached to buildings may enhance natural ventilation and promote convective cooling. Courtyards are well known from ancient times. They were used in ancient Greek and Roman architecture, as well as in Mesopotamia, the Indus Valley, the Nile Valley and in China. Their overall form and construction have endured over 6000 years with very few modifications, (Hinrichs, 1988).

Courtyards are associated with naturally ventilated buildings in hot climates. Atria and courtyards are generally hotter during the daytime but present a much lower temperature during the night (Chandra, 1989). Courtyards are transitional zones that improve comfort conditions by modifying the microclimate around the building and by enhancing the airflow in the building. Important research has been carried out to better understand the airflow processes and the cooling impact of courtyards (Bagneid, 1987; Etzion, 1988; Hoffman et al, 1994; Berger and Semega, 1995; Cadima, 2000; Majid et al, 2002; Feriadi, 2002b). A classification system of atria as well as an extensive literature review has been prepared by Eureca Laboratories (1982).

In a recent paper, Rajapaksha et al (2003) have studied the potential of a ventilated courtyard to provide passive cooling in buildings located in warm, humid climates. It was found that the overall performance depends upon the flow patterns. Better performance is reported when the courtyard acts as an air funnel, discharging indoor air into the sky.

Methods and tools to design atria and estimate the airflow through them have been proposed by Hunt and Holford (1998), Holford and Hunt (2000), Gage et al

(2001), Todorovic et al (2002) and Holford and Hunt (2003).

Solar chimneys

Solar chimneys have been extensively studied as a configuration to implement natural ventilation in buildings where solar energy is available. Solar chimneys are natural draught components that utilize solar energy to build up stack pressure and, thus, drive airflow through the chimney channel. Solar chimneys are similar to conventional chimneys except that the south wall is replaced by a glazed surface, which enables it to collect solar radiation (Haisley, 1981; Kumar et al, 1998; Afonso and Oliveira, 2000). Such a technique, called 'Scirocco room', is well known from traditional Italian architecture of the 16th century (Cristofalo et al, 1989). Quite recently, important experimental, numerical and theoretical research has contributed to a better understanding of solar chimneys.

The ability of solar chimneys to improve the ventilation rate in naturally ventilated buildings was studied by Bansal et al (1993, 1994). It was found that the impact of solar chimneys is substantial in inducing natural ventilation for low wind speeds. Gan and Riffat (1998) and Shao et al (1998) have also studied a solar-assisted technique to enhance natural ventilation, coupled with a heat-pipe heat-recovery system. They report a heat recovery efficiency of about 50 per cent. The performance of solar chimneys when integrated with air-conditioned buildings has been studied by Khedari et al (2003). It was reported that the solar chimney could reduce the average electrical consumption of the building by 10–20 per cent. The contribution of solar chimneys to improving ventilation and cooling in hot climates was studied by Bouchair, (1987, 1989, 1994) and Tan (2000). Theoretical models and simulation techniques to calculate the performance of solar chimneys have been proposed by Pedki and Sherif (1999), Rodrigues et al (2000) and Letan et al (2003).

Passive and active stacks have been incorporated within a high-rise residential building in Singapore (Priyadarsini et al, 2003). It was found that passive stacks cannot change the air velocity in the building, while the use of active stacks leads to a substantial increase in air velocity within the rooms. Similar results on the use of active stacks in naturally ventilated buildings are reported by Hien and Sani (2002).

Solar chimneys with a uniform heat flux on a single wall were investigated experimentally for different chimney gaps, heat flux inputs and different chimney inclinations by Chen et al (2003). No optimum gap was found, while it was reported that the airflow rate reached a maximum at a chimney inclination angle of around 45 degrees. This is about 45 per cent higher than that for a vertical chimney under otherwise identical conditions.

The integration of solar chimneys with cooling cavities to enhance both ventilation and cooling has been studied by various researchers (Barozzi et al, 1992; Aboulnaga and Abdrabboh, 1998, 2000; Hamdy and Fikry, 1998; Pasumarthi and Sherif, 1998; Hunt and Linden, 1999; Khedari et al, 2000a, 2000b; Li, 2000; Raman et al, 2001; Day et al, 2003). Cooling cavities induce downward buoyancy airflow in a vertical cavity, where the air is cooled using mainly evaporative cooling techniques. It has been found that such a combination leads to increased airflow rates and an important reduction of indoor temperature depending upon the characteristics of the system.

The ventilation performance of light/vent pipes has been studied by Oliveira et al (2001). Light/vent pipes are composed of two concentric tubes. The channel space between the tubes is allowed for airflow. Air is flowing either because of the temperature difference or because of the wind pressure. Experiments have shown that light/vent components enhance the airflow by 44 per cent.

Wind towers

Wind towers or cooling towers are well known and have traditionally been used in Middle Eastern and Persian architecture (Bahadori, 1978, 1985). Air enters the towers at the windward face, its higher part, and leaves at the lower part, which is in communication with the building. The air may be cooled by evaporative or convective cooling through the tower. Research has shown that inlet and outlet opening areas for wind towers have to be 3–5 per cent of the floor area that they serve (Nielsen, 2002).

New active developments, particularly coolers where the air was forced by a fan through wetted pads, have been used during the past in desert areas of the US. In natural downdraught coolers, the air is not forced through the pads and air is provided simply by gravity flow (Cunningham and Thompson, 1986). The performance of natural downdraught coolers is studied by Badran (2003), while the necessary pressure coefficients to evaluate the airflow in wind towers are provided by Karakatsanis et al (1986) and Bahadori (1981). In parallel, the performance of downdraught evaporative coolers has been studied in detail by various authors and design tools have been proposed (Givoni, 1991; Sodha et al, 1991; Chalfoun, 1992; Thompson et al, 1994).

A high-efficiency and innovative development of the downdraught evaporative cooler is achieved through the PDEC (Passive Downdraft Evaporative Cooling) research programme of the European Commission, (PDEC, 1995). The improvement consists of replacing the wetted pads with rows of atomizers – nozzles that produce an artificial fog by injecting water at high-pressure through minute orifices. This feature produces much better regulation of the system, a significant reduction of the pressure losses and a lower size of equipment.

Innovative components

Various innovative ventilation components have been proposed for integration within buildings. Fairey and Bettencourt (1981) have proposed a roof-top ventilation component known as 'La Sucka' that is based on the use of dampers on two or four sides of a roof-top cupola. The component permits the airflow through the leeward part of the cupola by closing the dampers in the windward façade. Fuller (1973) has proposed a 'dymaxion dwelling machine' that is a rotating roof which aligns to the wind direction, while Givoni (1968) has proposed a double-ceiling system for cross-ventilation.

Control of naturally ventilated buildings

In naturally ventilated buildings, controls have to be used to modify the indoor environment. Control can be automatic or manual by the building's occupants. Appropriate control, like window opening or use of blinds, may reduce the need for mechanical cooling. The importance of control has been clearly shown by various studies and research (Baker and Stadeven, 1995; Leaman and Bordass, 1995; Nicol et al, 1999).

Liem and van Paaseen (1998) have found that by controlling a naturally ventilated building, an improvement in the established comfort is observed; but the exact type of simple control algorithm has a marginal influence on the improvement. Kolokotroni et al (2001) have studied the performance of a naturally ventilated educational building in the UK and have concluded that although thermal mass and natural ventilation can reduce the effect of external hot weather and establish comfort in the building, manual or automatic control should be set in place so that the benefits are not offset by overcooling the building during cold spells.

Experimental surveys trying to identify the control actions undertaken by individuals in naturally ventilated buildings (Raja et al, 1998, 2001) have shown that controls are used in response to discomfort and, in general, occupants who have greater access to controls (for example, those close to a window) report less discomfort than those who have less access.

Various automatic control strategies can be used to achieve comfort in naturally ventilated buildings. Pitts

and Abro (1991) have developed an intelligent controller to optimize night ventilation in a building through an active solar chimney. The controller could calculate the cooling needs, the comfort conditions in the building and the cooling capacity of the solar chimney. Thus, appropriate decisions can be taken. La Roche and Milne (2001, 2002, 2003) have designed and tested an intelligent controller to optimize the use of mechanical ventilative cooling using a whole-house fan. The controller is based on a set of decision rules that takes into account indoor and outdoor temperatures, and experimental testing has shown that this can improve indoor comfort and reduce energy consumption.

In particular, artificial intelligence techniques seem to offer numerous advantages compared to classical control systems. Such a controller, for naturally ventilated buildings, was proposed by Dounis et al (1995a, 1995b). A comparison of ON-OFF, PID and PI with deadband and fuzzy controllers for naturally ventilated buildings has been performed by Dounis et al (1996a, 1996b), and it has been shown that fuzzy controllers present important advantages compared to other conventional strategies. Eftekhari and Marjanovic (2003) have proposed a fuzzy logic controller to optimize the opening position in naturally ventilated buildings. It was found that such a controller is capable of providing better thermal comfort inside the room than a manual control of openings or seasonal operation. Kolokotsa (2001) has designed and tested a prototype fuzzy controller for naturally ventilated buildings using local operating networks and smart cards technology. The research has been carried out in the frame of the European research project BUILTECH, and has resulted in the design of a prototype controller that significantly improves indoor environmental quality in naturally ventilated buildings. Kolokotsa (2003) has also tested five different fuzzy controllers for naturally ventilated buildings, particularly fuzzy P, fuzzy PID, fuzzy PI, fuzzy PD and adaptive fuzzy PD, and has concluded that all controllers achieve important energy and comfort improvements.

Indirect ventilative cooling techniques

Principles of night ventilative cooling

Night-time ventilation is associated with the circulation of low-temperature ambient air in a building and the reduction of the temperature of indoor air, but mainly of storage mass. Thus, indoor thermal conditions in a building are more positive during the following day. Night-time ventilation is suitable for areas with a high diurnal temperature range and where night-time temperature is not so cold as to create discomfort.

Night ventilation systems are classified as direct or indirect, depending upon the procedure used to transfer heat between the thermal storage mass and the conditioned space. In direct systems, cool air is circulated inside the building zones and heat is stored in the exposed opaque elements of the building. The reduced temperature mass of the building contributes to reducing the indoor temperature of the following day through convective and radiative procedures. Circulation of the air can be achieved by natural or mechanical ventilation. In direct systems, the mass of the building has to be exposed and the use of coverings or false floors or ceilings has to be avoided (Santamouris, 2003).

In indirect systems, cool air is circulated during the night through a thermal storage medium where heat is stored and is recovered during the following day period. In general, the storage medium is a slab covered by a false ceiling, a floor or a phase-change material storage, while the circulation of the air is always forced. It is evident that during the day period, the temperature of the circulated air has to be higher than the corresponding temperature of the storage medium. Direct and indirect night ventilation systems are used many times in a combined way.

Thus, night ventilation affects indoor conditions during the next day in four ways by: (Kolokotroni and Aronis, 1999):

1 reducing peak air temperatures;
2 reducing air temperatures throughout the day and, in particular, during the morning hours;
3 reducing slab temperatures; and
4 creating a time lag between the occurrence of external and internal maximum temperatures.

It is evident that the performance of night cooling systems depends upon three main parameters:

1 the temperature and the flux of the ambient air circulated in the building during the night period;
2 the quality of the heat transfer between the circulated air and the thermal mass;
3 the thermal capacity of the storage medium.

Important theoretical and experimental research has been carried out to better understand the phenomena, to evaluate the cooling potential of night ventilation techniques, and to develop computational and design tools and codes.

Extended experimental work on night ventilation techniques are reported by Baer, (1983, 1984); Agas et al (1991); Van der Maas and Roulet (1991); Geurra et al (1992); Barnard (1994); Givoni (1994, 1998a); Hassid

(1994); Van der Maas et al (1994); Blondeau et al (1995a, 2002); Ren (1995); Kolokotroni et al (1996a, 1997); Meierhans (1996); Santamouris and Assimakopoulos (1996); Santamouris et al (1996); Behne (1996); Feustel and Stetiu (1997); Aboulnaga and Abdrabboh (1998); Burton (1998); Dascalaki and Santamouris (1998a); Demeester et al (1998); Geros et al (1999); Nicol et al (1998); Wouters et al (1998); Zimmerman and Anderson (1998); Roucoult et al (1999); CEC (2000b); Liddament (2000); Shaviv et al (2000); Turpenny et al (2000a); Barnard et al (2001); Blake (2001); Axley and Emmerich (2002); Herkel et al (2002) and Todorovic et al (2002).

Cooling potential of night ventilative techniques applied to free-floating buildings

Various important theoretical and experimental works have been performed in order to assess the efficiency of night-cooling techniques applied to free-floating buildings. Most of the research shows that it is possible to reduce the peak indoor temperature of the next day between 0.5–3° C as a function of the airflow rate, the thermal storage capacity of the building and the daytime and night-time temperature of the ambient air.

Research carried out in southern climates has shown that the reduction of the indoor temperature during the following day may be between 2–3° C. Geros et al (1999) performed measurements in free-floating office buildings in Athens, Greece. The authors reported that under free-floating conditions, the use of night ventilation decreases the next-day peak indoor temperature by up to 3° C. Results of sensitivity analysis have shown that the expected reduction of the overheating hours varies between 39 per cent and 96 per cent for airflow rates for 10 and 30 air changes per hour (ACH), respectively (see Figure 8.4).

Shaviv et al (2001) have studied the cooling potential of night ventilation techniques in Israel for different levels of thermal mass. They showed that it is possible to achieve a reduction of 3–6° C in a heavy constructed building without operating an air-conditioning unit. Similar results are reported by Becker and Paciuk (2002) regarding the application of night ventilation techniques in office buildings in Israel.

Solaini et al (1998) have studied the performance of night ventilation techniques in an experimental building in Italy and found that these techniques play a very important role in its cooling needs. Silvestrini and Alessandro (1988) report that for Italian conditions, 3 ACH during the night may provide a good perception of comfort; when 10–15 ACH are applied, the building presents its minimum energy consumption for cooling.

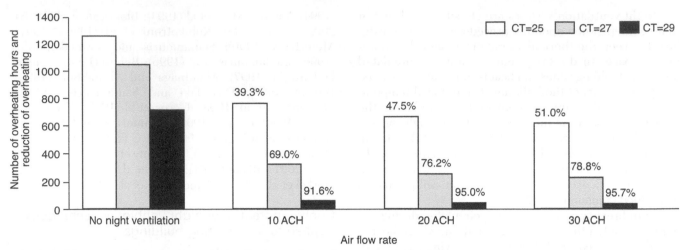

Note: The building is considered to operate under free-floating conditions. CT = constant temperature.

Source: Geros et al (1999)

Figure 8.4 *Average overheating hours and reduction due to the use of night ventilation in the Meletitiki Building, Athens, Greece*

In California, Givoni (1998) measured the potential of night ventilation to reduce maximum daytime temperatures in buildings with different mass levels. It was found that night ventilation has almost no effect on the low mass building, but has an important effect in heavy buildings. When the outdoor temperature was 38° C, the indoor maximum temperature of the high mass building was only 24.5° C.

Rainaweera and Hestnes (1994) have calculated the impact of night ventilation techniques when applied in typical dwellings in Sri Lanka. They found that an increase of the flow rate from 8 ACH to 14 ACH decreases the maximum indoor temperature of the next day by 0.5° C.

Golneshan and Yaghoubi (1985, 1990) have simulated the contribution of night ventilation techniques in Iranian residential buildings. They report that the use of 12 ACH per hour during the night with 1 ACH during the day may provide comfortable indoor conditions.

When applied in mild climates, night ventilation may reduce indoor temperatures by up to 1–2° C; but this may be sufficient to cover a very high part of comfort needs. Birtles et al (1996), as well as Kolokotroni et al (1998), have performed simulations to study the potential of night ventilation and thermal mass to cover the cooling needs of office buildings in the UK. They concluded that these techniques can provide the required cooling in most cases, while for the rest, a high percentage of the cooling requirements can be met by night ventilation.

Blondeau et al (1995b) have measured the potential of night ventilated offices in France and found that it is possi-

ble to reduce the maximum next-day indoor temperature by 1.5–2° C.

Neeper and McFarland (1982) and Kammerud et al (1984) have simulated the impact of night-ventilated massive passive solar houses in the US. They concluded that increasing the airflow by up to 10 ACH does not contribute to additional energy savings. Further studies by Chandra and Keresticioglu (1984) for the same building, considering a variable heat transfer coefficient, have shown that under specific conditions the optimum ACH is close to 25. Parker (1992) has measured the efficiency of whole-house fans to provide night cooling in a free-floating house in Florida. They report reductions of the total daily average interior temperature by over 2.5° F due to the removal of heat from the thermal mass of the building.

Cooling potential of night ventilative techniques applied to air-conditioned buildings

Night ventilation applied to air-conditioned buildings may reduce the required energy for cooling, as well as the peak electricity demand. Experiments, theoretical studies and real applications have shown that there is a very high savings potential that may exceed 50 per cent of the cooling load of massive buildings for an indoor temperature of close to 26° C and a high airflow rate. In parallel, the reduction of the peak electricity demand may decrease by up to 40 per cent. Different results are reported as a function of the climate and building characteristics, as well as of the airflow and operational conditions.

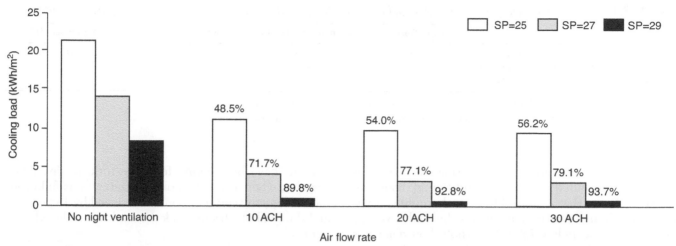

Note: The building is considered to operate under air-conditioning conditions. SP = set point.

Source: Geros et al (1999)

Figure 8.5 *Cooling load reduction due to the use of night ventilation in the Meletitiki Building, Athens, Greece*

Geros et al (1999) performed measurements in air-conditioned night-ventilated office buildings in Athens, Greece. They found that the early morning indoor air temperature can be reduced by 0.8–2.5° C depending upon the considered set-point temperature. Sensitivity analysis that attempts to identify the impact of airflow rates, as well as of the set-point temperature, has shown that the expected energy conservation varies between 48 per cent and 94 per cent for set-point temperatures of between 26–29° C and airflow rates of between 10 and 30 ACH, respectively (see Figure 8.5).

A similar analysis has been performed by Blondeau et al (1995), who have studied the cooling potential of night ventilation techniques when applied to an air-conditioned building in France. It is found that the lower the set-point temperature during daytime, the lower the contribution of night cooling. For a 22° C temperature set-point, night ventilation covers almost 12 per cent of the cooling load, while when the set-point temperature rises to 26° C, the corresponding contribution increases to 50 per cent of the load.

Carrilho da Graca et al (2002) have simulated the effect of night ventilation in a six-storey apartment in Beijing and Shanghai. They found that night cooling may replace air conditioning for about 90 per cent of the time in Beijing and 66 per cent in Shanghai. Olsen and Chen (2003) have studied the performance of night ventilation techniques coupled with a variable air volume flow rate (VAV) and a displacement ventilation system for an office building in the UK. They found that when night ventilation is associated with a VAV system, it contributes to

reducing the energy consumption for cooling by 12 per cent. In parallel, the peak chiller load is reduced by about 20 per cent for both the displacement ventilation and VAV systems when night cooling is used.

Kolokotroni and Aronis (1999) have studied the impact of night ventilation techniques in air-conditioned offices in the UK. They reported that application of night ventilation is beneficial and results in an energy saving of about 5 per cent and an installed capacity saving of about 6 per cent, while for heavyweight buildings the figures increase to about 15 and 12 per cent, respectively.

Martin et al (1984) have simulated the impact of night ventilation in air-conditioned non-residential buildings in Los Angeles, Atlanta and New York. They found that in Los Angeles, cooling energy consumption can be reduced by up to 7 per cent by night ventilation with 3 ACH. Increasing the airflow to 30 ACH does not contribute to additional benefits. In Atlanta, night ventilation has no impact in low-mass buildings, but it contributes to reducing the cooling load of high-mass buildings by 6–7 per cent for 3 ACH during the night. The performance is not improved significantly when 30 ACH are applied. Finally, in New York, small benefits can be achieved in high-mass buildings.

Various studies have considered the use of whole-house fans to provide night ventilation. Studies by Kusuda (1981) in air-conditioned buildings have shown that the use of whole-house fans to provide night-time ventilation may contribute to reducing the air-conditioning load by up to 56 per cent. Similar studies are reported by Burch and Treando (1979) and Ingley et al (1983). Burch and

Table 8.4 *Reduced electrical peak power demand of a building for different night ventilation strategies (percentage)*

		Berlin, Germany	Locarno, Switzerland	Red Bluff, US	San Francisco, US
No chiller	Natural night ventilation	−40[2]	−52[1]	–	−31
	Mechanical night ventilation	−38[2]	−51[1]	–	−29
With chiller	Mechanical night ventilation	−30	−28	0	−9

Notes: 1 Room temperatures and humidity levels are frequently beyond the thermal comfort range.
2 Indoor air humidity might exceed 60 per cent relative humidity (RH) for about 200h/a (hours per annum).

Source: Behne (1996)

Treando (1979) measured the performance of whole-house fans installed in an air-conditioned house in Houston, Texas, that was used to provide night ventilation. They found that on days when the daily average temperature was below 75° F, the whole-house fan was able to satisfy all of the cooling requirements. Savings in air-conditioning consumption from use of the whole-house fan varied between 6.5–10 per cent. Ingley et al (1983) performed a similar experiment in Florida and reported a 22 per cent electricity saving for the house with the whole-house fan while the daily air-conditioning use was reduced by 44 per cent on milder summer days.

Behne (1996) has performed a detailed study to evaluate the potential of night ventilated buildings under various operational conditions. The study has been performed for Berlin, Germany; Locarno, Switzerland; Red Bluff, US; and San Francisco, US. Naturally and mechanically ventilated free-floating and mechanically ventilated air-conditioned buildings have been considered. As shown in Table 8.4, under natural ventilation conditions, peak power gains vary between 31 per cent for San Francisco and 52 per cent for Locarno, while the peak power conservation in air-conditioned buildings varies from 9 per cent for San Francisco to 30 per cent for Berlin.

Slab cooling

In indirect systems, cool night air is circulated through a thermal storage medium where heat is stored and recovered during the day period. In general, the storage medium is a slab covered by a false ceiling or floor, while circulation of the air is always forced. During the following day, the temperature of the circulated air has to be higher than the corresponding temperature of the storage medium. A design analysis of this concept for office buildings is given by Barnaby et al (1980). Slab cooling combined with phase change materials (PCM) storage has been applied in a real building in the UK by Barnard (1994; Barnard et al, 2001).

Fleury (1984) simulated the impact of night cooling through hollow-core concrete floor slabs in a small office building in Los Angeles and New York. No important benefits have been found for New York, while in Los Angeles, when a suitable control strategy was followed, the total energy consumption was reduced by 13 per cent and the total electricity peak demand was reduced by 7 per cent.

Use of phase change materials (PCM)

Phase change materials can be used to store energy during the night and to recover it during the day. Cool air is circulated during the night at the PCM store and is stored under the form of latent heat. During the following day, the high-temperature ambient air is circulated through the PCM, where the latent heat offered to the material cools the air. The efficiency of the system deals primarily with the phase change temperature of the material, the temperature of the ambient air during the night period and the airflow rate. Phase change materials can be paraffin, eutectic salts, etc., and can be embedded in microcapsules, thin heat exchangers, plaster, gypsum board or other wall-covering materials.

During recent years, many research studies and experimental applications have been carried out. Kang Yanbing et al (2003) studied the use of an external PCM store associated with night ventilation techniques. A fatty acid was used as a phase change material. They found that the use of a PCM store decreases the maximum room temperature of the next day by almost 2° C, compared to a commonly night-ventilated building.

Turpenny et al (2000a, c) have proposed and tested a PCM storage with embedded heat pipes, coupled with night ventilation. A high heat-transfer rate was measured and it was concluded that the system can ameliorate the performance of night cooling techniques.

The use of PCM wallboard coupled with mechanical night ventilation in office buildings has been studied by Stetiu and Feustel (1996). They concluded that PCM storage associated with night ventilation techniques offers the opportunity for system downsizing in climates where the outside air temperature drops below 18° C at night. Calculations for a prototype IEA (International Energy Agency) building located in California show that PCM wallboard could reduce the peak cooling load by 28 per cent.

Modelling night ventilation

Proper design of buildings using night-ventilation cooling techniques needs to consider all of the parameters that define the energy and environmental performance of buildings. The use of detailed simulation programmes using well-validated algorithms is the more appropriate method to achieve the best possible efficiency and global performance.

When detailed simulation codes are not used, more simplified assessment methods may be employed. During recent years, several codes have been prepared to calculate the specific performance of night ventilation techniques. These tools are designed to help architects and engineers to consider in a more simplified but accurate way the sizing of night cooling techniques. In what follows, information on some of the data is provided.

NiteCool (Tindale et al, 1995) was developed especially for assessing a range of night-cooling ventilation strategies and was designed under the Energy Related Environmental Issues in Buildings (EnREI) Department of the Environment (DOE) programme in the US. The tool is based on single zone ventilation. LESOCOOL is another simple computer tool to evaluate the potential of night ventilative cooling (Roulet et al, 1996). LESOCOOL calculates the cooling potential and the overheating risk in a naturally or mechanically ventilated building, showing the temperature evolution, the airflow rate and the ventilation heat transfer. It can also take into account convective or radiative heat gains.

Santamouris et al (1996) have proposed a detailed methodology to calculate the performance of air-conditioned as well as free-floating night-ventilated buildings. The method is based on the principle of modified cooling degree days and is extensively compared against theoretical and experimental data. The method is integrated within the simulation tool SUMMER (Santamouris et al, 1995) and calculates the variation of the balance point temperature of a free-floating or air-conditioned night-ventilated building, as well as the overheating hours and the cooling load. In parallel, it performs comparisons with a conventional free-floating or air-conditioned building.

Givoni (1992, 1998a), has proposed a formula to predict the expected indoor maximum temperature with different amounts of mass and insulation in a night venti-lated building. Shaviv et al (2001) has proposed a method to estimate the decrease of the maximum temperature from the diurnal temperature swing as a function of the amount of thermal mass and the night ventilation rate.

(Millet, 1997) has proposed a simplified resistance capacitance model that takes into account the thermal inertia of the building and the impact of night ventilation. Attention is paid to the impact of the outdoor noise (related to the windows opening at night). This model was validated by comparing its results to a more detailed one (TRNSYS) and was used to produce guidance rules.

Finally, Stein and Reynolds (1992) have proposed calculation methods and rules of thumb to estimate the amount of heat that can be removed from the building for given boundary conditions in a night ventilated building.

Constraints and limitations of ventilative cooling techniques

Ventilative cooling is a very powerful technique, but it presents important limitations. Main problems are associated with noise and pollution, reduction of the wind speed in the urban environment and moisture control. In fact, moisture and condensation control is necessary, particularly in humid areas. Pollution and acoustic problems, as well as problems of privacy, are associated with the use of natural ventilation techniques.

Outdoor pollution presents a serious limitation for naturally ventilated buildings, especially in urban areas. Stanners and Bourdeau (1995) have estimated that in 70 to 80 per cent of European cities with more than 500,000 inhabitants, the levels of air pollution, regarding one or more pollutants, exceeds the World Health Organization (WHO) standards at least once per year. Air cleaning through filtration can be applied when mechanical ventilation or flow-controlled natural-ventilation components are used. Noise can be a serious limitation for naturally ventilated buildings. Stanners and Bourdeau (1996) reported that unacceptable noise levels of more than 65dBA affect between 10–20 per cent of urban inhabitants in most European cities. In parallel, as estimated by the Organisation for Economic Co-operation and Development (OECD) (OECD, 1991), almost 130 million people in OECD countries are exposed to noise levels that are unacceptable.

Recent research has shown that the most significant limitation of natural ventilation techniques is due to the specific climatic conditions of cities. Because of specific urban characteristics, there is a serious increase in ambient temperature because of the heat island effect, as well as a serious decrease in wind speed in urban canyons. Both reasons seriously decrease the cooling potential of natural and night-ventilative cooling techniques.

Geros et al (2001) have carried out specific experiments in ten urban canyons in Athens to study the reduction of the airflow in single-sided and naturally

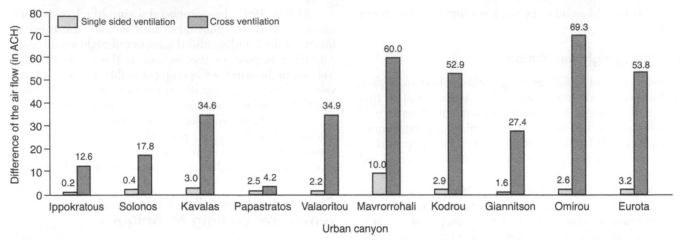

Source: Geros et al (2001)

Figure 8.6 *Reduction of air change rate for single-sided and cross-ventilated buildings in ten urban canyons*

cross-ventilated buildings. They found that because of the reduced wind speed, the airflow through the buildings can decrease by up to 90 per cent (see Figure 8.6). Thus, efficient integration of natural and night ventilation techniques in dense urban areas requires full knowledge of wind characteristics, as well as adaptation of ventilation components to local conditions.

Geros et al (1999) have compared the cooling load of a night ventilated building when located in ten specific urban canyons against the load of the same building located in a non-obstructed site. They reported that because of the reduced wind speed in the canyons, the

cooling load of urban buildings increases by between 6–89 per cent for the single-sided ventilation, and by between 18–72 per cent for the cross-ventilated building depending upon the characteristics of the canyon (see Figure 8.7)

In parallel, a similar comparison has been performed for a free-floating night-ventilated building (Geros et al, 1999). It has been calculated that the maximum indoor temperatures of the urban buildings increase between 0.0–2.6° C for single-sided buildings and between 0.2–3.5° C for cross-ventilated buildings, depending upon the characteristics of the canyon. Figure 8.8 illustrates the specific differences for the ten urban canyons. Thus, a

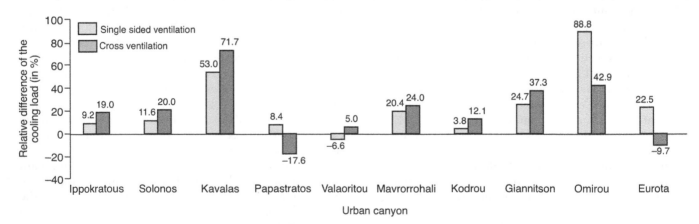

Note: The analysis refers to ten urban canyons where experiments have been performed and results are given for the single-sided and cross-ventilated buildings.

Source: Geros et al (1999)

Figure 8.7 *The difference of the cooling load calculated for a night ventilated building located in a canyon and in a non-obstructed site*

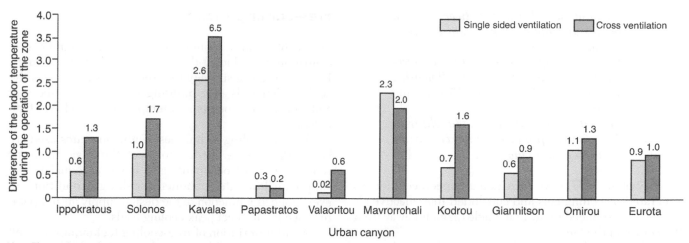

Note: The analysis refers to ten urban canyons where experiments have been performed and results are given for the single-sided and cross-ventilated buildings.

Source: Geros et al (1999)

Figure 8.8 *The difference between the maximum indoor air temperature calculated for a night ventilated building located in a canyon and in a non-obstructed site*

correct sizing and design of night-cooled buildings using natural ventilation techniques has to be based on data appropriate for urban locations.

Use of fans and mechanical ventilation systems to provide thermal comfort

Use of fans for comfort

Box fans, oscillating or ceiling fans can increase the interior air speed and improve comfort (Chand, 1973b; Chandra, 1985). Higher air speeds permit the building to be operated at a higher set-point temperature and thus to reduce its cooling needs. As reported by Chandra et al (1986), for every degree Fahrenheit increase of the thermostat during the summer, the cooling load is decreased by 7–10 per cent. Air-circulation fans allow the thermostat to increase by 4° F; thus, fans can contribute up to 40 per cent of the cooling needs of buildings under the assumption that the occupants are always close to the fan. James et al (1996) have shown that the additional use of ceiling fans in air-conditioned buildings contributes to substantial savings of energy if the air-conditioning set-point is lowered.

Ceiling fans have dominated the US market. According to a study by Ecos Consulting and the Natural Resources Defense Council of the USA (2001), two out of every three homes in the US have at least one ceiling fan, and, on average, each fan consumes about 130kWh per year. In total, there are almost 193 million ceiling fans in the US. Ceiling fans can save energy when users raise air-conditioning thermostats. Rohles et al (1983) and Scheatzle et al (1989), have shown that ceiling fans can extend the comfort zone outside the typical ASHRAE comfort zone. In particular, at an air velocity of 1.02m/sec, comfort may be achieved at 27.7° C for 73 per cent relative humidity, 29.6° C for 50 per cent humidity and 31° C for 39 per cent relative humidity. Fairey et al (1986) have shown that the use of ceiling or oscillating fans may significantly contribute to reducing the cooling load of buildings in the southern US if the thermostat settings are raised accordingly. As reported, energy savings of about 30 per cent are calculated for typical frame buildings in Orlando and Atlanta by increasing the thermostat setting from 25.56° C to 27.78° C. The energy savings may increase by up 50 per cent for heavy-mass buildings.

In the Florida climate, savings are roughly 14 per cent for a 2° F increase, according to the Florida Solar Energy Center. Although studies suggest a 2–6° F increase in the thermostat set-point, James et al (1996) report that in 386 surveyed Florida households, they have not identified any statistically valid differences in thermostat settings between houses using fans and those without them, although fans were used an average of 13.4 hours per day.

Chand (1973b), in his pioneering work, has studied the air motion produced by a ceiling fan and has concluded that:

- *the minimum clearance between the fan blades and the ceiling should be about 30cm;*
- *the capacity of a fan to meet the requirement of a room with a longer dimension L metres should be about 55 Lm³/min; while*
- *the reduction of the ceiling height from 2.9m to 2.6m produces an increase in the air movement in the zone.*

Aynsley et al (1977) have provided air speed contours as generated by ceiling fans and concluded that effective air speeds are produced up to 1 fan blade diameter away from the centre of the fan.

Schmidt and Patterson (2001) have designed a new high-efficiency ceiling fan that can decrease the power consumption and, therefore, electricity charges by a factor of between two and three. A new very efficient ceiling fan of improved aerodynamics blades has been designed and tested by Parker et al (1999). The new ceiling fan presents a much higher airflow performance than existing fans and uses advanced control technology.

Finally, Wu (1989) has demonstrated the potential of oscillating fans to extend the comfort zone. In particular, for an air speed of 1.52m/sec, comfort is achieved at 31° C at 50 per cent relative humidity (RH), at 32° C at 39 per cent RH, or at 33° C at 30 per cent RH.

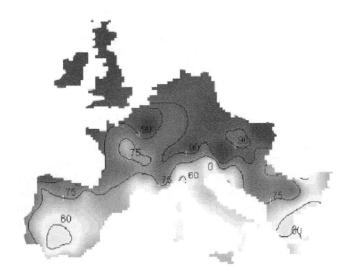

Source: Maldonado (1999)

Figure 8.9 *Percentage of hours in which the free cooling is applicable for the south zone of a reference building*

Free-cooling techniques

Free-cooling and economizer cycles can significantly contribute to reducing the cooling demand of buildings. Free cooling is a strategy that reduces or minimizes the cooling demands of a building by using an excess of ambient air when outdoor air temperatures are lower than indoors.

In free-cooling techniques, ventilation rates used are larger than those needed to meet the basic fresh air requirements of occupants and lower or equal to the supply airflow rates obtained for design conditions in every zone as a function of the design supply air temperature needed to meet peak cooling loads.

A full description of free-cooling techniques, as well as an assessment of the energy potential of free-cooling techniques for Europe, is given by Maldonado (1999) (see Figure 8.9). The design and details of free cooling are also described in ASHRAE (1980) and Perkins (1984).

Recently, Olsen et al (2003), have studied the performance of various low-energy cooling techniques coupled with a VAV and a displacement ventilation system for an office building in the UK. They found that the annual energy cost for displacement ventilation and VAV systems that use free cooling is about 20 per cent less than for the existing building, which uses a fixed minimum supply air rate that does not take advantage of free cooling.

Hybrid systems

Hybrid ventilation can be described as 'systems that provide a comfortable internal environment using both natural ventilation and mechanical systems but using different features of these systems at different times of the day or season of the year and within individual days' (Heiselberg, 2002). Compared to conventional ventilation systems, hybrid ventilation systems vary because they are based on an intelligent control system that permits switching between natural or mechanical modes in order to minimize energy consumption.

The main advantages of the hybrid ventilation systems are summarized by Heiselberg (2002):

- Hybrid ventilation results in higher user satisfaction as it permits a higher degree of individual control of the indoor climate.
- These systems optimize the balance between indoor air quality, thermal comfort, energy use and environmental impact and thus fulfil the needs for a better indoor environment and reduced energy consumption.

- Hybrid ventilation systems have access to natural and mechanical ventilation modes and exploit the benefits of each mode in the best way.
- These systems are also very appropriate solutions for complex buildings since they are associated with more intelligent systems and control.

Various hybrid ventilation systems have been proposed and applied in different types of buildings. An extended review of the systems, as well as of their existing applications, is provided by Delsante and Vik (2001). Olsen et al (2003) have also found that in the UK, while natural ventilation alone cannot maintain appropriate summer comfort conditions, the use of a hybrid system employing natural ventilation, together with a VAV system to maintain comfort during extreme periods, is the best choice, using at least 20 per cent less energy than any purely mechanical system.

Results from the first-generation hybrid-ventilated buildings show that such a technique has a very high cooling potential. It has been found that it is quite effective in providing good indoor air quality (IAQ) and thermal comfort, while energy performance is good, though requiring further improvements.

Summary

The energy consumption of buildings is high and is expected to increase further because of improved standards of living and the increase in world population During the last 20 years, air conditioning has presented a very high penetration rate and this has contributed significantly to increasing absolute energy consumption, as well as the peak electricity load of the building sector.

Convective cooling techniques have proven to be extremely energy efficient under specific climatic conditions. Extensive experimental and theoretical studies have shown that the application of convective cooling techniques may substantially reduce or neutralize the cooling load of buildings.

Over the last 20 years, important basic and industrial research has been carried out that has resulted in the development of new high-efficiency strategies, systems, control devices and tools. However, the continuing increase of energy consumption, primarily because of the important increase in air-conditioning installations, demands a more profound examination of convective cooling techniques and their impact upon buildings.

References

Aboulnaga, M. M. (1998) 'A roof solar chimney assisted by cooling cavity for natural ventilation in buildings in hot arid climates: An energy conservation approach', *Renewable Energy*, vol 14, no1–4, pp357–363

Aboulnaga, M. M. and Abdrabboh, S. N. (1998) 'Improving night ventilation into low-rise buildings in hot-arid climates exploring a combined wall-roof solar chimney', in *Proceedings of Renewable Energy: Energy Efficiency, Policy and the Environment*, World Renewable Energy Congress V, 20–25 September, Florence, Italy Sayigh, A. A. M. (ed), Pergamon, Elsevier Science Ltd, vol 3, pp1469–1472

Aboulnaga, M. M. and Abdrabboh, S. N. (2000) 'Improving night ventilation into low-rise buildings in hot-arid climates exploring a combined wall-roof solar chimney', *Renewable Energy*, vol 19, pp47–54

Adnot, J. (ed) (1999) *Energy Efficiency of Room Air-Conditioners*, EERAC, Study for the Directorate General for Energy (DGXVII) of the Commission of the European Communities, May, Brussels

AFEAS (2001) *Issue Areas: Production and Sales of Fluorocarbons*, AFEAS, Paris

Afonso, C. and Oliveira, A. (2000) 'Solar chimneys: Simulation and experiment', *Energy and Buildings*, vol 32, pp71–79

Agas, G., Matsaggos, T., Santamouris, M. and Argyriou, A. (1991) 'On the use of the atmospheric heat sinks for heat dissipation', *Energy and Buildings*, no 17, pp321–329

Allard, F., Dorer, V. B. and Feustel, H. E. (1990) *COMIS: Fundamentals of the Multizone Air Flow Model*, Comis Technical Note 29, AIVC, Coventry, UK

Allard F. (ed) (1998) *Natural Ventilation of Buildings*, James & James, London, UK

Arens, E. A., Blyholder, A. G. and Schiller, G. E. (1984) 'Predicting thermal comfort of people in naturally ventilated buildings', *ASHRAE Transactions*, vol 90, 1B, pp272–279

Arens, E. A. and Watanabe, N. S. (1986) 'A method for designing naturally ventilated buildings using bin climate data', *ASHRAE Transactions*, vol 90, 2B, pp773–792

ASHRAE (American Society of Heating, Refrigerating and Air-Conditioning Engineers) (1980) *Systems Handbook*, ASHRAE, Atlanta

ASHRAE (1992) *ASHRAE Standard 55: Thermal Environmental Conditions for Human Occupancy*, ASHRAE, Atlanta

Auliciems, A. (1989) 'Air conditioning in Australia III: Thermobile controls', *Architectural Science Review*, vol 33, pp43–48

Auliciems, A. and De Dear, R. (1986) 'Air conditioning in Australia I: Human thermal factors', *Architectural Science Review*, vol 29, pp67–75

Axley, J. W. and Emmerich, S. J. (2002) 'A method to assess the suitability of a climate for natural ventilation of commercial buildings', in *Proceedings of Indoor Air 2002 (9th International Conference on Indoor Air Quality and Climate)*, Monterey, California, vol 2, 30 June–5 July, pp854–859

Aynsley, R. (1999) 'Estimating summer wind driven natural ventilation potential for indoor thermal comfort', *Journal of Wind Engineering and Industrial Aerodynamics*, vol 83, pp515–525

Aynsley, R. M., Melbourne, W. and Vickery, B. J. (1977) *Architectural Aerodynamics*, Applied Science Publishers, London

Badran, A. A. (2003) 'Performance of cool towers under various climates in Jordan', *Energy and Buildings*, vol 35, pp1031–1035

Baer, S. (1983) 'Raising the open U value by passive means', in *Proceedings of the Eighth National Passive Solar Conference*, ASES, Glorieta, New Mexico, Boulder, Colorado, pp839–842

Baer, S. (1984) *Cooling with Night Air*, Alburquerque, New Mexico

Bagneid, A. A. (1987) *Courtyard Bioclimates: Comparative Experiments*, MSc thesis, University of Arizona, Arizona

Bahadori, M. N. (1978) 'Passive cooling in Iranian architecture', *Scientific American*, p238

Bahadori, M. N. (1981) 'Pressure coefficients to evaluate air flow pattern in wind towers', in *Proceedings of the International Passive and Hybrid Cooling Conference*, American Section of ASES, Miami Beach, Florida, pp206–210

Bahadori, M. N. (1985) 'An improved design of wind towers for natural ventilation and passive cooling', *Solar Energy*, vol 35, no 2, pp119–129

Baker, N. (1993) 'Thermal comfort evaluation for passive buildings – a PASCOOL task', in *Proceedings of the Conference on Solar Energy in Architecture and Planning*, H. S. Stephens and Associate, Florence

Baker, N. and Standeven, M. (1994) 'Thermal comfort in free running buildings', in *Proceedings of the Conference PLEA 94, Architecture in the Extremes*, Israel

Baker, N. and Standeven, M. (1995) *PASCOOL Comfort Group*, Final Report, February, EEC, Brussels

Bansal, N. K., Mathur, R. and Bhandari, M. S. (1993) 'Solar chimney for enhanced stack ventilation', *Building and Environment*, vol 28, no 3, pp373–377

Bansal, N. K., Mathur, R. and Bhandari, M. S. (1994) 'Study of solar chimney assisted wind tower system for natural ventilation in buildings', *Building and Environment*, vol 29, no 4, pp495–500

Barnaby, C. S., Hall, D. H. and Dean, E. (1980) 'Structural mass cooling in a commercial building using hollow core concrete plank', in *Proceedings of Passive Solar Conference*, Amherst, MA, AS/ISES, pp747–751

Barnard, N. (1994) 'Fabric energy storage of night cooling', in *Proceedings of CIBSE National Conference 1994*, Brighton Conference Centre, vol 1, pp88–100

Barnard, N., Concannon, P. and Jaunzens, D. (2001) *Modelling the Performance of Thermal Mass*, BRE Information Paper IP 6/01, Construction Research Communications Ltd, Garston

Barozzi, G. S., Imbabi, M., Nobilel, E. et al (1992) 'Physical and numerical modeling of a solar chimney-based ventilation system for buildings', *Building and Environment*, vol 27, no 4, pp433–445

Becker R. and Paciuk, M. (2002) 'Inter-related effects of cooling strategies and building features on energy performance of office buildings', *Energy and Buildings*, vol 34, pp25–31

Becker, S., Potcher, O. and Yaakov, Y. (2003) 'Calculated and observed human thermal sensation in an extremely hot and arid climate', *Energy and Buildings*, vol 35, pp747–756

Behne, M. (1996) *Alternatives to Compressing Cooling in Non Residential Buildings to Reduce Primary Energy Consumption*, Deutsche Forschungsgemeinschaft (DFG) and the US Department of Energy, May

Belayat, E. K, Nicol, J. F. and Wilson, M. (2002) 'Thermal comfort in Algeria: Preliminary results of field studies', in *Proceedings of EPIC 2002 Conference*, Lyon, France

Berger, X. and Semega, Y. (1995) 'Cold air chimneys in the old city of Nice', in Santamouris, M. (ed) *Proceedings of the International Workshop on Passive Cooling*, Athens

Besant-Jones, J. and Tenenbaum, B. (2001) *The California Power Crisis: Lessons for Developing Countries*, Energy and Mining Sector, Board Discussion Paper Series, Paper No 1, April, World Bank, Washington, DC

Birtles, A. B., Kolokotroni, M. and Perera, M. (1996) *Night Cooling and Ventilation Design for Office Type Buildings*, WREN, London

Blake, T. G. (2001) 'Thermal mass storage and energy saving using underfloor air conditioning', *Proceedings of the Seventh REHVA World Congress and Clima 2000 Naples 2001 Conference*, Naples, Italy, 15–18 September

Blondeau, P., Sperandio, M. and Allard, F. (1995a) 'Night accelerated ventilation and air conditioning for buildings cooling in summer', in Santamouris, M. (ed) *Proceedings of International Workshop on Passive Cooling*, Athens

Blondeau, P., Sperandio, M. and Allard, F. (1995b) 'Night ventilation for building cooling in summer', *Solar Energy*, vol 61, pp327–335

Blondeau, P., Sperandio, M. and Allard, F. (2002) 'Multicriteria analysis of ventilation in summer period', *Energy and the Environment*, vol 37, pp165–176

Bouchair, A. (1987) 'Passive solar induced ventilation', in *Proceedings of the Miami International Conference on Alternative Energy Sources*, Miami Beach, Florida, vol 1, part VIII

Bouchair, A. (1989) *Solar Induced Ventilation in the Algerian and Similar Climates*, PhD thesis, Leeds University, UK

Bouchair, A. (1994) 'Solar chimney for promoting cooling ventilation in southern Algeria', *Building Services Engineering Research and Technology*, vol 15, no 2, pp81–93

Bouden, C. and Ghrab, N. (2001) 'Thermal comfort in Tunisia: Results of a one year survey', in *Proceedings of the Conference on Moving Thermal Comfort Standards into the 21st Century*, Windsor, UK, pp197–206

Brager, R. and De Dear. J. (1998) 'Thermal adaptation in the built environment: A literature review', *Energy and Buildings*, vol 27, no 1, pp83–96

Brager, R. and De Dear. J. (2000) 'A standard for natural ventilation', *ASHRAE Journal*, vol 42, no 10, pp21–28

BREEZE (1993) *BREEZE 6.0: User Manual*, Building Research Establishment, Garston, UK

BSI (British Standards Institution) (1980) *BS 5925: Code of Practice for Design of Buildings: Ventilation Principles and Designing for Natural Ventilation*, BSI, London

Burch, D. M. and Treando, S. J. (1979) 'Ventilating residences and their attics for energy conservation', *Summer Attic and Whole House Ventilation*, NBS Special Publication 548, National Bureau of Standards, Washington, DC

Burton, S. (1998) 'Energy comfort 2000 – the application of low energy technologies to seven new non-domestic buildings', in *Proceedings of the EPIC 1998*, vol 2, pp473–478

Busch, J. (1992) 'A tale of two populations: Thermal comfort in air conditioned and naturally ventilated offices in Thailand', *Energy and Buildings*, vol 18, pp235–249

Cadima, P. (2000) 'The effect of design parameters on environmental performance of the urban patio: A case study in Lisbon', in *Proceedings of PLEA 1998*, Lisbon, Portugal, June

Cardinale N., Minucci, M. and Ruggiero, F. (2003) 'Analysis of energy saving using natural ventilation in a traditional Italian buildings', *Energy and Buildings*, vol 35, pp153–159

Carrol, W. C. et al (1982) *Passive Cooling Technology Assessments*, Synthesis Report, LBL-15184, Berkeley, California

Carrilho da Graca G., Chen, Q., Glicksman. L. R. and Norfold, L. K. (2002) 'Simulation of wind driven ventilative cooling systems for an apartment building in Beijing and Shanghai', *Energy and Buildings*, vol 34, pp1–11

CEC (California Energy Commission) (2000a) *Alternatives to Compressor Cooling*, P600-00-003, January, CEC, California

CEC (2000b) *Keeping Your Cool with Ceiling and Whole House Fans*, CEC, California

Chalfoun, N. (1992) *CoolT, V. 1.4*, Copyright Cool Tower Performance Program, Environmental Research Laboratory, University of Arizona, Tuscon, Arizona

Chand, I. (1973a) 'Effect of veranda on room air motion', *Civil Engineering Construction and Public Works Journal*, November–December

Chand, I. (1973b) 'Studies of air motion produced by ceiling fans', *Research and Industry*, vol 18, no 3, pp50–53

Chand, I. and Bhargava, P. K. (1975) 'A quantitative study of the air deflecting characteristics of horizontal sashes', *The Indian Architect*, May

Chand, I., Bhargava. P. K. and Krishak, N. L. V. (1975) 'Study of the influence of a pelmet type wind deflector on indoor air motion', *Building Science*, vol 10, pp231–235

Chand, I., Bhargava, P. K. and Krishak, N. L. V. (1998) 'Effect of balconies on ventilation inducing aeromotive force on low-rise buildings', *Building and Environment*, vol 33, no 6, pp385–396

Chandra, S. (1985) *Fans to Reduce Cooling Cost in the South East*, Florida Solar Energy Center, Florida

Chandra, S. (1989) 'Ventilative cooling', in Cook, J. (ed) *Passive Cooling*, MIT Press, Cambridge, MA

Chandra, S., Fairey, F. W. and Bowen, A. (1982a) *Passive Cooling by Ventilation: A Literature Review*, FSEC-CR-37-82, January, Florida Solar Energy Center, Florida

Chandra, S., Fairey, P., Houston, M. and Kerestecioglu, A. A. (1983) 'Wing walls to improve natural ventilation: Full scale results and design strategies', in *Proceedings of the Eighth National Passive Solar Conference*, ASES, Boulder, CO

Chandra, S, Fairey, P. W. and Houston, M. M. (1986) *Cooling with Ventilation*, SERI/SP-273-2966, DE86010701, Solar Energy Research Institute, Colorado

Chandra, S. and Kerestecioglu, A. A. (1984) 'Heat transfer in naturally ventilated rooms, data from full-scale measurements', *ASHRAE Transactions*, vol 90, no 1b, pp221–225

Chandra, S., Kerestecioglu, A. A. and Cromer. W. (1982b) 'Performance of naturally ventilated homes' in *Proceedings of the Seventh National Passive Conference, Passive 82*, Knoxville, Tenn, US

Chen, Z. D., Bandopadhayay, P., Halldorsson, J., Byrjalsen, C., Heiselberg, P. and Li, Y. (2003) 'An experimental investigation of a solar chimney model with uniform wall heat flux', *Building and Environment*, vol 38, pp893–906

Chiang, C. M., Lai, P. C., Chou, Y. and Li, Y. (2000) 'The influence of an architectural design alternative (transom) on indoor air environment in conventional kitchen in Taiwan', *Building and Environment*, vol 35, no 7, pp579–585

CIBSE (Chartered Institution of Building Services Engineers) (1997) *Natural Ventilation in Non-Domestic Buildings: Applications Manual AM10*, CIBSE, London

CICA (Confederation of International Contractors Association) (2002) 'Industry as a partner for sustainable development', *Construction*

Cristofalo, S. D., Orioli, S., Silvestrini, G. and Alessandro, S. (1989) 'Thermal behavior of "Scirocco rooms" in ancient Sicilian villas', *Tunnelling and Underground Space Technology*, vol 4, no 4, pp471–473

CSTB (Centre Scientifique et Technique du Batiment) (1992) *Guide sur la climatisation naturelle de l'habitat en climat tropical humide – Methodologie de prise en compte des parametres climatiques dans l' habitat et conseils pratiques*, vol 1, CSTB, Paris, France (in French)

Cunningham, W. and Thompson, T. (1986) 'Passive cooling with natural draft cooling towers in combination with solar chimneys', *Proceedings of the PLEA 86, Passive and Low Energy Architecture*, Pecs, Hungary

Day, Y., Sumathy, K., Wang, R. Z. and Li, Y. G. (2003) 'Enhancement of natural ventilation in a solar house with a solar chimney and a solid adsorption cooling cavity', *Solar Energy*, vol 74, pp65–75

Dascalaki, E. and Santamouris, M. (1998a) 'The Meletitiki Case Study Building', in Allard, F. (ed) *Handbook on Natural Ventilation of Buildings*, James & James, London

Dascalaki, E. and Santamouris, M. (1998b) *Manual of AIOLOS*, Department of Applied Physics, University of Athens, Greece

De Dear, R. J. (1991) 'Diurnal and seasonal variation in the human thermal climate of Singapore', *Singapore Journal of Tropical Geography*, vol 10, pp13–25

De Dear, R. J. (1998) *ASHRAE Transactions*, vol 104, no 1b, pp1141–1152

De Dear, R. J. and Brager, G. S. (1998) *ASHRAE Transactions*, vol 104, no 1a, pp145–167

De Dear, R. J. and Brager, G. S. (2002) 'Thermal comfort in naturally ventilated buildings: Revisions to ASHRAE Standard 55', *Energy and Buildings*, vol 34, pp549–561

Delsante, A. and Vik. T. A. (2001) *Hybrid Ventilation – State of the Art Review, IEA Energy Conservation in Buildings and Community Systems Programme*, Annex 35, Hybrid Ventilation and New and Retrofitted Office Buildings

Demeester, J., Wouters, P. and Ducarme, D. (1998) 'Natural ventilation in office-type buildings – results and conclusions of monitoring activities', in *Proceedings of Ventilation Technologies in Urban Areas, 19th Annual Conference*, Oslo, Norway, 28–30 September, pp407-413

Der Petrossian, B. (2001) 'Conflicts between the construction industry and the environment', *Habitat Debate*, vol 5, no 2, p5

Dounis, A. I., Bruant, M. and Santamouris, M. (1995a) 'A fuzzy logic system as airflow estimator', in Santamouris, M. (ed) *Buildings: Proceedings of the International Workshop on Passive Cooling*, Athens, Greece

Dounis, A. I., Bruant, M., Guarraccino, G., Michel, P. and Santamouris, M. (1996a) 'Indoor air quality control by a fuzzy reasoning machine in naturally ventilated buildings', *Applied Energy*, vol 54, no 1, pp11–28

Dounis, A. I., Bruant, M., Santamouris, M., Guarraccino, G. and Michel, P. (1996b) 'Comparison of conventional and fuzzy control of indoor air quality in buildings', *Journal of Intelligent and Fuzzy Systems*, vol 4, pp131–140

Dounis, A. I., Santamouris, M., Lefas, C. C. and Argiriou, A. (1995a) 'Design of a fuzzy set environment comfort system', *Energy and Buildings*, vol 21, pp81–87

Ecos Consulting and the Natural Resources Defense Council (2001) 'New ceiling fan takes flight', *Environmental Building News*, vol 10, no 3, March

Eftekhari, M. and Marjanovic, L. D. (2003) 'Application of fuzzy control in naturally ventilated buildings for summer conditions', *Energy and Buildings*, vol 35, pp645–655

EIA (Energy Information Administration) (1978–1982, 1984, 1987, 1990, 1993, 1997) *Residential Energy Consumption Surveys*, EIA, National Oceanic and Atmospheric Administration, Air-Conditioning and Refrigeration Institute, Paris

Emmerich, S. J., Dols, W. S. and Axley, J. W. (2001) *Natural Ventilation Review and Plan for Design and Analysis Tool*, NISTIR 6781, Prepared for Architectural Energy Corporation Boulder, Colorado, August

Ernest, D. R. (1991) *Predicting Wind – Induced Air Motion, Occupant Comfort and Cooling Loads in Naturally Ventilated Buildings*, PhD thesis, University of California at Berkeley, Berkeley, California

Etheridge, D. W. (2002) 'Non dimensional methods for natural ventilation design', *Building and Environment*, vol 37, pp1057–1072

Etzion, Y. (1988) 'The thermal behaviour of closed courtyards: Planning, physics and climate technology for healthier buildings', in *Proceedings of the Healthy Buildings Conference*, Stockholm, Sweden

Eureca Laboratories (1982) *Experimental Investigation of Thermal Induced Ventilation in Atria*, Eureca Laboratories, Sacramento, California

Fairey, P. W. and Bettencourt, W. (1981) 'La Sucka – a wind driven ventilation augmentation and control device', in *Proceedings of the International Passive and Hybrid Cooling Conference*, AS/ISES, Boulder, CO, pp196–200

Fairey, P., Chandra, S. and Kerestecioglou, A. (1986) *Ventilative Cooling in Southern Residences: A Parametric Analysis*, FSEC-PF-108-86, Florida Solar Energy Center, Miami

Fanger, P. O. (1972) *Thermal Comfort. Analysis and Applications in Environmental Engineering*, McGraw Hill, New York

Fathy, H. (1986) *Natural Energy and Vernacular Architecture*, The University of Chicago Press, Chicago

Feriadi, H. (2002a) 'Thermal comfort for naturally ventilated houses in Indonesia', in *International Symposium Building Research and Sustainability of the Built Environment in the Tropics*, Jakarta, 14–16 October

Feriadi, H. (2002b) 'Natural ventilation via courtyard for tropical buildings', in *International Symposium Building Research and Sustainability of the Built Environment in the Tropics*, Jakarta, 14–16 October

Feustel, H. E. and Stetiu, C. (1997) *Thermal Performance of Phase Change Wallboard for Residential Cooling Application*, LBL-Report 38320, LBL, Berkeley, California

Fleury, B. (1984) *Ventilation Cooling through Hollowcore Concrete Floor Slabs in a Small Office Building*, MSc thesis, Mechanical Engineering, University of California, Berkeley, California

Fleury, B. (1990) 'Ventilative cooling: State of the art', in *Proceedings of the Workshop on Passive Cooling*, Ispra, Italy, 2–4 April

Flourentzou, F., Van der Maas, J. and Roulet, C. A. (1998) 'Natural ventilation for passive cooling: Measurement of discharge coefficients', *Energy and Buildings*, vol 27, pp283–292

Fracastoro, G. V., Mutani, G. and Perino, M. (2002) 'Experimental and theoretical analysis of natural ventilation by window openings', *Energy and Buildings*, vol 34, pp817–827

Fuller, B. (1973) *The Dymaxion World of Buckminister Fuller*, Anchor Press, Doubleday, New York

Gagge, A. P. (1973) 'Rational temperature indices of man's thermal environment and their use with a two node model of his temperature regulation', *Federation Proceedings*, vol 32, no 5, pp1572–1582

Gagge, A. P., Fobelets, A. P. and Berglund, L. G. (1986) 'A standard predictive index of human response to the thermal environment', *ASHRAE Transactions*, vol 92, no 2b, pp709–731

Gage, S. A., Hunt, G. R. and Linden, P. F. (2001) 'Top down ventilation and cooling', *Journal of Architecture and Planning Research*, vol 18, no 4, pp286–301

Gan, G. and Riffat, S. B. (1998) 'A numerical study of solar chimney for natural ventilation of buildings with heat recovery', *Applied Thermal Engineering*, vol 18, pp1171–1187

Germano, M., Roulet, C. A., Allard, F. and Ghiaus, C. (2002) 'Potential for natural ventilation in urban context: An assessment method', in *Proceedings of the EPIC 2002*, Lyon, France

Geros, V. (1998) *Ventilation Nocturne: Contribution à la réponse thermique des bâtiments*, PhD thesis, INSA, Lyon, France

Geros, V., Santamouris, M., Tsangrassoulis, A. and Guarracino, G. (1999) 'Experimental evaluation of night ventilation phenomena', *Energy and Buildings*, vol 29, pp141–154

Geros, V., Santamouris, M., Papanikolaou, N. and Guarraccino, G. (2001) 'Night ventilation in urban environments', in *Proceedings of the AIVC Conference*, Bath, UK

Geurra J., Molina, J. L., Rodriguez, E. A. and Velazquez, R. (1992) 'Night ventilation in industrial buildings: A case study', in *Proceedings of the Conference Indoor Air Quality, Ventilation and Energy Conservation, Fifth International Jacques Cartier Conference*, Montreal, Canada, 7–9 October, Center for Building Studies, Concordia University, Montreal, Canada, pp476-483

Ghiaus, C. and Allard, F. (2002) 'Assessment of natural ventilation potential of a region using degree hours estimated on global weather data', in *Proceedings of the Conference on EPIC 2002*, Lyon, France

Givoni, B. (1968) *Ventilation Problems in Hot Countries*, Research report to Ford Foundation, Building Research Station, Israel Institute of Technology, Haifa, Israel

Givoni, B. (1976) *Man, Climate and Architecture*, Second edition, Applied Science Publishers, London

Givoni, B. (1991) 'Modeling a passive evaporative cooling tower', in *Proceedings of the 1991 Solar World Congress*, Denver, Colorado, US

Givoni, B. (1992) 'Comfort, climate analysis and building design guidelines', *Energy and Buildings*, vol 18, pp11–23

Givoni, B. (1994) *Passive and Low Energy Cooling of Buildings*, Van Nostrand Reinhold, Los Angeles, CA

Givoni, B. (1998a) *Climate Considerations in Buildings and Urban Design*, Van Nostrand Reinhold, Los Angeles, CA

Givoni, B. (1998b) 'Effectiveness of mass and night ventilation in lowering the indoor daytime temperatures. Part 1: 1993 experimental periods', *Energy and Buildings*, no 28, pp25-32

Golneshan, A. A. and Yaghoubi, M. A. (1985) 'Natural cooling of a residential room with ventilation in hot arid regions', in *Proceedings of the Conference on CLIMA 2000*, Copenhagen, Denmark

Golneshan, A. A. and Yaghoubi, M. A. (1990) 'Simulation of ventilation strategies of a residential building in hot arid regions in Iran', *Energy and Buildings*, vol 14, pp201–205

Hagelskjaer, S. (2002) 'Natural ventilation : Experience and results in office buildings', in *Proceedings of the EPIC 02*, Lyon, France

Haghighat, F., Li, Y. and Megri, A. C. (2001) 'Development and validation of a zonal model – POMA', *Building and Environment*, vol 36, pp1039–1047

Haisley, R. (1981) 'Solar chimney theory: Basic Precepts', in *Proceedings of the International Passive and Hybrid Cooling Conference*, Miami Beach, Florida

Heiselberg, P. (ed) (2002) *Principles of Hybrid Ventilation*, IEA Energy Conservation in Buildings and Community Systems Programme, Annex 35, Hybrid Ventilation and New and Retrofitted Office Buildings, Paris

Hamdy, I. F. and Fikry, M. A. (1998) 'Passive solar ventilation', *Renewable Energy*, vol 14, no 1–4, pp381–386

Hardiman, P. (1992) *Utersuchung nat ürlicher Lüftungssysteme zur Verbesserung des Raumklimas von konteng ünstigen Wohnhausern auf Java/Indonesien*, PhD thesis, Universitat Stuttgart, Germany

Hassid, S. (1994) 'Evaluation of passive cooling strategies for Israel', in Etzion, Y., Erell, E., Meir, I. A. and Pearlmutter, D. (eds) *Proceedings of 11th PLEA International Conference on Architecture of the Extremes*, Dead Sea, Israel, The Desert Architecture Unit, J. Blaustein Institute for Desert Research, Ben-Gurion University of the Negev, 3–8 July, pp162–169

Hensen, J. L. M. and Centrenova, L. (2001) 'Energy simulations of traditional versus adaptive thermal comfort for two moderate climate regions', in *Proceedings of the Conference Moving Comfort Standards into the 21st Century*, Windsor, UK, pp78–91

Herkel S., Pfafferot, J. and Wambsganb, M. (2002) 'Design, monitoring and evaluation of a low energy office building with passive cooling by night ventilation', in *Proceedings of the EPIC 2002 Conference*, Lyon, France

Hien, W. N. and Sani, H. (2002) 'The study of active stack effect to enhance natural ventilation using wind tunnel experiment and computational fluid dynamics', in *Proceedings of the International Symposium on Building Research and the Sustainability of Built Environment in the Tropics*, Jakarta, Indonesia

Hien, W. N. and Tanamas, J. (2002) 'The effect of wind on thermal comfort in the tropical environment', in *Proceedings of the International Symposium on Building Research and the Sustainability of Built Environment in the Tropics*, Jakarta, Indonesia

Hinrichs, C. L (1988) 'The courtyard housing form as a traditional dwelling in the Mediterranean region', in *Proceedings of the PLEA 88 Conference*, Pergamon Press

Hoffman, M. E., Mossen, A., De Urritia, O. and Wertheim, I. (1994) 'The courtyard as a climate design element: A quantitative approach', in *Proceedings of the Conference on PLEA94*, Israel

Holford, J. M. and Hunt, G. R. (2000) 'When does an atrium enhance natural ventilation?', in *Proceedings of Innovations in Ventilation Technology, 21st AIVC Annual Conference*, The Hague, The Netherlands, 26–29 September, Air Infiltration and Ventilation Centre, UK

Holford, J. M. and Hunt, G. R. (2003) 'Fundamental atrium design for natural ventilation', *Building and Environment*, vol 38, pp409–426

Howarth, A T. (1985) 'The prediction of air temperature variations in naturally ventilated rooms with convective heating', *Building Service Engineering Research and Technology*, vol 64, pp169–175

Humphreys, M. (1975) 'Field studies of thermal comfort compared and applied', *Journal of Installation Heating and Ventilation Engineers*, vol 44, pp5–27

Humphreys, M. (1978) 'Outdoor temperature and comfort indoors', *Building Research and Practice*, vol 6, no 2, pp92–105

Humphreys, M. and Nicol, J. F. (2000) 'Outdoor temperature and indoor thermal comfort – raising the precision of the relationship for the 1998 ASHRAE database of field', *ASHRAE Transactions*, vol 206, no 2, pp485–492

Humphreys, M. and Nicol, J. F. (2002) 'The validity of ISO-PMV for predicting comfort votes in every day thermal environments', *Energy and Buildings*, vol 34, no 6, pp667–684

Humphreys, M., Nicol, J. F. and McCartney, J. K. (2002) 'An analysis of some subjective assessments of indoor air quality in five European Countries', in Levin, H. (ed) *Proceedings of the Ninth International Conference on Indoor Air*, Santa Cruz, US, pp86–91

Hunt, G. R. and Holford, J. M. (1998) 'Top-down ventilation of multi-storey buildings', in *Proceedings of the 19th AIVC Conference on Ventilation in Urban Areas*, Oslo, Norway, pp197–205

Hunt, G. R. and Linden, P. F. (1999) 'The fluid mechanics of natural ventilation – displacement ventilation by buoyancy-driven flows assisted by wind', *Energy and the Environment*, vol 34, pp707–720

Hyde, R. (2000) *Climate Responsive Design*, E. and F. Spon, London

IIR (International Institute of Refrigeration) *Industry as a Partner for Sustainable Development – Refrigeration*, IIR, Paris

Inard, C., Bouia, H. and Dalicieux, P. (1996) 'Prediction of air temperature distribution in buildings with a zonal model', *Energy and Buildings*, vol 24, pp125–132

Inard, C. and Buty, D. (1991) 'Simulation of thermal coupling between a radiator and a room with zonal models', *Proceeding of 12th AIVC Conference*, vol 2, Ottawa, Canada, pp125–131

Ingley, H. A., Dixon, R. W. and Buffington, D. E. (1983) *Residential Conservation Demonstration Program*, Prepared for the Florida Public Service Commission, University of Florida, Gainesville, Florida

ISO (International Organization for Standardization) *ISO International Standard 7730: Moderate Thermal Environments – Determination of the PMV and PPD Indices and Specification of the Conditions of Thermal Comfort*, Second edition, ISO, Geneva

James, P. W., Sonne, J. K., Vieire, R., Parker, D. and Anello, M. (1996) 'Are energy savings due to ceiling fans just hot air?', in *Proceedings of ACEEE Summer Study on Energy Efficiency in Buildings*, US

JARN (Japan Air Conditioning and Refrigeration News) and JRAIA (Japan Refrigeration and Air Conditioning Industry Association) (2002) *Air Conditioning Market*, JARN and JRAI, Japan

Jokl, M. V. (1987) 'A new COMECON standard for thermal comfort within residential and civic buildings', *Indoor Air 87*, Berlin, vol 3, pp457–460

Kammerud R., Anderson, B., Place, W., Ceballos, E. and Curtis, B. (1984) 'Ventilation cooling of residential buildings', *ASHRAE Transactions*, vol 90, no 1b, pp226–252

Kang Yanbing, Jiang Yi and Zhang Yinping (2003) 'Modeling and experimental study on an innovative passive cooling system – NVP system', *Energy and Buildings*, vol 35, pp417–425

Karakatsanis. C., Bahadori, M. and Vickery, B. (1986) 'Evaluation of pressure coefficients and estimation of air flow rates in buildings employing wind towers', *Solar Energy*, vol 37, no 5, pp363–374

Khedari, J., Boonsri, B. and Hirunlabh, J. (2000a) 'Ventilation impact of a solar chimney on indoor temperature fluctuation and air change in a school building', *Energy and Buildings*, vol 32, no 1, pp89–93

Khedari, J., Mansirisub, W., Chaima, S., Pratinthong, N. and Hirunlabh, J. (2000b) 'Field measurements of performance of roof solar collector', *Energy and Buildings*, vol 31, pp171–178

Khedari J., Rachapradit, N. and Hirunlabh, J. (2003) 'Field study of performance of solar chimney with air conditioned building', *Energy*, vol 28, pp1099–1114

Kindangen, G., Krauss, P. and Depecker, P. (1997) 'Effects of roofs shape on wind-induced air motion inside buildings', *Building and Environment*, vol 32, no 1, pp1–11

Kindangen, G., Krauss, G. and Rusaquen, G. (1996) 'Influence of architectural parameters on air flow distribution inside a naturally ventilated single unit house', *Proceedings of the Conference on PLEA 96*, Lounain La Neuve, France

Klitsikas, N., Balaras, C., Argiriou, A. and Santamouris, M. (1995) 'Comfort field studies in the frame of PASCOOL in Athens, Hellas', in Santamouris, M. (ed) *Proceedings of the International Workshop on Passive Cooling*, Athens, Greece

Kolokotroni, M. and Aronis, A. (1999) 'Cooling-energy reduction in air-conditioned offices by using night ventilation', *Applied Energy*, vol 63, pp241–253

Kolokotroni, M., Kukadia, V. and Perera, M. (1996a) 'NATVENT – European project on overcoming technical barriers to low energy natural ventilation', in *Proceedings of the CIBSE/ASHRAE Joint National Conference*, Part 2, CIBSE, London, pp36–41

Kolokotroni, M., Perera, M. D., Azzi, D. and Virk, G. S. (2001) 'An investigation of passive ventilation cooling and control strategies for an educational building', *Applied Thermal Engineering*, vol 21, pp183–199

Kolokotroni, M., Tindale, A. and Irving, S. J. (1997) 'NiteCool: Office Night Ventilation Pre-Design Tool', in *Proceedings of the 18th AIVC Conference: Ventilation and Cooling*, Athens, Greece, 23–26 September

Kolokotroni M., Watkins, R., Santamouris, M., Niachou, K., Allard, F., Belarbit, R., Ghiaus, C., Alvarez, S. and Salmeron, J. M. (2002) 'Lissen: Passive ventilation cooling in urban buildings: An estimation of potential environmental impact benefits', in *Proceedings of the EPIC 2002 Conference*, Lyon, France

Kolokotroni, M., Webb, B. C. and Hayes, S. D. (1996b) 'Summer cooling for office-type buildings by night ventilation', in *Proceedings of the 17th AIVC Conference on Optimum Ventilation and Air Flow Control in Buildings*, Gothenburg, Sweden, 17–20 September, vol 2, pp591–599

Kolokotroni, M., Webb, B. C. and Hayes, S. D. (1998) 'Summer cooling with night ventilation for office buildings in moderate climates', *Energy and Buildings*, vol 27, pp231–237

Kolokotsa, D. (2001) *Design and Implementation of an Integrated intelligent Building Indoor Environment Management System Using Fuzzy Logic, Advanced Decision Support Techniques, Local Operating Network Capabilities and Smart Card Technology*, PhD thesis, Technical University of Crete, Chania, Greece

Kolokotsa, D. (2003) 'Comparison of the performance of fuzzy controllers for the management of the indoor environment', *Building and Environment*, vol 38, pp1439–1450

Kukreja, C. P. (1978) *Tropical Architecture*, McGraw-Hill, New Delhi

Kumar, S., Sinha, S. and Kumar, N. (1998) 'Experimental investigation of solar chimney assisted bioclimatic architecture', *Energy Conversion and Management*, vol 39, pp5–6

Kusuda, T. (1981) *Savings in Electric Cooling Energy by the Use of Whole House Fan*, NBS TN 1138, National Bureau of Standards, Gaithersburg, MD

La Roche, P. and Milne, M. (2001) 'Smart controls for whole house fans', in *Proceedings of Cooling Frontiers: The Advanced Edge of Cooling Research and Applications in the Built Environment*, College of Architecture and Environmental Design, Arizona State University, Tempe, Arizona

La Roche, P. and Milne, M. (2002) 'Effects of shading and amount of mass in the performance of an intelligent controller for ventilation', in *Proceedings of the ACES 2002 Solar Conference*, Reno, Nevada

La Roche, P. and Milne, M. (2003) 'Effects of window size and thermal mass on building comfort using an intelligent ventilation controller', *Solar Energy*

Leaman, A. J. and Bordass, W. T. (1995) 'Comfort and complexity: Unmanaged bedfellows?', in *Proceedings of the Work Place Comfort Forum*, RIBA, London, 22–23 March

Lebrun, J. (1970) *Exigences physiologiques et modalites physiques de la climatisation par source statique concentree*, PhD thesis, University of Liege, France

Letan, R., Dubovsky, V. and Ziskind, G. (2003) 'Passive ventilation and heating by natural convection in a multi-storey building', *Building and Environment*, vol 38, pp197–208

Li, Y. (2000) 'Buoyancy-driven natural ventilation in a thermally stratified one-zone building', *Energy and Environment*, vol 35, pp207–214

Liddament, M. W. (2000) 'Making ventilation work for cooling', in Sayigh, A. A. M. (ed) *Proceedings of Renewable Energy: The Energy for the 21st Century – World Renewable Energy Congress VI*, Brighton, UK, 1–7 July, Pergamon Press, London, Part 1, pp420–425

Liddament, M. W. and Orme, M. (1998) 'Energy and ventilation', *Applied Thermal Engineering*, vol 18, pp1101–1109

Liem, S. H. and van Paassen, A. H. C. (1998) 'Hardware and controls for natural ventilation cooling', in *Proceedings of* the 18th AIVC Conference on Ventilation and Cooling, Athens, Greece, vol 1

Lin Borong, Tan Gang, Wang Peng, Song Ling, Zhu Yingxin and Zhai Guangkui (2004) 'Study on the thermal performance of the Chinese traditional vernacular dwellings in summer', *Energy and Buildings*, vol 36, no 1, January, pp73–79

Madsen, T. L. (1987) 'Measurement and control of thermal comfort in passive solar systems', in Faist, A., Fernandes, E. and Sagelsdorff, R. (eds) *Proceedings of the Third International Congress on Building Energy Management ICBEM 87*, Presses Polytechniques Romandes, Lausanne, vol IV, pp489–496

Majid, N. H. A., Razak Saphan, A. and Denan, Z. (2002) 'Towards a sustainable environment: An analysis on courtyards microclimate in the tropical region', in *International Symposium Building Research and Sustainability of the Built Environment in the Tropics*, Jakarta, Indonesia, 14–16 October

Maldonado, E. (ed) (1999) *Efficient Ventilation Systems for Buildings*, European Commission, Directorate General for Energy and Transport, Brussels

Mallick, F. H. (1994) 'Thermal comfort in tropical climates: An investigation of comfort criteria for Bangladesh subjects', in *Proceedings of the Conference on PLEA 1994*, Israel

Mallick, F. H. (1996) 'Thermal comfort and building design in the tropical climates', *Energy and Buildings*, vol 23, pp161–167

Martin, A. J. (1995) *Control of Natural Ventilation*, BSRIA Technical Note TN 11/95, The Building Services Research and Information Association, Atlanta, GA

Martin, M., Fleury, B., Kammerud, R. and Webster, T. (1984) *Parasitic Power Requirements for Night Ventilated Non Residential Buildings*, Report of LBL, Berkeley, California, August

Matthews, J. and Nicol, J. F. (1995) 'Thermal comfort of factory workers in Northern India', in Nicol, J. F., Humphreys, M. A., Sykes, O. and Roaf, S. (eds) *Standards for Thermal Comfort: Indoor Air Temperature Standards for the 21st Century*, E. and F. Spon, London

McCartney, K. J. and Humphreys, M. A. (2002) 'Thermal comfort and productivity', in Levin, H. (ed) *Proceedings of the Ninth International Conference on Indoor Air*, Santa Cruz, US, vol 1, pp822–827

McCartney, K. J. and Nicol, F. (2002) 'Developing an adapting comfort algorithm for Europe: Results of the SCATS project', *Energy and Buildings*, vol 34, no 6, pp623–635

Meierhans, R. A. (1996) 'Room air conditioning by means of overnight cooling of the concrete ceiling', *ASHRAE Transactions*, vol 102, part 1

Millet, J. R. (1997) 'Summer comfort in residential buildings without mechanical cooling', in *Proceedings of the Second International Conference on Buildings and the Environment*, Paris, 9–12 June, vol 1, pp307–315

Milne, G. R. (1995) 'The energy implications of a climate-based indoor air temperature standard', in Nicol, J. F., Humphreys, M. A., Sykes, O. and Roaf, S. (eds) *Standards for Thermal Comfort: Indoor Air Temperature Standards for the 21st Century*, E. and F. N. Spon, London

Neeper, D. A and McFarland, R. D. (1982) 'Some potential benefits of fundamental research for the passive solar heating and cooling of building', LA 9423-MS-Los Alamos, Los Alamos National Laboratories, US

Nicol, J. F. (1973) 'An analysis of some observations of thermal comfort in Roorkee, India, and Bagdad, Iraq', *Annals of Human Biology*, vol 1, no 4, pp411–426

Nicol, J. F. (2002) 'Why international thermal comfort standards don't fit tropical buildings', in *International Symposium Building Research and Sustainability of the Built Environment in the Tropics*, Jakarta, 14–16 October

Nicol, J. F. (2003) 'Thermal comfort: State of the art and future directions', in Santamouris, M. (ed) *Solar Thermal Technologies – The State of the Art*, James & James, London

Nicol, J. F. and Humphreys, M. (1973) 'Thermal comfort as part of a self regulating system', *Building Research and Practice*, vol 6, no 3, pp191–197

Nicol, J. F. et al (1995) Standards for Thermal Comfort: Indoor Air Temperature Standards for the 21st Century, Chapman and Hall, London, p247

Nicol, J. F. and Raja, I. A. (1995) 'Time and thermal comfort in naturally ventilated buildings', in Santamouris, M. (ed) *Proceedings of the International Workshop on Passive Cooling*, Athens, Greece

Nicol, J. F., Raja, I. A., Allandin, A. and Jamy, G. N. (1999) 'Climatic variations in comfort temperatures: The Pakistan projects', *Energy and Buildings* vol 30, pp261–279

Nicol, J. F. and Roaf, S. (1994) 'Pioneering new indoor temperature standards – The Pakistan Project', in *Proceedings of the Conference PLEA 94*, Israel

Nicol, J. F., Robinson, P. and Kessler, M. R. (1998) 'Using night cooling in a temperate climate', in *Proceedings of Environmentally Friendly Cities, PLEA 98 (Passive and Low Energy Architecture) Conference*, Lisbon, Portugal, June, James & James, London, pp467–470

Nielsen, H. K. (2002) *Stay Cool: A Design Guide for the Built Environment in Hot Climates*, James & James, London

OECD (Organisation for Economic Co-operation and Development) (1991) *Fighting Noise in the 1990s*, OECD, Paris

Oliveira, A., Silva, A. R., Afonso, C. F. and Varga, S. (2001) 'Experimental and numerical analysis of natural ventilation with combined light/vent pipes', *Applied Thermal Engineering*, vol 21, pp1925–1936

Olgay, F. (1973) *Design with Climate – Bio-climatic Approach to Architecture Regionalism*, Princeton University Press, US, pp94–112

Olsen, E. L. and Chen, Q. Y. (2003) 'Energy consumption and comfort analysis for different low-energy cooling systems in a mild climate', *Energy and Buildings*, vol 35, pp561–571

Parker, D. S. (1992) *Measured Natural Cooling Enhancement of a Whole House Fan*, FSEC-PF-273-92, Florida Solar Energy Center, Cocoa, Florida

Parker, D. S., Michael, P. and Callahan, J. K. S. (1999) *Development of a High Efficiency Ceiling Fan 'The Gossamer Wind'*, FSEC-CR-1059-99, Florida Solar Energy Center, Cocoa, Florida,

Parsons, K. C., Webb, L. H., McCartney, K. J., Humphreys, M. A and Nicol, J. F. (1997) 'A climatic chamber study into the validity of the Fangers PMV/PPD thermal comfort index for subjects wearing different levels of clothing insulation', in *Proceedings of the CIBSE National Conference*, London, Part 1, pp193–205

Pasumarthi, N. and Sherif, S. A. (1998) 'Experimental and theoretical: Performance of a demonstration solar chimney model – Part I: Mathematical model development', *International Journal of Energy Research*, vol 22, no 3

PDEC (1995) *EU DGXII JOR3CT950078 PDEC Project*

Pedki, M. M. and Sherif, S. A. (1999) 'On a simple analytical model for solar chimneys', *International Journal of Energy Research*, vol 23, no 4

Pelletret, R., Allard, F., Haghighat, F. and Van der Maas J. (1991) 'Modelling of large openings', Presented at the 21st AIVC Conference, Canada

Perkins, D. (1984) 'Heat balance studies for optimising passive cooling with ventilation air', *ASHRAE Journal*, vol 26, no 2, pp27–29

Pitts, A. C. and Abro, R. S. (1991) 'Intelligent control of night time ventilation', in Alvarez, S. (ed) *Architecture and Urban Space*, Kluwer Academic Press, The Netherlands, pp595–600

Prianto, E., Bonneaud, F., Depecker, P. and Peneau, J. P. (2000) 'Tropical-humid architecture in natural ventilation efficient point of view – a reference of traditional architecture in Indonesia', *International Journal on Architecture Science*, Hong Kong, vol 1, no 2, pp80–95

Prianto, E. and Depecker, P. (2002) 'Characteristic of airflow as the effect of balcony, opening design and internal division on indoor velocity: A case study of traditional dwelling in urban living quarter in tropical humid region', *Energy and Buildings*, vol 34, pp401–409

Priyadarsini, R., Cheong, K. W. and Wong, N. H. (2003) 'Enhancement of natural ventilation in high-rise residential buildings using stack system', *Energy and Buildings*

Rainaweera, C. and Hestnes, A. G. (1994) 'Enhanced cooling in typical Sri Lankan Dwellings', *Proceedings of the Conference PLEA 94*, Israel

Raja, A., Fergus, J., Nicol, J. F., McCartney, K. J. and Humphreys, M. A. (2001) 'Thermal comfort: Use of controls in naturally ventilated buildings', *Energy and Buildings*, vol 33, pp235–244

Raja, A., Nicol, J. F. and McCartney, K. J. (1998) 'Natural

ventilated buildings: Use of control for changing indoor climate', in *Proceedings of the World Renewable Energy Congress*, Pergamon Press, Oxford, UK, vol V, pp391–394

Rajapaksa, I., Nagai, H. and Okumiya, M. (2003) 'A ventilated courtyard as a passive cooling strategy in the warm humid tropics', *Renewable Energy*, vol 28, pp1755–1778

Raman, P., Sanjay, M. and Kishore, V. V. N. (2001) 'A passive solar system for thermal comfort conditioning of buildings in compo site climates', *Solar Energy*, vol 70, no 4, pp319–329

Ren, J. (1995) *Night Ventilation for Cooling Purposes Part I: Reference Building and Simulation Model IBPSA, 4*, International Conference in Madison, Wisconsin, US

Rijal, H. B, Yoshida, Y. and Umeriya, N. (2002) 'Investigation of the thermal comfort in Nepal', in *International Symposium Building Research and Sustainability of the Built Environment in the Tropics*, Jakarta, 14–6 October

Rodrigues, A. M., Canha da Piedade, A., Lahellec, A. and Grandpeix, J. Y. (2000) 'Modelling natural convection in a heated vertical channel for room ventilation', *Building and Environment*, vol 35, pp455–469

Rodriguez, E. A., Alvarez, S. and Coronel, J. F. (1994) 'Modeling stratification patterns in detailed building simulation codes', in *Proceedings of European Conference on Energy Performance and Indoor Climate in Buildings*, Lyon, France

Rohles, F. H., Konz, S. A. and Jones, B. W. (1983) 'Ceiling fans as extenders of the summer comfort envelope', *ASHRAE Transactions*, vol 89, no 1a, pp245–263

Rosenbaum, M. (1999) 'Passive and low energy cooling survey', *Environmental Building News*, pp1–14

Roucoult, J. M., Douzane, O. and Langlet, T. (1999) 'Incorporation of thermal inertia in the aim of installing a natural nighttime ventilation system in buildings', *Energy and Buildings*, no 29, pp129–133

Roulet, C., van der Maas, A. and Flourentzos, F. (1996) 'A planning tool for passive cooling of buildings', in *Proceedings of the Seventh International Conference on Indoor Air Quality and Climate*, Nagoya, Japan, 21–26 July

Santamouris, M. (2003) *Night Ventilation Strategies*, Air Information and Ventilation Center, (AIVC), Brussels

Santamouris, M. (2004) 'Passive cooling techniques', in *Advances of Solar Energy*, ASES, London

Santamouris, M. and Argiriou, A. (1997) 'Passive cooling of buildings – results of the PASCOOL Program', *International Journal of Solar Energy*, vol 18, pp231–258

Santamouris, M. and Assimakopoulos, D. (eds) (1996) *Passive Cooling of Buildings*, James & James, London

Santamouris, M. and Fleury, B. (1989) *The Cooling Potential of Passive Cooling Techniques in Greece*, Report prepared by PROTECHNA on behalf of the Centre for Renewable Energy Sources of Greece (CRES), Athens, Greece

Santamouris, M., Geros, V., Klitsikas, N. and Argiriou, A. (1995) 'SUMMER: A computer tool for passive cooling applications', in Santamouris, M. (ed) *Proceedings of the International Symposium: Passive Cooling of Buildings*, Athens, Greece, 19–20 June

Santamouris, M., Mihalakakou, G., Argiriou, A. and Asimakopoulos, D. (1996) 'The efficiency of night ventilation techniques for thermostatically controlled buildings', *Solar Energy*, vol 56, no 6, pp479–483

Scheatzle, D., Wu, H. and Yellot, J. (1989) 'Extending the summer comfort with ceiling fans in hot arid climates', *ASHRAE Transactions*, vol 95, no 1

Schmidt, K. and Patterson, D. J. (2001) 'Performance results for a high efficiency tropical ceiling fan and comparisons with conventional fans: Demand side management via small appliance efficiency', *Renewable Energy*, vol 22, pp169–176

Shao, L., Riffat, S. B. and Gan, G. (1998) 'Heat recovery with low pressure loss for natural ventilation', *Energy and Buildings*, vol 28, pp179–184

Sharma, M. R. and Ali, S. (1986) 'Tropical Summer Index – a study of thermal comfort in Indian subjects', *Building and Environment*, vol 21, no 1, pp11–24

Shaviv, E., Capeluto I. G. and Yezioro, A. (2000) 'A simple design tool for determining the effectiveness of thermal mass and night ventilation as passive cooling design strategy', in *Proceedings of Roomvent 2000, Air Distribution in Rooms: Ventilation for Health and Sustainable Environment*, 9–12 July, vol 2, Elsevier Science Publishers, London, Reading, UK, pp881–886

Shaviv, E., Yezioro, A. and Capeluto, I. G. (2001) 'Thermal mass and night ventilation as passive cooling design strategy', *Renewable Energy*, vol 24, pp445–452

Silvestrini, G. and Allessandro, S. (1988) 'Cooling of ancient buildings through night forced ventilation', *Proceedings of the Conference PLEA 88*, Pergamon Press, Oxford

Sobin, H. J. (1981) 'Window design for passive ventilative cooling : An experimental model scale study', in *Proceedings of the International Passive and Hybrid Cooling Conference*, Miami Beach, Florida

Sodha, M. et al (1991) 'Thermal performance of a room coupled to an evaporative cooling tower', in *Proceedings of the 1991 Solar World Congress*, Denver, Colorado, US

Solaini, G., Dallo, G. and Scansani, S. (1998) 'Simultaneous application of different natural cooling technologies to an experimental building', *Renewable Energy*, vol 15, pp277–282

Stanners, D. and Bourdeau, P. (eds) (1995) *Europe's Environment – The Dobris Assessment*, European Environmental Agency, Denmark

Standeven, M. A. and Baker, N. (1994) 'Diurnal variation in comfort conditions observed in the PASCOOL comfort surveys', in *Proceedings of the Conference on Energy Performance and Indoor Climate in Buildings*, Lyon, France

Stein, B. and Reynolds, J. (1992) *Mechanical and Electrical Equipment for Buildings*, John Wiley and Sons, New York

Stetiu, C. and Feustel, H. E. (1996) 'Phase change wallboard as an alternative to compressor cooling in California residences?', in *Proceedings of the ACEEE 1996 Summer Study on Energy Efficiency in Buildings*, American Council for an Energy-Efficient Economy, Washington, DC, vol 10, pp157–163

Stoops, J., Pavlou, C., Santamouris, M. and Tsangrassoulis, A. (2000) *Report to Task 5 of the SCATS project (Estimation of Energy Saving Potential of the Adaptive Algorithm)*, European Commission, Brussels

Taki. A. M, Ealiwa, M. A, Howarth, A. T. and Seden, M. R. (1999) 'Assessing thermal comfort in Ghadames, Libya: Application of the adaptive model', *Building Services Engineers Research Technology*, vol 20, no 4, pp205–210

Tan, C. C. (2000) *Solar Induced Ventilation*, PhD thesis, School of Design and Environment, National University of Singapore, Singapore

Tanabe, S. and Kimura, K. (1994) 'Effects of temperature, humidity and air movement on thermal comfort under hot and humid conditions', *ASHRAE Transactions*, vol 100, no 2, pp953–960

Tindale, A. W., Irving, S. J., Concannon, P. J. and Kolokotroni, M. (1995) 'Simplified method for night cooling', in *Proceedings of the CIBSE, National Conference 1995*, Eastbourne, UK, 1–3 October, vol 1, pp8–13

Thompson, T. Chalfoun, N. and Yoklic, M. (1994) 'Estimating the thermal performance of natural down-draught evaporative coolers', *Energy Conversion and Management*, vol 35, no 11, pp909–915

Todorovic, B., Randjelovic, I. and Krstic, A. (2002) 'Air pressure – A potential force for night cooling of atrium buildings', in *Proceedings of the EPIC 2002 Conference*, Lyon, France

Toffum, A., Melikov, K., Rasmussen, L. W., Kuciel, A. A., Cinalska, E. A., Tynel, A., Bruzda, M. and Fanger, P. O. (2000) *Human Response to Air Movement: Part I. Preference and Draught Discomfort*, DTU International Center for Indoor Environment and Energy, Lyngby, Denmark

Togari, S., Arai, Y. and Miura, K. (1993) 'A simplified model for predicting vertical temperature distribution in a large space', *ASHRAE Transactions*, vol 99, pp84–99

Turpenny, J. R., Etheridge, D. W. and Reay, D. A. (2000a) 'Novel ventilation cooling system for reducing air conditioning use in buildings', in *Proceedings of Roomvent 2000, Air Distribution in Rooms: Ventilation for Health and Sustainable Environment*, Reading, UK, 9–12 July, Elsevier Science Publishers, London, vol 2, pp875–880

Turpenny, J. R., Etheridge, D. W. and Reay, D. A. (2000b) 'Novel ventilation cooling system for reducing air conditioning in buildings. 1. Testing and theoretical modelling', *Applied Thermal Engineering*, vol 20, pp1019–1038

Turpenny, J. R., Etheridge, D. W. and Reay, D. A. (2000c) 'Novel ventilation system for reducing air conditioning in buildings. Part 2: Testing of prototype', *Applied Thermal Engineering*, vol 21, pp1203–1217

UNCHS (United Nations Centre for Human Settlements – Habitat) (1993) *Development of National Technological Capacity for Environmental Sound Construction*, HS/293/93/E, UNCHS, Nairobi

Van der Maas, J., Florentzos, F., Rodriguez, J.-A. and Jaboyedoff, P. (1994) 'Passive cooling by night ventilation', in *Proceedings of the European Conference on Energy Performance and Indoor Climate*, Lyon, France, 24–26 November

Van der Maas, J. and Roulet, C.-A. (1991) 'Night time ventilation by stack effect', *ASHRAE Technical Data Bulletin*, vol 7, no 1, Ventilation and Infiltration, pp32–40

Vieira, R. K. and Parker, D. S. (1991) *Energy Use in Attached and Detached Residential Developments*, FSEC-CR-381-91, Florida Solar Energy Center, Cape Canaveral, Florida

Vollebregt, R., Boonstra, C. and Santamouris, M. (1998) 'Natural ventilation and cooling strategies in new office designs', *Energy Comfort 2000*, European Commission, Directorate General For Energy, Brussels

Walton, G. N. (1994) *CONTAM93: User Manual (NISTIR 5385)*, National Institute of Standards and Technology, US

Webb, C. G. (1959) 'An analysis of some observations of thermal comfort in an equatorial climate', *BJIM*, vol 16

Wilkins, J. (1995) 'Adaptive comfort control for conditioned buildings', in *Proceedings of the CIBSE National Conference*, Eastbourne, UK, Chartered Institute of Building Service Engineers, London, part 2, pp9–16

Wouters, P., Demeester, J., Ducarme, D., Kofoed, P. and Zaccheddu, E. (1998) 'Overview and synthesis of the monitoring activities carried out in the framework of the NATVENT EC Joule Project', in *Proceedings of the EPIC 1998 Conference*, vol 3, pp958–963

Wu, H. (1989) 'The use of oscillating fans to extend the summer comfort envelope in hot arid climates', in *Proceedings of the ASHRAE Far East Conference on Air Conditioning in Hot Climates*, Malaysia, October

Wu, H. (1990) 'The effect of using nighttime ventilation and building thermal mass for passive cooling', in *Proceedings of the Solar 90 Conference*

Wurtz, E., Nataf, J. and Winkelmann, F. C. (1996) *Two and Three-Dimensional Natural and Mixed Convection Simulation Using Modular Zonal Model*, LBNL Report, Berkeley, California

Zimmerman, M. and Andersson, J. (1998) *Low Energy Cooling – Case Study Buildings*, International Energy Agency, Energy Conservation in Buildings and Community Systems Programme, Annex 28 Final Report, Paris

9

The Effect of Ventilation on Health and Other Human Responses

Olli Seppänen

Introduction

The effects of ventilation on indoor air quality and health are a complex issue. It is known that ventilation is necessary to remove indoor-generated pollutants from indoor air or to dilute their concentration to acceptable levels. But since the limit values of all pollutants are not known, the exact determination of required ventilation rates based on pollutant concentrations is seldom possible. The selection of ventilation rates has also to be based on epidemiological research (see, for example, Seppänen et al, 1999), laboratory and field experiments (for example, CEN, 1996; Wargocki et al, 2002a) and experience (see, for example, ECA, 2003).

Ventilation may also have harmful effects on indoor air quality and climate if not properly designed, installed, maintained and operated as summarized by Seppänen (2003). Ventilation may bring harmful substances indoors or deteriorate the indoor environment. Ventilation also interacts with the building envelope and may deteriorate the structures of the building. Ventilation changes the pressure differences over the structures of a building and may cause or prevent the infiltration of pollutants from structures or adjacent spaces. Ventilation is also, in many cases, used to control the thermal environment or humidity in buildings.

Ventilation can be implemented with various methods that may also affect health (see, for example, Seppänen and Fisk, 2002; Wargocki et al, 2002a). In non-residential buildings and hot climates, ventilation is often integrated with air conditioning, which makes the operation of ventilation more complex. Since ventilation is used for many purposes, its health effects are also various and complex. This chapter summarizes the current knowledge of the positive and negative effects of ventilation on health and other human responses. The focus of the chapter is on the office-type working environment and on residential buildings. In industrial premises, the problems of air quality are usually more complex and very much case specific. They are subject to occupational safety legislation and are not discussed here.

Ventilation and indoor air quality

Exposure to pollutants in indoor air may cause a variety of effects. The severity of the effects covers a wide spectrum, from perception of unwanted odours to cancer. The effects may be acute or may develop over a longer time. Some examples of health effects related to indoor air are as follows (ECA, 2003):

- Ventilation may dilute the concentration or disperse airborne virus or bacteria, causing infectious diseases.
- Some micro-organisms can grow in air humidifiers and may result in pneumonia (*Legionella*) and 'humidifier fever'.
- High humidity indoors is associated with an increased growth of micro-organisms such as mould and bacteria. It increases allergies and other types of symptoms. Most allergic and asthmatic people react on exposure to mould.
- An increased risk of developing lung cancer has been linked to exposure to environmental tobacco smoke (ETS) and to radon decay products.

For some effects, clear relationships with exposure to indoor air pollution have been reported. Among these are respiratory diseases (particularly among children), allergies (particularly to house dust mites) and mucous membrane irritations (particularly due to formaldehyde). Large numbers of people have been, and are still being, affected.

Many chemicals encountered in indoor air are known or are suspected to cause sensory irritation or stimulation at least at high concentrations. As pointed out by the World Health Organization (WHO, 1989), many different sensory systems that respond to irritants have receptors situated on or near the body surface. Some of these systems tend to facilitate the response rather than habituate, and their reactions are delayed. On the other hand, in the case of odour perception, the reaction is immediate but is also influenced by olfactory fatigue on prolonged exposures. In general, sensory systems are tuned towards registering environmental changes rather than the absolute levels. Sensory effects are important parameters in indoor air quality control for several reasons. They may appear as (ECA, 2003):

- adverse health effects on sensory systems (for example, environmentally induced sensory dysfunctions);
- adverse environmental perceptions that may be harmful *per se* or constitute precursors of disease to come on a long-term basis (for example, annoyance reactions and triggering of hypersensitivity reactions);
- sensory warnings of exposure to harmful environmental factors (for example, odour of toxic sulphides and mucosal irritation due to formaldehyde); and
- important tools in sensory bioassays for environmental characterization (for example, using the odour criterion for general ventilation requirements or for screening building materials to find those with low emissions of volatile organic compounds).

It is important to realize that the sensory effects of pollutants are not necessarily linked to their toxicity. Indeed, some harmful air pollutants, such as radon or carbon monoxide (CO), are not sensed at all. Therefore, perceived air quality is not a universal measure of adverse effects.

Sensory effects reported to be associated with indoor air pollution are, in most cases, multi-sensory, and the same perceptions or sensations may originate from different sources. Humans integrate different environmental signals to evaluate the total perceived air quality and to assess comfort or discomfort. However, it is not known how this integration occurs. Perceived air quality is, for example, mainly related to stimulation of both the trigeminus and olfactorius nerves.

Comfort and discomfort by definition are influenced by more complex psychological factors, and for this reason the related symptoms, even when severe, cannot be documented without 'perceptional' assessments.

The role of ventilation in controlling pollutants in the indoor environment

The basic equation (1) to calculate the required ventilation rate for pollutant control is simple and relates the generation of pollutant, concentration differences of indoor and outdoor air and ventilation efficiency if ventilation is the only way to remove the pollutants from the room's air (CEN, 1996; Seppänen, 1998):

$$Q_h = \frac{G_h}{C_{h,i} - C_{h,o}} \cdot \frac{1}{\epsilon_v} \tag{1}$$

where:

Q_h = the airflow needed for selected air quality with respect to any contaminant in the air,

G_h = the generation of contaminant;

$C_{h,i}$ = acceptable contaminant concentration in indoor air;

$C_{h,o}$ = the contaminant concentration of intake air;

ϵ_v = the ventilation efficiency ($\epsilon_v = 1$ for complete mixing to $\epsilon_v = 2$ for piston flow).

Use of equation (1) in design means that the ventilation airflows in buildings are rationally selected and distributed to all rooms, depending upon the pollution loads. The problems, however, are in the application since:

- The acceptable concentration of various pollutants in indoor air is not known, especially for the mixtures of hundreds of the compounds found in indoor air.
- The generation rate of pollutants indoors is not usually known.
- The contaminant concentration of intake air is not known with respect to all pollutants.
- The concentration of contaminants in the supply air may be different from the outdoor air due to processes in the air handling system or structures through which the supply air is flowing.
- Only a limited amount of information is available on the ventilation efficiency of various air distribution systems.

Equation (1) can be also used to calculate required ventilation rates for moisture control or thermal control if the term for contaminant generation is replaced with moisture generation or heat generation, and concentration with moisture contents or enthalpy of the air, respectively. Ventilation efficiency is dependent upon the pollutant type, location and the way in which it is generated.

Since the exact mathematical relations or threshold values for pollutants in the indoor air are not available, information on the health effects of ventilation is, in many cases, based on experimental research. Parameters in the experiments may include ventilation rates, ventilation systems, contaminants in indoor air, and physical characterization of the indoor environment. These factors affect human responses through each other but also independently. This is illustrated in Figure 9.1.

Primary human responses to ventilation that are dealt with in this chapter are:

- infectious diseases;
- sick building syndrome (SBS) symptoms;
- task performance and productivity;
- perceived air quality; and
- respiratory allergies and asthma.

Ventilation may affect several other parameters of the indoor environment that may also have health effects. These include:

- thermal conditions;
- effects through moisture control;
- effects through the control of pressure differences;
- draught; and
- noise.

Ventilation rates and human responses in the office environment

Ventilation and infectious diseases

The review of ventilation rates and human responses (Seppänen et al, 1999) summarizes the results of four studies available at that time on the health effects of ventilation rates. These were performed in a jail, barracks, a home for the elderly and offices. All of them reported significant association between low ventilation rates and an increase in health problems: pneumonia, upper respiratory illnesses, influenza and short-term sick leave, respectively (Brundage et al, 1988; Hoge et al, 1994; Drinka et al, 1996; Milton et al, 2000). Even though the ventilation rates were estimated and not measured, the consistent findings are a strong indication of the association of ventilation rates with health effects. The strongest evidence is provided by the most recent study of these (Milton et al, 2000). The association with sick leave was analysed for 3720 employees in 40 buildings using 115

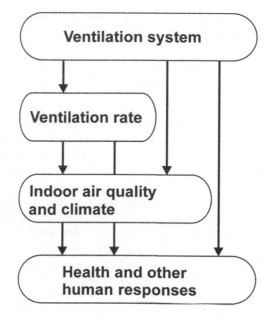

Source: chapter author

Figure 9.1 *Ventilation affects health and other human responses through several pathways*

independently ventilated ventilation areas. Among office workers, the relative risk for short-term sick leave was 1.53 (1.22–1.92 confidence interval) with the estimated ventilation of 12l/s per person compared with a ventilation rate of 24l/s per person.

A quantitative relationship between ventilation rate and sick leave was further developed by Fisk et al (2003). They used the Wells-Riley model – equation (2) – of airborne disease transmission (Nardell et al, 1991) and fitted the data from epidemiologic studies mentioned above in it:

$$P = \frac{D}{s} = 1 - \exp\left[-\frac{ipqt}{Q} \right] \qquad (2)$$

where P = proportion of new disease cases among the susceptible persons; D = number of new disease cases; s = number of susceptible persons; i = number of infectors; p = breathing rate; q = the rate at which an infector disseminates infectious particles; t = time that infectors and susceptibles share a confined space or ventilation system; and Q = rate of supply of outdoor air.

Equation (2) neglects the removal of infectious particles by filtration and by deposition on room surfaces, which are significant processes in removing airborne particles from room air. These removal processes can be expressed with effective removal rates per unit volume n_f and n_d, yielding the equation:

$$P = \frac{D}{s} = 1 - \exp\left[\left(-\frac{ipqt}{V}\right)\Big/(n_v + n_f + n_d)\right] \quad (3)$$

where n_v is the ventilation rate; n_f is the removal rate of infectious particles by filtration, equal to the product of the recirculation airflow rate and the filter efficiency; and n_d is the removal rate of particles due to deposition on room surfaces.

In this equation the term $ipqt/V$ is the unknown. It was calculated from the experimental data and applied in the equation. Figure 9.2 plots the calculated values of illness or short-term sick leave versus ventilation rate, normalized by the illness or sick leave rate predicted with no ventilation. All predictions show the expected decrease in illness over time; however, the rate of decrease varies dramatically for low ventilation rates, with the prediction based on the data of Drinka et al (1996) appearing as an outlier.

For comparison to the disease transmission model, a much simpler model was used in which the disease prevalence is proportional to the reciprocal of the total infectious particle removal rate:

$$P \propto 1/(n_v + n_f + n_d) \quad (4)$$

This model is consistent with the assumption that the disease prevalence in the building is proportional to the indoor concentration of infectious particles. The simple particle concentration model – equation (4) – provides a mid-range prediction (see Figure 9.2).

Fisk et al (2003) draw the conclusion that the majority of existing literature indicates that increasing ventilation rates will decrease respiratory illness and associated sick leave. A disease transmission model, calibrated with empirical data, can be used to estimate how ventilation rates affect sick leave; however, the model predictions have a high level of uncertainty.

The association of ventilation rates with sick building syndrome (SBS) symptoms in commercial and institutional buildings

Reviews by Seppänen et al (1999) and Wargocki et al (2002a) on the association of ventilation rates and human responses show that ventilation rates below 10l/s per person are associated with a significantly inferior prevalence or value of one or more health or perceived air quality outcomes.

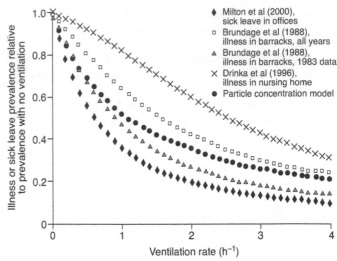

Source: adapted from Fisk et al (2003)

Figure 9.2 *Predicted trends in illness of sick leave versus ventilation rate*

Seppänen et al (1999) reviewed current literature on the associations of ventilation rates in non-residential and non-industrial buildings (primarily offices) with health and other human outcomes. Twenty studies, with close to 30,000 subjects, investigated the association of ventilation rates with human responses, and 21 studies, with over 30,000 subjects, investigated the association of carbon dioxide concentration with these responses. Almost all studies including ventilation rates below 10l/s per person found these ventilation rates to be associated in all building types with statistically significant worsening in one or more health or perceived air quality outcomes. Some studies determined that increases in ventilation rates above 10l/s per person, up to approximately 20l/s per person, were associated with further significant decreases in the prevalence of SBS symptoms or with further significant improvements in perceived air quality. The ventilation rate studies reported relative risks of 1.1–6.0 for sick building syndrome symptoms for low compared to high ventilation rates.

Each assessment included in the reviewed studies with sick building symptoms as an outcome are presented in Figure 9.3. The references to the papers cited are provided in Seppänen et al (1999). When outcomes at two levels of ventilation rate are compared, each level is represented with a circle, with multiple comparisons within single studies displayed separately. If the study compared outcomes among groups of workers experiencing different ranges of ventilation rate (for example, < 10l/s per person versus > 10l/s per person), Figure 9.3 presents the approximate mean ventilation rate within each range.

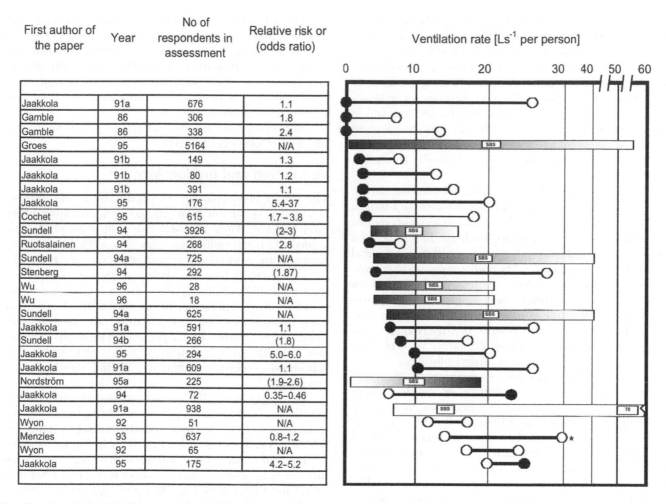

First author of the paper	Year	No of respondents in assessment	Relative risk or (odds ratio)
Jaakkola	91a	676	1.1
Gamble	86	306	1.8
Gamble	86	338	2.4
Groes	95	5164	N/A
Jaakkola	91b	149	1.3
Jaakkola	91b	80	1.2
Jaakkola	91b	391	1.1
Jaakkola	95	176	5.4–37
Cochet	95	615	1.7–3.8
Sundell	94	3926	(2–3)
Ruotsalainen	94	268	2.8
Sundell	94a	725	N/A
Stenberg	94	292	(1.87)
Wu	96	28	N/A
Wu	96	18	N/A
Sundell	94a	625	N/A
Jaakkola	91a	591	1.1
Sundell	94b	266	(1.8)
Jaakkola	95	294	5.0–6.0
Jaakkola	91a	609	1.1
Nordström	95a	225	(1.9–2.6)
Jaakkola	94	72	0.35–0.46
Jaakkola	91a	938	N/A
Wyon	92	51	N/A
Menzies	93	637	0.8–1.2
Wyon	92	65	N/A
Jaakkola	95	175	4.2–5.2

Note: The circles in the chart denote mean ventilation rates compared in the assessment.

Source: adapted from Seppänen et al (1999)

Figure 9.3 *Summary of studies with measured ventilation rates per person and sick building syndrome (SBS) symptoms*

Statistically significant differences in outcomes at different ventilation rates are illustrated graphically within the table, with a shaded circle indicating at least one significantly worsened health or perception outcome at that ventilation rate. All circles being unshaded indicate lack of a statistically significant increase in any outcome with ventilation. In general, the criteria for statistical significance are $p < 0.05$, or a 95 per cent confidence interval that excludes unity. For each assessment, the range of relative risks (essentially the symptom prevalence at a lower ventilation rate divided by symptom prevalence at a higher ventilation rate) is included in the tabulated information. Several assessments treated the measured ventilation rates as a continuous variable in the statistical

model used to analyse study data, and the results are illustrated with shaded or unshaded horizontal bars. If the ventilation rate variable was a statistically significant parameter in the model, suggesting a dose–response relationship, the horizontal bar is shaded with the darker shading, denoting more symptoms.

Almost all of the studies included in the review, which included ventilation rates below 10l/s per person, found that these rates were associated with a significantly worse prevalence or value of one or more health or perceived air quality outcomes. Most assessments included multiple statistical tests – for example, for the association of multiple symptoms with ventilation rate – and in 25 of 34 assessments with information provided, 50 per cent or more of the statistical tests indicated a significant association with ventilation rate. Available studies further show that increases in ventilation rates above 10l/s per person, up to approximately 20l/s per person, are sometimes associated with a significant decrease in the prevalence of SBS symptoms. The less consistent findings for relationships in the range above 10l/s per person are compatible with the prediction that benefits per unit increase in ventilation would be likely to diminish at higher ventilation rates and, thus, be more difficult to detect epidemiologically.

However, some studies that collected outcome data at several levels or ranges of ventilation assessed with a statistical model whether there was a dose–response relationships. Data from many such studies indicate a dose–response relationship between ventilation rates and health and perceived air quality outcomes of up to approximately 25l/s per person. Nevertheless, available data are not sufficient to quantify an average dose–response relationship. Only five studies were conducted in hot, humid climates; thus, the results of this review apply primarily for moderate and cool climates. Most of these studies have been conducted in office buildings.

Based on these results, Seppänen et al (1999) concluded that in office buildings or similar spaces constructed using current building practices, increases in ventilation rate in the range of between 0–10l/s per person will, on average, significantly reduce occupant symptoms and improve perceived air quality. Increases in ventilation rate above 10l/s per person, and up to 20l/s per person, may further reduce symptoms and improve air quality, although these benefits are currently less certain, based on available data. No threshold for effects is evident at 10l/s per person or at any other specific ventilation rate. As ventilation rates increase, benefits gained for occupants per additional unit of ventilation are likely to decrease in magnitude and to require larger studies for convincing demonstration. Benefits that have yet to be consistently demonstrated in this way (for example, for ventilation rates above 10l/s per person) may still be of substantial public health importance. Ventilation standards may thus need to periodically revisit the available evidence for occupant benefits at particular ventilation rates, as well as the magnitude of these benefits, weighed against the current incremental costs of increasing ventilation. This process would be new since minimum ventilation rates in existing codes and standards do not substantially reflect health data, such as is reviewed here.

Dose–response relations between measured average carbon dioxide (CO_2) concentration and several typical SBS symptoms were also reported from the analysis of the US BASE study data (Apte et al, 2000).

The effect of ventilation on SBS symptoms was reported in an intervention study in the tropics (Tham et al, 2003a). The authors reported a statistically significant reduction in intensity of headaches and difficulty in concentration when ventilation was increased from 9.8–22.7l/s per person, and an increase when the intervention was reversed.

SBS symptoms are also related to productivity and sick leave (Niemelä et al, 2003).

Ventilation and productivity

The effect of ventilation on productivity was demonstrated by Wargocki et al (2000a) in a simulated office environment. They exposed five groups of six female subjects to three ventilation rates (3l/s, 10l/s and 30l/s per person), one group and one ventilation rate at a time. The performance of four simulated office tasks improved monotonically with increasing ventilation rates, and the effect reached significance in the case of text typing. For each twofold increase in ventilation rate, performance improved on average by 1.7 per cent. The study indicates the benefits of ventilation at rates well above the minimum levels prescribed in existing standards and guidelines.

Wargocki et al (2003) reported that the increase of ventilation from 2.5–25l/s per person caused performance to increase as indicated by reduced average talk time (P < 0.055) in a call centre with a new filter in the air handling system, and to decrease (P < 0.05) with a used filter so that talk time was about 10 per cent lower with a new filter than with a used filter at high outdoor air supply rate.

Tham et al (2003b) also reported that the performance of office workers increased in the tropics when the outdoor air supply rate was increased from 9.8–22.7l/s per person at a temperature of 24.4° C.

Ventilation and perceived air quality

The effect of ventilation on perceived air quality has been well documented. Many ventilation standards have been based on perceived air quality (for example, ASHRAE 62, 2001). Perceived air quality has been related to predicted percentage of dissatisfaction with and acceptability of air quality in several studies and has been summarized in CEN (1996) and ECA (2003). The summary of the field measurements (Seppänen et al, 1999) on perceived air quality and ventilation rates also shows the connection (see Figure 9.4).

In some laboratory tests with ventilation rates and pollution loads (Wargocki et al, 2000b) the perceived air quality (PAQ) has been correlated with performance. The estimated effect on performance based on typing, addition

First author of the paper	Year	No of respondents in assessment	Relative risk or (odds ratio)
Bluyssen	95a	5164	N/A
Groes	95	5164	N/A
Nordström	95b	225	(03–0.4)
Palonen	90	580	1.9–2.8
Ruotsalainen	94	268	2.8
Cochet	95	522	N/A
Zweers	90	12	N/A
Jaakkola	94	72	0.47

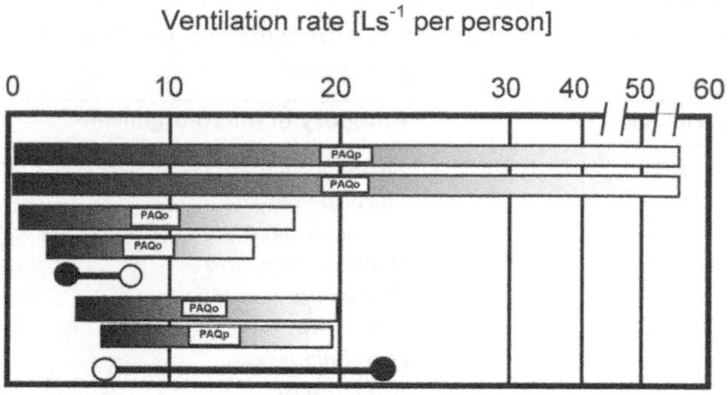

Note: The circles in the chart denote mean ventilation rates compared in the assessment. A black circle denotes a significantly worse perceived air quality. The bars denote assessments of dose–response relationships. The shaded end of a bar denotes a dose–response relation between ventilation and perceived air quality in the indicated ventilation range.

Source: adapted from Seppänen et al (1999)

Figure 9.4 *Summary of studies with measured ventilation rates per person and perceived air quality*

and proof-reading tests was 1.5 per cent for each 10 per cent of people dissatisfied with air quality. This linkage, however, may overlap with the linkage between SBS and productivity since PAQ may also affect the prevalence of SBS as both are subjectively reported.

Ventilation in residences

In residential buildings, the role of ventilation to control indoor-generated pollutants depends largely upon the indoor sources, which may vary significantly between homes. The removal of pollution sources is, in general, a more effective way of controlling indoor air quality than diluting the concentration of pollutants by ventilation.

Only little information is available on the health effects of measured ventilation rates in residential buildings. However, a review by a European group (Wargocki et al, 2002a) concluded that ventilation rates below 0.5 air changes per hour (ACH) are a health risk in Nordic residential buildings. This is quite a low value, and the minimum ventilation rates may be much higher in moderate climates where the outdoor air humidity is higher and more ventilation is needed to carry away the indoor-generated moisture. In residential buildings, ventilation rates depend upon the air tightness of the building envelope and the ventilation system. In Nordic countries, most new residential buildings use balanced ventilation with mechanical supply and exhaust of the air with heat recovery from the ventilation air. With mechanical ventilation, the required ventilation rates are easily provided to each room. The drawbacks have been poor technical implementation of these systems.

Ventilation, however, affects and can control several pollutants in residences. They are discussed in the following sections.

Mould and mites

Low ventilation may lead to high indoor humidity and moisture accumulation within building structures or materials. This may lead to microbial growth and, subsequently, to microbial contamination and other emissions in buildings. In epidemiological studies, moisture damage and microbial growth in buildings have been associated with a number of health effects, including respiratory symptoms and diseases and other symptoms (Bornehag et al, 2001; Nevalainen, 2002). The health effects associated with moisture damage and microbial growth seem to be consistent in different climates and geographical regions (ISIAQ, 1996).

It has been shown with relatively good certainty that building-related moisture and microbial growth increase the risk of respiratory symptoms, respiratory infections, allergy and asthma. The underlying mechanisms are irritation of mucous membranes, allergic sensitization and non-specific inflammation. Toxic mechanisms may also be involved, especially in connection with toxin-producing fungi and bacteria. Certain building materials seem to support the growth of potentially toxic microbes and even induce toxin production more readily than other materials (Nevalainen, 2002).

The primary controlling factor of the mould growth is moisture, especially the relative humidity indoors and in building structures. The limit values of the relative humidity of some species of mould are presented in Table 9.1.

Table 9.1 *Moisture levels required for the growth of selected micro-organisms from construction, finishing or furnishing materials*

Moisture level	Category of micro-organism
High (a_w > 0.90; ERH > 90%)	Tertiary colonizers (hydrophilic)
	Mucor plumbeus
	Alternaria alternata
	Stachyrobotrys atra
	Ulocladium consortiale
	Yeasts – for example, *Rhodotorula*
	sporobolomyces
	Actinomyces
Intermediate (a_w 0.80–0.90; ERH 80–90%)	Secondary colonizers
	Cladosporium cladosporioides
	C. sphaerosperum
	Aspergillus flavus
	A. versicolor[*]
Low (a_w < 0.80–0.90; ERH < 0.80–90%)	Primary colonizers (xerophilic)
	Aspergillus versicolor[**]
	A. glaucus group
	A. penicillioides
	Penicillium brevicompactum
	P. chrysogenum
	Alemia sebi

Notes: a_w = water activity; ERH= equilibrium relative humidity

[*] At 12° C

[**] At 25° C

Source: ISIAQ (1996)

In most cases, ventilation reduces the moisture content indoors (most of the time in most industrialized countries the indoor absolute humidity is higher indoors than outdoors). The effect of ventilation is twofold. Ventilation can remove indoor-generated moisture directly and dilute the water content of the air to a lower level. In some climatic conditions (summer in some coastal areas), the outdoor moisture contents may be high. In those conditions, ventilation is not effective. However, those conditions do not necessarily last very long.

Several studies have been performed on the effect of improved ventilation with or without dehumidification on the population of dust mites in residences (Harving et al, 1993; Chivato et al, 1997; Crane et al, 1998; Jones, 1999; Niven et al, 1999; Warner et al, 2000). Most of them have reported a reduction in the mite population but not an improvement in symptoms or allergen levels. Some researchers (Howieson, 2002) have pointed out the short-comings in these studies. It is natural that the effect of ventilation and dehumidification is not seen in symptoms and other human outcomes if the indoor environment is not cleaned from the accumulated allergens at the same time. The mite population generates many times more fragments of dead mites and faecal pellets than the mass of the living population. If these allergens are not removed at the same time, it is obvious that the symptoms will not be alleviated.

Indoor-generated particulate matter

The health effects of particulate matter depend upon particle size and composition. The particles with a diameter smaller than 5µm do not settle by gravity (settling velocity about 0.15mm/s) but stay airborne with air currents. The most harmful, in general, are those with small diameters (< 0.3µm). They penetrate deep in the lungs and are the most difficult to remove from the air. Inert particles may also carry bacteria and virus, as well as allergens. Typical sizes for the particles carrying bacteria is 1–2µm; particles with fungal spores are larger and are in the range of 2–3µm. Particles with allergens (such as animal dander) vary widely in their size range. Since particle size varies substantially between harmful particles, the efficiency of various removal measures varies as well.

Dust concentration in the air is also affected by the dust accumulated on room surfaces from where they become airborne through the activities and air currents in the room. An effective method against exposure to dust is to keep surfaces clean and dustless. This means frequent dusting or cleaning surfaces with effective vacuum cleaners. As the finest particles easily penetrate the dust collection reservoir of vacuum cleaners and pollute the room air, it is important that cleaners filter the fine particles from the air returned back to the room or exhaust the air outdoors, as is done in central vacuum cleaning systems. As the act of cleaning itself temporarily increases the dust concentration in the air, cleaning should be done during unoccupied periods. Adequate ventilation should also be provided during cleaning.

Environmental tobacco smoke

Scientific evidence has shown the adverse health effects of passive smoking (see, for example, Jones, 1999). Many of the thousands of compounds in smoke are carcinogenic (Gold, 1992). The size of the particles of the smoke is relative small, which makes cleaning difficult.

The most effective method of controlling exposure to tobacco smoke is the control of smoking. It is almost impossible to control exposure to safe levels with other means in a room where smoking takes place. Control with ventilation by diluting smoke requires very high ventilation rates. If the tobacco smoke is diluted to an acceptable level from a health standpoint, in a long-term exposure it would require a ventilation rate of 555l/s per smoked cigarette in an hour (a nicotine concentration of 1µg/m^3

and 2mg of nicotine per cigarette). Such ventilation rates are impossible in practice. However, ventilation dilutes the concentration of harmful compounds of environmental tobacco smoke, and makes the air quality acceptable in short-term exposure, with considerably lower ventilation rates (see, for example, Leaderer and Cain, 1983).

Pollen

Pollen is a common allergen outdoors and indoors as pollen is carried inside with ventilation air and clothing. Pollen particles are large and can be removed from incoming air easily. Even in buildings with exhaust ventilation, filters can be installed in the air inlets. Control is easiest in buildings with mechanical exhaust and supply ventilation, commonly used in Nordic countries. Germination of pollen begins in humid conditions and results in small allergenic particles that are more difficult to remove from the air than pollen itself.

Room air cleaners are used to remove particles from the air effectively. Their effect on the total dose depends upon their total cleaning effect. Effective cleaning airflow of such devices is the removal efficiency multiplied by the airflow through the device. The effective airflow divided by the air volume of the space to be cleaned should be more than 1 before cleaning is effective. Removal efficiency of air cleaners depends upon particle size. Small particles are more difficult to clean than large ones. Electrical air cleaners also remove small particles from the air. Their drawback is potential ozone generation, which may be harmful.

Nitrogen oxides

Scientific evidence shows the negative health effects of exposure to nitrogen oxides (Jones, 1999). Sources of nitrogen oxides are all kinds of combustion, ranging from internal combustion engines to gas cooking and heating appliances. Scientific evidence shows more adverse health effects of the indoor climate in residences with open-flame cooking with gas or solid fuel (Burr, 2001). It is thus obvious that all open-flame cooking or heating without proper chimneys or flue gas pipes should be avoided.

The use of range hoods will help as they can capture part of the flue gases; but their typical capture efficiency is only 60 per cent, and the effectiveness is highly dependent upon their proper use and cooking practice. All cooking, space heating and water heating that can be done with proper control of combustion and removal of flue gases will improve air quality. In most cases, this technology is also more cost effective than many others.

Ventilation may also cause malfunctioning of heating appliances. Exhaust ventilation may create underpressure in an apartment or house, which could cause back-draught in a heating system. This should, of course, be avoided.

Formaldehyde and other volatile organic compounds (VOCs)

Material emissions have been recognized to have an influence on the total pollution load of buildings. Research and development activities in the area of material emissions began with formaldehyde emissions from particle boards about 30 years ago. Labelling schemes and quality control have solved the problem with particle boards manufactured in Europe, but not with those imported from some other countries. During the last decade, research has shown that almost all materials emit chemical pollutants. The focus has been on paints, varnishes and flooring materials. Unfortunately, harmful emissions are not limited to the finishing materials, but also include furniture and partitions. In some cases, sealants and injection putties have also created problems due to high volatile organic compound (VOC) emissions. Even though the emission rates of materials have been significantly reduced due to labelling systems (for example, Seppänen, 2003), ventilation is needed to dilute VOC concentrations to acceptable levels, particularly in new and renovated buildings.

Carbon monoxide (CO)

Carbon monoxide is extremely poisonous and odourless. The source of carbon monoxide is incomplete combustion in which CO is generated instead of CO_2. The adequate supply of air for combustion will prevent the incomplete combustion and the generation of carbon monoxide. The most potential sources of CO are room heaters with gas, oil or solid fuel. The only way to control the exposure is to control the source and to ensure the proper use of appliances where the generation of CO is possible (Jones, 1999). Use of these appliances should be avoided and proper ventilation provided.

Carbon dioxide (CO$_2$)

The indoor source of carbon dioxide comes from the metabolism of a building's occupants and pets. The outdoor concentration of CO_2 is fairly constant, but varies depending upon the location and the time of day. Typical outdoor air concentrations are 350 to 450 parts per million (ppm – that is, cm^3/m^3). Indoor air concentration depends upon ventilation and indoor sources (number of occupants). Typical indoor air concentrations of CO_2 are between 500–1500 ppm. CO_2 in these concentrations is not harmful; however, it is an indicator of other pollutants in the air and ventilation rate. The CO_2 concentration depends upon the number of occupants, and duration of occupancy, ventilation rate and room volume. Since occupancy (the generation of CO_2) varies, the concentration of CO_2 is seldom constant in a building and should be evaluated accordingly. Control of CO_2 is the same as control of ventilation (Seppänen et al, 1999).

Ozone

Ozone is an example of a pollutant with a higher outdoor than indoor concentration. Ventilation, particularly through windows, will increase ozone levels indoors and cause harmful effects. Ozone may react with indoor-generated compounds and result in more harmful compounds than the original chemicals in the reaction (Wolkoff et al, 2000)

Ventilation and temperature control

Ventilation is also commonly used for temperature control. In many countries, the outdoor temperature is, for most of the year, below the indoor temperature and ventilation can be used to reduce high room temperatures. The adverse effects of high room temperatures are well documented. High room temperature increases the prevalence of sick building syndrome symptoms, deteriorates perceived air quality, increases the sensation of air dryness in winter, and affects performance and productivity at work (ECA, 2003; Seppänen et al, 2003).

Night-time ventilative cooling in both hot and moderate climates provides an attractive opportunity for indoor temperature control with a low environmental impact. Its principle is based on daily temperature swings during hot periods. A typical daily temperature swing is around 12° C. The cool night-time air can be used to cool the building during the night and, in this way, to decrease daytime temperatures.

Ventilation can also be used to increase the room air velocity and, in this way, to increase the convective heat transfer and decrease the thermal stress in high temperatures.

The effect of air temperature on the thermal comfort is well known; but its effect on air quality is not so widely recognized. Studies have shown that warm and humid air is stuffy (Berg-Munch, 1980), and warm room air temperature in the winter causes a higher number of typical sick building symptoms than cooler air (Seppänen and Jaakkola, 1989). The relationship between the number of symptoms and temperature is close to linear in the temperature range from 20° to 26° C. Laboratory experiments (Fang et al, 1998) have suggested that perceived quality of polluted air depends upon the enthalpy of the air. In these tests, air was polluted with emissions from typical building materials. Studies with a whole body exposure did not show as strong effects of the enthalpy of the air on perceived air quality (Fang et al, 1997). However, the influence was still very significant. Humphreys et al (2002) found that the thermal state of the subjects (as recorded by their comfort vote) was by far more influential than any particular characteristic of the environment (including enthalpy) in deciding indoor air quality. These findings suggest the use of low room air temperature and low relative humidity in the winter from a standpoint of good indoor air quality (IAQ) and energy economy.

Cleaning with recirculating and filtering air

Several classes of health effects are linked to particle exposures that may be reduced via filtration. The sizes of the particles linked with these health effects vary widely. Since filter efficiency and rates of particle deposition to surfaces also vary with particle size, the concentration reductions attained from filtration, and the incremental benefits of using higher-efficiency filters, will vary markedly with particle type. Despite the widespread use of filtration systems, the influence of different air filtration options on indoor concentrations of particles has not been well documented.

Fisk et al (2002) used a model and data on particle size distributions, filter efficiencies and particle deposition rates to estimate the reductions in the indoor mass concentrations of particles attainable from use of filters in heating, ventilating and air-conditioning (HVAC) supply airstreams. Predicted reductions in cat and dust mite allergen concentrations range from 20–60 per cent. Increasing filter efficiencies above approximately ASHRAE Dust Spot 65 per cent did not significantly reduce predicted indoor concentrations of these allergens. For environmental tobacco smoke particles and outdoor fine mode particles, calculations indicate that relatively large (for example, 80 per cent) decreases in indoor concentrations are attainable with practical filter efficiencies. Increasing the filter efficiency above ASHRAE Dust Spot 85 per cent results in only modest predicted incremental decreases in indoor concentration.

Association of heating, ventilating and air-conditioning (HVAC) system types with sick building syndrome (SBS) symptoms

Summaries (Seppänen and Fisk, 2002; Wargocki et al, 2002a) showed that most studies on the association of ventilation systems and human responses indicate that relative to natural ventilation, air conditioning (with or without humidification) was consistently associated with a statistically significant increase in prevalence by approximately 30–200 per cent of one or more SBS symptoms. In two of three analyses from a single study (assessments), symptom prevalence was also significantly higher in air-conditioned buildings than in buildings with simple mechanical ventilation and no humidification. The available data also suggest, with less consistency, an increase in the risk of symptoms with simple mechanical ventilation relative to natural ventilation. The statistically significant association of mechanical ventilation and air conditioning with SBS symptoms is much more frequent than expected by chance and, moreover, is not likely to be a consequence of being confounded by several potential personal, job or building-related confounders. A group of European scientists (Euroven group) elaborated and tested several hypotheses on the reasons for improper performance of mechanical systems but was only able to find support for some of the hypotheses. The group concluded that potential causes of adverse health effects due to HVAC systems comprise poor maintenance and hygiene in HVAC systems; intermittent operation of HVAC systems and lack of moisture control; and lack of control of HVAC system materials and loaded filters (Wargocki et al, 2002a).

The results of the review by Seppänen and Fisk (2002) are presented in Figures 9.5 and 9.6. Figure 9.5 presents the assessments comparing symptoms among occupants of air-conditioned buildings with those in naturally ventilated or simple mechanically ventilated buildings. Figure 9.6 presents the assessments comparing symptoms associated with simple mechanical ventilation to symptoms associated with natural ventilation. Figures 9.5 and 9.6 provide the following data:

- the number of symptoms or symptom groups in the analyses;
- the number of symptoms that were statistically significantly associated with HVAC system type; and
- when available, the range of relative risks or odds ratios for statistically significant associations.

Additionally, the presence or absence of statistically significant associations of HVAC system types with outcomes is illustrated graphically within the tables using an adaptation of the format of Mendell (1993). HVAC system types are indicated by circles located in the appropriate columns. When the type of humidification was uncertain or included multiple types, the circle is replaced with a horizontal bar extending across the applicable columns. Within these tables, shading of a circle or horizontal bar (relative to no shading) indicates that the study found a statistically significant increase in prevalence of one or more symptoms among occupants with that HVAC system type relative to buildings with the reference type of HVAC. Unshaded circles at both ends of a connecting line indicate that the subjects served by different types of HVAC systems did not have significantly different symptom prevalences. The numbers adjacent to the circles denote the number of buildings in the assessment with that type of HVAC system. Blank spaces in the tables indicate that the information was not reported.

Referring to Figure 9.5, 16 of 17 assessments found a statistically significant increase in the prevalence of one or more symptoms with air conditioning relative to natural ventilation. Nine of these assessments controlled for two or more types of confounding factors, and eight of the nine found a significant increase in symptoms with air conditioning. Two of three assessments found a statistically significant increase in the prevalence of symptoms with air conditioning relative to simple mechanical ventilation without air conditioning; however, no significant increase in symptom prevalences was found in the assessment with the largest number of buildings. Air conditioning with or without humidification was associated with significant increases in symptom prevalences. The studies provided minimal information to assess the hypothesized increase in risks with various types of humidification. In 12 of 20 assessments, air conditioning was associated with a significant increase in the prevalence of a majority of the symptoms or symptom groups. Most of the relative risks or odds ratios were between 1.3 and 3.0, indicating roughly up to 30–200 per cent increases in symptom prevalences in the air-conditioned buildings.

The results of the nine assessments that did not involve air-conditioned buildings are provided in Figure 9.6. In five of seven assessments that compared simple mechanical ventilation to natural ventilation or to sets of buildings with both natural and exhaust ventilation, prevalences of one or more symptoms were statistically significantly higher with simple mechanical ventilation.

Reference		Study and Building Characteristics					Ventilation System Type								Results	
							Mechanical Without AC				Air Conditioning					
First Author	Year	Controlled confounders	No of respondents in comparison	Sealed (S) or openable (O) windows	Smoking	Recirculation *	Natural Ventilation	Mechanical exhaust	Simple mechanical, no humidification	Simple mechanical, with humidification	No Humidification	Steam Hum.	Evaporative Hum.	Spray Hum.	Number of Symptoms with signicantly higher prevalences in assessment^^	Range of risk ratio or (odds ratio) for outcomes
Jaakkola	95	P,W,B	868	O		Y/N	7				9				2 of 14 S	1.5–2.6
Mendell	96	P,W	710	S	N	Y	3				6				6 of 7 S	1.6–5.4
Burge^	87	none	1459	S/O			11				10				10 of 10 S	(1.3–2.1)
Harrison^	87	none	1044	S		Y/N	8				6				6 of 6 S	(1.7–2.9)
Zweers	92	P,W,B	2806	S/O	Y		21				●				5 gr. of S	1.5–1.7
Jaakkola	95	P,W,B	335	O		Y	7					2			3 of 14 S	(1.9–2.5)
Burge^	87	none	863	S/O			11					4			8 of 10 S	(1.3–2.1)
Zweers	92	P,W,B	3573	S/O	Y		21					●			5 of 5 gr. of S	1.3–1.9
Jaakkola	95	P,W,B	559	O		Y/N	7						3		3 of 14 S	(2.0–2.7)
Teeuw	94	none	927	S/O		Y/N	7						7		5 of 8 S	1.4–2
Burge^	87	none	1991	S/O			11							15	10 of 10 S	(1.4–2.2)
Finnegan^	87	none	787	S	Y	Y/N	3							3	6 of 11 S	(2.5–4.8)
Harrison^	87	none	2080	S		Y/N	8							13	5 of 6 S	(2.1–3.2)
Hedge^	84	none	1214				2							2	2 of 2 S	(2.7–3.0)
Zweers	92	P,W,B	3846	S/O	Y		21							●	5 of 5 gr. of S	1.5–2.1
Brasche	99	P,W													3 of 7 S	(1.4–1.4)
Hawkins	91	P	255		N	Y	6							6	S score	
Jaakkola	95	P,W,B	1828	O		Y/N			18		9				2 of 14 S	(1.3–1.7)
Jaakkola	95	P,W,B	1295	O		Y/N			18				2		1 of 14 S	(1.8–1.8)
Jaakkola	95	P,W,B	1519	O		Y/N			18					3		

^ as reanalyzed by Mendell (1990)   P = personal factors, W = work factors, B = building factors
*In mechanically ventilated buildings   Hum = Humidification   gr = groups

Key: ○──○ } No statistically significant difference in symptoms   ○──● } Statistically significant difference in symptoms

Source: adapted from Seppänen and Fisk (2002)

Figure 9.5 *Comparison of SBS symptom prevalences with and without air conditioning*

Reference		Study and Building Characteristics					Ventilation System Type				Results	
First Author	Year	Controlled confounders	No of respondents	Sealed (S) or openable (O) windows*	Smoking	Recirculation*	Natural Ventilation	Mechanical exhaust	Simple mechanical	Simple mechanical, with humidification	Symptoms with significantly higher prevalences in assessment	Range of risk ratio or (odds ratio) for outcomes
Jaakkola	95	P,W,B	456	O		N	7	2				
Skov	90	P,W	2369	O		Y/N	9		5		2 of 2 gr. of S	1.4 – 1.8
Jaakkola	95	P,W,B	1460	O		Y/N	7		18		1 of 14 S	2.2
Mendell	96	P,W	300	O	N	Y	3		3		4 of 7 S	1.5 – 5.4
Burge^	87	none	1386	S/O			11	7			3 of 10 S	(0.7 – 0.8)
Sundell	94	P,W,B	778	S/O								
Zweers	92	P,W,B	3009	S/O	Y		21				2 gr. of S	1.3 – 1.5
Sundell	94	P,W,B	788									
Zweers	92	P,W,B	2879	S/O	Y		21				4 gr. of S	1.4 – 2.1

^as reanalyzed by Mendell (1990)   P = personal factors, W = work factors, B = building factors
*In mechanically ventilated buildings   gr. = group

Key ○—○ } No statistically significant difference in symptom prevalences   ○—● } Statistically significant difference in symptom prevalences

Source: adapted from Seppänen and Fisk (2002)

Figure 9.6 *Comparisons of symptom prevalences among buildings without air conditioning*

The study with the largest number of buildings (Sundell, 1994) did not find a significantly higher symptom prevalence with simple mechanical ventilation; however, only 10 of 540 rooms in this study had natural ventilation. In one of the five assessments (Skov et al, 1990) with increased symptoms in buildings with simple mechanical ventilation, two buildings with mechanical ventilation had humidifiers – a possible risk factor. When prevalences were significantly higher with simple mechanical ventilation, the odds ratios or relative risks ranged from 1.4 to 2.3, with one outlier of 6.0. One of these seven assessments had the opposite finding (significantly more symptoms with natural ventilation) and one had no statistically significant findings. In two other assessments in Figure 9.6, prevalences of symptoms with mechanical exhaust ventilation did not differ significantly from prevalences with natural or simple mechanical ventilation.

The results portrayed in Figures 9.5 and 9.6 provide minimal information on the potential additional risks of humidification. Hedge (1984) compared symptom prevalences among three sets of air-conditioned buildings: buildings without humidification, buildings with steam humidification and buildings with evaporative humidification. The prevalences of five of ten symptoms differed significantly among the three HVAC types, suggesting that humidification type may affect symptom prevalences. For eight of ten symptoms, prevalences were highest with evaporative humidification. The results of Zweers et al (1992) reported in Figure 9.6, comparing symptom prevalences with simple mechanical ventilation (independently with and without humidification) to symptom prevalences with natural ventilation, also suggest that humidification may be associated with higher prevalences of two out of five symptom groups.

The reasons for the consistent increases in symptom prevalences with mechanical ventilation and particularly with air conditioning remain unclear. Multiple deficiencies in HVAC system design, construction, operation or maintenance may contribute to the increases in symptom prevalences, including deficiencies that lead to pollutant emissions from HVAC systems. These are discussed in Chapter 10.

Pollutants in air-handling equipment and systems

As reported in the previous section, several studies have shown that the prevalence of SBS symptoms is usually higher in air-conditioned buildings than in buildings with natural ventilation. One explanation for the association of SBS symptoms and mechanical HVAC systems is VOCs and other chemical pollutants that are emitted by HVAC components and ductworks. It has been shown that chemical and sensory emissions of building materials, ventilation systems and HVAC components are also significant and play a major role in the indoor air quality of a space (Fanger, 1988). The emissions may originate from any component in the HVAC system. Measurements of chemical emission from typical materials used in HVAC systems are sparse. Measurements indicate that emission rates of VOCs emitted by the materials vary considerably (Morrison and Hodgson, 1996; Morrison et al, 1998). High emitting materials in their measurements were used in duct liner, neoprene gaskets, duct connectors and duct sealant. High surface area materials such as sheet metal had lower emission rates.

The European audit project on indoor air quality (Bluyssen et al, 1996), the European Database Project on Air Pollution Sources and the European AIRLESS project (Bluyssen et al, 2001) have shown that the perceived quality of supply air is not always the best possible, and is often even worse than the perceived quality of outdoor air quality. The perceived air quality of the air supplied to the rooms, however, was usually not as bad as it was immediately after passing through a filter. This may be due to absorption in duct systems or chemical reactions in the air. A recent study (Wargocki et al, 2002b) has shown a slightly lower but still significant pollution load from building sources, including the air handling systems (0.04–0.27olf/m^2).

The emission of VOCs may increase when the components and surfaces become dirty due to inferior maintenance. This hypothesis is supported by several field studies that have reported the association between the indoor air problems and cleanliness of HVAC systems. Crandall et al (1996) reported that poor HVAC cleanliness was significantly related to elevated multiple respiratory symptoms with a risk ratio (RR) of RR = 1.8, dirty filter with RR = 1.9, debris inside air intake with RR = 3.1, and dirty ductwork with RR = 2.1. These are all indicators of sources of chemical pollutants in the HVAC system.

Operation and maintenance

The strength of pollution sources may increase when dirt accumulates. Thus, the proper maintenance of the HVAC system is important to keep components clean. Several studies have found a significant relation between the indicators of poor maintenance and sick building symptoms. Crandall et al (1996) report a risk ratio of 2.0 between multiple lower respiratory symptoms and a lack of scheduled air handler inspection, and RR = 2.8 with air ductwork that has never been cleaned. Burge et al (1990) found that low symptom buildings tend to have better operation and maintenance, including manuals and instructions for maintenance. A study in the US (Dorgan et al, 1999) based on a survey of 96 office buildings reported a statistically significant trend between indoor air quality and level of maintenance. Specifically, as the level of system maintenance was increased (both depth and frequency), the IAQ level improved. The association of SBS and improper maintenance is supported by findings in the Swedish mandatory HVAC inspection programme (Engdahl, 1998). Unsatisfactory maintenance instructions were the reason for failing to pass the inspection in more than half of the buildings with a mechanical ventilation system. Angell and Daisey (1997) found that inferior HVAC maintenance was related in 39 per cent of 49 schools with poor air quality, and dirty supply air ducts and air handlers in 18 per cent of cases.

Evidence from epidemiological studies and laboratory measurements support the hypothesis that contaminated HVAC systems may be a source of pollutants and increase exposure to pollutants, which increase prevalence of sick building symptoms in office buildings. There is also a substantial amount of evidence that suggests neglected maintenance as being a major reason for these problems. Evidence also suggests that inferior maintenance is a common problem in both North America and Europe.

Severe problems are also created with condensation if the components are not properly maintained, drained and cleaned. Improperly maintained condensing cooling coils may be a major source of microbial pollution in buildings. Several studies and guidelines (for example, ISIAQ, 1996) have pointed out the importance of the cleanliness of cooling coils. For example, a study in southern California discovered that one third of the cooling coils in the large air handling units and two-thirds in the small ones were contaminated in the US (Byrd, 1996).

The importance of the cleanliness of air handling systems has been already recognized in national guidelines and standards in many countries (VDI 6022, 1997; ASHRAE, 2001; FiSIAQ, 2001; CEN, 2003).

Ventilation and pressure differences

Ventilation also affects pressure differences over building structures. This can be beneficial or harmful. One of the most important issues with respect to healthy buildings is to keep building structures dry and prevent condensation in and on structures. In cold climates, the water content of air is usually higher indoors than outdoors. If pressure is higher indoors, air with high moisture content may flow into the cold structure and water vapour may condense and cause mould and other harmful effects. To decrease the risk of condensation, buildings should have higher exhaust flow rates than supply flow rates in cold climates. This decreases the potential convection of humid indoor air to the structures. In hot climates, the problem is reversed, and supply airflow should be greater than the exhaust airflow.

Pressure difference may cause other harmful effects. If the pressure inside is smaller than outside, the pollutants in the structures or in the other side of the structures may be drawn in. One example of this is the entry of radon into buildings from the ground in houses with slab on the ground construction or in poorly ventilated crawl spaces. This is a common problem in buildings with exhaust ventilation built on the ground with high concentrations of radon in soil gas. A similar problem is faced if the house is constructed on polluted ground, such as old dump site.

Moisture damages in structure often cause the growth of mould. If the air flows through the polluted structure inside, it may also carry harmful pollutants inside. It has been shown that mould spores can be carried inside through a base floor in a building with mould growth in the crawl space.

Future research needs

The complex relationship between ventilation rate and indoor air quality greatly complicates research on the associations of ventilation rates and systems with health outcomes, productivity and perceived air quality. Many of the epidemiological studies have failed to control for important potential confounders or have incompletely characterized the study buildings and study methods. Difficulties and inaccuracies in ventilation rate measurements have also served as a barrier to this area of research.

Limitations in existing data make it essential that future studies better assess health, productivity and PAQ changes in the ventilation rate range of between 10–25l/s per person in office-type environments. Future research should be based on well-controlled cross-sectional studies or well-designed blinded and controlled experiments. The most effective studies will include high-quality measurements of ventilation rates, ample study power to detect the effects considered of public health importance, and, if possible, improved measures of adverse occupant outcomes – for example, more sensitive or more objective assessment tools.

Future research, to be optimally useful for policy efforts, should place an increased emphasis on dose–response relations useful for quantitative risk assessment, associations of health outcomes with ventilation rates per unit floor area (to assess the effects of pollutants from building sources, as well as those from occupant source), and buildings that are not office-type environments.

There is great demand for research into ventilation in residences, schools and other environments with susceptible occupants, such as the young, the elderly and unhealthy people.

The following research items were identified by a European expert group (ECA, 2003):

- Research must be conducted on the actual ventilation rates and energy use in different types of existing buildings, as well as the simultaneous effects on IAQ, health, well-being and productivity.
- There should be focused research on the linkage between properties of air-handling systems and human responses (air conditioning versus natural ventilation).
- A measured evaluation should be undertaken of 'good' naturally ventilated buildings and their properties, including human responses.
- Improved strategies and systems to control ventilation rates should be developed, including demand-controlled systems using pollutant sensors for indoor and outdoor air quality and more accurate measurement of airflow within the system.
- Research and new technology should focus on cleaning indoor air and outdoor air for ventilation.
- Research and development should highlight the importance of design tools for ventilation and IAQ calculations.

- There should be research on how to get maximum benefit from climate and natural forces in providing adequate ventilation, and how to integrate this within building design.
- Research must provide answers on how to protect building occupants against the sudden release of toxic substances in buildings.
- There should be more research on the characterization of pollution sources in buildings, and technologies to control the sources and their effects on health, well-being and productivity.
- There should be an increase in the effectiveness of ventilation at removing contaminants by using computer simulations and measurements.
- The effects of IAQ on health and productivity should be studied further and existing laboratory results validated. How performance improvements are being attributed to IAQ should also be studied in more detail.

References

Angell, W. J. and Daisey, J. (1997) 'Building factors associated with school indoor air quality problems: A perspective', in *Proceeding of Healthy Buildings Conference 1997*, vol 1, pp143–148

Apte, M., Fisk, W. and Daisey, J. (2000) 'Association between indoor CO_2 concentrations and sick building syndrome symptoms in US office buildings: An analysis of the 1994–1996 BASE study data', in *Proceedings of Indoor Air 2000*, vol 10, pp246–257

ASHRAE 62. (2001) *ANSI/ASHRAE Standard 62-2001: Ventilation for Acceptable Indoor Air Quality*, American Society of Heating Refrigerating and Air-Conditioning Engineers, Atlanta

Berg-Munch, B. (1980) 'The influence of ventilation, humidification and temperature on sensation of freshness and dryness of air', in *Proceedings of International Conference on Building Energy Management*, Portugal

Bluyssen, P., de Oliviera Fernandes, E., Groes, L., Clausen, G., Fanger, P. O., Valbjørn O., Bernhard, C. and Roulet, C. (1996) 'European indoor air quality audit project in 56 office buildings', *International Journal of Indoor Air Quality and Climate*, vol 6, no 4, pp221–238

Bluyssen, P., Seppänen, O., Fernandes, E., Clausen, G., Müller, B., Molina, J. and Roulet, C, A. (2001) 'AIRLESS: A European project to optimise indoor air quality and energy consumption of HVAC systems', in *Proceedings of CLIMA 2000*, Naples

Bornehag, C.-G., Blomquist, G. and Gyntelberg, F. et al (2001) 'Dampness in buildings and health', in *Proceedings of Indoor Air 2001* vol 11, pp72–86

Brundage, J., Scott, R. M. and Wayne, M. et al (1988) 'Building-associated risk of febrile acute respiratory diseases in army trainees', *JAMA*, vol 259, no 14, pp2108–2112

Burge, S., Jones, P. and Robertson, A. (1990) 'Sick building syndrome', in *Proceedings of Indoor Air '90*, vol 1, pp479–484

Burr, M. L. (2001) 'Combustion products', in Spengler, J. D., Samet, J. M. and McCarthy, J. F. (eds) *Indoor Air Quality Handbook*, McGraw-Hill, New York, Chapter 29

Byrd, R. (1996) 'Prevalence of microbial growth in cooling coils of commercial air-conditioning systems', in *Proceedings of Indoor Air '96*, Seec Ishibashi Inc, Japan, vol 3, pp203–207

CEN (European Committee for Standardization) (1996) *Technical Report CR 1752: Ventilation for buildings: Design criteria for indoor environment*, CEN, Brussels

CEN (2003) *European Standard. Draught, prEN 13779: Ventilation for non-residential buildings – Performance requirements for ventilation and room conditioning systems*, CEN, Brussels

Chivato, T., Montoro, A., Martinez, D., Gill, P., Zubeldia, J., De Barrio, M., Baeza, M. L., Rubio, M. and Laguna, R. (1997) 'Clinical tolerance, parasitological efficacy and environmental effects of dehumidifiers in stable asthmatics sensitized to house dust mites', *Allergol. et Immunopathol.*, vol 25, no 2, pp67–72

Crandall, M., Sieber, W. and Malkin, R. (1996) 'HVAC and building environmental findings and health symptoms associations in 80 office buildings', in *Proceedings of IAQ 1996*, pp103–108

Crane, J., Ellis, I., Siebers, R., Grimmet, D., Lewis, S. and Fitzharris, P. (1998) 'A pilot study of the effect of mechanical ventilation and heat exchange on house-dust mites and Der p 1 in New Zealand homes', *Allergy*, vol 53, pp755–762

Drinka, P., Krause, P. and Schilling, M. (1996) 'Report of an outbreak: Nursing home architecture and influenza-A attack rates', *Journal of American Geriatric Society*, vol 44, pp910–913

Dorgan, C. B., Dorgan, C. E. and Linder, R. J. (1999) 'The link between IAQ and maintenance', in *Proceedings of IAQ and Energy*, American Society of Heating, Refrigerating and Air-Conditioning Engineers Inc, Atlanta, GA, pp47–62

ECA (European Collaborative Action) (2003) *Urban Air, Indoor Environment and Human Exposure: Ventilation, Good Indoor Air Quality and Rational Use of Energy*, Report 23, European Collaborative Action, Joint Research Centre –Institute for Health and Consumer Protection, European Commission, Office of Official Publications, Luxemburg

Engdahl, F. (1998) 'Evaluation of Swedish ventilation systems', *Building and Environment*, vol 33, no 4, pp197–200

Fang, L., Clausen, G. and Fanger, P. O. (1997) 'Impact of temperature and humidity on the perception of indoor air quality during immediate and longer whole-body exposures', in *Proceedings of Healthy Buildings '97*, Healthy Buildings/IAQ´97, Washington, DC, vol 2, pp231–236

Fang, L., Clausen, G. and Fanger, P. O. (1998) 'Impact of temperature and humidity on the perception of indoor air quality', *International Journal of Indoor Air Quality and Climate*, vol 8, no 2, pp80–90

Fanger, P. O. (1988) 'Introduction of the olf and the decipol units to quantify air pollution perceived by humans indoors and outdoors', *Energy and Buildings*, vol 12, pp106–112

FiSIAQ (Finnish Society of Indoor Air Quality and Climate) (2001) *Classification of Indoor Climate 2000*, FiSIAQ, Espoo, Finland, Publication 5 E

Fisk, W. J., Faulkner, D., Palonen, J. and Seppanen, O. (2002) 'Performance and costs of particle air filtration technologies', *Indoor Air Journal*, vol 12, pp223–234

Fisk, W. J., Seppänen, O., Faulkner, D. and Huang, J. (2003) 'Economizer system cost effectiveness: Accounting for the influence of ventilation rate on sick leave', in *Proceedings of Healthy Buildings 2003*, National University of Singapore, Singapore, 7–11 December

Gold, D. (1992) 'Indoor air pollution', *Clinics in Chest Medicine*, vol 13, pp215–229

Harving, H., Korsgaard, J. and Dahl, R. (1993) 'House-dust mites and associated environmental conditions in Danish homes', *Allergy*, vol 48, pp106–109

Hedge, A. (1984) 'Evidence of a relationship between office design and self-reports of ill health among office workers in the United Kingdom', *Journal of Architectural Planning and Research*, vol 1, pp163–174

Hoge, C. W., Reichler, M. R., Dominiguez, E A. et al (1994) 'An epidemic pneumococcal disease in an overcrowded, inadequately ventilated jail', *New England Journal of Medicine*, vol 331, no 10, pp643–648

Howieson, S., Lawson, A., McSharry, C., Morris, G. and McKenzie, E. (2002) 'Indoor air quality, dust mite allergens and asthma', in Levin, H. (ed) *Proceedings of the Ninth International Conference on Indoor Air Quality and Climate*, vol 1, pp113–118

Humphreys, M. A., Nicol, J. F. and McCartney, K. J. (2002) 'An analysis of some subjective assessments of indoor air-quality in five European countries', in Levin, H. (ed) in *Proceedings of the Ninth International Conference on Indoor Air Quality and Climate*, vol 5, pp86–91

ISIAQ (International Society of Indoor Air and Climate) (1996) *Control of moisture problems affecting biological indoor air quality: ISIAQ guideline*, Task Force I, Flannigan, B. and Morey, P. (eds), Ottawa. Canada

Jones, A. P. (1999) 'Indoor air quality and health', *Atmospheric Environment*, vol 33, pp4535–4564

Leaderer, B. and Cain, W. (1983) 'Air quality in buildings during smoking and nonsmoking occupancy', *ASHRAE Transactions*, vol 89, pp601–611

Mendell, M. J. (1993) 'Non-specific symptoms in office workers: A review and summary of the epidemiologic literature', *International Journal of Indoor Air Quality and Climate*, vol 3, pp227–236

Milton, K., Glenross, P. and Walters, M. (2000) 'Risk of sick leave associated with outdoor air supply rate, humidification, and occupant complaint', *International Journal of Indoor Air Quality and Climate*, vol 10, pp211–221

Morrison, G. and Hodgson, A. (1996) 'Evaluation of ventilation system materials as sources of volatile organic compounds', in *Proceedings of the Seventh International Conference of Indoor Air Quality and Climate*, vol 3, pp585–590

Morrison, G., Nazaroff, W., Cano-Cruiz, A., Hodgson, A. and Modera, M. (1998) 'Indoor air quality impacts of ventilation ducts: Ozone removal and emissions of volatile organic compounds', *Journal of the Air and Waste Management Association*, vol 48, pp941–952

Nardell, E. A., Keegan, J. and Cheney, S. A. (1991) 'Theoretical limits of protection achievable by building ventilation', *American Review of Respiratory Disease*, vol 144, pp302–306

Nevalainen, A. (2002) 'Of microbes and men', in Levin, H. (ed) *Proceedings of the Ninth International Conference on Indoor Air Quality and Climate*, vol 3, pp1–9

Niemelä, R., Seppänen, O. and Reijula, K. (2003) 'Prevalence of SBS-symptoms as an indicator of health and productivity in office buildings', in Tham, K. W., Sekhar, C. and Cheong, D. (eds) *Proceedings of Healthy Buildings Conference Singapore 2003*, vol 3, pp251–256

Niven, R. M., Fletcher, A. M., Pickering, A. C., Custovic, A., Sivour, J., Preece, A. R., Oldham, L. A. and Francis, H. (1999) 'Attempting to control mite allergens with mechanical ventilation and dehumidification in British houses', *Journal of Allergy and Clinical Immunology*, vol 103, pp756–762

Seppänen, O. (1998) 'Ventilation strategies for good indoor air quality and energy efficiency', *IAQ and Energy*, American Society of Heating Refrigeration and Air Conditioning, Atlanta, GA, pp257–276

Seppänen, O. (2003) 'Healthy buildings – from science to practice', Plenary lecture, in *Proceedings of Healthy Buildings Conference 2003*, Singapore

Seppänen, O. and Fisk, W. (2002) 'Association of ventilation type with SBS symptoms in office workers', *International Journal of Indoor Environment and Health*, vol 12, no 2, pp98–112

Seppänen, O., Fisk, W. J. and Faulkner, D. (2003) 'Cost benefit analysis of the night time-ventilative cooling in office building', in *Proceedings of Healthy Buildings Conference 2003*, Singapore

Seppänen, O. A., Fisk, W. J. and Mendell, M. J. (1999) 'Association of ventilation rates and CO_2-concentrations with health and other responses in commercial and institutional buildings', *International Journal of Indoor Air Quality and Climate*, vol 9, pp252–274

Seppänen, O. and Jaakkola, J. (1989) 'Factors that may affect the results of indoor air quality studies in large office buildings', in Nagda, N. and Harper, J. (eds) *Design and Protocol for Monitoring Indoor Air Quality*, ASTM STP 1002, American Society of Testing Materials

Skov, P., Valbjørn, O., Pedersen, B. and DISG (1990) 'Influence of indoor climate on the sick building syndrome in an office environment', *Scandinavian Journal of Work, Environment and Health*, vol 16, pp363–371

Sundell, J. (1994) 'On the association between building ventilation characteristics, some indoor environmental exposures, some allergic manifestations and subjective symptom reports', *Indoor Air*, Supplement no 2/94, pp1–49

Tham, H. C., Willem, H. C., Sekhar, S. C., Wyon, D. P., Wargocki, P., Fanger, P. O. (2003a) 'The SBS-symptom and environmental perceptions of office workers in the Tropics at two air temperatures and two ventilation rates', in *Proceedings of Healthy Buildings 2003 Conference*, Singapore, vol 1, pp182–187

Tham, H. C., Willem, H. C., Sekhar, S. C., Wyon, D. P., Wargocki, P., Fanger, P. O. (2003b) 'Temperature and ventilation effects on the work performance of office workers (study of a call center in tropics', in *Proceedings of Healthy Buildings 2003 Conference*, Singapore, vol 3, pp280–286

VDI (Verein Deutscher Ingenieure) 6022 (1997) *Hygienic Standards for Ventilation and Air-Conditioning Systems – Offices and Assembly Rooms*, VDI (Association of German Engineers), Germany

Wargocki, P., Bako-Biro, Z., Clausen, G. and Fanger, P. (2002b) 'Air quality in a simulated office environment as a result of reducing pollution sources and increasing ventilation', *Energy and Buildings*, vol 34, pp775–783

Wargocki , P., Sundell, J., Bischof, W., Brundrett, G., Fanger, O., Gyntelberg, F., Hanssen, S. O., Harrison, P., Pickering, A., Seppänen, O. and Wouters. P. (2002a) 'Ventilation and health in non-industrial indoor environments: Report from a European Multidisciplinary Scientific Consensus Meeting', *International Journal of Indoor Environment and Health*, vol 12, pp113–128

Wargocki, P., Wyon, D. and Fanger, P. (2000b) 'Pollution source control and ventilation improve health, comfort and productivity', in *Proceedings of Cold Climate HVAC '2000*, pp445–450

Wargocki, P., Wyon, D. and Fanger, P. (2003) 'Call-centre operator performance with new and used filters at two outdoor air supply rates', in Tham, K. W., Sekhar, C. and Cheong, D. (eds) *Proceedings of Healthy Buildings Conference Singapore 2003*, vol 3, pp213–218

Wargocki, P., Wyon, D., Sundell, J., Clausen, G. and Fanger, P. (2000a) 'The effects of outdoor air supply rate in an office on perceived air quality, sick building syndrome (SBS) symptoms and productivity', in *Proceedings of Indoor Air 2000*, vol 10, pp222–236

Warner, J. A., Frederick, J. M., Bryant, T. N., Weich, C., Raw, G., Hunter, C., Frank, R., McIntyre, D. A. and Warner, J. O. (2000) 'Mechanical ventilation and high-efficiency vacuum cleaning: A combined strategy of mite and mite allergen reduction in the control of mite-sensitive asthma', *The Journal of Allergy and Clinical Immunology*, vol 105, pp75–82

WHO (World Health Organization) (1989) *Guidelines for Community Noise*, Berglund, B., Lindvall, T. and Schwela, D. H. (eds), WHO, Geneva

Wolkoff, P., Clausen, P. A., Wilkins, C. K. and Nielsen, G. D. (2000) 'Formation of strong airways irritants in terpene/ozone mixtures', *International Journal of Indoor Air Quality and Climate*, vol 10, pp82–91

Zweers, T., Preller, L., Brunekreef, B. and Boleij, J. S. M. (1992) 'Health and comfort complaints of 7043 office workers in 61 buildings in the Netherlands', *Indoor Air*, vol 2, no 3, pp127–136

10
Advanced Components for Ventilation

W. F. de Gids

Introduction

Since the Air Infiltration and Ventilation Centre's (AIVC's) publication *Advanced Ventilation Systems* (Knoll, 1992), the development of systems and components has been accelerated. For a description of the principles and mechanisms involved, this technical report is still very valuable. This chapter may be seen as an update of *Advanced Ventilation Systems* in terms of new developments on components for ventilation systems.

A ventilation system is a composition of components for ventilation. Ventilation means air supply and exhaust. For some people, only mechanical ventilation exists and natural ventilation is not considered a ventilation system at all. Within the international community, a ventilation system is the combination of all components intended for ventilation, from inlet to (roof) outlet. This means that windows or even vent holes are considered as components for ventilation. A chapter about advanced components is difficult to write. What is advanced, today, may not be advanced in the near future; what is advanced today in a certain country may not at all be advanced in other countries. Apart from climate and building practices, strategies in building regulations and standards and historical and cultural aspects all play an important role in ventilation. As a result, types of ventilation system components are described in this chapter. Although many examples given in this chapter are from residential ventilation practices, there is no reason not to apply them in small offices and other utility-type buildings.

The role of infiltration

The tightness of the building envelope is critical to energy performance, the correct functioning of a ventilation system and comfort. Draught and temperature control may be disturbed by poor air tightness within a building. The general rule is 'build tight, ventilate right'. This general rule is explained, together with background and ventilation strategies, in the AIVC *Guide to Energy Efficient Ventilation* (Liddament, 1996).

Low air tightness levels may cause infiltration at a rate that is larger than the flow through an air inlet. This is, of course, not the right way to ventilate. Relying upon infiltration to satisfy indoor air quality requirements is not advisable because:

- One needs an air leakage level of the building, which over time will lead to too high airflow rates and, hence, unwanted energy use.
- Within a leaky building there might be rooms with almost no infiltration; because of its nature, the air tightness of buildings is not necessarily well distributed over the building envelope.

Nevertheless, there are still countries where infiltration is considered as an air supply for ventilation purposes. In the following section the purpose for air supplies is described.

Air inlets

Air inlets are, in most cases, openings between the inside and outside. Through natural forces, wind and buoyancy, air is driven through the openings to ventilate the room

Source: chapter author

Figure 10.1 *An example of a window frame in a dwelling*

behind. Mechanical air supply is also used in many buildings, not only through a network of ducts, but also through decentralized supply units for each room. In natural ventilation, air supply is a key part of the design of buildings. In cases where air supplies are designed in the wrong way, one might expect complaints by occupants regarding draught and stuffiness. Substantial differences can be found in the capacity or size of inlets available on the European market. The reason for these differences is primarily the large difference in the requirements of varying national building regulations. For example, for the same size of room, the difference in the capacity of air inlets expressed in square centimetre net openings is as follows:

- France = 20cm^2;
- UK = 40cm^2;
- Belgium = 70cm^2;
- The Netherlands = 100cm^2.

Because of these differences, the way in which air inlets are manufactured in different countries varies substantially. Most of the data on controlled air inlets has been taken from the NATVENT project (Perera, 1999).

Windows

Relatively large windows are frequently used for airing and, in some countries, for indoor air quality control. They can be tilted, turning and sliding windows or combinations of these. The most advanced windows are windows with an automatic control on the opening to ensure good indoor air-quality levels, at the same time as reasonable conditions of comfort. Comfort conditions and the use of windows are very dependent upon weather conditions, such as outside temperature, wind and rain. In Figure 10.1 an example is given with a window frame consisting of a side-hung window and a fixed non-movable window in which an air inlet is built in at the top. In mild climates, windows are used for airing purposes, as well as for ventilation in relation to indoor air quality control. In moderate climates, the comfort problem already plays an important role when windows are used for ventilation. Large, low openings may cause draught and, hence, comfort problems. The control of the opening is normally manual and is quite crude. There are components available to control the opening of these large windows, in wide as well as ajar positions, as shown in Figure 10.2. The window in the lower left of Figure 10.2 is not completely closed; but because of a double clamp, the window is in an ajar position, so is slightly open (8mm). Also reasonably advanced are the so-called double-sliding windows (see Figure 10.3). The advantage of a double-sliding window is twofold: one can control the flow rate easily by adjusting the position of the two opposite window panes. At the same time, one can adjust air velocity to the room by adjusting the opening to the room in such a way that comfort problems during the heating season will not occur as frequently as in the case of side-hung turning windows. Side-hung windows are used quite frequently in moderate climates. People do not use them only for ventilation or indoor air quality control, but also for temperature control. Figure 10.4 provides information on the use of windows from the AIVC report *Inhabitants' Behaviour with Regard to Ventilation* (Dubrul, 1988). From Figure 10.4 it might be concluded that a high relation between window use and outside temperature exists. The average opening time is three to four hours, mainly during daytime, even for bedrooms. This is not very energy efficient. Airing for 20 to 30 minutes is normally sufficient. The conclusion is that side-hung windows are not suitable for indoor air quality control in climates with low outside temperatures. Another disadvantage of larger windows is the problem of burglary, especially in cases where these windows are situated at ground level.

Vents

Vent lights

In some window frames a vent light has been incorporated (see Figure 10.5). When vent lights are situated about 1.8m above floor level, they might work sufficiently for indoor air quality control over a wide range of temper-

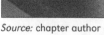
Source: chapter author

Figure 10.2 *Some control elements for adjusting side-hung windows*

atures. The problem is manual controllability. As a result, they work better than side-hung windows, causing less draught problems; but they may still cause draught problems at low outside temperatures. The use of vent lights is less dependent upon the time of day. Like side-hung windows, their use is very dependent upon outside temperature (see Figure 10.6)

Inlet grilles

Manually controlled inlets can fulfil indoor air quality control with all types of extraction. The application of passive extraction is possible, as is the integration of cross-ventilation within a system. Integration with mechanical exhaust systems is also relatively easy. In cases where the natural driving force fails, the mechanical extract system guarantees a minimum required flow through the building. The distribution may not be adequate, but a flow is at least guaranteed (see Figure 10.7)

In Figure 10.7, the upper figure is a typical trickle vent with a slit in the upper part of the window frame; on the inside there is a diffuser with a manual control. Typical sizes of trickle vents' ventilation openings are in the range of 15–40cm^2.

The lower picture comprises an inlet grille for installation above the double glazing in a window. Typical sizes of grilles are 70–140cm^2. These types of inlets feature continuous manual control possibilities.

Over the last number of years, a wide range of options for improving inlets have been developed, such as:

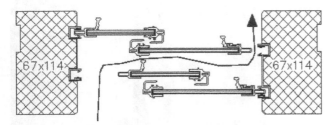

Source: De COCK van Gelder BV

Figure 10.3 *An example of a double-sliding window*

- possibilities for demand control based on sensors or presence detectors;
- sound attenuation; and
- air cleaning or filtering.

In areas with high outside noise levels due to air, rail or car traffic, these openings may cause significant problems when open. Occupants have a tendency to keep them closed to avoid hearing outside noise. As a result, sound-attenuated inlets are available, featuring a sound reduction up to about 25 dBA.

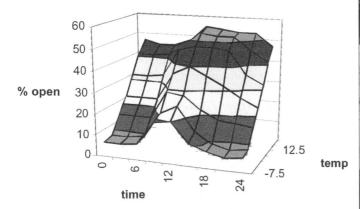

Source: TNO Building and Construction Research

Figure 10.4 *The relation between percentage of windows open over the time of the day against the 24-hour mean outside temperature*

Source: chapter author

Figure 10.5 *An example of a vent light in an open and closed position, including control mechanism*

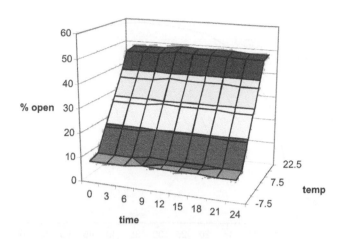

Source: TNO Building and Construction Research

Figure 10.6 *The relation between percentage of vent lights open over the time of the day against the 24-hour mean outside temperature*

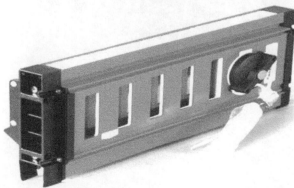

Source: ALUSTA BV

Figure 10.7 *Two examples of manually controlled inlet grilles*

Pressure-controlled air inlets

The objective of pressure-controlled inlets can be expressed as a constant natural supply of airflow independent of the pressure difference caused by wind and buoyancy. Their main feature is the prevention of over-ventilation, resulting in less comfort problems while saving energy.

Passive air inlets

Low pressure

Pressures differences across façades are very often in the range of 1–5Pa. The idea is to keep the flow rate independent of the pressure difference for most of the time. An example from The Netherlands with a constant flow rate of 1Pa is given in Figure 10.8. The size of this inlet is about 10cm high and 18cm deep. The principle is a balancing vane, which controls the opening. Air comes in on the right-hand side, causing an over-pressure on the vane, which tilts and closes the opening gradually. The performance of the inlet given in Figure 10.8 is shown in Figure 10.9. The performance curve is not flat or horizontal, which one might expect, but declining because there is a fixed built-in compensation for the air tightness of the façade.

High pressure

The high-pressure air inlet typically starts to control at around 10Pa reaching a more stable airflow rate at somewhere around 20Pa. The percentage occurrence of exceeding time pressures of 10–20Pa is quite small. An example of this performance is given in Figure 10.10.

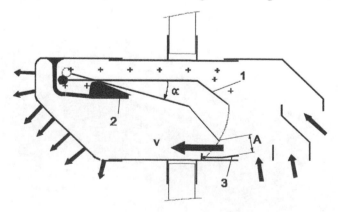

Source: EMPA

Figure 10.8 *An example of a passive pressure-control inlet*

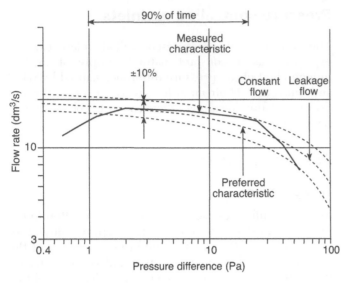

Source: TNO Building and Construction Research

Figure 10.9 *The relation between flow rate through and pressure difference across the passive air inlet*

Industry in France has already produced these type of inlets for many years. The idea is that mechanical extract ventilation may deliver a pressure difference across the inlet grille larger than 15Pa. This is valid for airtight dwellings in the event that all windows are closed. This approach is typical for France.

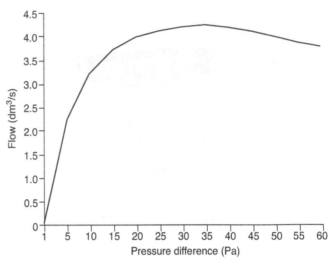

Source: TNO Building and Construction Research

Figure 10.10 *An example of a pressure-controlled air inlet with an almost constant airflow at 20Pa*

Active air inlets

Active pressure-controlled inlets are available in The Netherlands and, as with passive air inlets, they control from 1Pa onwards. An example of such an inlet is given in Figure 10.11. The performance curve is presented in Figure 10.12. A small electronic motor controls the opening position of the inlet. The airflow is constant over

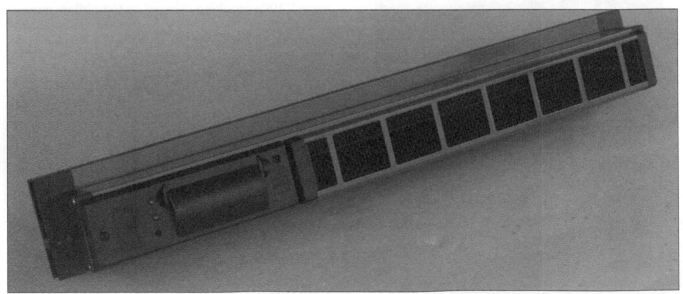

Source: EU NATVENT

Figure 10.11 *An example of an active inlet working at 1Pa*

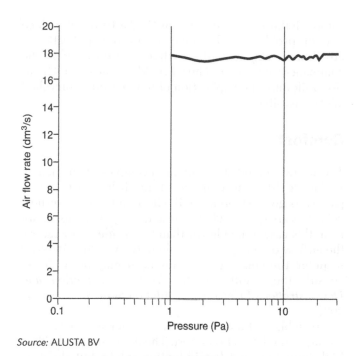

Source: ALUSTA BV

Figure 10.12 *An example of the performance of an active inlet controlling at 1Pa*

Humidity-controlled inlets

Most humidity-controlled inlets act on the relative humidity (RH) of the air. The calibration of the available inlets is quite reproducible and accurate. The principle of these inlets is primarily based on the change in length of a tape, due to a change in RH. An example of the characteristic for an inlet from France is given in Figure 10.13. As can be seen, the characteristic shows a hysteresis effect. For instance, at 50 per cent RH coming from a higher RH, the flow rate is 15m^3/h, while 50 per cent RH coming from a lower RH gives an airflow rate of about 25m^3/h. This difference is quite remarkable.

Pollutant-controlled inlets

Pollutant-controlled inlets are rarely available on the market. Nevertheless, it is good to realize that pollutant-controlled ventilation is applied in several types of buildings, such as schools, theatres, shopping malls, congress halls and parking lots. A signal from the sensor can control the extract fan and/or the inlets with electronic controls. CO_2 sensors that are coupled to a central control unit are the most commonly applied pollutant-control inlets today. These units control the exhaust fan as well as the inlets. An illustration of a CO_2 sensor is given in Figure 10.14. The types of sensors applied today are:

- carbon monoxide (CO) for parking lots; and
- CO_2 and mix gas for all other buildings with a variation in occupancy over time.

a wide range of pressure differences, which frequently occur. These inlets are also available with sound attenuation. This is the most advanced option, and its application is in great demand in controlled hybrid ventilation. Systems that are now on the market are, in most cases, based on carbon dioxide (CO_2) or timer controlled.

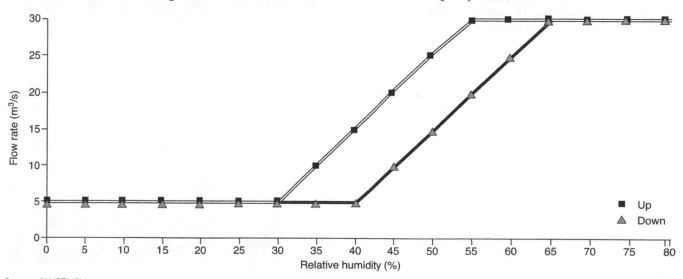

Source: ALUSTA BV

Figure 10.13 *The relation between relative humidity (RH) and flow rate*

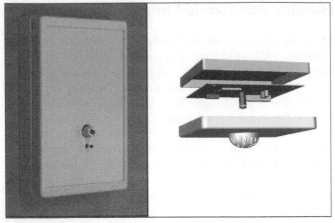

Figure 10.14 *A photo and a schematic view of a CO_2 sensor*

The number and positions of the sensors, together with the control algorithm, determine whether or not the 'set-point flow rate value' will be maintained.

Temperature-controlled inlets

Temperature-controlled inlets are on the market in countries where the outside temperature is more important as a driving force for ventilation. An example of the characteristics of a temperature-controlled air inlet from Sweden is given in Figure 10.15. The lower the outside temperature, the larger is the pressure difference and the lower the airflows will be. The control of the inlet is a bi-metal sensor which, by bending, closes the inlet. It is unclear how these inlets are calibrated. Most of these inlets also have considerable 'dead time' and hysteresis

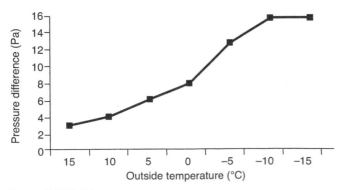

Figure 10.15 *An example of the performance of a temperature-controlled air inlet*

(although this is not presented in Figure 10.15). To reduce draught problems and to save energy, the application is possible in climates where thermal buoyancy is the dominant driving force. Thus, in colder climates and high-rise buildings, the application of temperature-controlled inlets is feasible.

Comfort

The airflow behind an air inlet is a complex phenomenon. In Figure 10.16 an example of an airflow distribution pattern is given (Dörer, 2004). Close to the inlet, airflow velocities are quite high. Because the supply temperature is, in this case, much lower than the inside temperature, the airflow direction is down along the window. In order to judge the comfort of air inlets, it may be useful to consult the results of the AIVC's *Evaluation and Demonstration of Domestic Ventilation Systems* (Månsson, 2002).

In Table 10.1 an overview is given for several types of air inlets in relation to comfort. The inlets are divided into high-induction and low-induction inlets; but the size, shape, aspect ratio and low or high position also play a role.

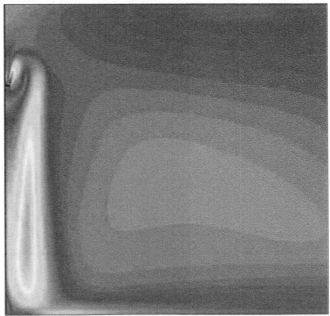

Figure 10.16 *A computational fluid dynamics (CFD) result of the airflow distribution behind a natural air inlet*

Table 10.1 *Comfort due to air inlets for several outside temperatures*

Type of ventilation opening		Outside temperature in degrees Celsius (°C)						
General	Specific	−15	−10	−5	0	5	10	15
High induction	Direction supply upwards	-	-	0	+	++	++	++
	Circular opening, radial flow direction	−	−	-	-	0	++	++
	Horizontal supply	−	−	−	−	-	+	++
Low induction or windows ajar	Horizontal supply, high position	−	-	-	0	+	+	++
	Vertical opening, medium position	−	−	−	−	−	-	-
	Horizontal opening, low position	−	−	−	−	−	-	-
Infiltration	Good air tightness plus mechanical exhaust	−	-	-	-	-	0	0
	Medium air tightness plus mechanical exhaust	−	−	−	-	-	+	++

Notes: The scores describe the percentage of the occupied zone that fulfils the comfort conditions according to Fanger's comfort theory:
++ 100–95%; + 95–85%; 0 85–75%; - 75–50%; − 50–0%

Source: IEA

Remarks

The philosophical difference in building regulations worldwide regarding ventilation and the way in which requirements are described have a large influence on the design and performance of air inlets. For example:

- In France air inlets may not be fully closed, while in The Netherlands inlets must be completely closable.
- In The Netherlands the controllability of these inlets must be between 1–25Pa, whereas in France the control may start at about 15–20Pa.

For all types of inlets, it is important to be able to clean them regularly. For controlled inlets, calibration is especially needed, even occasionally during yearly maintenance.

Integration within the ventilation system

Inlets cannot be seen as isolated components for ventilation. They can only fulfil their task when they are opened or controlled or when enough pressure difference across the inlet is available. They also depend upon the exhaust part of the system. This might involve passive vents

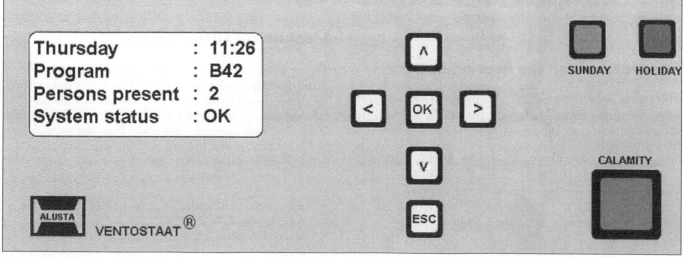

Source: ALUSTA BV

Figure 10.17 *A so-called 'ventostat' for time programming the demand per room*

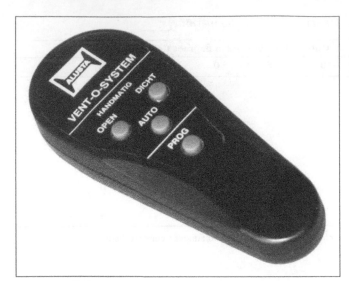

Source: ALUSTA BV

Figure 10.18 *A remote control for inlets*

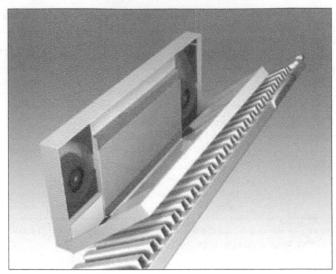

Source: CLIMARAD BV

Figure 10.19 *An air inlet combined with a warm water radiator*

outside where one might expect an under-pressure (mostly at roof level), or a fan with ducting to the rooms to be extracted. Systems with controlled air inlets are now available for dwellings and office-type buildings. Some systems have a central control unit where, depending upon the occupation of rooms, the desired airflow rate can be programmed. Instead of having a sensor, time programming is used to determine ventilation demand. An example of such a control unit is given in Figure 10.17. The same system has, for each of the rooms, the possibility of a remote control to overrule the central control setting of an inlet manually. After a specified period of time (from 3 to 12 hours), the central control unit of the system takes over again (see Figure 10.18).

Integration with the heating system

The latest development is the combination of air inlets with a warm water radiator. Incoming air can be preheated, minimizing draught problems. Although the idea is quite old, the application of such units in reality, including the control for preheating, has taken some time to achieve. A more complex control system is necessary. Some of these radiator vents have for the possibility of filtering, and even CO_2-controlled vents are on the market. An example is provided in Figure 10.19.

Mechanical supply units per room

Mechanical supply units per room are sometimes applied in cases where either natural forces cannot sufficiently bring enough air in, or in cases where other reasons may prevail, such as sound attenuation and or filtering of air. Relatively small units ($0.5 \times 0.5 \times 0.15$m, with capacities of up to 60dm^3/s) are available. Most of them feature sound attenuation. Some of them also have the possibility of filtering, as well as CO_2 sensor control (see Figure 10.20)

Mechanical supply valves

These valves are connected to ductwork. There has been little development over the last few years concerning mechanical supply valves. A frequent problem is the dust forming either on the valve itself or on the wall or ceiling around it. This dust does not come from the supplied air but is caused by the induction of air. The dust from the internal environment is deposited on and/or around the supply valves. Possible solutions for this occurrence are:

- reduce the induction, which, in fact, means lower supply velocities; or
- bring the air in at a certain distance from the wall or ceiling.

Source: ALUSTA BV

Figure 10.20
*An example of a
local mechanical
supply unit*

Source: Bergschenhoek BV

Figure 10.21
*An example of a
mechanical
air-supply valve*

An example is given in Figure 10.21. The pressure difference across these valves is typically in the range of 10–30Pa at flow rates varying from about 5–25dm³/s.

Although it is, in principle, possible to apply the same type of pressure control on these inlets, there are (as far as is known) no products available with these options.

Extract valves

Fixed extract valves

Fixed valves mean, in this context, non-manual adjustable valves for normal use in combination with mechanical extraction. Most valves can be adjusted; but the purpose of that adjustment is to commission the system. Low-pressure valves have resistances that are almost negligible; they are more or less just aesthetical components. An example is provided in Figure 10.21. The pressure difference across the valve is normally less than 1Pa at nominal flow rates. A more advanced one is a result from RESHYVENT (de Gids, 2004) (see Figure 10.22). An example of a high-pressure valve is given in Figure 10.23. Typical pressure differences across these valves are 20–40Pa, at flow rates varying from 5–25dm³/s. These valves are widely applied in mechanical extraction.

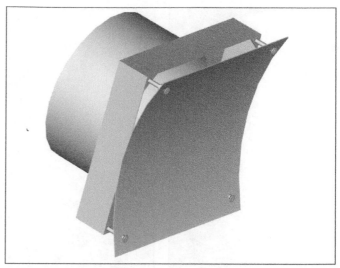

Source: Bergschenhoek BV

Figure 10.22 *A fixed low-pressure extract valve*

Pressure-controlled valves

Pressure-controlled extract valves have been available for many years in France. The principle behind them is a flexible, soft plastic cylinder that is blown up by the inflowing air. When the extracted airflow increases, the cylinder slowly closes the extract opening (see Figure 10.24). The flow stabilizes a pressure difference of about 20Pa.

Ductwork

Ducts

Chapter 3 of this book is devoted to ductwork. In this chapter, only some of the latest developments are described concerning sound attenuation in ductwork, special fittings, odd shapes and materials. The most applied type of ducting in buildings is probably the round spiro-type duct. Rectangular ducting is also widely used. In special cases, rectangular ducting is integrated within a concrete floor. Ducting is no longer an obstacle. Maximum height of ductwork is mostly around 80mm. The rectangular ducting integrated within the floor can easily be crushed, which leads to high resistances. Ducts are also available as concrete blocks. A Belgian manufacturer makes storey-high concrete ducts. Contrary to what one might expect, concrete ductwork may also be quite airtight. In the case of retrofitting, this is done by a plastic coating or lining on the inside of the duct. However, even

Source: Bergschenhoek BV

Figure 10.23 *An example of a high-pressure extract valve*

Source: AERECO

Figure 10.24 *A pressure-controlled extract valve*

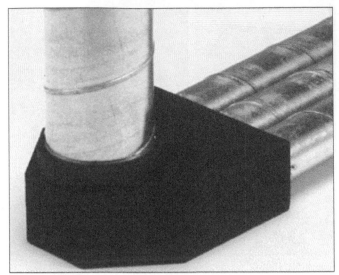

Source: EU TIP-Vent

Figure 10.25 *Special low-pressure fittings for domestic ventilation systems*

newly manufactured 'floor storey-high' concrete ducts may reach class B (de Gids, 1996) This can only be realized when the ducts are constructed in an industrial environment and assembled on site by good craftsmen. The resistance of these ducts is comparable with metal-type ducting (de Gids, 2003) Some special duct shapes are available on the market, such as oval (metal) ducting. These ducts can be used in the case of small height for installations.

Fittings

Some special fittings have been developed within the framework of the European Union (EU) TIP-Vent project (Wouters, 2000) in order to minimize the resistance of ductwork. From analyses on existing systems, it became clear that an important part of the total pressure drop in a typical domestic mechanical extract system was caused by the distribution from the vertical ducts to the horizontal ducts. Figure 10.25 provides an example of such fittings. A reduction of about 10–5Pa resulted at flow rates of around 40–0dm³/s.

Active sound attenuators for small ducts

In the TIP-Vent project, an attempt was made to develop an active sound attenuator for ducts and flow rates used in domestic ventilation. Active noise control is based on the principle of a counter-noise, which is a sound wave with the same amplitude and same frequency, but in phase opposition with the primary wave to be attenuated.

The principle and some results are shown in Figure 10.26. Sound attenuation of 10–15dB is obtained between 100 hertz (Hz) and 600Hz, with some fluctuations.

Fans

Even for domestic ventilation, fans vary over a wide range of capacities; types of fans are also quite different. An overview of fans used for domestic ventilation is given in Figure 10.27. Fan power is the second most important factor responsible for energy consumption due to ventilation, after conditioning of air. On a percentage basis, it was verified that fan energy consumption could generally be low if fans are properly selected, especially in residential buildings (usually bellow 10 per cent of the total heating, ventilating and air-conditioning, or HVAC, energy consumption). In other types of buildings, fan consumption can exceed 20 per cent of total needs.

Fan power for reasonable designs typically ranges in non-residential buildings from 1–5W/(dm³/s). In the TIP-Vent project, the energy for fans in domestic ventilation 0.5W/(dm³/s) was used as a reference situation. The improvements made in TIP-Vent are around a factor of 3. The fan power was reduced from 0.5–0.15W/(dm³/s).

A unique type of fan specially developed for hybrid ventilation systems is shown in Figure 10.28. This fan has a flat curve instead of the steep curves of most ordinary ventilation. The reason for this is because these fans are used in hybrid ventilation.

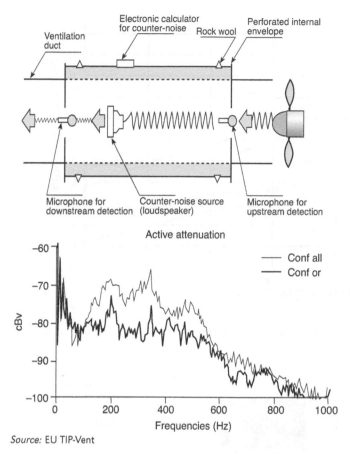

Figure 10.26 *An active sound attenuator for domestic ventilation*

Source: EU TIP-Vent

Source: StorkAir BV

Figure 10.27 *An overview of fans used in domestic ventilation: (a) duct fan; (b) axial fan; (c) radial fan in a fan box*

Heat recovery units

Heat recovery units can be a component of a balanced ventilation system. They transfer heat from the extracted airflow to the incoming air for preheating. This is a very energy-efficient way of achieving ventilation. Today, high efficiencies are possible. Efficiencies measured under laboratory conditions of about 90 per cent are possible. Heat recovery in balanced systems requires a supply ducting and distribution system. This means more crafts-manship than ordinary extract ventilation systems. The design of the inlet position and flow direction has to be carried out with special care, otherwise draught problems may occur. Possibilities for filtering and humidification are available. An example is provided in Figure 10.29. Typical pressure differences across these high-efficiency units are 50–100Pa at flow rates from 42–70dm³/s.

Roof outlets

Roof outlets or cowls are the ultimate destination of ventilation air before it is delivered to the outside. There are different types of cowls:

- passive;
- wind driven;
- active mechanical; and
- hybrid.

An example is given in Figure 10.30.

The main goal of the cowl is to overcome back-draughting of the extracted flow. It might also be seen as a protection against the negative effect of the wind. Some people expect an extra under-pressure or suction effect from these cowls. Although this is, in some cases, true, the real positive effect in terms of under-pressure is relatively small. The only cowl that almost always produces an under-pressure is the turning cowl, which always finds a position as a wind vane. The wind-driven fan that rotates due to wind features a unique performance. For low wind velocities, these fans normally work as a resistance; but they always ensure a certain flow to the outside. The performance is, however, not proportional to wind velocity. For non-rotating cowls, one might be very pleased to reach a performance where for all wind attack angles an over-pressure can be prevented (see Figure 10.31).

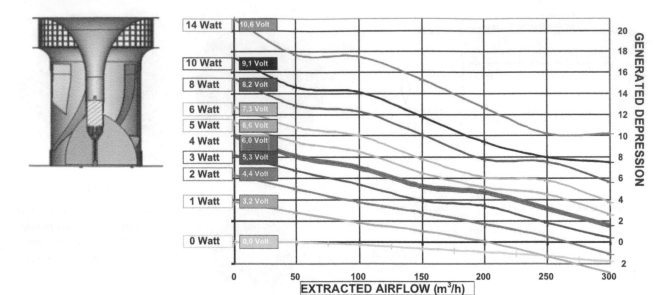

Source: AERECO

Figure 10.28 *A very low-pressure but energy-efficient fan*

Source: StorkAir BV

Figure 10.29 *An example of an open heat-recovery unit
for domestic-balanced ventilation*

Source: TNO Building and
Construction Research

Figure 10.30 *An example of
a roof outlet or cowl*

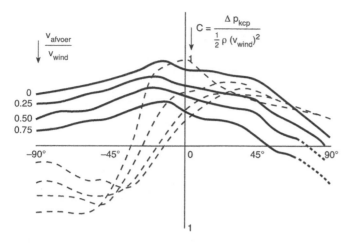

$$C = \frac{\Delta p_{kcp}}{\frac{1}{2}\,\rho\,(v_{wind})^2}$$

$\dfrac{v_{afvoer}}{v_{wind}}$

Source: TNO Building and Construction Research

Figure 10.31 *The performance curve of a cowl*

Photovoltaic applications

Photovoltaic (PV) cells became more common in situations where components were close to the perimeter of the building. The latest developments involve using PV for the energy supply of fans and control components for ventilation. In Airlit PV (Palmer, 1999), an approach was chosen to power the control for the opening of windows used for night cooling. In RESHYVENT, attempts are made to use PV for the fan power of hybrid ventilation systems. The size and capacity of the batteries, as well as price, are still barriers to applying photovoltaics (see Figure 10.32)

Source: Building Research Establishment

Figure 10.32 *An application of photovoltaics (PV) integration in ventilation*

Acknowledgements

We gratefully thank the several industrial manufacturers of ventilation components or systems for providing the information upon which the illustrations for this chapter are based.

References

de Gids, W. F. (1974) *Drie onderzoeken naar de werking van natuurlijke ventilatie IG-TNO*, Rapport C330, TNO, Delft, The Netherlands

de Gids, W. F. (1996) *Luchtdichtheid van betonnen kanalen*, TNO 96-BBI-R0374, TNO, Delft, The Netherlands

de Gids, W. F. (2003) *Bepaling van de weerstand van betonnen kanalen*, TNO 2003 GGI-R046, TNO, Delft, The Netherlands

de Gids, W. F. (2004) *RESHYVENT EU Project on Demand Controlled Hybrid Ventilation System*, Cauberg Huygen, Maastricht

Dörer, V. (2004) *Design Parameters*, EU RESHYVENT project, Technical report, EMPA, Duebendorf

Dubrul, C. (1988) *Inhabitants' Behaviour with Regard to Ventilation*, IEA ECBCS Annex 8, AIVC, Bracknell, UK

Knoll, B. (1992) *Advanced Ventilation Systems*, TN 35, AIVC, Coventry, UK

Liddament, M. W. L. (1996) *Guide to Energy Efficient Ventilation*, AIVC, Coventry, UK

Månsson, L. G. M. (2002) *Evaluation and Demonstration of Domestic Ventilation Systems*, IEA ECBCS Annex 27, AIVC, Coventry, UK

Palmer, J. (1999) *Airlit PV, EU Project V: The Development of a Façade Unit to Provide Daylight and Ventilation with Integrated Photovoltaic Power*, BRE, Watford, UK

Perera, E. (1999) *NATVENT EU Project on Natural Ventilation in Offices*, BRE, Watford, UK

Wouters, P. (2000) *TIP-Vent EU Project: Towards Improved Performances of Mechanical Ventilation*, BBRI, Brussels

11
Ventilation Standards and Regulations

Peter Wouters, Christophe Delmotte and N. Heijmans

Introduction

The construction of buildings and the installation of ventilation systems are economic activities that often involve large budgets and major responsibilities. It is important that the client and the supplier (building contractor, ventilation firms, etc.) clearly agree on the expected performances of the building or system. In order to create a building environment that aims to meet the requirements of the client as well as of the society, it is preferable that for each project the needs are explicitly stated and that one counts not too much on so-called implied needs.

Standards and regulations can be a very good tool for handling the requirements of society. Regulations are mandatory and, therefore, applicable to all projects covered by the regulation, whereas this is not always the case with standards.

Reasons for ventilation

Standards should be based on a clear philosophy. In the past, the primary and almost the only reason for ventilation was to remove or dilute the indoor-generated pollutants and to supply fresh air for human beings.

There are several types of pollutants that must be removed from the building: pollutants related to human occupancy (human effluents) and to human activities (combustion, smoking, cooking, etc.), as well as pollutants related to the building itself (building materials), its content (furniture), its maintenance (cleaning products) and its environment (ingress of soil gases such as Radon).

Ventilation is also needed to reduce the exposure to airborne microbes causing infectious diseases, and to control pressure levels in the building to prevent pollutants from spreading.

Other reasons for ventilation may be, for example, humidity control to:

* prevent growth of dust mites;
* prevent microbiological growth in building structures (walls, floors and ceilings); and
* prevent building constructions from damage.

Since the beginning of the 1990s, ventilation has received increased attention in the context of summer comfort control by so-called 'free cooling'. In practice, one observes an increased need for treating this kind of cooling strategy in standards and regulations.

Ventilation, indoor air quality (IAQ) and energy use

Ventilation is an important parameter for the indoor air quality in buildings. In general, the more ventilation, the lower is the exposure to pollutants from inside. But ventilation also has its energy consequences. If heating or cooling is required, the energy penalty is the most important reason to minimize the amount of ventilation air (see Figure 11.1).

Ventilation and infiltration

Apart from purpose-provided ventilation, there is also transport of air through the air leakage of the building envelope (due to the fact that buildings are never perfectly airtight). This is normally called infiltration. In some standards, the infiltration may be part of the required ventilation, whereas in other standards infiltration is explicitly excluded. From the energy point of view, the reduction of infiltration is important. For indoor air quality considerations, there is a clear distinction between

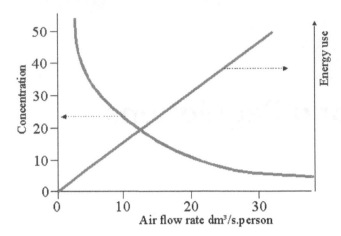

Source: chapter authors

Figure 11.1 *Typical relation between ventilation rate, concentration of pollutants and the energy use for ventilation*

ventilation and infiltration. Infiltration is uncontrolled, so one cannot guarantee the flow rate.

Policy in standards and regulations related to ventilation

Descriptive versus performance-based approach

Most standards and regulations are descriptive. This basically means that the requirements are not expressed in terms of maximum allowable dose or exposure levels, but in terms of variables that are assumed to have a certain link with the allowable dose or exposure levels – for example, flow rates in case of mechanical ventilation systems and cross-sections of natural air-supply openings.

There clearly is a need to develop more performance-oriented regulations and standards. The process of evaluating health effects due to pollutant sources in buildings is quite complex. On the one hand, we have the pollutant originating from several sources in and outside the building. On the other hand, there is the human being, who is partly exposed to contaminated outdoor air and partly to contaminated indoor air.

The total exposure depends, among other things, upon:

- source strength, varying in time and location; and
- the distribution of time people spend outdoors and indoors in different locations.

Most existing standards and regulations require flow rates based on assumed pollutants and their acceptable levels of concentration. The absolute highest level of performance approach may be to define the maximum effect of pollutants on the health and well-being of people. This last high-level approach is almost practically impossible to exploit.

The best practice today is probably the exposure of people as the underlying model for a performance-based approach. Therefore, during the design and use of the building in practice, one should consider:

- all possible indoor pollutant sources; and
- the time different people spend in different buildings indoors and within the building in the different rooms, and outdoors on several locations with different concentrations.

From that data, the total exposure should be evaluated.

Evolution in philosophies and needs concerning assessment of ventilation-related performances

Indoor air quality requirements

It is clear that the boundary conditions for specifying the requirements in relation to ventilation have clearly evolved during the last decades (see Figure 11.1). For decades, the needs concerning indoor air quality (IAQ) and required ventilation rates were based on laboratory experiments – for example, von Pettenkofer's research in 1860. Carbon dioxide (CO_2) was recognized as an appropriate marker of IAQ due to body odour.

The majority of today's standards and regulations are still based on this approach and limit values for IAQ levels were directly related to pollution by people – for example, in terms of maximum CO_2 levels.

During the 1980s and 1990s, field research has gained in importance. Various studies have highlighted the importance of other pollutions sources in many types of buildings.

Although there is much discussion concerning the most appropriate approach for handling these other sources, there is, at present, a consensus that these sources are important. As a result, the resulting airflow rates for achieving a 'similar level of IAQ' will be higher.

In order to deal with the non-occupancy-related pollution sources, it is important to quantify these sources. It is interesting to observe two tendencies:

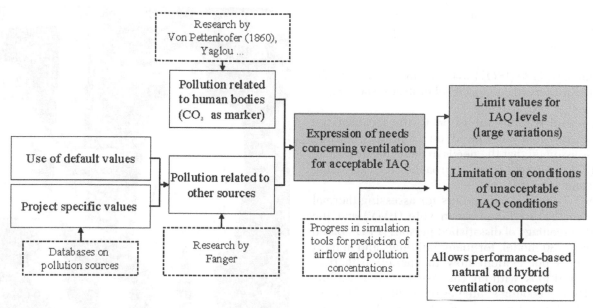

Source: chapter authors

Figure 11.2 *Schematic representation of various approaches for expressing ventilation needs*

- One approach is to handle these pollution sources by adopting default values. The motivation for this approach is the fact that it is far from evident how to quantify the various sources.
- The other approach is to quantify the various pollution sources. This requires appropriate databases.

A further development (and quite similar to thermal comfort assessment) is the use of a probabilistic approach. It assumes that it is allowed, during limited periods, to have IAQ conditions below the target values. In order to allow such an approach, it is important to have appropriate prediction tools which allow one to estimate the IAQ conditions in non-steady state conditions.

Such developments in assessment methods (and related requirements in standards and regulations) are important for enabling performance-based approaches – for example, natural and hybrid ventilation concepts. In the future, it will be essential to have performance-based approaches in order to enable a correct comparison between various systems and to develop innovative systems (for example, hybrid ventilation and demand-controlled systems).

Thermal comfort requirements

As is indicated in 'Reasons for ventilation', ventilation is becoming increasingly used as part of a strategy for improving thermal comfort in summer and/or to avoid/eliminate the use of active cooling. Therefore, it is also important to pay attention to the characterization of thermal comfort conditions in standards and regulations.

For characterizing thermal comfort conditions (which, at the same time, allows one to express needs), indices of thermal comfort have been derived. These are mainly based on laboratory research. Some of these studies were already being carried out in the beginning of the 20th century (as is shown in Figure 11.3). The most common indices are:

- *Predicted mean vote* (PMV): this predicts the mean response of a large group of people according to the International Organization for Standardization (ISO)/American Society of Heating, Refrigerating and Air-Conditioning Engineers (ASHRAE) thermal sensation scale (+3 = hot; +2 = warm; +1 = slightly warm; 0 = neutral; –1 = slightly cool; –2 = cool; –3 = cold).
- *Effective temperature* (ET*): the temperature of an environment at 50 per cent relative humidity (RH) that results in the same total heat loss from the skin as in the actual environment. The ET* value depends upon clothing and activity.
- *Standard effective temperature* (SET): the equivalent air temperature of an isothermal environment at 50 per cent RH in which a subject, while wearing clothing standardized for the activity concerned, has the

same heat stress (skin temperature) and thermoregulatory strain (skin wetness) as in the actual environment.

In ISO standard 7730 (ISO, 1994), equations are given for evaluating the thermal comfort level in steady-state conditions.

It is useful to draw attention to the fact that the derived expressions are based on laboratory testing, where it is assumed that the needs in daily life are similar. Figure 11.4 illustrates some approaches for expressing thermal comfort.

In ISO 7730, key parameters for assessing thermal comfort are the predicted mean vote (PMV) and the predicted percentage of dissatisfied persons (PPD). Based on the ISO 7730 model, for many buildings and in most climates, it is not evident how to guarantee thermal comfort conditions without the use of active cooling systems. Passive cooling strategies (including night ventilation) will often not be sufficient to meet the required comfort conditions as stated in ISO 7730.

Concerns have been expressed about the relation between the stated needs in laboratory conditions and the needs in the real world, and therefore also about the equations used in ISO 7730.

Therefore, and in parallel with laboratory studies, there has been a whole range of thermal comfort field studies (for example, Humphreys and Nichol, 1998). The comfort temperatures found from the various field studies

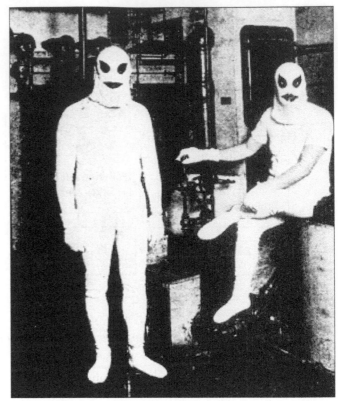

Source: Donaldson and Nagengast (1994)

Figure 11.3 *Suits to test the reaction of the human body to 'wet bulb' conditions*

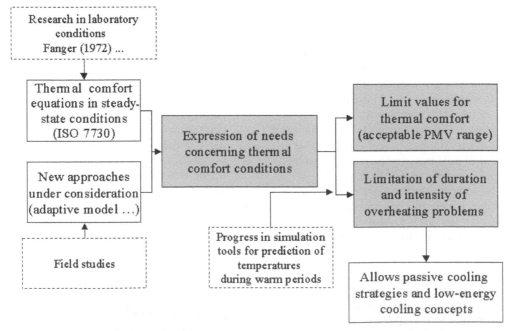

Source: chapter authors

Figure 11.4 *Schematic representation of various approaches for expressing thermal comfort*

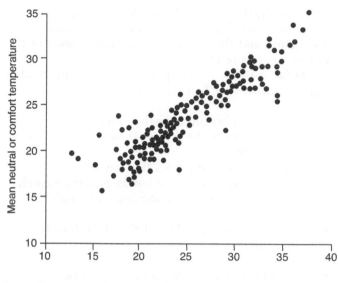

Figure 11.5 *Correlation between the comfort temperature and the mean indoor temperature*

vary notably one from another and are sometimes difficult to reconcile with the temperatures calculated from ISO standard 7730 or ASHRAE 55. Consideration of these

phenomena led to viewing thermal comfort as part of a self-regulating system. This way of regarding thermal comfort has become generally known as the adaptive model, where the adaptive principle is described as: 'if a change occurs such as to produce discomfort, people react in ways that tend to restore their comfort'. This approach is currently receiving much interest. At present, the following conclusions can be drawn:

- There are some doubts as to whether it is possible to directly translate needs identified in laboratory conditions into normal working situations.
- The various studies seem to indicate that people worldwide accept a much more diverse set of thermal environments than in laboratory-based studies. If so, then the application of the present standards leads to unnecessary energy consumption for heating and cooling.

This chapter will not provide a detailed discussion of the adaptive model. However, and in order to understand the potential impact of the adaptive model on indoor climate needs and induced energy use, the following observations are important (see Figure 11.5).

Based on a collection of available surveys, there seems to be a good correlation between the comfort temperature

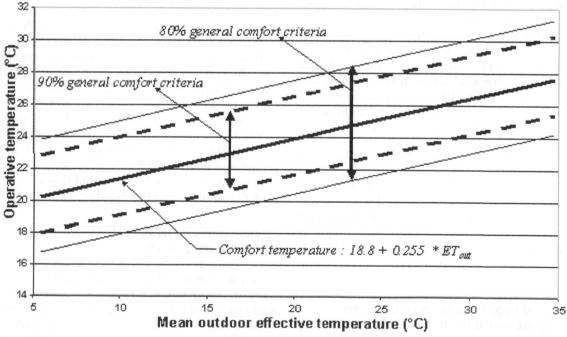

Figure 11.6 *Prediction of optimum comfort temperature and acceptable temperature ranges for buildings in naturally ventilated buildings (80–90 per cent general comfort criteria)*

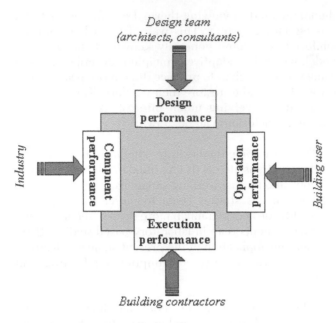

Source: chapter authors

Figure 11.7 *Various performance-related aspects of ventilation systems*

of the occupants and the average indoor temperature of the accommodation. Each point in the Figure 11.5 corresponds with a distinct survey.

Although still debated, there are many people who believe that this concept may lead to new comfort criteria, where different targets may be fixed for different operation modes. A proposal for such new criteria has been developed by De Dear (1998), who makes a distinction between 'centralized HVAC [heating, ventilating and air conditioning]' and 'natural ventilated' buildings (see Figure 11.6).

Further research will probably allow one to draw conclusions about the relevance and impact of modifying the comfort equations as stated in, for example, ISO 7730. If modifications will be needed, they can be considered as an example of new knowledge (= results from field research) resulting in new needs (concerning thermal comfort), where the application of this new need will have a considerable impact upon another need (limiting energy use in buildings).

Standards and regulations in relation to energy-efficient ventilation

Standards and regulations can be an important instrument in stimulating energy-efficient ventilation systems. The performances of ventilation systems are not only deter-

mined by the design, but also by the performances of the components used in the installation, the quality of the installation and the operation and maintenance of the system (see Figure 11.7).

As far as the type of requirements are concerned, a major distinction can be made between:

*   standards and regulations that directly focus on the energy efficiency of the ventilation systems and/or its components; and
*   standards and regulations that pay attention to the overall energy efficiency of buildings; this approach is also called the 'energy performance approach'.

Examples of standards and regulations that directly focus on energy-efficient ventilation are:

*   the air tightness performances of ductwork;
*   the air tightness performances of the building envelope;
*   specific fan power; and
*   the efficiency of heat exchangers.

Standards that focus on the total energy performance of a building are not directly stimulating energy-efficient ventilation but, as far as measures in energy-efficient ventilation are competitive with energy-efficiency measures in other areas, can strongly stimulate the use of energy-efficient ventilation technology (such as demand-controlled ventilation; energy-efficient fans; good air tightness of buildings and ductwork; and correct control of airflow rates).

For European Union (EU) countries, this latter approach will become mandatory in 2006. Indeed, the European Energy Performance in Buildings Directive (EPBD) (European Commission, 2003) requires that all member states will have assessment methods in line with this philosophy whereby minimum performance requirements have to be met by new buildings, as well as for major renovations of large buildings (see Appendix 2).

Standards and regulations in the European Union

Given the fact that there is a clear tendency in Europe to have a major common European component in the national standards and regulations regarding indoor climate and the energy efficiency of buildings, and since most of the Air Infiltration and Ventilation Centre (AIVC) and International Energy Agency (IEA) countries are directly concerned with this, this section draws specific attention to the European context.

European legislation and European directives

The European Union has, according to the Single European Act and the Maastricht Treaty, the ability to take certain legislative measures in relation to the energy efficiency of buildings. Various possibilities exist, such as directives, mandates, regulations and decisions.

A directive is a legislative instrument within the EU that is binding for member states with regard to the objective to be achieved, but which leaves to the national authorities the choice of form and methods used to obtain the objectives agreed at EU level within their domestic legal systems.

For this chapter, two directives are of particular importance:

- the Construction Product Directive (CPD); and
- the Energy Performance of Buildings Directive (EPBD).

Construction Product Directive (CPD)

Paragraph 71 of the White Paper on completing the internal market, approved by the European Council in June 1985, states that, within the general policy, particular emphasis will be placed on certain sectors, including construction. The removal of technical barriers in the construction field, to the extent that they cannot be removed by mutual recognition of equivalence among all the member states, should follow the new approach set out in the council resolution of 7 May 1985, which calls for the definition of essential requirements on safety and other aspects that are important for the general well-being, without reducing the existing and justified levels of protection in the member states.

The CPD dates from 1989. The six essential requirements specified in the CPD provide the basis for the preparation of harmonized standards at European level for construction products:

1 mechanical resistance and stability;
2 safety in case of fire;
3 hygiene, health and the environment;
4 safety in use;
5 protection against noise; and
6 energy economy and heat retention.

Several of these requirements directly (for example, number 3) or indirectly deal with indoor air quality and ventilation.

Energy Performance of Buildings Directive (EPBD)

This directive was adopted in December 2002 and requires from the member states a series of actions in relation to the energy efficiency of buildings. The full text is given in Appendix 2. The requirements are specified in Article 1:

This Directive lays down requirements as regards:
(a) *the general framework for a methodology of calculation of the integrated energy performance of buildings;*
(b) *the application of minimum requirements on the energy performance of new buildings;*
(c) *the application of minimum requirements on the energy performance of large existing buildings that are subject to major renovation;*
(d) *energy certification of buildings; and*
(e) *regular inspection of boilers and of air-conditioning systems in buildings and, in addition, an assessment of the heating installation in which the boilers are more than 15 years old.*

The EPBD mainly deals with energy-related aspects. However, explicit attention is also paid to aspects of the indoor climate. In Article 4, the following specification is given:

Member States shall take the necessary measures to ensure that minimum energy performance requirements for buildings are set, based on the methodology referred to in Article 3. When setting requirements, Member States may differentiate between new and existing buildings and different categories of buildings. These requirements shall take account of general indoor climate conditions in order to avoid possible negative effects such as inadequate ventilation.

European standards

Three bodies are responsible for the planning, development and adoption of European standards:

1 European Committee for Electrotechnical Standardization (CENELEC) for electro-technical issues (www.cenelec.be);

Table 11.1 *Overview of possible European Committee for Standardization (CEN) documents*

	EN European standard	CEN/TS technical specification	CEN/TR technical report	CWA CEN workshop agreement	CEN guide
Technical body	TC	TC	TC or BT	Workshop	BT or CA
Principal participation	National delegations	National delegations	National delegations	No restriction	National delegations
Interests represented	All	All	All	Specific	All
Standstill*	Yes	No	No	No	No
Public comments	6 months	No†	No	Optional	No
Level of approval	National CEN members	TC	TC or BT	Workshop	BT or AG or CA
Rule for approval	Weighted vote	Weighted vote	Simple majority	Consensus	Simple majority
Ratification	BT	BT	BT	None	CA or AG
National announcement	Yes	Yes	Optional	Yes	Optional
National availability	Yes	Yes	Optional	Optional	Optional
Publication	National standard	Optional	Optional	Optional	Optional
Withdrawal of conflicting standards	Yes	No	No	No	No
Languages	Three official languages	At least one official language	At least one official language	At least one official language	At least one official language
Review	Five years maximum	Three years maximum	No limit	Three years maximum	No limit

Notes: * Standstill is an obligation accepted by the national members of CEN/CENELEC not to take any action, either during the preparation of an EN (and harmonization document (HD) for CENELEC) or after its approval, which could prejudice the harmonization intended, and, in particular, not to publish a new or revised national standard that is not completely in line with an existing EN (and HD for CENELEC).

† Unless the CEN/TS results from a draft developed as a prEN up to and including the CEN inquiry

Abbreviations: EN = European standards; prEN = pre-European standards; AG = General Assembly (supreme authority of CEN); CA = administrative board (authorized agent of the General Assembly (AG) to direct CEN's business and accomplish all administrative actions and measures to achieve its objectives); BT = technical board (technical body that controls the full standards programme and promotes its speedy execution by the technical committees (TCs), the CEN Management Centre (CMC) and other bodies); TC = technical committees (technical structure with precise title, scope and work programme, established in the CEN System by the technical board (BT) essentially to manage the preparation of European standards (EN) in accordance with an agreed business plan).

Source: chapter authors

2   European Telecommunications Standards Institute (ETSI) for most of the information and communications technologies (www.etsi.org);
3   European Committee for Standardization (CEN) for all other sectors (www.cenorm.be).

CEN is a legal association, the members of which are the national standards bodies of 28 European countries and 8 associates (organizations representing social and economic interests at European level).

The CPD mandates the following organizations for carrying out specific activities:

• CEN and CENELEC for preparing and adopting the standards; and
• European Organization for Technical Approval (EOTA) (www.eota.be) for preparing the technical approvals.

As a result of the CPD, a large number of actions have started up in the framework of the European Committee

for Standardization (CEN). Ventilation-related aspects are mainly handled in Technical Committee 156, whereas some of the energy-related aspects are covered by Technical Committee 89.

The procedure for approving European standards is rather complicated, and a detailed description falls outside the objective of this chapter. The adoption of standards does not require unanimity but is based on a weighted voting principle (see www.cenorm.be).

The most well-known CEN documents are the European standards. However, CEN can produce various kinds of documents; a schematic overview is given in Table 11.1.

The following observations are important. The CPD and the EPBD are part of European legislation. European standards resulting from the CPD have to be implemented at national level. It is an obligation for the member states to implement these standards at national level.

The European standards themselves do not automatically have an impact on the building sector and on the

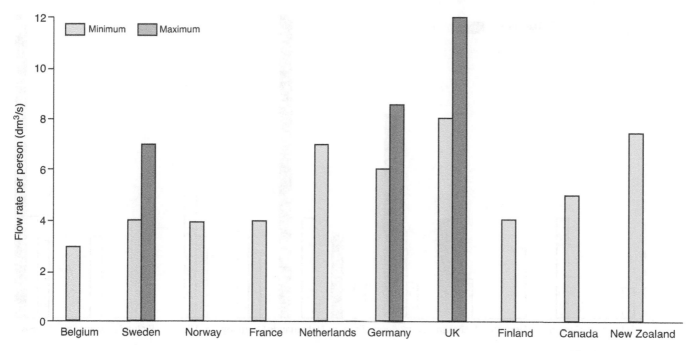

Source: Wouters et al (2001)

Figure 11.8 *Required flow rates per person for dwellings in standards (TIP-Vent 2000)*

individuals in the member states. Only if a standard is used for specifying a certain performance level (for example, in the framework of a national legislation or by being part of project specific requirements) does it becomes part of the requirements for quality.

The EPBD does not contain requirements for European citizens but for the member states. Each member states must set up the required regulations and related activities.

The increased importance of European standards puts a high pressure on CEN to deliver all required assessment procedures. An example is the generally recognized role of night ventilation as part of an overall strategy for passive cooling. At present, there is no sufficiently refined procedure at CEN level, nor is such a procedure in preparation.

Types and levels of requirements in standards and regulations

Variation between countries in airflow requirements

Ventilation standards throughout Europe, but also worldwide, differ considerably in their philosophies and in expressing ventilation targets as comparable values. In Figure 11.8 the variation in requirements is presented for dwellings, whereas in Figure 11.9 the variation is presented for office buildings.

Another example is the approach developed in the framework of CEN TC 156. The uncertainty in real indoor air quality needs is clearly illustrated by the fact that the IAQ and ventilation targets in relation to occupants-related pollution varies by 400 per cent from $5–20dm^3/s$ per person. If one takes into consideration smoking, the variation is even larger (see Table 11.2).

Variation of ventilation requirements as a function of time

The airflow requirements found in standards are the result of weighting various concerns – for example, air quality, health, energy and productivity. This weighting process can be strongly influenced by external factors (for example, the oil crisis in the 1970s) and new findings of research (for example, the impact on health and productivity).

Given the lack of precise knowledge on some of these facts, it is not surprising that the airflow requirements in standards may vary considerably as a function of time. ASHRAE 62, of which each revision is the outcome of an intensive review process, is a good example. In Figure

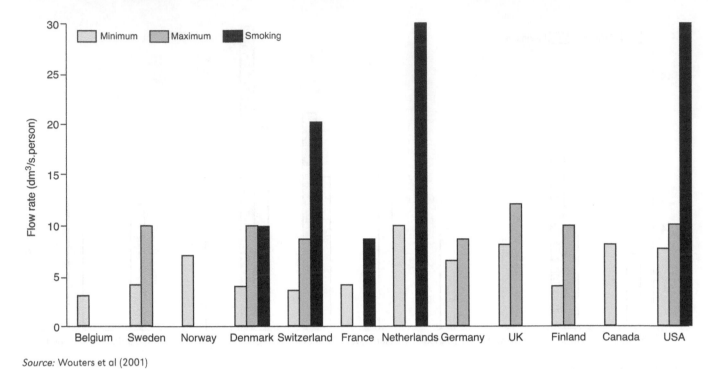

Source: Wouters et al (2001)

Figure 11.9 *Required flow rates per person for offices in standards (TIP-Vent 2000)*

11.10, the evolution of the airflow requirements per person are presented and the major discussion items are indicated. Major variations occur as a function of time.

Different priorities or concerns

In most cases, standards and regulations reflect more or less generally agreed procedures for characterizing certain processes and needs. Therefore, it seems logical to expect a large degree of similarity in the standards and regulations of various countries. However, this is not always the case. This section focuses on natural ventilation.

Natural ventilation is one of the possible strategies for controlling indoor air quality. The aim of natural ventilation devices is to guarantee that there is, within certain limits, the possibility of acceptable indoor air quality. It is important to recognize that requirements concerning natural ventilation devices are not only determined by expectations with respect to indoor air quality (see Figure 11.11, left-hand side) but also by a whole range of other assumptions:

- First, several assumptions (see Figure 11.11, middle) have to be made for translating IAQ requirements into airflow rate requirements. These assumptions are the same for mechanical and natural ventilation strategies.

Table 11.2 *Rates of outdoor air per person*

Indoor air quality (IAQ) classes	CO_2 concentration above outdoor concentration	Outdoor airflow Non-smoking area		Outdoor airflow Smoking area	
		Typical value (l/s per person)	Default value (l/s per person)	Typical value (l/s per person)	Default value (l/s per person)
IDA1	< 400 ppm	> 15	20	> 30	40
IDA2	Between 400 ppm and 600 ppm	10–15	12.5	20–30	25
IDA3	Between 600 ppm and 1000 ppm	6–10	8	12–20	16
IDA4	> 1000 ppm	< 6	5	< 12	10

Source: prEN 13779

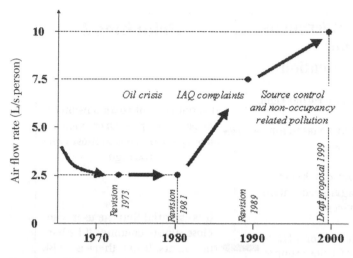

Source: chapter authors

Figure 11.10 *Evolution of requirements in ASHRAE 62*

- Then, a number of specific assumptions (see Figure 11.11, right-hand side) are needed in order to derive requirements concerning natural ventilation devices. These assumptions (stated or implied) include:
  - assumptions concerning boundary conditions:
    - climate-related data (such as temperature, wind and local shielding);
    - building data (such as air tightness and leakage distribution);

- occupants (for example, occupancy profile, window use and reaction to draught);
- assumptions concerning acceptable periods of rather poor indoor air quality.

It is clear that it is impossible to guarantee excellent indoor climate conditions under all weather conditions with a natural ventilation system. One has, therefore, to define the maximum allowable deviations.

Application of design criteria for natural ventilation devices

In order to illustrate the assumptions previously discussed, the philosophy is applied to the requirements found for natural air-supply openings. In Figure 11.12, one finds in the middle of the figure a series of possible considerations and/or points of attention. Basically, all of them are quite reasonable; but they don't lead to the same requirements. On the left-hand side of Figure 11.12, a series of considerations is used for defining a set of priorities (which will result in a series of requirements). On the right-hand side of Figure 11.12, the same is done where the other considerations are taken as priorities. Both sets of priorities look reasonable; but the consequences for ventilation characteristics are completely different. The left set is well in line with the underlying assumptions for building regulations in France, whereas the right set is quite close to the approach adopted in The Netherlands and Belgium.

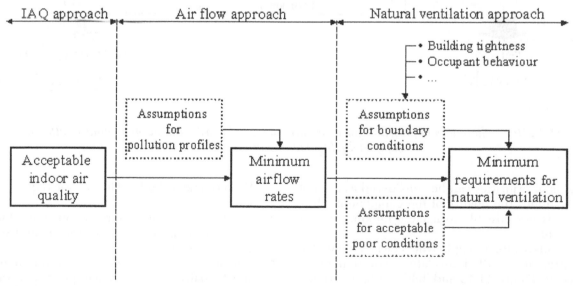

Source: chapter authors

Figure 11.11 *From an indoor air quality (IAQ) approach to requirements concerning natural ventilation:*
A succession of assumptions

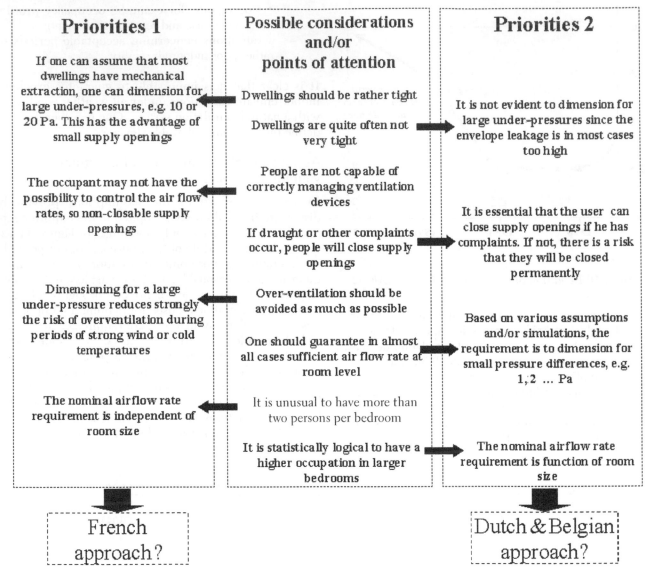

Source: Wouters (2000)

Figure 11.12 *How differences in priorities for a given set of logical considerations and points of attention lead to completely different quality requirements*

Both sets of priorities can be considered as a set of needs that defines the quality of the components. It is clear that it is not possible to make a ranking of both quality concepts.

Another interesting example which clearly shows the wide variation in ventilation-related requirements is synthesized in Figure 11.13 and Table 11.3, where the requirements for natural air-supply openings are expressed as a function of the floor area.

Various tendencies have been observed:

• In certain countries, the requirement is a function of the floor area, whereas in other countries, it is a constant value.

• The units for expressing the requirements differ substantially: the reference pressures vary from 1–20Pa; in the UK, the requirement is still expressed in square millimetre open sections.

Table 11.3 *Comparison of air supply requirements in various countries*

	Belgium	France	The Netherlands	UK
Air supply closable?	Obligatory	Not allowed	Obligatory	Required according to guidance in approved document 1
Dimensioning principles	Airflow proportional with floor area 1 dm³/s.m² floor area for ΔP = 2Pa	Fixed value	Airflow proportional with floor area 0.9 dm³/s.m² floor area for ΔP = 1Pa	Fixed value
Mechanical exhaust ≤ 7m² bedroom	7 dm³/s at 2Pa	30 m³/h at 20Pa	7 dm³/s at 1Pa	8000mm²
14m² bedroom	14 dm³/s at 2Pa	30 m³/h at 20Pa	12.6 dm³/s at 1Pa	8000mm²
Natural exhaust ≤ 7m² bedroom	7 dm³/s at 2Pa	45 m³/h at 20Pa	7 dm³/s at 1Pa	8000mm²
14m² bedroom	14 dm³/s at 2Pa	45 m³/h at 20Pa	14 dm³/s at 1Pa	8000mm²

Source: chapter authors

- Certain countries impose non-closable supply openings whereas other do not allow it.
- In one country, the air supply requirements are a function of the type of air extraction used.
- In absolute terms, very large differences are found: for a 7m² bedroom, there is a variation of a factor 3.7, whereas for a 14m² bedroom, the variation is a factor 6.8.

North American standard for residential ventilation (ASHRAE 62)

Requirements and guidance to make the air in homes healthier and safer without adding significant costs is provided in a newly approved standard from the American Society of Heating, Refrigerating and Air-Conditioning Engineers (ASHRAE). ASHRAE, founded in 1894, is an international organization of 55,000 individuals. Its objective is to advance – through research, standards writing, publishing and continuing education – the arts and sciences of heating, ventilation, air conditioning and refrig-

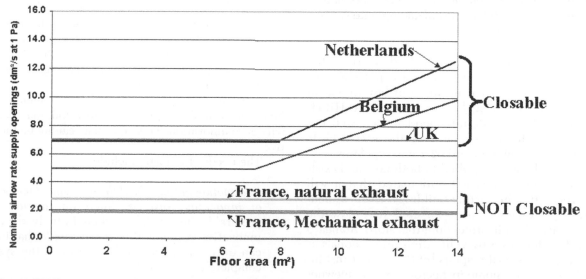

Source: Wouters (2000)

Figure 11.13 *Comparing requirements for air inlets in bedrooms*

eration to serve the evolving needs of the public (see www.ashrae.org). ASHRAE Standard 62.2-2003, *Ventilation and Acceptable Indoor Air Quality in Low-Rise Residential Buildings*, is the first and only recognized indoor air quality standard for residences in the US.

The standard provides the minimum requirements necessary to achieve acceptable indoor air quality for dwellings. Acceptable IAQ means that the indoor air will not likely pose a significant health hazard and will not be irritating or have unacceptable odours.

Residential ventilation traditionally has not been a major concern in North America because it was felt that between operable windows and envelope leakage, occupants were getting enough air. This has changed as homes are built tighter to save energy. The need for the standard is driven by today's tighter-built houses, new knowledge about the relationship of IAQ and health, and a public that is placing more importance on IAQ and health, in general.

The standard provides a variety of measures to improve IAQ at minimal cost:

- source control of moisture and other specific pollutants through the use of local exhaust fans;
- criteria to minimize back-draughting and other combustion-related contaminants;
- provisions for reducing contamination from attached garages;
- minimum whole-house ventilation rates;
- performance criteria for air-moving equipment to help ensure that equipment will operate as intended and will not be unacceptable to the occupants; and
- guidance on how to select, install and operate systems for maximum benefit.

ASHRAE writes consensus standards that are in the public interest; but these standards are not codes or regulations since they have no force of law in and of themselves. Standard 62.2, like many other ASHRAE standards, is written in code-intended language so that it can be easily adopted by regulatory authorities. Efforts are under way to adopt Standard 62.2 into regulations.

The standard may be applied to both new and existing houses, although it is not anticipated that it will become a regulation in existing buildings. Even without legal force, professionals are expected to follow the standard of care of their profession; therefore, Standard 62.2 should be followed by ventilation professionals as part of their professional responsibilities. Other institutions, such as utilities, home inspectors and homeowner associations, can use Standard 62.2 to evaluate existing construction.

New challenges in the context of ventilation standards

Assessment of innovative systems

Standards and regulations cannot cover all kind of possible technologies for various reasons:

- There are always new systems coming on the market.
- A description of the assessment procedures for all possible systems would lead to extremely voluminous documents, which the building industry finds difficult to handle.

Typically, but not always, the more innovative systems are not always covered by the standards or regulations. In the framework of this chapter, 'innovative technologies' are defined as technologies:

- which, in most cases, give a better performance in terms of the energy performance of buildings than the commonly used technologies; *and*
- whose performance cannot be assessed by the procedure in the energy performance calculation method.

Energy performance regulation (EPR) approaches in different European countries, as well as the EPBD, have increasingly become driving factors for industry in relation to product development. One of the key questions for industry is: *how can we improve and optimize our products with respect to the performance assessment made in EPR approaches?*

In order to encourage such an approach, it is crucial that there is transparency in relation to the assessment methodology. Our impression, as well as the reaction from several major manufacturers who are active throughout Europe, is that this transparency is lacking for the majority of the innovative systems.

As such, EPR approaches, which, in principle, should be strong stimuli for product innovation, may become a barrier to innovation. This, of course, cannot be the objective.

The needs of industry include:

- *Information in time about the assessment methodology of innovative systems.* Such information should preferably be available when initially developing innovative systems. It is *unacceptable* that industry can only ask for an evaluation once the system is completely developed.
- *A minimum level of coherence in the assessment procedures by the different member states.* It is extremely

inefficient to be confronted with completely different assessment schemes in different countries.

- *A sufficiently open assessment scheme.* This allows for optimization of systems and products.

Work on the assessment of innovative ventilation systems in relation to energy performance regulation has been done in the IEA HYBVENT project (Wouters et al, 2002) and in the European Community (EC) RESHYVENT project. Moreover, a more general study has been done in the framework of the SAVE ENPER project (ENPER, 2003). In general, the use of detailed simulations seems to be crucial.

Climate data

The availability of more precise climate data is becoming more and more important for several reasons:

- The increased interest in natural and hybrid ventilation systems for IAQ control, in combination with a performance-based approach, requires the use of (rather) detailed simulation tools and, therefore, of (rather) detailed climate data (such as temperature and wind).
- The increased interest in passive cooling strategies in combination with more refined thermal comfort models often requires the use of advanced simulation tools. Hourly values of outdoor temperature, wind speed and direction are then important. For urban areas, it may not be appropriate to use the data of meteorological stations since the specific urban characteristics (for example, the heat island effect and the canyon effect) may have a strong impact on local outdoor climate conditions.
- The assessment of innovative ventilation systems in the context of energy performance regulations probably requires the use of hourly simulation tools. Again, detailed climatic data are needed.

The level of detail of climatic data can vary:

- *The time scale of the data (for instance, monthly, hourly and minutely data).* In most cases, hourly data are sufficient. However, for the assessment of, for instance, demand-controlled ventilation systems, it may be necessary to use data with a short interval.
- *The geographical scale.* In urban areas or areas with large changes in climate within short distances (for example, coastal areas and mountains), it may be important to have access to precise local data.

As far as standards and regulations are concerned, it is not evident how to take all of these aspects into account.

Product characterization of detection systems

In several countries, there is an increased use of advanced ventilation components: high-efficiency heat exchangers, direct current fans, presence detectors, CO_2 detectors, IAQ sensors, etc. Quality control and a correct characterization of the component performances are then of increased importance.

France has a long experience with humidity-controlled ventilation systems (more than 1 million buildings) and there are specific procedures for assessing the various systems on the market, including technical agreements. Given the growing interest in other kinds of demand-controlled ventilation systems (such as CO_2 and presence detection), research is under way to better characterize such control systems, where technical agreements are foreseen. Two ongoing research projects are briefly presented in the following subsections.

CO_2 sensors

CO_2 control has, in principle, a very large energy savings potential. However, it is not evident how to guarantee the long-term accuracy of CO_2 sensors. Therefore, a detailed assessment method has been developed (Villenave et al, 2003) for characterizing the performances of CO_2 sensors. Results obtained for five CO_2 sensors are shown in Table 11.4.

Some observations are as follows:

- For several sensors, the average deviation (column 2) is larger than the value claimed by the manufacturer (last column).
- It is probably important to clearly define the various deviations – for example, how to deal with maximum deviations and dynamic effects.
- What level of deviation is acceptable? Is a deviation of 100ppm or 150ppm acceptable if the allowable increase in CO_2 level is limited (for example, CEN CR 1752 class C allows only an increase of 1190ppm; a deviation of 150ppm would mean an error of 13 per cent).
- How can one handle the variation of outdoor CO_2 concentrations?

Presence detectors

It is well known that there are large differences in the detection levels of presence detectors. In France, a study has been carried out (Bernard et al, 2003) to develop an

Table 11.4 *Accuracy of measurements of five types of CO_2 sensors*

Sensor	Average deviation		Maximal deviation		Manufacturer value
	ppm	percentage	ppm	percentage	
A	124	10	198	17	± 100 ppm
B	80	8	114	16	± 50 ppm
C	213	23	483	66	± 140 ppm
D	72	10	189	19	± 150 ppm
E	143	12	481	44	± 90 ppm

Note: According to this study, there is no major drift as a function of time.

Source: Villenave et al (2003)

assessment scheme of presence detectors. Figure 11.14 shows the detection field of a specific infrared detector. It is expected that there will be technical agreement schemes for presence detectors whereby such kinds of performance characterization will be mandatory, including minimum performance levels.

Occupancy-related aspects

There is an increased need for more detailed information on occupancy-related data:

- It is well known that occupants can have a very strong impact on the performance of ventilation systems and on the IAQ level in buildings and the related energy use.
- There is an increased interest in and market take-up of demand-controlled ventilation systems whereby the occupants play a crucial role.

Therefore, it is important to have reliable and/or representative information on, for example:

- occupancy profiles of buildings (at what time of the day there are a specified number of individuals in the various rooms);
- window and door use; and
- household activities.

Given the large variation between households, it may be necessary to make use of probabilistic models.

Summary

At present, the majority of standards are directly or indirectly still based on the early research findings of von Pettenkofer and Yaglou whereby human bio-effluents (with CO_2 as a tracer) are the basis for applying standards. Despite this common starting point, there are wide variations in the different standards with respect to airflow specifications. This can be partly explained by the relative weight given to various considerations, where external factors (for example, oil crises and environmental considerations), as well as new research findings (for example, impact on health and economic impact of poor indoor air quality) are not receiving the same importance in all countries.

Source control and/or ventilation for non-occupancy-related pollution sources is today the focus of attention in most standardization committees. However, and although nearly everyone agrees that these pollution sources should be considered, there is not yet a consensus for dealing with this kind of pollution. The fact that interest in this kind of pollution is only a recent phenomena can be explained, one the one hand, by the increased occurrence in buildings of materials and systems with high pollution loads (such as carpets, cleaning products and HVAC systems) and, on the other hand, by increased scientific knowledge.

The majority of the current standards and regulations are primarily descriptive and rarely performance based. One of the challenges for the future is to replace these descriptive approaches by performance-based concepts.

As far as energy-related requirements are concerned, there are clear benefits to having requirements in terms of the overall energy efficiency of buildings. The new European Energy Performance Directive will impose such an approach for all European Union member states.

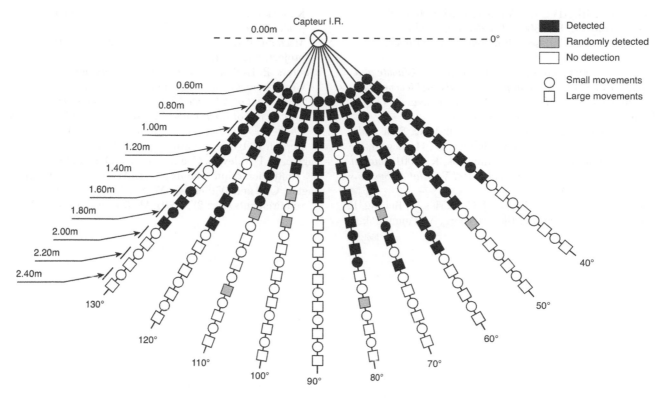

Source: Bernard et al (2003)

Figure 11.14 *Detection field of a specific infrared detector*

References

ASHRAE 62-1989 (American Society of Heating Refrigerating and Air-Conditioning Engineers) (1989) *Ventilation for Indoor Air Quality*, ASHRAE, Atlanta, US

ASHRAE 62n (2001) *Proposed Addendum Ventilation Rate Procedure to ASHRAE Standard 62.2-2001: Ventilation for Acceptable Indoor Air Quality*, Public review draft, ASHRAE, Atlanta, US

Bernard, A. M., Villenave, J. G. and Lemaire, M. C. (2003) *Demand Controlled Ventilation (DCV) Systems: Performances of Infrared Detection*, 24th AIVC Conference, Washington, DC

CEN CR 1752 (European Committee for Standardization) (1998) *Ventilation for Buildings: Design Criteria for the Indoor Environment*, CEN, Brussels

CEN EN 13779 (2004) *Ventilation for Non-residential Buildings: Performance Requirements for VEntilation and Room-Conditioning Systems*, CEN, Brussels

De Dear, R. (1998) 'A global database of thermal comfort field experiments', *ASHRAE* winter meeting January 1998, *ASHRAE Technical Data Bulletin*, vol 14, no 1, Atlanta, US

Donaldson, B. and Nagengast, B. (1994) *Heat and Cold: Mastering the Great Indoors*, ASHRAE, Atlanta, GA

ENPER (2003) *Task B2: Energy Performance of Buildings – Assessment of Innovative Technologies*, May, www.enper.org/

European Commission (2003) 'Directive 2002/91/EC of the European Parliament and of the Council of 16 December 2002 on the energy performance of buildings', *Official Journal of the European Communities*, Brussels, 4 January

Fanger, P. O. (1972) *Thermal Comfort*, McGraw-Hill, New York

Humphreys, M. and Nicol, F. (1998) 'Understanding the adaptive approach to thermal comfort', *ASHRAE* winter meeting January 1998, *ASHRAE Technical Data Bulletin*, vol 14, no 1, Atlanta, US

ISO (International Organization for Standardization) (1994) *Standard 7730 – Moderate Thermal Environments: Determination of the PMV and PPD Indices and Specification of the Conditions for Thermal Comfort*, second edition, ISO, Geneva

Villenave, J. G., Bernard, A. M., and Lemaire, M. C. (2003) *Demand Control Ventilation Systems: Performances of CO_2*

Detection, 24th AIVC Conference, Washington, DC

von Pettenkofer, M. S. (1860) *Über den Luftwechsel in Wohngebäuden*, Cottasche Buchhandlung, Munich, Germany

Wouters, P. (2000) *Quality in Relation to Indoor Climate and Energy Efficiency: An Analysis of Trends, Achievements and Remaining Challenges*, PhD thesis, University of Louvain, Louvain-La-Neuve, Belgium

Wouters, P., Barles, P., Blomsterberg, Å., Bulsing, P., de Gids, W., Delmotte, C., Faÿsse, J. C., Filleux, C., Hardegger, P., Leal, V., Maldonado, E. and Pennycook, K. (2001) *Towards Improved Performances of Mechanical Ventilation Systems*, EC JOULE TIP-Vent project

Wouters, P., Delmotte, C., Faysse, J. C., Barles, P., Bulsing, P., Filleux, C., Hardegger, P., Blomsterberg, A., Pennycook, K., Jackman, P., Maldonado, E., Leal, V. and de Gids, W. (2000) *Towards Improved Mechanical Performances of Mechanical Ventilation Systems*, EC JOULE TIP-Vent project, August, Brussels

Wouters, P., Heijmans, N., de Gids, W., Van der Aa, A., Guarracino, G. and Aggerholm, S. (2002), *Performance Assessment of Advanced Ventilation Systems in the Framework of Energy and IAQ Regulations: Critical Issues, Challenges and Recommendations*, HybVent Technical Report, HybVent, www.hybvent.civil.auc.dk/

Yaglou, C. P., Riley, E. C. and Coggins, D. I. (1936) 'Ventilation requirements', *ASHRAE Transactions*, vol 42

Yaglou, C. P. and Witheridge, W. N. (1937) 'Ventilation requirements (Part 2)', *ASHRAE Transactions*, vol 43

Appendix 1
European Committee for Standardization (CEN) Documents TC 156

These are limited to TC 156 (status: December 2003).

A1.1 *Published European Committee for Standardization (CEN) documents*

WI (working item) number	Reference	Title
156042	CR 14378	Ventilation for buildings – ductwork; determination of mechanical energy loss in ductwork components (2002)
156056	CR 1752	Ventilation for buildings – design criteria for the indoor environment (1998)
156037	EN 12220	Ventilation for buildings – ductwork; dimensions of circular flanges for general ventilation (1998)
156036	EN 12236	Ventilation for buildings – ductwork hangers and supports; requirements for strength (2002)
156041	EN 12237	Ductwork – circular sheet metal air ducts; requirements for testing strength and leakage (2003)
156047	EN 12238	Air terminal devices – aerodynamic testing and rating for mixed-flow applications (2001)
156060	EN 12239	Air terminal devices – aerodynamic testing and rating for displacement flow applications (2001)
156049	EN 12589	Air terminal units – aerodynamic testing of constant and variable rate terminal units (2001)
156059	EN 12599	Ventilation for buildings – test procedures and measuring methods for handing overinstalled ventilation and air-conditioning systems (2000)
156076	EN 12792	Ventilation for buildings – symbols, units and terminology (2003)
156050	EN 13030	Terminals – performance testing of louvres subject to simulated rain (2001)
156053	EN 13053	Air handling units – ratings and performance for units, components and sections (2001)
156038	EN 13180	Ductwork – dimensions and mechanical requirements for flexible ducts (2001)
156061	EN 13181	Terminals – performance testing of louvres subject to simulated sand (2001)
156063	EN 13182	Ventilation for buildings – instrumentation requirements for air velocity measurements in ventilated spaces (2002)
156065	EN 13264	Ventilation for buildings – floor-mounted air terminal devices; test for structural classification (2001)
156074	EN 13403	Non-metallic ducts – ductwork made from insulation duct boards (2003)
156034	EN 1505	Ventilation for buildings – sheet metal air ducts and fittings with rectangular cross-section; dimensions (1997)
156039	EN 1506	Ventilation for buildings – sheet metal air ducts and fittings with circular cross-section; dimensions (1997)
156046	EN 1751	Ventilation for buildings – air terminal devices; aerodynamic testing of dampers and valves (1998)
156052	EN 1886	Ventilation for buildings – air handling units; mechanical performance (1998)
156045	ENV 12097	Ventilation for buildings – ductwork; requirements for ductwork components to facilitate maintenance of ductwork systems (1997)

Source: Belgian Building Research Institute (BBRI)

A1.2 *Ratified CEN documents*

WI number	Reference	Title
156030	prEN 13141-1	Ventilation for buildings – performance testing of components/products for residential ventilation; Part 1: Externally and internally mounted air transfer devices
156031	prEN 13142	Ventilation for buildings – components/products for residential ventilation; required and optional performances characteristics
156032	prEN 14134	Ventilation for buildings – performance testing and installation checks of residential ventilation systems
156033	prEN 13465	Ventilation for buildings – calculation methods for the determination of airflow rates in dwellings
156035	prEN 14239	Ventilation for buildings – ductwork; measurement of ductwork surface areas
156066	prEN 13141-2	Ventilation for buildings – performance testing of components/products for residential ventilation; Part 2: Exhaust and supply air terminal devices
156067	prEN 13141-3	Ventilation for buildings – performance testing of components/products for residential ventilation; Part 3: Range hoods for residential use
156068	prEN 13141-4	Ventilation for buildings – performance testing of components/products for residential ventilation; Part 4: Fans used in residential ventilation systems
156070	prEN 13141-6	Ventilation for buildings – performance testing of components/products for residential ventilation; Part 6: Exhaust ventilation system packages used in a single dwelling
156075	prEN 13141-7	Ventilation for buildings – performance testing of mechanical supply and exhaust units (including heat recovery) for mechanical ventilation systems intended for single-family dwellings
156081	prEN 14240	Ventilation for buildings – chilled ceilings; testing and rating

Note: These documents are work items on a stage between ratification and publication.

Source: BBRI

A1.3 *Under-approval documents*

WI number	Reference	Title
156040	prEN 1507	Ventilation for buildings – ductwork; rectangular sheet metal air ducts – requirements for testing strength and leakage
156057	prEN 13779	Ventilation for non-residential buildings – performance requirements for ventilation and air-conditioning systems
156064	prEN 14788	Ventilation for buildings – design and dimensioning of residential ventilation systems
156069	prEN 13141-5	Ventilation for buildings – performance testing of components/products for residential ventilation; Part 5: Cowls and roof outlet terminal devices
156071	prEN 14277	Ventilation for buildings – air terminal devices; method for measurement by calibrated sensors in or close to air terminal devices (ATDs)
156082	prEN 14518	Ventilation for buildings – chilled beams; testing and rating of passive chilled beams
156087	prEN 1886 rev	Ventilation for buildings – air handling units; mechanical performance
156088	prEN 13053 rev	Ventilation for buildings – air handling units; ratings and performance for units, components and sections

Note: These are active work items at a stage between the beginning of the enquiry and the end of the formal vote.

Source: BBRI

Mandate for new European Committee for Standardization (CEN) standards

The European Energy Performance of Buildings Directive (EPBD) requires member states to implement energy performance calculation procedures. In principle, it is assumed that maximum use of CEN standards is made. However, a whole range of CEN standards is missing; therefore, at the end of 2003, the European Commission gave a series of mandates to CEN for preparing new CEN standards.

A1.4 *Under-development documents*

WI number	Reference	Title
156051	–	Ventilation for buildings – terminals; comfort criteria
156058	–	Ventilation for buildings – cooling load
156073	–	Ventilation for buildings – fire precautions for air distribution systems in buildings
156077	–	Ventilation for buildings – calculation methods for airflow rates in commercial buildings
156078	–	Ventilation for buildings – calculation methods for energy losses due to ventilation and infiltration in commercial buildings
156079	–	Ventilation for buildings – calculation methods for energy losses due to ventilation and infiltration in dwellings
156080	–	Ventilation for buildings – air terminal devices; Part I: Aerodynamic testing and rating for mixed-flow applications for non-isothermal testing
156083	–	Testing and rating of active chilled ceilings
156084	–	Ventilation for buildings – flat oval ductwork; dimensions
156085	–	Ventilation for buildings – material specification of surface treatment for ductwork
156086	prEN 12097 rev	Ventilation for buildings – ductwork; requirements for ductwork components to facilitate maintenance of ductwork systems
156090	prEN 1506 rev	Ventilation for buildings – sheet metal air ducts and fittings with circular cross-section; dimensions
156091	–	Ventilation for buildings – flat oval ductwork; leakage and strength
156092	–	Ventilation for buildings – air terminal devices; aerodynamic testing and rating for mixed-flow applications for non-isothermal testing – Part 2: Hot jets
156093	–	Technical report (TR) on basic data used to produce prEN 14240 *Ventilation for Buildings – Chilled Ceilings: Testing and Rating* and WI 00156082 *Ventilation for Buildings – Chilled Beams: Testing and Rating of Passive Chilled Beams*
156094	prEN 13141-8	Ventilation for buildings – performance testing of components/products for residential ventilation; Part 8: Performance testing of inducted mechanical supply and exhaust ventilation units (including heat recovery) for mechanical ventilation systems intended for a single room
156095	–	Airflow measurements in ventilation systems

Note: These are active work items that have not yet reached the stage of enquiry.

Source: BBRI

A1.5 List of mandates to CEN within the context of the Energy Performance of Buildings Directive (EPBD)

No	Work item	Present stage	Publication	Which technical committee (TC)?	Stage end 2004	Comment
1	Energy performance of buildings – energy certification of buildings	New	EN	TC 89	Stage 40	Stage 64 2006-06
2	Energy performance of buildings – overall energy use, primary energy and carbon dioxide (CO_2) emissions	New	EN	TC228	Stage 40	Stage 64 2006-09
3	Energy performance of buildings – ways of expressing energy performance of buildings	New	EN	TC 89	Stage 40	Stage 64 2006-06
4	Energy performance of buildings – applications of calculation of energy use to existing buildings	New	EN	TC 89	Stage 40	Stage 64 2006-06
5	Energy performance of buildings – applications of calculations for building types in EPBD Annex, Part 3	New	EN	TC 89	Stage 40	Stage 64 2006-06
6	Energy performance of buildings – additional applications of calculations for the inclusion of the positive influences of day-lighting, solar shading, passive cooling, position and orientation, renewables, district heating and cooling, combined heat and power (CHP) (including on-site) and for modular inclusion of future technologies	New	EN	TC 89	Stage 40	Stage 64 2006-06
7	Energy performance of buildings – certification systems	New	EN	TC 89	Stage 40	Stage 64 2006-06
8*	Energy performance of buildings – systems and methods for the inspection of boilers and heating systems	New	EN	TC 89	Stage 40	Stage 64 2006-06
9*	Energy performance of buildings – systems and methods for the inspection of air-conditioning systems	New	EN	TC 89	Stage 40	Stage 64 2006-06
10	Heating systems in buildings – method for calculating system energy requirements and system efficiencies; Part 1: General	46	EN	TC 228	49	Stage 64 2006
11	Heating systems in buildings – method for calculating system energy requirements and system efficiencies; Part 2.1: Space heating emission systems	40	EN	TC 228	46	Stage 64 2006
12	Heating systems in buildings – method for calculating system energy requirements and system efficiencies; Part 2.2: Space heating generation systems: • Part 2.2.1: boilers • Part 2.2.2: heat pumps • Part 2.2.3: thermal solar systems (renewable energy system) • Part 2.2.4: the use of CHP electricity and heat • Part 2.2.5: the use of district heating • Part 2.2.6: the use of other renewables, heat and electricity	31 ? ? ? ?	EN	TC 228	40	Stage 64 2007
13	Heating systems in buildings – method for calculating system energy requirements and system efficiencies; Part 2.3: Space heating distribution systems	31	EN	TC 228	40	Stage 64 2007
14	Heating systems in buildings – method for calculating system energy requirements and system efficiencies; Part 3.1: domestic hot water systems, including generation efficiency parts for boilers, heat pumps and renewable energy systems as thermal solar systems	31	EN	TC 228	40	Stage 64 2007

No	Work item	Present stage	Publication	Which technical committee (TC)?	Stage end 2004	Comment
15	Dynamic calculation of room temperatures and of load and energy for buildings with room conditioning systems (including solar shading, passive cooling and position and orientation)	31 ?	prEN	TC 156	Stage 40	Stage 64 2007-09
16	Energy performance of buildings – energy requirements for lighting (including day-lighting)	New ?	prEN	TC 169	Stage 40	
17	Thermal performance of buildings – calculating energy use for space heating and cooling; simplified method	New	prEN	TC 89	Stage 40	Stage 64 2007 Based on EN ISO 13790
18	Thermal performance of buildings – calculating energy use for space heating; simplified method	64	EN13790	TC89	Stage 64 by 2003	existing
19	Thermal performance of buildings – sensible room cooling load calculation; general criteria and validation procedures	32	prEN	TC 89	Stage 40	Stage 64 2006-02
20	Energy performance of buildings – calculating energy use for space heating and cooling; general criteria and validation procedures	32	prEN	TC 89	Stage 40	Stage 64 2006-02
21	Ventilation for buildings – calculation methods for the determination of airflow rates in dwellings	49	EN	TC 156	64	FV ends 2003-10
22	Ventilation for buildings – calculation methods for the determination of airflow rates in buildings (items 21 and 22 could merge)	11	prEN	TC 156	40	Or part 2 of 13465
23	Ventilation for buildings – calculation methods for energy requirements due to ventilation and infiltration in buildings	11	prEN	TC 156	40	Stage 64 2007
24	Ventilation for buildings – calculation methods for energy requirements due to ventilation and infiltration in dwellings (to be merged with WI 23)	11	prEN	TC 156	40	Merging with 23
25	Calculation methods for energy efficiency improvements through the application of integrated building automation products and systems	31	prEN	TC247	Stage 40	
26	Review of standards dealing with calculation of heat transmission in buildings – first set	Review	EN ISO	TC 89	Stage 40 2004-03	Revision led by ISO
27	Review of standards dealing with calculation of heat transmission in buildings – second set	Review	EN ISO	TC 892	Stage 40	Revision led by ISO
28	Ventilation for non-residential buildings – performance requirements for ventilation and room conditioning systems	49	EN	TC 156	64	May be subject to immediate revision
29	Design of embedded water-based surface heating and cooling systems to facilitate renewable low-temperature heating and high-temperature cooling	30	EN	TC228	40	Stage 64 2006
30	Performance requirements for temperature calculation procedure without mechanical cooling	46	EN	TC89	49	Stage 64 2005
31	Performance requirements for temperature calculation procedure with mechanical cooling	46	EN	TC89	49	Stage 64 2005
32	Requirements for standard economic evaluation procedures	11	PrEN	TC228	40	Stage 64 2007

Notes: * Included under 'overall energy use' to provide TC 89 with responsibility for coordination and assignment to appropriate TC.

Source: CEN (European standards intended to support the EPBD)

Appendix 2

Energy Performance of Buildings Directive (EPBD)

Global context

On 4 January 2003, the *Official Journal of the European Communities* published 'Directive 2002/91/EC of the European Parliament and of the Council of 16 December 2002 on the Energy Performance of Buildings'. This directive provides major legal boundary conditions for future developments in Europe in the area of the energy efficiency of buildings and measures related to the indoor climate in buildings.

The directive and the links to ventilation-related aspects

The official text of the EPD follows, with chapter authors' comments given in bold italics.

Article 1: Objective

The objective of this Directive is to promote the improvement of the energy performance of buildings within the Community, taking into account outdoor climatic and local conditions, as well as indoor climate requirements and cost-effectiveness.

This Directive lays down requirements with regard to:

(a) the general framework for a methodology of calculation of the integrated energy performance of buildings;
(b) the application of minimum requirements on the energy performance of new buildings;
(c) the application of minimum requirements on the energy performance of large existing buildings that are subject to major renovation;
(d) energy certification of buildings; and
(e) regular inspection of boilers and of air-conditioning systems in buildings and in addition an assessment of the heating installation in which the boilers are more than 15 years old.

Article 2: Definitions

For the purpose of this Directive, the following definitions shall apply:

1 *Building*: a roofed construction having walls for which energy is used to condition the indoor climate; a building may refer to the building as a whole or parts thereof that have been designed or altered to be used separately.

2 *Energy performance of a building*: the amount of energy actually consumed or estimated to meet the different needs associated with a standardized use of the building, which may include, *inter alia*, heating, hot water heating, cooling, ventilation and lighting. This amount shall be reflected in one or more numeric indicators which have been calculated, taking into account insulation, technical and installation characteristics, design and positioning in relation to climatic aspects, solar exposure and influence of neighbouring structures, own-energy generation and other factors, including indoor climate, that influence the energy demand.

3 *Energy performance certificate of a building*: a certificate recognized by the Member State or a legal person designated by it, which includes the energy performance of a building calculated according to a methodology based on the general framework set out in the Annex.

4 *Combined heat and power (CHP)*: the simultaneous conversion of primary fuels into mechanical or electrical and thermal energy, meeting certain quality criteria of energy efficiency.

5 *Air-conditioning system*: a combination of all components required to provide a form of air treatment in which temperature is controlled or can be lowered, possibly in combination with the control of ventilation, humidity and air cleanliness.

6 *Boiler*: the combined boiler body and burner-unit designed to transmit to water the heat released from combustion.

7 *Effective rated output* (expressed in kW): the maximum calorific output specified and guaranteed by the manufacturer as being deliverable during continuous operation while complying with the useful efficiency indicated by the manufacturer.

8 *Heat pump*: a device or installation that extracts heat at low temperature from air, water or earth and supplies the heat to the building.

Article 3: Adoption of a methodology

Member States shall apply a methodology, at national or regional level, to calculate the energy performance of buildings on the basis of the general framework set out in the Annex. Parts 1 and 2 of this framework shall be adapted to technical progress in accordance with the procedure referred to in Article 14(2), taking into account standards or norms applied in Member State legislation.

This methodology shall be set at national or regional level.

The energy performance of a building shall be expressed in a transparent manner and may include a carbon dioxide (CO_2) emission indicator.

Article 4: Setting energy-performance requirements

1   Member States shall take the necessary measures to ensure that minimum energy performance requirements for buildings are set, based on the methodology referred to in Article 3. When setting requirements, Member States may differentiate between new and existing buildings and different categories of buildings. These requirements shall take account of general indoor climate conditions in order to avoid possible negative effects, such as inadequate ventilation, as well as local conditions and the designated function and the age of the building. These requirements shall be reviewed at regular intervals, which should not be longer than five years and, if necessary, updated in order to reflect technical progress in the building sector.

In this article, particular attention is given to indoor climate conditions in order to avoid possible negative effects, such as inadequate ventilation.

2   The energy performance requirements shall be applied in accordance with Articles 5 and 6.
3   Member States may decide not to set or apply the requirements referred to in paragraph 1 for the following categories of buildings:
   •   buildings and monuments officially protected as part of a designated environment or because of their special architectural or historic merit, where compliance with the requirements would unacceptably alter their character or appearance;
   •   buildings used as places of worship and for religious activities;
   •   temporary buildings with a planned time of use of two years or less, industrial sites, workshops, and non-residential agricultural buildings with low

energy demand and non-residential agricultural buildings that are in use by a sector covered by a national sectoral agreement on energy performance;
   •   residential buildings that are intended to be used for less than four months of the year;
   •   stand-alone buildings with a total useful floor area of less than 50m².

Article 5: New buildings

Member States shall take the necessary measures to ensure that new buildings meet the minimum energy performance requirements referred to in Article 4.

This article is very clear: energy performance requirements must be imposed for all buildings (see definition of building in §2) whereby the calculation method must meet the requirements as given in the Annex.

For new buildings with a total useful floor area over 1000m², Member States shall ensure that the technical, environmental and economic feasibility of alternative systems such as the following is considered and is taken into account before construction starts:

   •   decentralized energy supply systems based on renewable energy;
   •   CHP;
   •   district or block heating or cooling, if available;
   •   heat pumps, under certain conditions.

Article 6: Existing buildings

Member States shall take the necessary measures to ensure that when buildings with a total useful floor area over 1000m² undergo major renovation, their energy performance is upgraded in order to meet minimum requirements in so far as this is technically, functionally and economically feasible.

Member States shall derive these minimum energy performance requirements on the basis of the energy performance requirements set for buildings in accordance with Article 4. The requirements may be set either for the renovated building as a whole or for the renovated systems or components when these are part of a renovation to be carried out within a limited time period, with the above-mentioned objective of improving the overall energy performance of the building.

Article 7: Energy performance certificate

1   Member States shall ensure that, when buildings are constructed, sold or rented out, an energy perform-

ance certificate is made available to the owner or by the owner to the prospective buyer or tenant, as the case might be. The validity of the certificate shall not exceed ten years.

A certificate will be obligatory for all buildings when constructed, sold or rented. As described in the Annex, indoor climate and ventilation-related parameters have to be included. Ideally, such certificates should also correctly assess innovative technologies as demand-controlled ventilation, hybrid ventilation, etc.

Certification for apartments or units designed for separate use in blocks may be based on:

- a common certification of the whole building for blocks with a common heating system; or
- the assessment of another representative apartment in the same block.

Member States may exclude the categories referred to in Article 4(3) from the application of this paragraph.

2   The energy performance certificate for buildings shall include reference values such as current legal standards and benchmarks in order to make it possible for consumers to compare and assess the energy performance of the building.

The certificate must include reference values and this probably also means information on ventilation-related aspects.

The certificate shall be accompanied by recommendations for the cost-effective improvement of the energy performance.

Recommendations concerning ventilation can be part of the certificate.

The objective of the certificates shall be limited to the provision of information and any effects of these certificates in terms of legal proceedings or otherwise shall be decided in accordance with national rules.

3   Member States shall take measures to ensure that for buildings with a total useful floor area over 1000m^2 occupied by public authorities and by institutions providing public services to a large number of persons and, therefore, frequently visited by these persons, an energy certificate, not older than ten years, is placed in a prominent place clearly visible to the public.

The range of recommended and current indoor temperatures and, when appropriate, other relevant climatic factors may also be clearly displayed.

Article 8: Inspecting boilers

With regard to reducing energy consumption and limiting carbon dioxide emissions, Member States shall either:

(a) Lay down the necessary measures to establish a regular inspection of boilers fired by non-renewable liquid or solid fuel of an effective rated output of 20kW to 100kW. Such inspection may also be applied to boilers using other fuels.

Boilers of an effective rated output of more than 100kW shall be inspected at least every two years. For gas boilers, this period may be extended to four years.

For heating installations with boilers of an effective rated output of more than 20kW that are older than 15 years, Member States shall lay down the necessary measures to establish a one-off inspection of the whole heating installation.

On the basis of this inspection, which shall include an assessment of the boiler efficiency and the boiler sizing compared to the heating requirements of the building, the experts shall provide advice to the users on replacing the boilers, other modifications to the heating system and on alternative solutions.

(b) Take steps to ensure the provision of advice to the users on the replacement of boilers, other modifications to the heating system and on alternative solutions, which may include inspections to assess the efficiency and appropriate size of the boiler. The overall impact of this approach should be broadly equivalent to that arising from the provisions set out in (a). Member States that choose this option shall submit a report on the equivalence of their approach to the Commission every two years.

Article 9: Inspecting air-conditioning systems

With regard to reducing energy consumption and limiting carbon dioxide emissions, Member States shall lay down the necessary measures to establish a regular inspection of air-conditioning systems of an effective rated output of more than 12kW.

This inspection shall include an assessment of the air-conditioning efficiency and the sizing compared to the cooling requirements of the building. Appropriate advice shall be provided to the users on possible improvement or replacement of the air-conditioning system and on alternative solutions.

Advice on improvement or replacement of air conditioning is required, as well as on alternative solutions. This might include information on, for example, passive cooling when applying night ventilation concepts.

Article 10: Independent experts

Member States shall ensure that the certification of buildings, the drafting of the accompanying recommendations and the inspection of boilers and air-conditioning systems are carried out in an independent manner by qualified and/or accredited experts, whether operating as sole traders or employed by public or private enterprise bodies.

Each Member State must take the necessary measures to guarantee the required number of qualified and/or accredited experts (including their specifications). This requires very substantial resources. For Germany alone, during the first year more than 2 million certificates have to be delivered.

Article 11: Review

The Commission, assisted by the Committee established by Article 14, shall evaluate this Directive in the light of experience gained during its application, and, if necessary, make proposals with respect to, *inter alia*:

* possible complementary measures referring to the renovations in buildings with a total useful floor area less than $1000m^2$;
* general incentives for further energy-efficiency measures in buildings.

Article 12: Information

Member States may take the necessary measures to inform the users of buildings of the different methods and practices that serve to enhance energy performance. Upon Member States' request, the Commission shall assist Member States in staging the information campaigns concerned, which may be dealt with in Community programmes.

Article 13: Adapting the framework

Points 1 and 2 of the Annex shall be reviewed at regular intervals, which shall not be shorter than two years.

Any amendments necessary in order to adapt points 1 and 2 of the Annex to technical progress shall be adopted in accordance with the procedure referred to in Article 14(2).

Article 14: Committee

1   The Commission shall be assisted by a Committee.
2   Where reference is made to this paragraph, Articles 5 and 7 of Decision 1999/468/EC shall apply, having regard to the provisions of Article 8 thereof.
    The period laid down in Article 5(6) of Decision 1999/468/EC shall be set at three months.
3   The Committee shall adopt its Rules of Procedure.

Article 15: Transposition

Member States shall bring into force the laws, regulations and administrative provisions necessary to comply with this Directive at the latest on 4 January 2006. They shall forthwith inform the Commission thereof.

When Member States adopt these measures, they shall contain a reference to this Directive or shall be accompanied by such reference on the occasion of their official publication. Member States shall determine how such reference is to be made.

Member States may, because of lack of qualified and/or accredited experts, have an additional period of three years to apply fully the provisions of Articles 7, 8 and 9. When making use of this option, Member States shall notify the Commission, providing the appropriate justification, together with a time schedule with respect to the further implementation of this Directive.

Article 16: Entry into force

This Directive shall enter into force on the day of its publication in the *Official Journal of the European Communities*.

Article 17: Addressees

This Directive is addressed to the Member States.

Annex: General framework for calculating the energy performance of buildings (Article 3)

The methodology of calculating the energy performances of buildings shall include at least the following aspects:

(a) the thermal characteristics of the building (shell and internal partitions, etc.); these characteristics may also include air tightness;
(b) heating installation and hot water supply, including their insulation characteristics;
(c) air-conditioning installation;
(d) ventilation;

(e) built-in lighting installation (primarily the non-residential sector);

(f) position and orientation of buildings, including outdoor climate;

(g) passive solar systems and solar protection;

(h) natural ventilation;

(i) indoor climatic conditions, including the designed indoor climate.

> *Ventilation and indoor air quality (IAQ)-related aspects are very dominant in the specifications of the calculation method: 'air tightness', 'ventilation', 'natural ventilation', 'indoor climate conditions', etc.*

The positive influence of the following aspects shall, where relevant in this calculation, be taken into account:

(a) active solar systems and other heating and electricity systems based on renewable energy sources;

(b) electricity produced by CHP;

(c) district or block heating and cooling systems;

(d) natural lighting.

For the purpose of this calculation, buildings should be adequately classified into categories such as:

(a) single-family houses of different types;

(b) apartment blocks;

(c) offices;

(d) education buildings;

(e) hospitals;

(f) hotels and restaurants;

(g) sports facilities;

(h) wholesale and retail trade services buildings;

(i) other types of energy-consuming buildings.

Index

Printed and bound by CPI Group (UK) Ltd, Croydon, CR0 4YY

22/10/2024

01777611-0005